Creation, Separation, and the Mind
The Three Towers of Singularity

The Application of Universal Code in Reality

By

William E. Wismann, David E. Martin,

and Norbert Schwarzer

"Creation, Separation, and the Mind – The Three Towers of Singularity – The Application of Universal Code in Reality", RASA strategy book

1st Edition, June 2024

Publishing: RASA Energy Inc.

Print: IngramSpark

ISBN: 9798218444839 (Hardcover)

ISBN: 9798218444846 (eBook)

Acknowledgement

The Authors wish to acknowledge, with gratitude, the infinitely unfolding Universe and the exceptional minds that have sought to understand its breadth, depth, and phases. Further, we wish to honor our friends, partners, spouses, and children for the invaluable perspective that they have afforded us in both our thoughts as well as the insights without which our observations would be far less humane.

1 Contents

2 Preface

Since the mid-19th century, chemistry has rejected "vitalism" in favor of the secular atomized models of John Dalton and Antoine Lavoisier. With Dmitri Mendeleev's periodic table from 1869, the obsession with the particulate became hegemon for physics, chemistry and its associated mathematics calcified with Niels Bohr's Nobel Prize in Physics in 1922.

The understanding of the nature of everything narrowed to "tight-binding" concepts and first order interaction considerations, excluding many nonlinearities, but also all weaker forms of fundamental communications, compactified dimensionalities, their entanglements, and recessive overtones. Driven by the desire of seeing mathematical tools and concepts, even if those are often only partially successful in their first applications, not as what they truly are, namely only models and distorted illustrations of reality, but as reality itself, this error led to a disastrous evolution in what is sold as "science". "Far-too-early-simplification" and the dogmatic insistence on "a world of particles" was the inevitable result. Naturally, this development was often strongly supported by the ego of certain personalities and an omnipresent hunt for "public support", recognition and short-lived attention. Subsequently, the whole of reality evolved more into a science show rather than a truly truth-seeking science. By focusing conceptual ontologies on the assumption that particulate mass and 'energy' exist in thermodynamics, context, and sequential memory effects have been rendered subordinate to serial synthetic or lytic events.

The subsequent introduction of certain parameters via correcting fields, bringing in what the potential approach and the "zoo of particles" would miss otherwise, can only be seen as what these attempts are:

<u>Perturbative amendments inside an incomplete theory</u>.

The scalar Higgs field in Quantum Theory, the test masses in General Theory of Relativity (GTR), and the so-called inflation field in cosmology are prominent examples of postulates within incomplete theoretical frameworks.

"Energy" is thought to be required to alter the "atomic", and "thermodynamics" serves as the governing principle. And while this dogma has served to bring us countless military advances in mobility, communications, and apocalyptic ecosystem destruction in the last century, it has rendered us largely impotent to innovate our relationship with the dualism defined by matter and energy.

The established science is thereby so fixated on concepts like "particles", "forces", and "potentials" without being able to explain where these "elements" or "phenomena" come from (at least not in a metric consistent or illustratively graspable way), that many scientists no longer even realize that these concepts are anything but models and – since their introduction – always have been. As models, however, they automatically come with a built-in approximation and fundamental uncertainty, that modern science only all too often simply and generously ignores. And this is a grave mistake, because if there is one thing that is certain in this universe, it is the omnipresent uncertainty of everything... but most especially with all human concepts about the ontology of reality. Einstein's blunder with the cosmological constant can be seen quite small in comparison to a multitude of scientific dogmas and the stubbornness in which they are – all too often – applied to completely inadequate scenarios, experiments, and observations. It is almost an irony that especially in the very field, which only exists because of the universal uncertainty of everything, the "concept of uncertainty" is dogmatically opposed when it comes to question the theoretical foundations of this very field. The Nobel Prize in Physics in 1922 may be seen as the pivot element for the start of these developments.

On top of all that comes an almost religious obsession of modern science with a technical instrument called computer. One can observe in almost all fields how rather mediocre personalities, "experts" they consider themselves, start to treat problem solving not via a proper cognitive analysis of the thing of interest, but by an automated process which immediately requires the "assistance" of prefabricated tools like finite and boundary elements, molecular dynamics, random walk, renormalization group theory, and so on. This predilection imposes serial, static, dimension delimited processes devoid of the *in situ* context in which reality persistently, generatively, and infinitely orthogonally manifests and transforms. It is almost like an addiction… and perhaps, in many cases, in fact, it is an addiction. This not only leads to a permanent overlooking of important closed form and much better analytical solutions, but also to unnecessary numerical errors with often catastrophic consequences and the complete overshadowing and whitewashing of entire groups of natural phenomena and effects. Almost the same holds for the new fetish of science, artificial intelligence, or AI. Here, it is ultimately the AI programmers who make the final and decisive decision about what is thought and recognized, and what is not. As most of them program their artificial "intelligence" in the simplest possible way, namely via linear algebra, the result – naturally – will not be able to "think" lateral, nonlinear, dynamically flexible, dimensionally unlimited, and hence, naturally.

In this book, as bold as it may sound, we will not just address these problems, but also show how to overcome them. For illustration we will consider quite sophisticated "demonstrators" or simply – to put it dryly – experiments.

In the present experiments, here, because of their transformative power in what human beings realize as socioeconomic space-time often referred to as "towers" or key technologies, we demonstrate kinetic, intelligent, infinitely linear independent transformational behavior of statistical phase experiences which integrate context, sequence and "strange effects" into accretive and purification engineering.

We introduce the notion of "accretive" to not only describe classical combinations, but also to demonstrate that the effect of such associations alters the physical, chemical, and utilitarian properties of the resultant material.

We introduce the notion of "purification" to not only describe the classic separation of impurities, but also to show that the effect of such dissociation transforms the physical, chemical, and utilitarian properties of the resultant materials.

To tackle these problems – most surprisingly – does not require a "new physics", but only a consequent application of what was already there before 1922.

And this is what we will do in this book – no more and no less.

2.1 And Where Is the Proof of Our Bold Claims?

As our claims in the "preface" above are quite bold (at least they will be considered that way by those who are bound to oppose us in order to save their own comfortable position in a show being named "science"), we here are going to include the rather unusual but hopefully convincing "weapon" of direct evidence in this very preface. Thereby, we do not necessarily need mathematics straight away because we can easily refer to the existing proofs in our main book [P1], this one here, and our other publications. After all, this material is all publicly available and so the only thing we need to do here is to clearly reference where each of these building blocks can be found and briefly hint what they mean.

In this way – so we hope – we can now, perhaps even need to, very clearly criticize the concepts of particles, mass, potential and so on already in the preface and still not run the risk of being dismissed as charlatans, by the thousands who will be unwilling (or unable) to read the rest of the book [P1] (and the other publications) or handle the math. Anyone who cannot or does not want to believe us, potentially even intends to attack us straight away, is strongly advised to first study AND refute the mathematical proofs.

And as those are given AND clearly referred to, there will be no excuse. We will only accept reasonable, mathematically well-funded and falsifiable criticism.

So, here is a list of the boldest of our claims from above and their entanglement with our corresponding references presenting the necessary mathematical and therefore completely falsifiable proofs:

1. There are no particles → [P1], section "There Are No Particles":
   The proof is given via a so-called "proof by contradiction", where we assume that particles exist and then, by running this very particle-concept through the Hamilton extremal principle in the form of the most fundamental and general Einstein-Hilbert action [P2], we lead this concept to its own and rather drastic failure.
2. About the metric origin of mass → [P1], e.g., section "From Scalar Mass and Energy to a Mass-Energy-Vector or -Matrix?" and [P3, P4, P5, P6, P7]:
   It can be shown that mass is just a certain form of entangled dimensions within a given system or space-time.
   As a by-product, also the equivalence of mass and energy becomes clear.
3. About the metric origin of a potential → [P6, P8]:
   Just as with the property "mass", it can be shown that potentials are forms of entangled dimensions within a given system or space-time.
4. About the metric origin of energy → see point 2 of this list here.
5. About the "unification" of Quantum Theory and the General Theory of Relativity [P9] → [P1], section "The Theory" and [P3, P4, P5, P6, P10]:
   It was shown that the unified theory always resided in the so-called Einstein-Hilbert action [P2]. Thus, it, which is to say a "Theory of Everything", is and was there since 1915, when David Hilbert published his little piece about the derivation of the Einstein field equations

[P9] from a Hamilton extremal principle. We have now shown that Hilbert's approach already contained it all. The only recipe necessary to see this was a bit of scaling.

6. About the artificial character of "artificial intelligence" → e.g., [P1], see all sections about the "Third Tower" and [P5, P6].
7. About the metric origin of thermodynamics → see [P3, P6, P11, P12, P13, P14, P15]: Metric thermodynamics can directly be obtained from the Einstein-Hilbert action by proper statistics of quantum and gravity centers.
8. About the metric origin of evolution → see [P16, P17].
9. …

2.2 Preface References

[P1] this book; please cite as:
W. E. Wismann, D. E. Martin, N. Schwarzer, “Creation, Separation, and the Mind – The Three Towers of Singularity – The Application of Universal Code in Reality”, RASA strategy book, 2024, ISBN: 9798218444839

[P2] D. Hilbert, “Die Grundlagen der Physik”, Teil 1, Göttinger Nachrichten, 1915, pp. 395-407

[P3] N. Schwarzer, “The World Formula: A Late Recognition of David Hilbert’s Stroke of Genius”, Jenny Stanford Publishing, 2022, ISBN: 9789814877206

[P4] N. Schwarzer, “The Math of Body, Soul, and the Universe”, Jenny Stanford Publishing, 2022, ISBN: 9789814968249

[P5] N. Schwarzer, “The Quantum Gravity War – How will the Nearby Unification of Physics Change the Future Warfare?”, Jenny Stanford Publishing, 2024, ISBN: 9789814968584

[P6] N. Schwarzer, “Mathematical Psychology – The World of Thoughts as a Quantum Space-Time with a Gravitational Core”, Jenny Stanford Publishing, ISBN: 9789815129274

[P7] N. Schwarzer, “How Fermions Get Massy – Can Quantum Gravity Answer this Question?”, self-published, Amazon Digital Services, 2023, Kindle, ASIN: B0C5TM9ZCG

[P8] N. Schwarzer, “The Metric Potential – How can Quantum Gravity Help Us to Understand the Origin of Forces”, self-published, Amazon Digital Services, 2023, Kindle, ASIN: B0BXSHQZX1

[P9] A. Einstein, “Grundlage der allgemeinen Relativitätstheorie”, Annalen der Physik (ser. 4), 49, 769–822

[P10] N. Schwarzer, “The Theory of Everything – Quantum and Relativity is Everywhere – A Fermat Universe”, Pan Stanford Publishing, 2020, ISBN-10: 9814774472

[P11] N. Schwarzer, “Quantum Gravity Thermodynamics – And it May Get Hotter”, self-published, Amazon Digital Services, 2019, Kindle, ASIN: B07XC2JW7F

[P12] N. Schwarzer, “Science Riddles – Riddle No. 20: Second Law of Thermodynamics – Where is its Fundamental Origin?”, self-published, Amazon Digital Services, 2019, Kindle, ASIN: B07Y79BTT9,

[P13] N. Schwarzer, “Science Riddles – Riddle No. 21: Evolution – Where is its Fundamental Origin?”, self-published, Amazon Digital Services, 2019, Kindle, ASIN: B07YKL37DL

[P14] N. Schwarzer, “Quantum Gravity Thermodynamics II – Derivation of the Second Law of Thermodynamics and the Metric Driving Force of Evolution”, self-published, Amazon Digital Services, 2019, Kindle, ASIN: B07XWPXF3G

[P15] N. Schwarzer, “Brief Proof of Hilbert’s World Formula – Dirac, Klein-Gordon, Schrödinger, Einstein, Evolution and the 2nd Law of Thermodynamics all from one origin”, self-published, Amazon Digital Services, 2020, Kindle, ASIN: B08585TRB8

[P16] N. Schwarzer, “How Systems Evolve – A Red Pill Course on Evolution”, self-published, Amazon Digital Services, 2022, Kindle, ASIN: B0B6RGBKQ4

[P17] N. Schwarzer, "The Mathematical Darwin – How Quantum Gravity Explains the Miracle of Evolution", upcoming book-project with Pan Stanford Publishing

3 Abstract

There are apparently simple experiments, processes, and observations which have eluded proper explanation for significant periods of time. In this book we will consider a variety of such cases, namely:

- The Einstein-Podolsky-Rosen paradox and the spooky action at a distance
- The intrinsic action of the human mind in the metric picture
- Schrödinger's cat, which we are going to deliver from its unjust fate of an existence in an infinite limbo state
- The so-called 3-generation problem of elementary particles, which we are going to solve by the means of the Bianchi identity in a scalarized Quantum Gravity Physics realm
- The particle-wave dualism, which we will address via a fundamentally based mathematical approach leading to the destruction of the concept of particles
- The appearance of so-called Sigmoid or S-curves everywhere in this universe, especially in the processes of life
- The origin of evolution
- The 720° half spin oddity
- An innovative, flexible, and quite general deposition technology
- An almost alchemic decomposition experiment

The two latter experiments or innovations are of quite some socioeconomic relevance, because – when fully unraveling their potential – they are quantum leaps not only in their very own industrial segments, but also with respect to the impact they will have on the society as a whole. Thereby, most interestingly, it can be shown that the goal of the "full unraveling of the two innovations' potential" within a complex socioeconomic system requires the incorporation of the second point in our list above into a suitable financial and economic approach.

Close observation shows us that the two latter innovations – in principle – are opposites, namely, one process which creates (or synthesizes) and one which separates. It is due to this fact that the two "powers" combined and entangled with a third, namely the mathematical description of the intrinsic processes of the human mind need to be considered socioeconomic "gamechangers". We think that it is therefore quite understandable that in this book we cannot reveal any technical details or physical explanations for these very special experiments. Readers being interested in such are kindly asked to contact the authors directly.

With this said, however, we are going to work out a new concept for the constructive understanding of such experiments via a universal analysis technique. Interestingly, this technique follows from the failed attempt of a point-wise localization of "something", which could be "anything", within a universe governed by minimum principles. As a result, one obtains the contradiction of the concept of particles. There are only waves and thus, consequently, everything of higher complexity can be decomposed into basic oscillations and ground waves and the latter can be used to compose things of higher complexity and structure. In essence, this just means that the general recipe for composition and decomposition – of everything – requires nothing more than a generalized Fourier formation and a generalized Fourier analysis, respectively. Another interesting consequence pops up in connection with the so-called Einstein-Podolsky-Rosen or EPR paradox and the alleged "spooky action at a distance", which we are going to merit in a motivator story in order to illustrate some details of the theory being of significance for our general purposes here.

Even though all fundamental and basic ingredients for the analysis of any process will be given in this book, as said in the paragraph before, regarding the two experiments, we cannot reveal any process-relevant or theoretical essentials, which could potentially endanger the intellectual property. Interested readers may please contact the authors directly.

With respect to all other points listed above, however, we will try to come to – hopefully – suitable hypotheses with respect to a physical explanation and mathematical description of the experiments, processes, and observations by the means of a starting point as fundamental and universal as possible.

We think that the recipe for such a universal approach is to be found in Hamilton's extremal principle by the means of the Einstein-Hilbert action.

4 The Motivator

As to work out the importance of unbiased, non-ideological, and as holistic and fundamental as possible approaches when it comes to try and explain AND mathematically comprehend various aspects of reality, we resort to an unusual thought experiment.

Thereby, the topic shall be the so-called Einstein-Podolsky-Rosen (EPR) paradox and its explanation in the realm of a Quantum Gravity Theory. For entertainment and better understanding we created a little story, where the original protagonists of the paradox, namely, Einstein, Podolsky, and Rosen, are joined by Sir Arthur Conan Doyle's figures Sherlock Holmes and Dr. Watson.

The little exercise shall motivate us to broaden our view and reduce our intrinsic bias when addressing practical problems.

At first, only the story will be presented without any equations or formulae disturbing the plot.

Then, at the end of this book, which is to say, after a quite long and probably hard slog with some quite fundamental and complex equations, their cumbersome derivation and – partially also – their applications, we will repeat the story for the mathematically trained reader with all the necessary equations given.

The paradox was chosen as an example or demonstrator, because it gives insight into the matter of dimensional entanglement on a very fundamental level. It will be evaluated in this book that the consideration of the position of objects requires the observation of additional dimensions in connection with the main attributes of a given space-time. From there automatically follows a higher dimensionality, which allows us to understand the processes of creation and separation via changes of such dimensionality with entangled attributes (= properties = degrees of freedom = dimensions). In addition, we can extract an effective mathematical description (see this book).

Thus, entanglement, respectively dimensional entanglement is the cornerstone for the understanding of creation and separation processes in general and the following story will help with the comprehension in a – hopefully – entertaining manner. Applying this generality to "mysterious" processes like our "innovative, flexible and quite general deposition technology" and our "almost alchemic decomposition experiment" will help in their understanding and mathematical formulation, because here, too, dimensions are added or taken away.

Another aspect of the EPR paradox, which is important in connection with the second, the "almost alchemic decomposition experiment", in fact, is the spooky action at a distance. With the latter – hopefully – being not so spooky after the story below, we might have a different explanation for the astounding affinity-behavior between the essential components of this experiment, which are the target element and the separation-agent. Their interaction, namely, is also – apparently – happening at a rather spooky distance. There even seems to be some kind of memory-effect, which makes simple "molecules" remember certain events (processes) and repeat them. Classically, the chemical potential (be it covalent, ionic, complex, or whatever else) does not give a conclusive explanation for this "distant reactivity behavior", its selectivity, and its scale-invariance (reaching over target chains from 2C to 230C and more).

Sherlock, Watson, Einstein

and

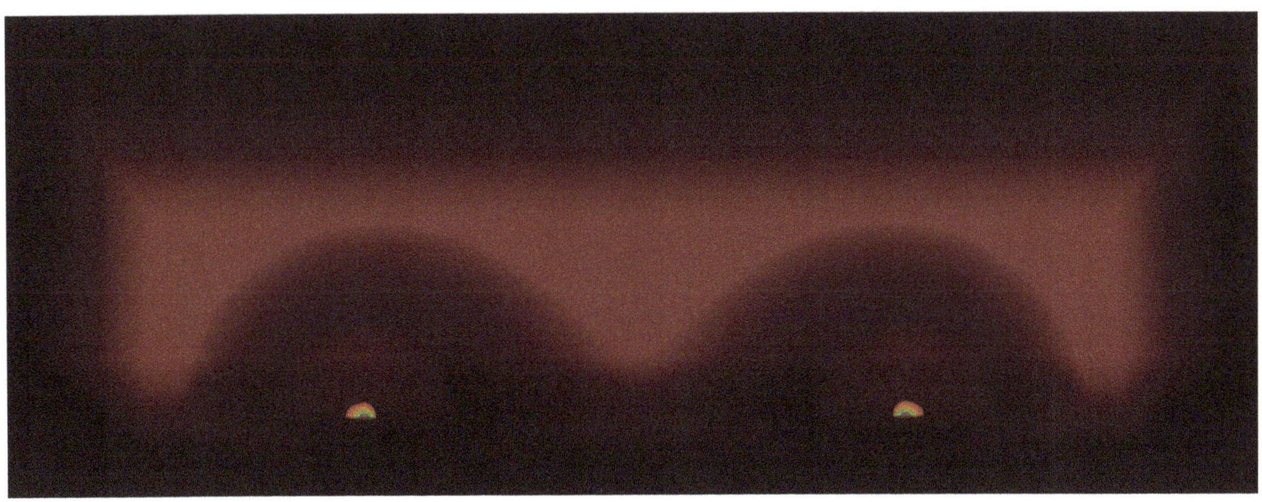

The Mystery of Entanglement and the Spooky Action at a Distance

By Norbert Schwarzer & T. Bodan (from [85] with thanks)
Illustrated by Livia & Fila Schwarzer

4.1 The Thought Experiment

The classical vacuum Einstein field equations give solutions of clear quantum character. Surprisingly, such solutions are extendable to forms which allow for entanglement. Thereby the omnipresent cosmologic constant guarantees the permanent connection of the so-called non-local parameters.

To make the theory understandable, the authors have resorted to a little thought experiment which turned into some kind of thought-meeting of a few of the greatest minds in history… fictive and non-fictive.

4.2 Introduction

"Come on, Watson," Holmes urged me, "you must have heard about these gentlemen."

"Of course, I have heard about the first one, this Einstein," I said. "A pipe-smoking genius who likes to stick out his tongue the moment he sees a camera pointing at him. However, much to my regret I have never heard, saw, or read anything about the other two… what were their names again, Holmes?" I asked.

At teatime, Holmes and Watson discuss the mystery of entanglement. [A]

"Podolsky and Rosen," Holmes answered. "To be precise, it is Professor Podolsky and Professor Rosen."

I merely nodded and in order to fight my desire to ask the question nagging so forcefully inside me, I bent forward and took the cup of tea in front of me. But my friend only knew me too well. He took his pipe out of his mouth and gave me one of his rare smiles.

"My dear Watson," he said, "of course I could enlighten you myself about the mystery these three most respected gentlemen are confronted with, but as there is a much better way, I suggest we just sit and enjoy the rest of our tea, while we are waiting."

"Waiting for what?" I asked.

"For the three gentlemen to arrive, of course!" Holmes replied.

I looked startled. My mouth must have fallen open, but Holmes – as polite as he was – generously ignored this quite obvious lack of self-control and pulled out his pocket watch. This brought me back to my senses.

"You still have this old thing?" I asked.

Holmes did not answer immediately. He turned the watch in his hand before he finally said, "Old habits, Doctor Watson, just old habits." Then he held up his pipe. "It is the same with this thing here." He twisted his hand slowly, as if to intensely observe the object in his hand. After all, an object he should know so well as he has had it for so many years. "As you have suggested about 70 years or so ago, I had stopped smoking."

I nodded vigorously. "And right I was in making such a suggestion."

"Yes, of course, you were right, Watson!" said Holmes, "but still, I think as if something very important would be missing if I wasn't having it in my mouth or in my hand."

I thought about this for a while. Then I nodded.

"More tea, Holmes?" I asked.

"Yes please!" Holmes answered, "… and now you may also place the question bothering you so profoundly."

"Well," I started, "how can we expect a visitor who is known to be dead?"

"Yes, Watson, Einstein is dead and so are the other two…" He paused and looked at me quite calmly. I could not make out a trace of irony neither in his words, nor in their intonation, nor in his face. Then, after it had felt like an endless stretch of time, he finally added, "… and so are we, Watson."

I stared at him.

"Yes Watson, we are dead, too. Well, to be precise, we never actually were alive, but this does not mean that we did not exist, you see?"

I shrugged.

"All is relative, Watson! All is relative!" Holmes said. Then he added, "Besides, Watson, Einstein is allowed to bring his pipe. He has it with him all the time, probably died with it… I don't know, but don't worry, he will not smoke it in here."

“So, he is going to use it for show then, just like you?” I asked.

“Not for show, my dear Doctor. As I just said, old habits, nothing but habits. What would we be without them.”

Then the bell rang.

Einstein – dead with his pipe still in his mouth. [A]

4.3 The Strangest Client There Could Be

"So, you are not the clients yourself, but you are here on someone else's behalf?" Still I could not comprehend it. It was the strangest thing I had ever heard and it already had taken us a while to get this far in the conversation.

Oh yes, the three men were quite eager to share information, but Holmes needed a while to make them all three understand that the information they were willing to give was not necessarily the one he and I needed to solve the problem or to help them solve it. At the beginning, it even seemed almost impossible to define the problem, even finding out who wanted that problem to be solved so very badly that he had organized this unusual gathering.

But let us move backwards in time a little bit.

It was obvious that the three men had no time to lose. They had barely entered the room when Einstein started to speak in a rather un-British briskly manner, "Only a very few moments of eternity were given to us, Mister Holmes," Einstein said after we had formally greeted each other and the usual amenities were – almost – hastily exchanged. Einstein looked imploringly at my friend when he added, "Only a few moments to solve the very problem we have already failed to solve a long time ago."

Podolsky added, "We have not been able to do so while we were alive, but this time with you, Mister Holmes…"

"And what or who gave you the idea that I could be of assistance?" Holmes asked.

"You and Doctor Watson!" insisted Podolsky, who was the only one of the three visitors who had bothered to take off his hat.

"My apologies, gentlemen," Holmes responded. "I meant to say 'us' of course." He corrected himself and gave a pronounced nod in my direction, his eyes winking. I understood the message and now it was I who repeated his question, "Still, Professor Einstein, we would both like to know what gave you the idea that WE could be of assistance?"

"It wasn't us," said Einstein almost a bit impatiently, "but, please, can we leave it for later? I don't think it matters here and I'd rather…"

"On the contrary Professor Einstein," Holmes interrupted him and Einstein looked startled, "it may matter very much… very much indeed."

There was a pause and I asked again, "So, who or what was it then? Who made you think…"

"The universe itself did!" he said with the same kind of impatience he had shown before.

"The universe? The universe has talked to you? How?" asked Holmes, who was the first who recovered from this peculiar answer. But when Einstein merely shrugged, he gave the answer himself. "It is what you call a gut's feeling, Professor, is it not?"

Einstein nodded and Rosen muttered an almost silent "Yes!" Then he added, "We concluded this, because all three of us have the same 'gut's feeling', as you just characterized it. We also think…" But Holmes took his hand in the air and silenced him easily. His gesture was very soft, but such a power emanated from the man that even a simple thing like a raised hand could change the climate in a room completely and instantly.

"So, the universe doesn't know!" he said distinctly. Everybody in the room stared at the man in the chair, who was deep in thought. Nobody said a word. Then quite suddenly, Holmes stood up and walked to the window. He looked out into the street for a while and when he finely turned and faced us, he was almost beaming with excitement.

"Now we know who the client is, gentlemen," he said.

"Who?" I asked, but Holmes didn't answer. Therefore, I faced the three professors and repeated my question in their direction, "So, you are not the clients yourself, but you are here on someone else's behalf?"

The three men nodded. Then we heard Holmes' voice. He had faced the window again and it was as if he spoke to himself. Nevertheless, his voice was loud and clear. It sounded through the room as if being amplified from every corner.

"The universe only knows what has been found out by the things it hosts. It is like a computer which only has a chance of knowing what one of his programs has evaluated. The universe is nothing but a huge computer and we are its programs.

This most specific problem you have brought with you, my dear Professors, must be of great importance to the universe, because it went to quite some length to construct a simulation like this."

"Simulation?" Einstein asked.

"Well, you could also give it another name, my dear Professor. What was it again, the one technique you always applied to understand such strange things like a bending space-time, like riding on a photon, or other rather strange adventures like this?"

"Are you by any chance referring to a 'thought experiment', Mister Holmes?" suggested Einstein.

"Yes, Professor, we might as well assume that we are all just parts – essential parts I hope – of a little thought experiment. So, now we, Mister Watson and I, are ready to listen to your story, gentlemen."

Are we all just programs running inside a huge universal computer machine? [A]

4.4 The Spooky Action at a Distance

"Originally," Rosen started, "we did this experiment..."

"This thought experiment!" rendered Einstein more precisely, thereby smiling.

"Yes, of course," Rosen added quickly, "at the time there was no chance... no chance to do such experiments in the real world. Thus, naturally, it only was a thought experiment. We constructed a certain quantum theoretical system with two objects in it..."

"Any particular objects?"

"No," Rosen said. "Just quantum objects, which is to say, small things which would follow the rules of Quantum Theory, that's the only condition."

"Almost!" hinted Einstein.

"Ah yes," Rosen agreed quickly again, "the two objects were assumed to be connected."

"Entangled!" Einstein said.

"You see, it is a very special form of connection. You simply assume that the two particles would have properties and that one of these properties kind of ties them together."

"Entangles them!" Einstein corrected. But Rosen seemed unperturbed by his interruptions, this obvious lack of education of his famous colleague. He took the hints as if they were coming from his own mind.

"All other properties are free to do as they please, only these specially connected..."

"Entangled!" said Einstein in the same almost indifferent tone as before. As much as he insisted on the correction as much one would have expected him say it a bit sharper when presenting it the second time, but there was no change to be detected. And as before, Rosen took it as if it were just coming from another part of his brain.

"... properties are kind of unfree," he finished his sentence.

"Could you please specify this for us?" asked Holmes.

"Of course!" Rosen replied. "Simply imagine the two objects having nothing more than the two properties, we might like to name them black or white, and positions in space for each of the objects."

"Space-time!" Einstein said.

"Pardon?" I asked.

"Space-time!" repeated Einstein, "it is positioned in space and time, and not just in space alone."

Rosen went on, "Quantum Theory allows us to formulate the system in such a way that we could make the black-and-white property coupled..."

"Entangled!" said Einstein.

"... while we leave the positions of the objects arbitrary."

"Which does mean what?" Holmes asked.

“It means that the quantum objects, let us imagine the two particles, could move apart quite easily. Nothing prevents them from traveling away from each other… farther and farther until they may be lightyears apart from each other. Then something or someone measures the black-and-white property of one of the two particles. Let us assume the coupling…”

“The entanglement!” said Einstein.

“… would bind the black for one particle to a white of the other. Then in the very instant that you have seen the color-state of the one which was measured, you will immediately also know the color-state of the other… even though it is several lightyears away and the two particles have no way to communicate with each other.”

“Well, what is so strange about that?” asked Holmes. “If I were taking coins out of a box, cut them into halves and were tossing them to my left and right, each piece in one direction and the other into the other one, then of course, if someone catches a coin half on my left, he would immediately know how its partner on the right does look like.”

The entangled halves of a coin. [A]

“Yes, Mister Holmes!” Podolsky said like somebody who had a lot of practice in making allowances to people who are less familiar with the funny laws of Quantum Theory. “But that is the macroscopic reality you are referring to. We, however, are talking about the quantum world. Here the particle does not have a defined black-or-white property until this very property is classified and measured. You might put it like ‘the particle has been black AND white until the very moment of its measurement’. It is like a fast-spinning coin and you have no way of knowing which side it will show until you have caught it.”

Holmes nodded. “I see!” he said softly.

“So, there is the same likelihood for the color white just as there also is for the color black to be measured at the first particle. There is also the same likelihood for the color white just as there is for the color black to be measured at the other particle… only, that…”

But suddenly Podolsky was interrupted by Holmes. His deep voice almost shook the room, “… only that the measurement should not have taken place with the first particle, before the second one is due with its own, otherwise the one measurement will already have fixed the result for the other, isn’t it so, Professor?”

The three scientists looked stunned.

"How do you know?" they asked in unison.

Holmes almost laughed. "Well, anything else would be no surprise, nothing out of the ordinary. So, I simply picked the most mysterious constellation as it was the most logical under the circumstances. After all, this gathering would not be, if it wasn't for some very great mystery."

Einstein smiled. "Yes!" he said simply, "but we aren't there yet. The real mystery is still to come."

"What is it?" I asked.

"What do you think makes these particles behave so strangely, I mean, what makes them appear so rigidly bound together?"

"There must be a connection!" I said simply. "This entanglement, as you named it?"

"Ok," Einstein said, "remember, the two particles are lightyears apart from each other. Nevertheless, the very instant the black-white property has been determined for one of the two particles, the thing, I mean the color, is also fixed for the other one. So, I ask you, what kind of connection could this be? What would allow such information to travel several billions of miles in no time at all?"

I was flabbergasted. "Oh!" was all I could say. But then I recovered quickly and said, "However, this was all just a thought experiment, right? I mean, it was your thought experiment?"

"Yes, that is quite true," admitted Einstein, "we came up with this funny idea, because we wanted to show the world, which is to say the scientific community, that Quantum Theory cannot be complete. There had to be something wrong or missing. If strange stuff like this was the outcome, surely there was a snag."

"And did the scientific community understand and accept your objection?" Holmes asked. His voice was almost a bit amused, as if he already knew the answer… and in fact, I was sure that he did.

"That's hardly the point. The community well understood what our problem was. So, all fine there. However, a few years back, from today, I mean, some clever scientists found a way to realize what we considered the perfect thought experiment, simply because we thought it would never be possible to perform it in practice. Yes, they did the one thing we were sure would never be possible at all. Not possible in the real world, but to our great dismay they found exactly what we had predicted. They measured the strangest thing there is. We name it

The Spooky Action at a Distance."

The spooky action at a distance. The two hands are throwing the dice, but the action in between them is unknown and kind of spooky. [A]

4.5 But Where Is the Problem

Everybody stared at Holmes. He had started to laugh. It was as if somebody had made a wonderful joke, but he was the only one who had gotten the gist.

"My apologies, gentlemen," Holmes said after a while, "but I couldn't help noticing that the very man who helped to create the paradox of this spooky action at a distance in Quantum Theory also is the one who has to be held responsible for making this a problem at all."

Everybody looked puzzled. Everybody, except Einstein, who understood, smiled, and said, "Yes, I got the point, Mister Holmes, but the constancy of the speed of light is not to be negotiated here. It is a solid and well-proven fact. Nothing can travel faster than light. This speed is about three hundred thousand kilometers per second and that is THE UNIVERSAL limit, end of story."

"And?" I threw in, in order to show that I simply could not see why there was a problem at all.

Einstein made an almost exasperated gesture, but Podolsky politely answered, "The thing is quite simple, Doctor Watson, if there is nothing faster than light, how can it be that the signal from one of the two particles…"

"… entangled particles…," added Einstein, his huge eyes rolling impatiently, but Podolsky took it just like Rosen had done and went on as if the add-on from Einstein, no matter how impolitely brought forward by the famous man, was just his own thought, "… reaches the other in an instant, with absolutely no time being elapsed? That is the riddle, Doctor Watson."

While the mass is puzzled, the genius chuckles amused. [A]

4.6 Finding the Starting Point

"So," I said, "then we are here because there is an effect, which allows some spooky action to travel huge distances in no time…"

"… no time at all…" Einstein said, but I ignored him.

"… but there is also a law which does not allow velocities faster than the speed of light."

"Precisely!" cried Rosen excitedly.

"So that's just an antagonism," I said. "One theory cannot be correct or must at least be somehow incomplete, right?"

"That is what we thought, too," Einstein said. "We guessed that it is Quantum Theory, which must be wrong, because it is too fantastic anyway. But then there were the experiments and the spooky action at a distance was observed in the real world."

I smiled at the famous scientist. I simply had to smile, I could not help it, because I knew that I would play the role of the devil's advocate now and I savored the moment before finally saying what had to be said anyway, "Well then, the solution is quite simple, is it not?" I started. "It must be the other theory, which is wrong, am I correct?"

"Yes," said Einstein a little bit sharper than necessary. "However, even ignoring the fact that this very theory is from me – we call it General Theory of Relativity, by the way – this theory has also been proven in reality. So, the riddle is, how can there exist two theories of such rather obvious antagonism inside the same world?"

"And you have no idea where to start in your efforts of finding an answer to that question?" I asked.

"No!" Einstein said with determination and Rosen and Podolsky nodded.

"But isn't that obvious?" Holmes said suddenly.

Everybody looked at him, the three professors puzzled, me merely interested. I knew my friend too well to be surprised by his power of deduction. He could draw conclusions when others were unable to even see the pieces, not to speak about the connection between. In the light of the topic of the day, I would even suggest that his brain must have made good use of the spooky action at a distance… taking the speed of his thoughts.

Holmes' speedy thoughts chasing themselves inside his neuronal network / brain. [A]

"You are here, gentlemen," Holmes said slowly. "I cannot see anybody from the quantum section. So, the starting point must have to do with you."

"Then we should start with our old thought experiment, the one thing they nowadays call the Einstein-Podolsky-Rosen or EPR paradox, right?" asked Rosen.

"No!" Holmes said sharply, "this is not what I had in mind. The fact that you are here does not automatically lead to the conclusion that your starting point from almost 90 years ago was correct. No, this only leads to the suggestion that you are able – in principle – to provide something which you must have forgotten to throw into the story 90 years ago. Now we need to find out what this missing piece could have been."

"But you just said that this is obvious?" Podolsky said and there was a tiny note of accusation in his voice.

"And so I still think it is," Holmes answered unperturbed.

"So why not just enlighten us then?" Einstein asked.

"Because you already know it yourself, don't you, Professor?" Holmes said. "You are just afraid of what might be the outcome, is it not so?"

Einstein inhaled deeply before he answered, "When doing the publication about this paradox, I was convinced that our doubts would settle the matter and that Quantum Theory in fact was incomplete. Then, a short while after my death, a certain Mister Bell found a way to design such an experiment and he also showed a mathematical way to prove that our own approach to address the problem was wrong… foolish even. The man succeeded, the experiments, which were performed later on, were most successful and as said, the three of us looked like fools."

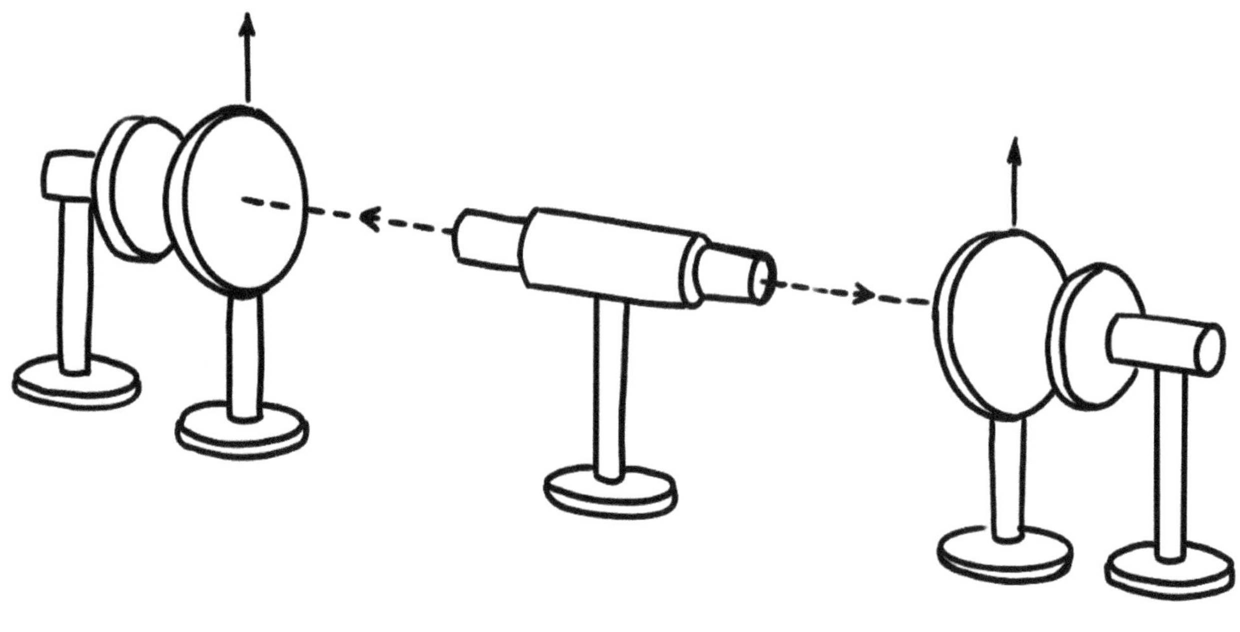

A Bell experiment. [A]

"Pardon me?" I asked. "But if the matter was decided, I mean with tests and everything, then what problem is there to be solved?"

Einstein shrugged.

"Because the universe was not satisfied with the answer that Mister Bell and all his colleagues had given. The whole understanding was still incomplete. So please, Professor Einstein, tell us everything!" Holmes urged.

Now, finally Einstein took off his hat. He drew a chair and sat down. Podolsky and Rosen followed his example.

"Everybody is of the opinion that I regretted the introduction of the cosmological constant as 'THE biggest blunder of my life'. In fact, I had made two such 'biggest blunders' and today I consider the second one even bigger than the first."

"You shouldn't be too hard on yourself," said Rosen, "after all, it was us who came up with the idea of the hidden parameters in the first place." Podolsky nodded vigorously in agreement.

Einstein made an impatient hand gesture.

God throwing dice. [A]

"Not really… I was so over consumed with the idea that there must be a God who does not throw dice that I would have taken anything. I would have grasped at straws to save my deep beliefs about the world. When you came with the idea that there are hidden parameters, assuring the connection of the two particles within our EPR paradox, I was so relieved that I couldn't see the obvious flaw… simply because I did not want to see… did not want to even think about it."

"What flaw?" asked Holmes.

"A theory like Quantum Theory, which was able to perfectly describe the world around us, including the tiny electrons and their peculiar behavior, should have deserved some more credit. It was not right of us to throw it overboard only because we did not like what this theory told us about our God."

"Instead of making up paradoxes in order to prove them wrong," added Rosen, "we should have gone deeper into the matter, question everything, but most of all, we should have questioned our beliefs, because… after all… they are only just…"

"… beliefs," ended Podolsky the sentence for him.

"So, what is the current state of the matter?" I asked.

Einstein took a deep breath, "Quantum theoreticians simply say that the two particles are not separated. No matter how far they are apart from each other, it is one system. If, however, it is one system, then the black-white-parameter is well-defined inside the system and as long as the system stays intact, the parameter may have oscillated as much as it likes, but it is always oscillating according to the boundary conditions of the whole system. Thus, if the boundary conditions demand the second particle to be black if the first was white, then this will be so all the time."

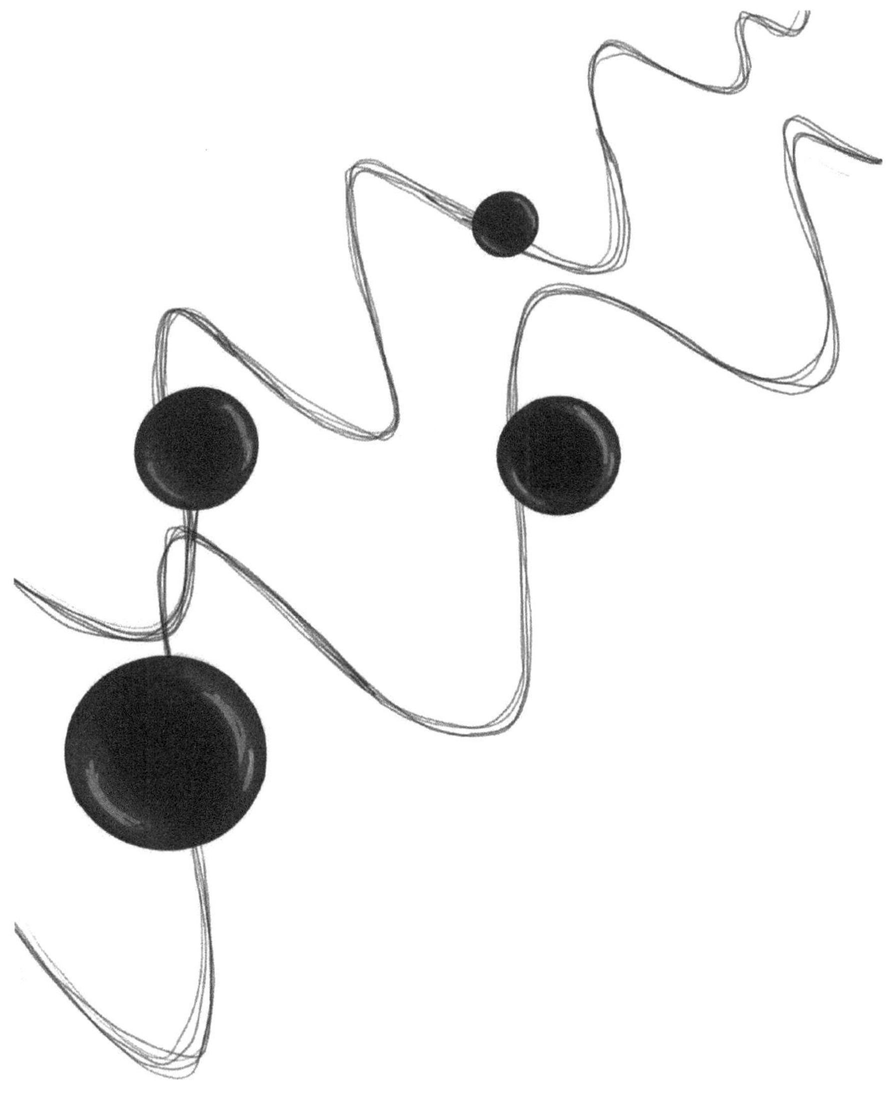

Oscillations between "particle manifestations". [A]

"All the time the system stays intact!" Rosen added softly.

"And what could destroy this state of intactness?" I asked.

"A measurement, for instance," said Podolsky.

There was silence for a while. Then a question occurred to me and I asked, "You do not like the explanation of the quantum guys, do you?"

"No!" said Rosen flatly and Holmes elaborated on his behalf, "You do not like it, because it is against the common sense to even consider something like a non-local black-and-white parameter, right?"

The three professors nodded and Holmes went on, "You have fought so hard to establish the fact that nothing can travel faster than the speed of light (an assumption derived from the assumption of entity resolution at a measured distance – both of which themselves are also assumptions) and that everything is relative and well-determined, which also includes that it has a certain position, that you almost detest the idea about a system, being spread over lightyears distances and still being just one system, one entity. Almost, … like being one cosmos."

They nodded again and Holmes went on, "You consider the fact unbearable that something like position loses its meaning, that locality does not matter with a thing like the black-and-white parameter."

"Yes!" said Einstein. Suddenly he got up and started pacing the room. "And this has nothing to do with my God," he said. "It has to do with the fact that I cannot imagine such a thing anymore. I cannot construct a proper association, I cannot do a…," he made a pause and I suggested, "… a thought experiment, Professor?"

Einstein nodded.

"Professor Einstein," Holmes started, "what kind of solution would console you?"

Einstein thought for a while and I assumed that he was seeking an answer, but he was already far beyond that point.

"It is very interesting that you think that an answer consoling us would also be the one satisfying the universe, Mister Holmes."

"Well, this is obvious, is it not?" Holmes replied.

"Yes indeed!" said Einstein, "because otherwise there wouldn't be much reason for us…," he pointed towards his colleagues, "… to be here at all."

"Precisely!" Holmes cried out, delightedly.

"But there is only one thing which would make me feel more comfortable with the EPR paradox, and that is impossible."

"Nevertheless, we are here and even more strangely, we are here together." Holmes made a waving movement with his right hand. "That fact alone should tell us that something or someone must be of the opinion that there is a chance."

But Einstein was skeptical. "The only thing that could convince me would be to show that the funny non-localized-parameter stuff would also reside in my own theory."

Now Rosen cut in and said, "This, however, requires the unification of Quantum Theory and the General Theory of Relativity. So, we are talking about a task that whole generations of scientists have failed to achieve."

"But why not try and then see how far we will get?" I suggested and I felt Holmes' gaze riveted on me. The warm smile of my dear friend was filling the room and his eyes were full of confidence. But there also was something else in his features. At first, I could not quite place it, but then I recognized it. Holmes was amused.

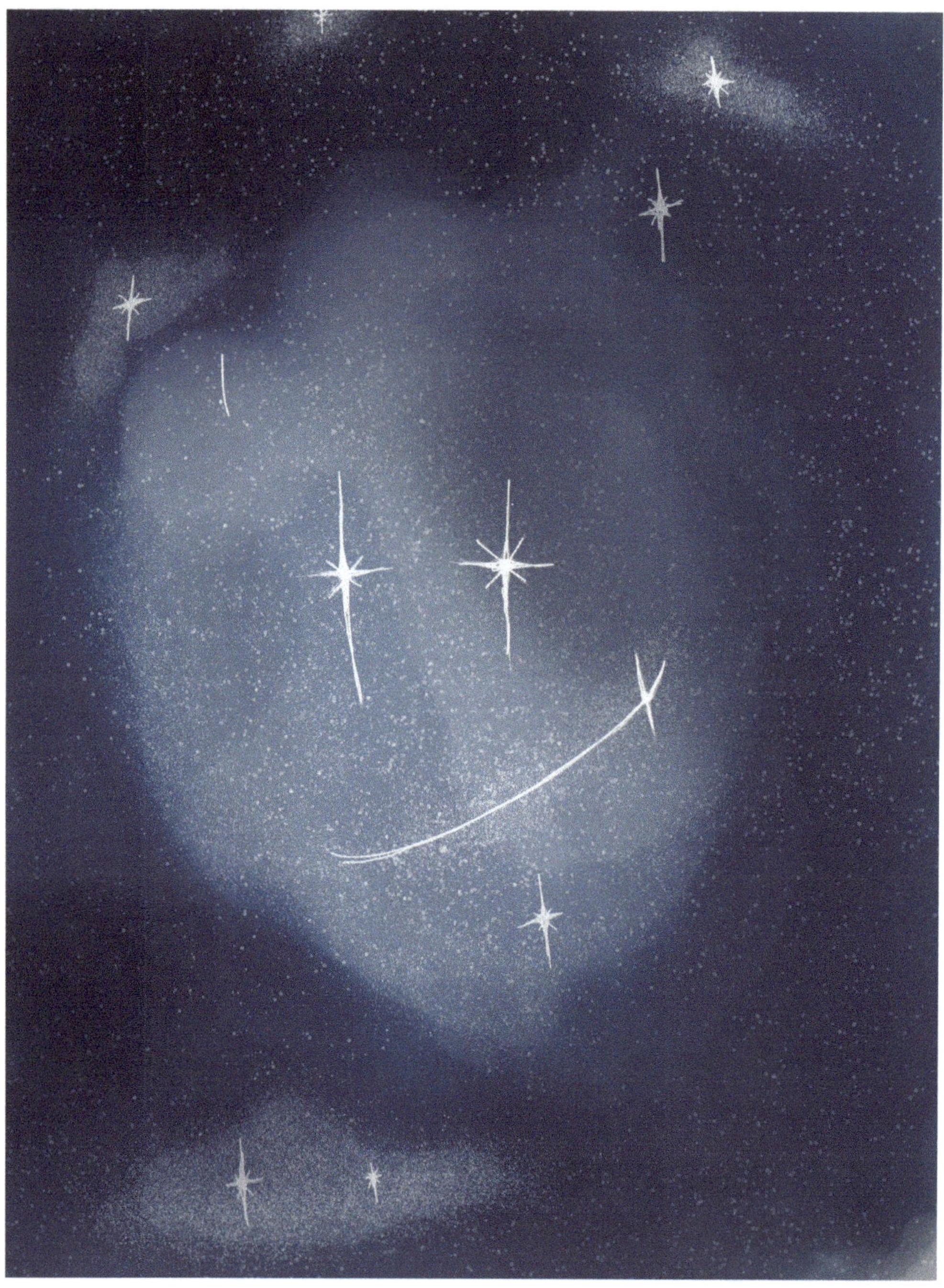

The smiling universe. [A]

4.7 Approaching the Problem

"Tell us about the other 'biggest blunder', Professor Einstein," Holmes said, and the person addressed cleared his throat at once. Obviously, this man had no problem admitting that he had made mistakes.

"Well, Mister Holmes, even though I have no idea how this could help us, that particular story is quite simple. After I had established my field equations, it was discovered that they, I mean the equations, give solutions to the whole universe. To my great surprise, however, the solutions demanded the universe to move… or evolve. I had always considered the universe as something static, something just being… existing. Now my own theory, my General Theory of Relativity suggested the universe to be most dynamic. This could not be. I could not accept this and so, I was looking for a way out and this is a way I found. I discovered that my equations were still correct if I added a constant term. I named this constant the 'cosmological constant' and gave it the symbol Λ, which is the big L from the Greek alphabet." Einstein made a short pause. Apparently, he was waiting for a remark, but nobody made any comment.

"Then Hubble, a famous astronomer in the United States, discovered the so-called red shift in the light of distant stars, which means he discovered that galaxies are moving away from us. The greater the distance to the moving galaxies, the greater the velocity with which these galaxies flee. It became immediately clear to me that this could only mean that the universe in fact was a dynamic entity and so I removed the cosmological constant. Which is to say I erased the one term I had just added from my equations and announced its introduction as the 'biggest blunder' of my life."

"And do you still think that it was a blunder?" Holmes asked.

"That depends…" Einstein said. Then he chuckled and said, "It is kind of relative, you know?"

"I see!" Holmes said. "Please, Professor Einstein. Would you be so kind and show us the full equation, I mean the one with this 'biggest blunder' in it?"

Einstein obliged. He stood up, strolled to the blackboard right next to the window and started to draw his famous equation.

"And now," Holmes demanded, "please, tell us what the various terms mean, Professor!"

"Certainly!" Einstein replied and pointed to the left-hand side of his equation. "This here is all just about curvature. It tells us how a certain space-time warps in order to find a stable state… a minimum in action, we physicists say."

"So, this whole thing comes out of a condition?" Holmes asked.

"Yes, of course," answered Einstein as if this was just obvious as anything, "the whole is nothing but the result of demanding that a space of a certain number of dimensions can only exist in states of minima."

"You mean, like finding the deepest point in a valley or even in a chain of mountains?" I asked.

"Precisely!" Einstein said.

"Interesting!" said Holmes, "and was it ever shown in a rigorous manner that your equations result from such a minimum principle?"

"Yes, of course!" Einstein said promptly. "A great mathematician of my time, David Hilbert, had shown this quite clearly. The only condition he needed was to demand a minimum for the curvature of the space. The curvature is this term here, this R. We call it the Ricci scalar. Then, when doing this minimum-evaluation-thing with the Ricci scalar, my equations come out automatically."

Einstein drawing another minimum. [A]

"That is quite fascinating," Holmes said. "So, there is nothing you need to postulate, except that you say:

A) Here is some space and the space has n dimensions. Perhaps if we set n=4 then we might speak about our ordinary space-time. Then…

B) … you say that this space could be warped somehow and you want to know the nature or geometry of this funny deformation.

C) So, you simply take the thing you called curvature and name it R in your equation…"

"The Ricci scalar!" said Einstein.

"… and then you do the Hilbert-trick and out comes your equation, right?"

"In a nutshell, yes!" said Einstein.

"What's the thing on the other side, this T-something?"

"That is matter," Einstein answered. "We called it the energy-momentum tensor and gave it the symbol T."

"Does it also come out of a minimum principle?" asked Holmes.

"No! Not really…" Einstein seemed uncertain, almost a bit uncomfortable, "To tell the truth, it had to be postulated."

"Why?" asked Holmes.

"Well, because we all thought there must be some matter in the universe and as the Ricci scalar alone does not show any matter, we had to bring it in somehow, didn't we?"

"In this case, discard it!" Holmes demanded.

The three professors looked flabbergasted.

"What?" Podolsky even cried out, "But where else should the matter come from then?"

"I don't care at the moment," answered Holmes, "but I most certainly do not want to bias my starting approach with something I have no idea where it comes from. Therefore, until there is no proper explanation about the origin of this T-term, would you please treat it as if it were not there… Thank you, Professor!"

"And what shall I put there instead?" Einstein asked almost a bit mockingly.

"A zero, of course!" Holmes answered simply.

Einstein obliged and drew a pronounced '0' after the '='-sign on the right-hand side of his equation. "But you should know that we then only have a vacuum equation, Mister Holmes."

"A vacuum is not necessarily nothing!" Holmes said simply and then added in a suddenly very busy tone, "And now we find the right solution for the 'spooky action mystery'!"

4.8 The Quantum Side of the General Theory of Relativity

"But there are already hundreds of solutions to Einstein's equations, if not thousands," Rosen said. "How shall we know in which direction to go? Where do we start?"

"Simple!" said Holmes, "there must be a set of solutions clearly sporting typical properties of quantum solutions."

"Like what?" Einstein asked, almost a bit snappishly.

"Like the solutions of Professor Dirac. I mean the ones where the time coordinate oscillates and can have two signs, namely one for matter and…"

"… the other one for antimatter," completed Podolsky. "My dear Mister Holmes, but the Dirac-solutions are solely solutions to the Dirac equation. Here, you might have missed that point, but trust me, it is an important one; we are talking about the Einstein field equations. This is something completely different. Or let me put it differently: We have two completely different equations, namely the one of Dirac and the one of Einstein and you expect us to find a solution satisfying both. This is impossible."

"Really?" said Holmes amused. Then he stood up and walked to the blackboard.

"How about this one here?"

With this, he wrote a rather simple expression on the blackboard and turned to his famous guests.

Sherlock Holmes and the metric Dirac. [A]

At first, they stared at him. Then, almost in perfect unison, they started to calculate. Einstein on the blackboard. Rosen and Podolsky on the table in front of them using sheets of paper and pencils apparently ready to be used for just this purpose. There was a bit of back and forth and a few exchanges between the blackboard and the table before the three faced Holmes and Einstein said, "Yes Holmes, that is a solution to my equation… and surprisingly…"

"It has the quantum properties of the Dirac solutions," added Rosen.

"And still a lot of degrees of freedom to be adjusted to the experimental observations," finished Podolsky.

Vivid discussion. [A]

"But how did you come up with it?" I asked.

"It was given to me." Holmes said and with this he opened a drawer and pulled out one sheet of paper. He handed it to Einstein, who only needed about three seconds to take in its content and then he asked, "Schwarzer? Never heard of this guy. Did you?" he addressed his colleagues.

Both shook their heads.

Holmes smiled. “That is quite understandable, gentlemen. After all, the person is a bit too recent to be known by someone who is as… well, as antique as you are. And what is more, he is not even a real scientist and this bit of paper isn’t even known among the scientific community today.”

“But the solution is correct, we just checked,” Podolsky said incredulously, “how come nobody knows this man or this solution?”

“Explaining this would lead us too far away. For now, it needs to suffice to tell you that nowadays the whole science and education system is often more a circus than a well-organized and fair community. Politicians and ideologists have taken over too many places and made it a farce, a mere show for an ever-dimmer audience to pretend importance and meaning where extremely often there is not a trace of either of the two.”

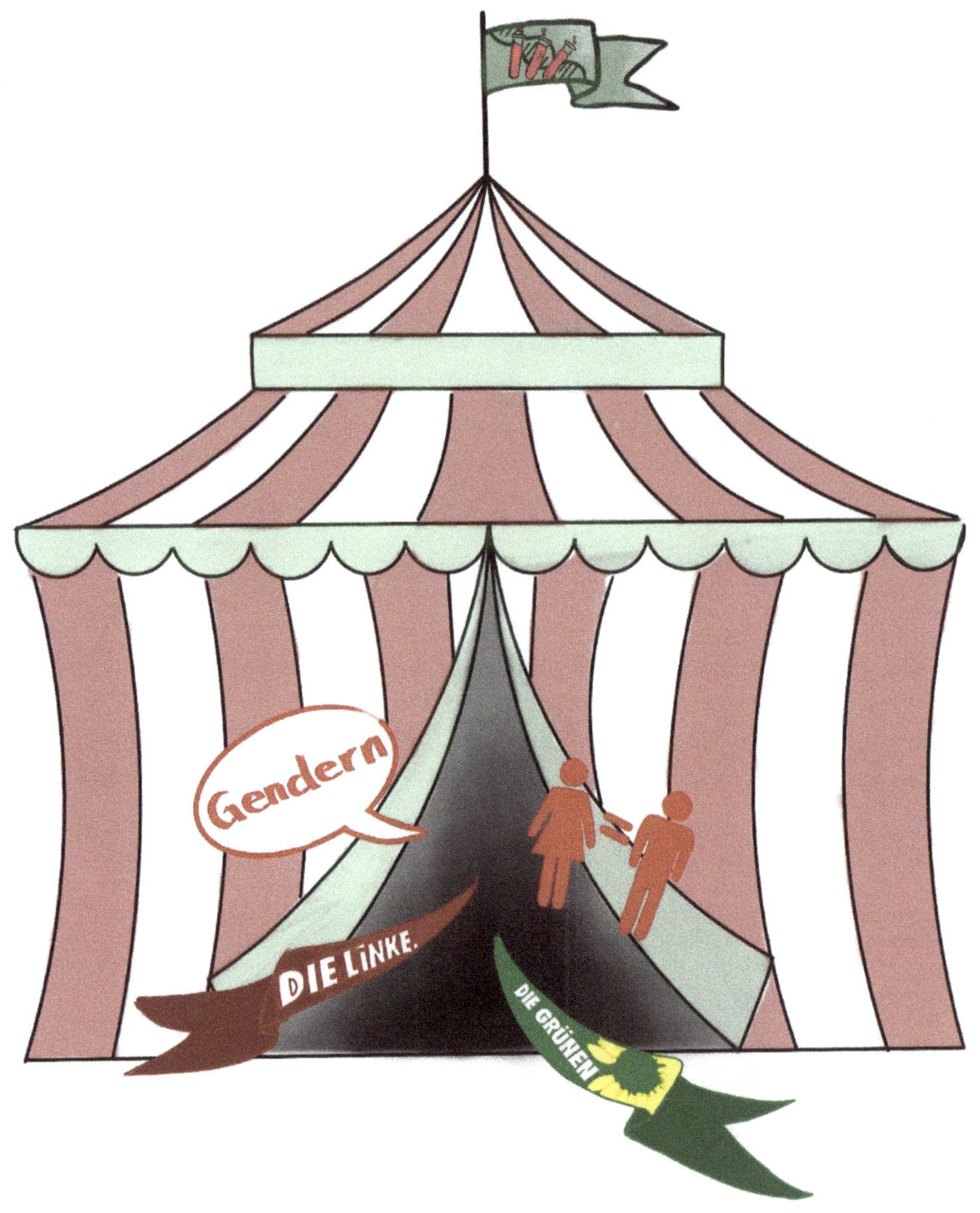

The science circus with only bad clowns… very bad and evil. [A]

“I’m very sorry to hear that!” said Einstein and his colleagues mumbled their agreement.

“Well, as bad as this may be, this can hardly be our concern right now. We have a different problem we want to solve and with the Dirac-like solution I think we do have a good start.” Holmes sounded extremely positive, but Einstein was not convinced.

“I’m sorry, but I simply do not see how this thing could be of help with the EPR paradox. It is great, of course, that we now have seen a solution to my equation, sporting quantum properties, but the EPR-thing is a different story, is it not?”

He looked straight at Holmes and when he found him smiling, he added, “You already have an idea, Mister Holmes?”

“Let’s call it a suggestion, Professor Einstein… just a suggestion. But before I will convey this to you, I need to ask you something. When reading papers about the General Theory of Relativity, one always finds hints about so-called test masses. What exactly does that mean?”

“Ha!” laughed Einstein loudly, “It means nothing else but that we have no true two-body solution. That’s just it.”

Holmes smiled. “Good!” he said, “This is what I thought it is and it suits me very well!”

“I’m sorry, gentlemen!” I cut in, “but I have to confess myself completely lost. What on earth is a two-body solution?”

“Oh, that is not a very complicated thing, Doctor,” said Rosen. “We have very nice solutions to the Einstein field equations as long as we are just talking about one body. Mister Holmes’ solution here, for instance…”

“It is not mine,” hinted Holmes, “I found it in a paper of a guy named Schwarzer and even he doesn’t claim it as his and therefore names it the ‘Dirac-Schwarzschild solution’.”

“… well, anyway, it also is a one-body or one-center solution and it seems to be so complicated to construct true two-center or two-body solutions in the General Theory of Relativity that we scientists resorted to the model of the test-mass. There we simply assume that a small mass, which is to say we have a mass small in comparison with another object, determining the warping of the space, comes near the space-warping object and, because the test-mass is so small, we can ignore the warping, which itself is producing. That’s just it.”

“So, you are cheating!” I said.

“Let’s better call it approximating,” Rosen suggested, smiling.

“Yes, and here, so I think, is the point,” Holmes cut in. “This approximation hinders you from getting the full picture. Without that little scientific skullduggery, you would probably be able to get the holistic view and find the missing something. Thus, you are missing exactly THE one thing you obviously need to understand the physical meaning about the peculiar effect which right now still appears as mystery.”

“The quantum guys will most certainly disagree at this point, Mister Holmes!” said Podolsky. “You know that their argument is that the entangled system is just one entity and that the black-white-property is non-local, which means, it is of such character that the position of the two particles and the huge distance they have from each other does not matter. For them, I mean quantum people, there is no communication and the EPR-treatment is not correct. In other words, they consider the problem solved, which is to say they consider it non-existent."

Holmes only laughed. "In other words: they ignore the observable distance, cannot give any explanation why this distance does not matter with the black-white-property, and forbid you to think about that, because they say it is the wrong question, right?"

"Yes, Mister Holmes," said Einstein, "it is a little hard to put it like you just did, but in essence, you… kind of… got the point."

"In other words," Holmes added, "the quantum believers say to us that the quantum world is too special to be understood anyway and that we simply have to accept its strangeness. Seeking an explanation for its strangeness is prohibited or at least counter argued with the hint, that Quantum Theory is nothing one can truly understand and therefore it is foolish to the extent to even try to do so."

"But you are not willing to accept this, Mister Holmes, are you?" asked Rosen.

"No!" Holmes said flatly, "99% of my cases I only solved because I never did accept such restrictions, such human made limitations, and I will not start to do it now."

"But Holmes," I blurted out. I simply could not restrain myself, "You say Quantum Theory should not forbid us to seek for reasonable explanations, but at the same time you accept the strange things that General Theory of Relativity gives out. Where is the difference?"

"Two reasons, my dear Watson, just two reasons:

a) The General Theory of Relativity comes out of a nice and clear mathematical principle. I am referring to the minimum principle of Hamilton in the form of Professor Hilbert's action integral. It goes like that: you start here and you end there… there is no fuzzy stuff in between, except for this postulated matter term. However, as you remember, I had just thrown it out of Einstein's theory, because I do not consider it genuine. Sorry for that, Professor Einstein!
b) I do not know about you, my dear Watson, but a curved space-time, my brain can handle, but non-local parameters and arbitrary, almost religiously dogmatic rules like this about 'it is just one system' or 'this is nothing the human brain could understand anyway' or even 'don't ask questions', I cannot bear."

"This is all very nice, Mister Holmes," interrupted Einstein, "but I don't see how to proceed from where we are now in order to come forward with this EPR-paradox or EPR-non-paradox. What is the 'sanity and reason' way to grasp the effect of entanglement?"

"I'm very glad you brought it down to this straightforward question, Professor," answered Holmes and smiled. "In my very humble opinion the key for the correct handling of entanglement or the EPR paradox…" Rosen snorted and Holmes added, "well, EPR-paradox or EPR-non-paradox then… Anyway, I think it all comes down to the two-body or two center problem. Without a consistent model for this thing, we will helplessly run in circles and this, I assure you, is not my preferred type of exercise."

Holmes lecturing. [B]

“But so far nobody has such a solution, Mister Holmes!” cut in Podolsky, “and please rest assured, we all have tried our best to find one.”

“Hm!” said Holmes thoughtfully, his hands under his chin, “then every one of you and the whole rest of the community, too, must have looked in the wrong direction.”

“With all due respect, Mister Holmes, but how on earth would you know where we have already looked?” Rosen almost sounded a little bit offended.

“Oh, that is obvious,” answered Holmes lightly, “pairs of bodies exist in reality and so there must be a way to mathematically describe such systems. Because, obviously the universe can handle them. However, as nobody has found the corresponding solution, they must all have sought at the wrong places… I mean as long as we exclude those blind dunderheads who would not see the solution even when having it right in front of them.”

Einstein laughed, while Rosen and Podolsky looked rather sour.

“Ok, Mister Holmes,” asked Rosen after Einstein had finished laughing, “and where should we start to search for the perfect solution then?”

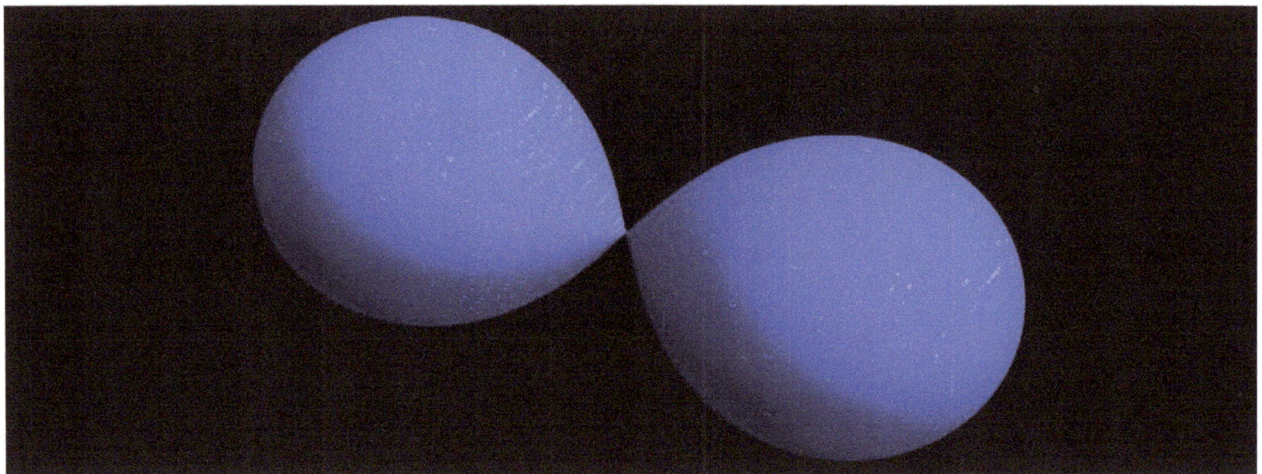

The two-body-problem.

“In the infinite space of a few more dimensions, preferably orthogonal ones, of course!” answered Holmes simply. Everybody in the room stared at my friend. We expected him to elaborate, but Holmes stood up, strolled to the blackboard again and simply extended his previous drawing. While adding term after term, the whole still looking relatively simple, he elaborated his idea. “Until today, you have always tried to mathematically treat the second body in the same space where you already had placed the first one. You thought that the coordinates of your mathematical space have to do with the spatial directions of your own experience, but this is not the case. You must separate the mathematical space from the real one, because the mathematical one does not stand for a spatial reality at all, it only gives the degrees of freedom of a certain object.”

“Good Lord,” said Einstein with dawning comprehension.

“As long as you have only one object,” Holmes went on, “you have the usual coordinates for space and time. Then, when talking about our normal space, for instance, you have the three dimensions for space and an additional one for time and these are also the degrees of freedom for your single object. However, when adding a second object, you should be aware of the fact that this object brings in its own degrees of freedom and this…”

“… this automatically increases the number of dimensions I have to treat the two-body system with!” Einstein finished the sentence for my friend.

“Precisely, Professor!” Holmes said.

Meanwhile Holmes had finished his drawing and stepped back from the blackboard. Einstein took one look at the new equation and cried, “So, it is eight dimensions now! You have simply doubled the number of degrees of freedom when going from one object to two.”

“That is correct, Professor,” said Holmes, “but may I attract your attention to the important fact that there is a peculiar coupling among the objects when solving the whole system. Which is to say, when introducing my little eight-dimensional approach here…”

"Metric!" said Einstein.

"Pardon me?" I asked.

"Metric!" repeated Einstein and then elaborated, "We call these systems, I mean the ones Mister Holmes has just drawn on the blackboard, metrics, Doctor Watson. No need to worry. It is only an expression which shows us that we deal with a certain mathematical tool which describes the state of a space."

"Ah!" I said and thought that his explanation was rather compact for a layman like me, but nevertheless, I thought to have gotten the gist about that thing on the blackboard and that was fine with me at the moment. After all, I would be able to ask Holmes about it later, when the "wise men" were gone.

While I was still mulling over the "tool" on the blackboard, the professors had set to work on Holmes' new "metric solution" and as they now had to consider eight dimensions instead of only four, they were considerably more "busy". Whereas I am using this funny description, which means the word "busy", here to point out that there was much more excitement among them than before.

Finally, Rosen cried, "They are coupled…"

"Entangled!" said Einstein.

"… via their cosmological constant." And Podolsky immediately added, "And as this cosmological constant is – well, just as the name says – a constant everywhere…"

"In the whole cosmos the system spans!" said Einstein.

"… we have found the coupling parameter holding the two objects firmly bound together."

"But it depends on the property the constant couples in. That is quite important!" elaborated Einstein, who had jumped to his feet and started to pace the room while he was speaking to us. "While here it is the two time-coordinates for each of the two objects and I'm sure, Mister Holmes, you have only chosen this, because this allowed us to reuse the results from the first one-body solution you had named a Dirac-Schwarzschild solution…? ..." Einstein stopped for a moment in his pacing and looked at my friend imploringly.

Holmes simply nodded.

"… we can couple…"

"Entangle!" I heard myself say.

"What?" Einstein was totally taken aback.

"Entangle, Professor Einstein!" I repeated. "I guess this is a bit more precise than 'couple', because I have realized that, so far, you had insisted quite intensely in using this term rather than anything else, have you not?"

There was a funny pause. Podolsky and Rosen chuckled and Holmes was almost beaming at me. Einstein took the pipe out of his mouth, smiled briefly, and then started pacing again.

"Where was I?" he said. "Ah yes. In this extremely simple example, it just happened to be that the entanglement was on time, because Mister Holmes was so kind to choose a close enough solution to the one-center case we had before – thank you for your thoughtfulness and consideration, Mister Holmes. But we clearly see that we could do the entanglement on every other pair or probably even triple or any tuple of degrees of freedom within this extended metric. The entanglement via the cosmological constant always assures the coupling. Be it with time, to perhaps couple matter and antimatter, be it spin, charge… it does not matter. I see no limits here. Every property… at least as far as I can see right now… could be made an entangled one by the cosmological constant… or is it constants?"

Einstein suddenly stopped. He faced us, looked everybody in the room straight in the eyes, and cried, "But this means…"

"… that we were right!" shouted Podolsky and Rosen in unison.

"Yes, my friends!" Einstein took over again. "There are 'hidden' parameters providing the entanglement, only that these are not truly hidden. As cosmological constants they are omnipresent and everywhere in the cosmos. They always assure the entanglement. There is no specific quantum-mystery after all. It was already inside my equations. All the time. The cosmological constant is the 'not so very much hidden parameter' and as it is just present… present like a celestial God, I'd almost say, the connection of the entanglement is guaranteed. Everywhere and everywhen, I'd say. But then… wait a moment." Einstein almost ran to Holmes' desk, stopped sharply before the edge, stared at my friend, and cried, "You knew!"

"Rest assured, Professor Einstein," my friend answered very calmly, "I did nothing more than guess."

"Sorry, gentlemen, but what is the sudden problem?" I asked.

But Einstein ignored me. He continued to stare at Holmes and said, "There are infinitely many of them. Entanglement is everywhere. Entangled systems, which are all little cosmoi, I mean, with their own private cosmological constant… Our whole universe is just brim-full with them."

"Yes!" said Holmes, "that is what comes out of your own equations, Professor Einstein."

"And the measurement?" asked Rosen. "What about the measurement, I mean when we determine such an entangled parameter with one particle? What happens to the system… to the…"

"… the little cosmos?" Einstein asked. "You mean, what happens to the little entangled system, which is forming its own small cosmos… a cosmos of two, of three, and so on? You would like to know what happens to this cosmos?"

Rosen nodded. That was his question.

"It will be destroyed during the measurement!" Holmes said softly.

There was silence for quite a while. Then Einstein mumbled, "The collapse of the wave function is the death of a minuscule universe… well, its coupling!"

"Sorry, gentlemen, but what is what?" I asked.

Podolsky looked at me in surprise, but then his eyes showed pity and he started to elaborate. "The quantum people saw that there is a problem in interpreting their funny combined black-and-white property, which should be either the one or the other, but in fact is always both… I mean it is both until the measurement. Then the quantum state becomes clear and its mathematical description takes on one specific form. As they named this mathematical description a 'wave function', they are speaking about the 'collapse of the wave function' in the moment of the measurement. Now your friend was just suggesting that this mysterious collapse is nothing but the destruction of a mini-cosmos. Fascinating, I'd say!"

"The perfect weapon," Rosen muttered and everybody looked at him in surprise. Everybody except Holmes.

"What did you say?" Podolsky asked.

"Can't you see the potential in this?" Rosen asked back. "Just look!"

He took a sheet of paper and wrote a few equations. Then he drew an arrow from his final equation to the metric Holmes had given on the blackboard and which they had copied from there. Podolsky took his hands over his mouth. Einstein looked shocked.

"God save us!" he said.

The space-time bomb.

"God save mankind!" Podolsky added.

There was silence. Then Rosen cleared his throat. In a hoarse voice, he addressed his colleagues, "What are we going to do about this?"

Silence again. Then Einstein jumped to his feet, "Nothing!" he said firmly. His two colleagues stared at him and he added, "I know what you expect me to do, but I'm not writing a letter to a president or any other leader, any politician in this world. Remember what happened last time! Politicians are bad, they are rotten from the inside and they are always dim. Otherwise, they would not have had the need to go into politics in the first place. Remember what I once said about things being infinite. Well, I'm still not so sure about the universe, but I'm definitely sure about the infinity of the stupidity of politicians, mass-media-people, and bureaucrats. The universe might be limited in some way, but the stupidity of politicians and their entourage most certainly is not. The only people you can trust to do things right in socioeconomic and political matters are those who are in politics but because of their qualification and not because of their political instinct (cunningness more like) or – even worse – their ideology. These trustworthy persons are people who usually came from the outside of politics and not from within this horrible, this evil, disgusting system… full of disgusting, evil people. Never trust a life-long politician or polit-professional. No, I will never again write a letter to these awful people."

A typical YGL politician. [A]

He hammered his fist on the table. Then he added, “We have to leave the matter in the hands of the holistic thinkers!” He looked intensely at my friend. “We have to leave it in the hands of those who are capable of seeing it all, those who are free of religious imbecility and ideological ignorance.”

He nodded in my friend’s direction who said nothing. Then Holmes simply stood up and made a small bow.

Are there holistic thinkers out there and will they save mankind? [A]

4.9 Good Bye

“What was it about this letter to the president?” I asked Holmes, after the three men had gone.

My friend did not answer. He bent forward and pulled two sheets of paper from the very drawer from which he had produced the Dirac-Schwarzschild solution earlier. Without a word, he handed the papers to me and I started to read. It was a letter and it ran like this:

Albert Einstein
Old Grove Rd.
Nassau Point
Peconic, Long Island

August 2nd, 1939

F.D. Roosevelt,
President of the United States,
White House
Washington, D.C.

Sir:

Some recent work by E.Fermi and L. Szilard, which has been communicated to me in manuscript, leads me to expect that the element uranium may be turned into a new and important source of energy in the immediate future. Certain aspects of the situation which has arisen seem to call for watchfulness and, if necessary, quick action on the part of the Administration. I believe therefore that it is my duty to bring to your attention the following facts and recommendations:

In the course of the last four months it has been made probable - through the work of Joliot in France as well as Fermi and Szilard in America - that it may become possible to set up a nuclear chain reaction in a large mass of uranium,by which vast amounts of power and large quantities of new radium-like elements would be generated. Now it appears almost certain that this could be achieved in the immediate future.

This new phenomenon would also lead to the construction of bombs, and it is conceivable - though much less certain - that extremely powerful bombs of a new type may thus be constructed. A single bomb of this type, carried by boat and exploded in a port, might very well destroy the whole port together with some of the surrounding territory. However, such bombs might very well prove to be too heavy for transportation by air.

The United States has only very poor ores of uranium in moderate quantities. There is some good ore in Canada and the former Czechoslovakia, while the most important source of uranium is Belgian Congo.

In view of this situation you may think it desirable to have some permanent contact maintained between the Administration and the group of physicists working on chain reactions in America. One possible way of achieving this might be for you to entrust with this task a person who has your confidence and who could perhaps serve in an inofficial capacity. His task might comprise the following:

a) to approach Government Departments, keep them informed of the further development, and put forward recommendations for Government action, giving particular attention to the problem of securing a supply of uranium ore for the United States;

b) to speed up the experimental work,which is at present being carried on within the limits of the budgets of University laboratories, by providing funds, if such funds be required, through his contacts with private persons who are willing to make contributions for this cause, and perhaps also by obtaining the co-operation of industrial laboratories which have the necessary equipment.

I understand that Germany has actually stopped the sale of uranium from the Czechoslovakian mines which she has taken over. That she should have taken such early action might perhaps be understood on the ground that the son of the German Under-Secretary of State, von Weizsäcker, is attached to the Kaiser-Wilhelm-Institut in Berlin where some of the American work on uranium is now being repeated.

Yours very truly,

A. Einstein

(Albert Einstein)

When I had finished reading, I looked questioningly at my friend and he finally spoke, “This letter, my dear Watson, led to one of the greatest research projects ever performed by mankind. They named it the ‘Manhattan Project’.”

“What was the goal of this project?”

“Well, more important than the goal probably is the question about the result. The project succeeded and so the goal also was the outcome,” Holmes elaborated.

“And what was the outcome then?” I asked.

Atom bomb explosion. [A]

“It was the atom bomb, Watson!”

Silence.

…

“What do you think, Holmes? Where did they go?” I asked.

“On!” Holmes answered simply and then he added, “However, in these modern times, some might rather formulate it as ‘they went to the next level’.”

Then he smiled and said, “I am of the opinion, Watson, that right now, it would be the perfect time for a nice cup of tea, don’t you think?”

A special tea. [A]

4.10 Epilog

I was almost bursting with the question, but pulled myself together and it was only after the tea was on the table, cups were filled, the honey was properly stirred into the hot, steaming liquid, that I dared to bring it forward.

"Holmes?" I started as casually as I could, but the moment the word was out of my mouth I knew that it already had betrayed me. Nevertheless, my friend sat rather quietly. He laid the spoon on the tray, lifted the saucer and cup, but did not drink. Instead, he only inhaled the steam wafting from the tea. "Fine Indian tea," he said as if he had not heard me. But then, he added, "Yes, Watson I know. You have asked yourself two questions…"

Sherlock Holmes with a cup of tea... or a cup of tea with Sherlock Holmes? [A]

"Ahhh..." I mumbled, because until this moment I had not noticed myself, that it actually was two questions, but when thinking about it...

"For one, which is the most obvious, you have asked yourself, why it was only us, I mean the three famous scientists plus you and me. Clearly, it would only have been most reasonable from the universe to also invite the very man who had recently come up with the idea, which turned out to provide the necessary prerequisite to solve the riddle, would it not, Watson?"

I nodded. Holmes sipped from his tea and while he was savoring it, I asked myself which part of his famous brain right now was busy with the analysis of the liquid in his mouth.

"Lime!" he said.

I concluded at once that he must refer to the honey, because the tea truly came from India and could not have anything to do with a lime tree.

"How can you taste this, Holmes?" I asked, "I mean this is a particular strong tea and the honey in it doesn't give much away, apart from its sweetness, but still, you…"

A glass of lime honey. [A]

"It says on the label, Watson!" my friend answered, pointing to the glass of honey on the tray. I felt embarrassed, but inwardly I had to smile nevertheless. My friend was not the type to harp and so he just drank a bit more from his tea and then he said, "There can be only one reason why he was not here today and I'm sure you already know, don't you, my dear Doctor Watson?"

Holmes did not wait for an answer.

"It is the subsequent, which is to say the second, question immediately following out of the answer of the first, which hinders you to make this very first answer manifest, is it not?"

I felt my mouth going dry. My friend was right, of course. It was not too difficult to give the only logical answer to the question why the man who had come up with the essential ingredient has not been with us today. I mean, in the same way as we were there. But what did this mean? Immediately, just as Holmes had predicted, that second question popped up in my head and I said, "But what are we then, Holmes?"

Holmes smiled and one could see that he was rather pleased with himself.

"I'd say that we are pretty good programs, Watson… or, as Einstein would probably have said, we are quite fascinating thought experiments, doing our job inside a nice 'computer'. A computer, which in fact happens to be a rather beautiful mind. It probably does belong to the one person you were missing and that itself is nothing but a thought experiment on an even greater level."

"And this awareness does make you feel… comfortable?" I asked.

But Holmes did not answer. He simply studied me. I felt like yet another interesting case to him, … like a small one, however, almost insignificant. Then I heard my friend say, "Just drink a bit more of that wonderful tea and then you only need to ask yourself one thing. It will immediately make you feel much, much better:

Why am **I** here?"

A thought in the author's brain. [A]

5 Creation and Separation – The Two Towers of Key Technology

In the abstract to this book, we announced a set of experiments, processes, or observations, which we intend to discuss and – except for protected intellectual property – also to explain. Splitting up this list into two main aspects, we might distinguish between:

A) Creation
   a. The Einstein-Podolsky-Rosen paradox and the spooky action at a distance
   b. The appearance of so-called Sigmoid or S-curves everywhere in this universe, especially in the processes of life
   c. The origin of evolution
   d. An innovative, flexible, and quite general deposition technology
   e. The 720° half spin oddity
   f. The intrinsic action of the human mind in the metric picture

B) Separation
   a. Schrödinger's cat, which we are going to deliver from its unjust fate of an existence in an infinite limbo state
   b. The so-called 3-generation problem of elementary particles, which we are going to solve by the means of the Bianchi identity in a scalarized Quantum Gravity Physics realm
   c. The particle-wave duality, which we will address via a fundamentally based mathematical approach leading to the destruction of the concept of particles
   d. An almost alchemic decomposition experiment
   e. The 720° half spin oddity
   f. The intrinsic action of the human mind in the metric picture

The attentive reader will have realized that the points e and f (bold) appear in both categories. This is because, with respect to point f, the human mind – in principle – as a very general "machine" is capable of both processes: creation and separation. We will therefore consider it separately in a third category (see further below, section "The Third Tower: The Mind").

With respect to the half spin symmetry oddity, something which has bothered many "ordinary thinkers" for over a century and which was always presented to us by the "quantum enlightened" as "that is the nature of quantum physics, something you don't understand because your brains are not quantum compatible, so shut up", we will see that both key technologies literally force us to rethink the problem and remove the oddity. Along the way – and this again couples back into point f about the human mind – we came to the conclusion that the self-exclaimed "quantum enlightened" had no idea about a proper explanation (a truly quantum gravity compatible one, we mean) themselves for the fact that something needs to be rotated twice before it is back to its original state before the rotation. Because this is what the half spin oddity is about. Concepts to explain this oddity, and which one does find here or there (e.g., [214]), trying to mitigate the half spin-strangeness via spatial entanglement only solve half of the problem as there never is an explanation about the character of the entanglement and what causes it. In this book we will offer such an explanation.

While from the other, which to say the non-mind- and non-spin-related, points the problems addressed in Aa to Ac and Ba to Bc are so well known that they need no explicit introduction, we will give the necessary basic information directly in connection with their mathematical formulation further below (with a few exceptions, mainly after the theory section) in this book.

Only the innovative experiments Ad and Bd require a proper phenomenological description. Thereby, we clearly state that for the reason of protecting the inventor's IP, we – of course – cannot give away the whole technology, but will only describe the observations – and hand over – the necessary tools, which are prerequisites for a general assessment, evaluation, and discussion.

5.1 Brief Introduction of an Innovative, Flexible, and Quite General Deposition Technology

There are many thin film deposition technologies and they are the cornerstones in many industrial fields. The reasons are clear: coatings increase the lifetime and reliability of all sorts of parts and tools and contribute considerably to productivity. Although coatings have been in production for many years, optimization of a coating-part and -tool design for a specific application is still a challenge due to the many influencing variables in the application process. Taking mechanical cutting, for instance, we have: temperature, cutting tool material and geometry, workpiece material, chemical interactions, and diffusion processes. The mechanical properties of the coating-tool system undergo changes during the cutting process due to the numerous influencing parameters. The same, respectively, something similar can be said about coatings and thin films in almost all other applications. The corresponding tests for a specific application are therefore still the fastest optimization process of the coating-tool or coating-part system. These tests reveal two typical weaknesses of almost all deposition techniques, which is the adhesion / cohesion of the coating and an often significant and most undesired deviation of the coating properties from the bulk properties of the coating material. The coating deposition technique that **we refer to here in this book as the first tower or T1 innovation**, does not have such weaknesses… at least not in the same severeness as this is known from almost all other deposition techniques. In addition, the T1-innovation shows an extremely high flexibility due to a huge number of degrees of freedom.

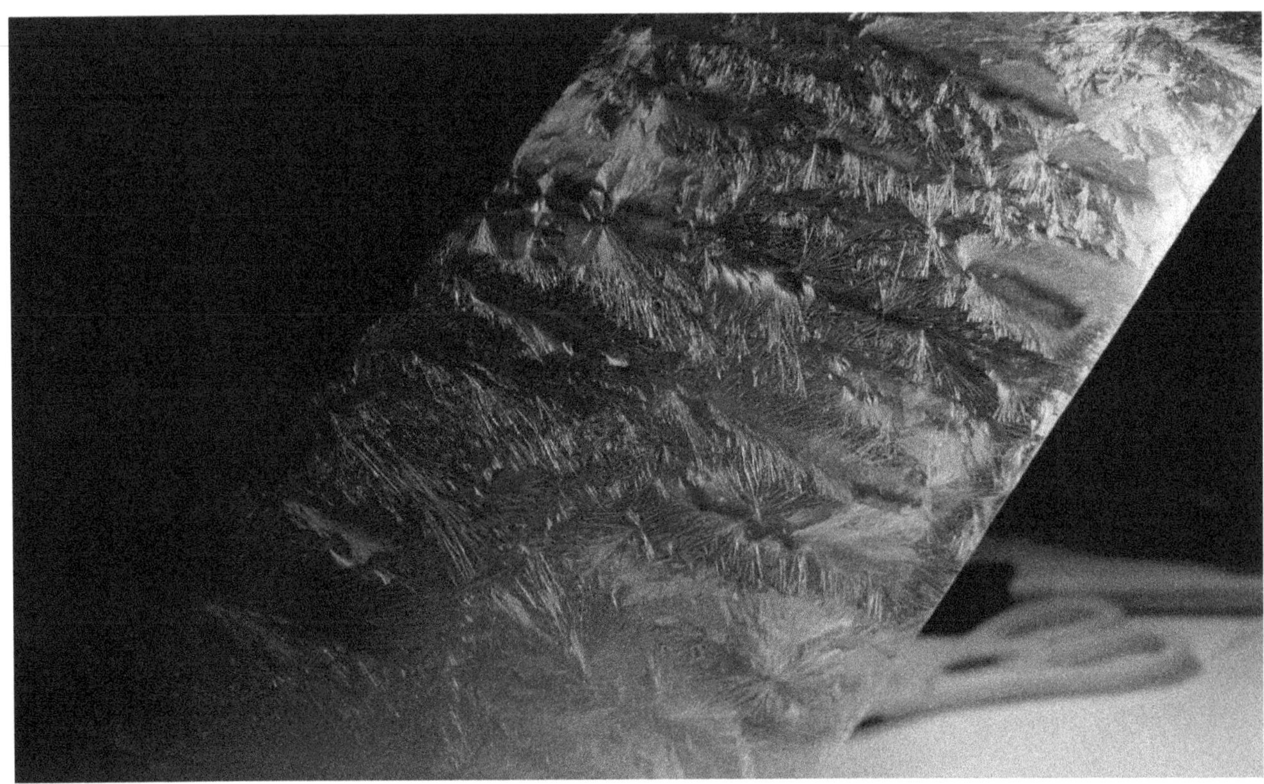

Fig. 1: A T1-coating example.

5.2 Brief Introduction of an Almost Alchemic Decomposition Experiment

Imagine mixing a bit of tar sand with a certain liquid, an, as we call it, Polar Selective Agent, or PSA.
Imagine not only to get the – sorry, not just the, but ALL – hydrocarbons out of the horrible mélange, but also result in completely clean sand and water, which resides as residual in the tar sands. Imagine, that you will observe no significant change in temperature regarding the essential reaction phase and that you will only require a little bit of heat for a final separation stage.

What sounds like a miracle is in fact **the second tower or T2 innovation we are going to consider in the book.**

Fig. 2: The second tower T2 at www.rasa.energy

Thereby, so it should explicitly be noted, is the set of various Polar Selective Agents so numerous and flexible that the same technology can principally be applied to a great variety of completely different separation processes. It is for this reason that one might have named the agent also "Poly Selective" or "Poly Polar…" instead of just "Polar…"

5.3 The Two Innovations with Respect to Their Dimensional Behavior

The following table orders the two processes or experiments Ad and Bd (see the beginning of this chapter) with respect to their dimensional behavior.

	1st Tower = T1	2nd Tower = T2
Preparation	Surface cleaning and environmental conditioning	Agent composition / reaction
Activation	Mainly electromagnetic-, EM-based affinity optimization for the deposition. Thereby activation mainly of target / substrate	Statistical electromagnetic mechanical, (STEMM), activation of the selective agent
Processing	Deposition	Separation and Removal of Target Elements
Dimensional Impact	Dimensional increase at target	Dimensional decrease of target due to separation into subsystems

Table 1: The dimensional behavior of our general deposition (1st Tower) and our separation (2nd Tower) technology.

As already hinted in the abstract of this book, we immediately see that the two innovations Ad and Bd are dimensional opposites, because one process adds additional dimensions to a given target, while the other one reduces the dimensionality of the target due to separation and/or extraction. It will be worked out in the section "The Third Tower: The Mind" that we can observe the same features or processes in the operation modes of the very system we consider as the human mind. It is due to this fact that we can conclude that understanding of the technical processes T1 and T2 on a sufficiently fundamental level will also give us access to the comprehension, modeling, and perhaps even optimizing and steering of mind-forming processes. In other words, as the recipes are in principle the same, respectively at least very similar to what we have at hand for the mathematical formulation of the innovations T1 and T2, we can also come to a mathematical description of the intrinsic processes of the human mind. Combining the latter then with the two T1-T2-innovations by using our own general entanglement techniques will result in no more than a socioeconomic "gamechanger". In order to protect this idea, we here refrain from revealing any technical details or physical-mathematical explanations for T1 and T2, but will give all necessary ingredients and a variety of examples with respect to the mind and a few other applications. We will see that there exists a universal analysis technique, providing us with a new concept for the constructive understanding of any experiment or process. This technique follows from the failed attempt of a point-wise localization of "something", which could be "anything" within a universe governed by minimum principles. As a result, one obtains the contradiction of the concept of particles. There are only waves and thus, consequently, everything of higher complexity can be decomposed into basic oscillations and ground waves and the latter can be used to compose things of higher complexity and structure. Mathematically, this just means that the general recipe for composition and decomposition – of everything – requires nothing more than a generalized Fourier formation and a generalized Fourier analysis, respectively. Another interesting consequence pops up in connection with the so-called Einstein-Podolsky-Rosen or EPR paradox and the alleged "spooky action at a distance" and it was for this reason that we have chosen the paradox for our motivator story at the beginning of this book. To illustrate some details of the theory being of significance for our general purposes here, we needed a figurative understanding of the state of entanglement and it is exactly the EPR paradox which provides quite some entertaining insight into the matter.

With such insight, the job of understanding and mathematically describing the above-mentioned experiments, processes, or observations will still be "a long hard slog", but it will probably be more fun. Thereby, please note: We cannot reveal any process-relevant or theoretical essentials, which could potentially endanger the intellectual property related to the T1 and T2 innovations. Here, interested readers may please contact the authors directly.

5.4 The Connection of the Two Towers to Certain Examples, Processes, and Theories from Apparently Completely Different Fields

Our selection of examples Aa to Ac and Ba to Bc from various fields is based on the following connection to the two innovations Ad and Bd:

5.4.1 The Einstein-Podolsky-Rosen (EPR) Paradox and the Spooky Action at a Distance

The paradox and its explanation give insight into the matter of dimensional entanglement on a very fundamental level. It is shown in this book that the consideration of the position of objects requires the observation of additional dimensions in connection with the main attributes of a given space-time. The subsequent higher orthogonal[1] dimensionality allows us to understand the processes of creation and separation via changes of such dimensionality with entangled dimensions. This also allows for an effective mathematical description (see this book).

Thus, entanglement, respectively dimensional entanglement, as we see, is the cornerstone for the understanding of creation and separation processes in general. Applying this generality to "mysterious" processes like our deposition technique T1 and our separation experiment T2 will help in their understanding and mathematical formulation, because here, too, dimensions are added or taken away.

Another aspect of the EPR paradox, which is important in connection with the second experiment T2, in fact, is the spooky action at a distance. With the latter now being not so spooky anymore, we might have a different explanation for the astounding affinity-behavior between the target element and the separation-agent, also – apparently – happening at a rather spooky distance. Classically, namely, the chemical potential (be it covalent, ionic, complex, or whatever else) does not give a conclusive explanation for this "distant interaction", its selectivity, and its scale-invariance (reaching over target chains from 2C to 230C and more).

[1] Thereby it is the orthogonality which assures the individuality of each dimension.

5.4.2 The Appearance of Sigmoid or S-Curves

… everywhere in this universe, especially in the processes of life.

Especially the deposition technique T1 as a growth process will profit from a thorough understanding of the functional dependencies coming with growth in general. However, these functions are also connected with one of the most important evolution equations there is, namely, the diffusion equation. Diffusion, on the other hand, is strongly connected with separation processes and as we will show in this book that a first principle-based derivation of this equation leads to an ordinary diffusion and an anti-diffusion, providing some insight into the strange behavior observable in the separation experiment T2 along the way, we definitively should cover the S-curves in here.

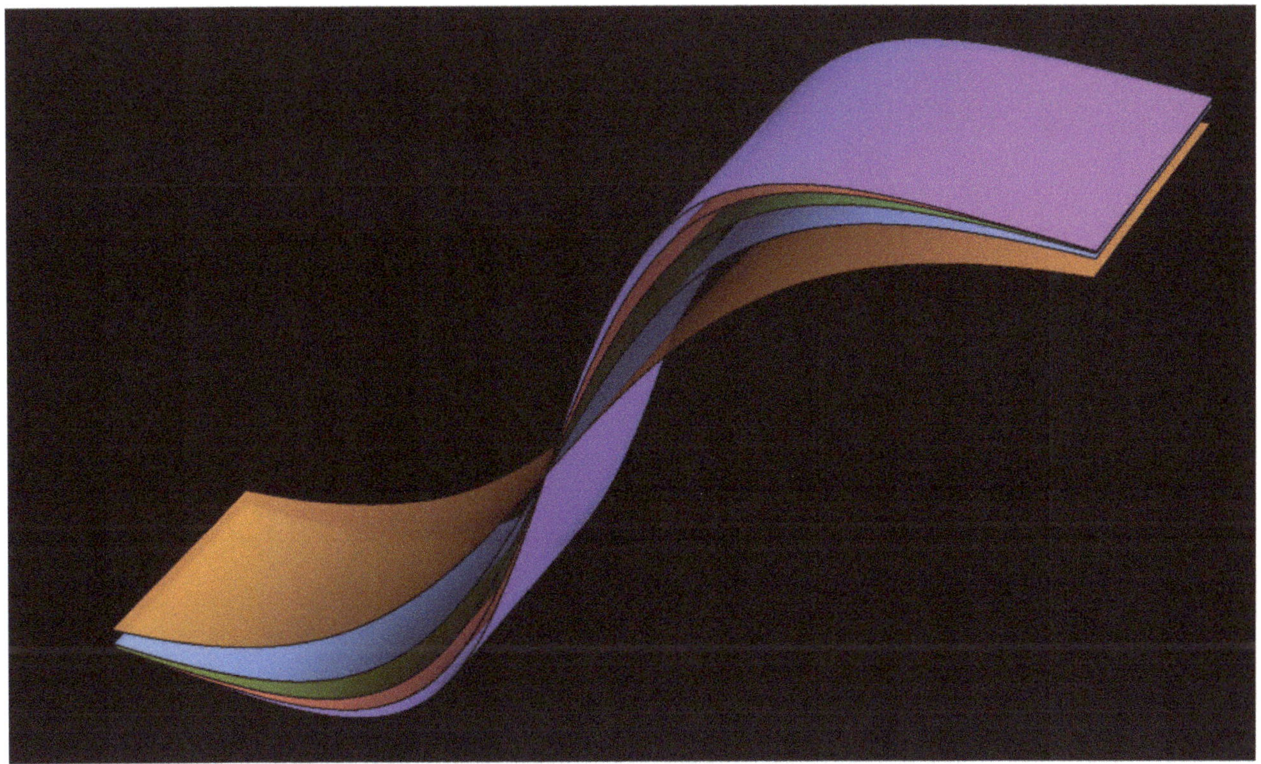

Fig. 3: Sigmoid or S-Curves – you find them everywhere.

5.4.3 The Origin of Evolution

Apart from what was already said in the subsection above about the importance of the understanding of Sigmoid curves in connection with our innovations T1 and T2, we also must consider the influence of self-organization which automatically leads us to evolution and its many derivatives and appearances.

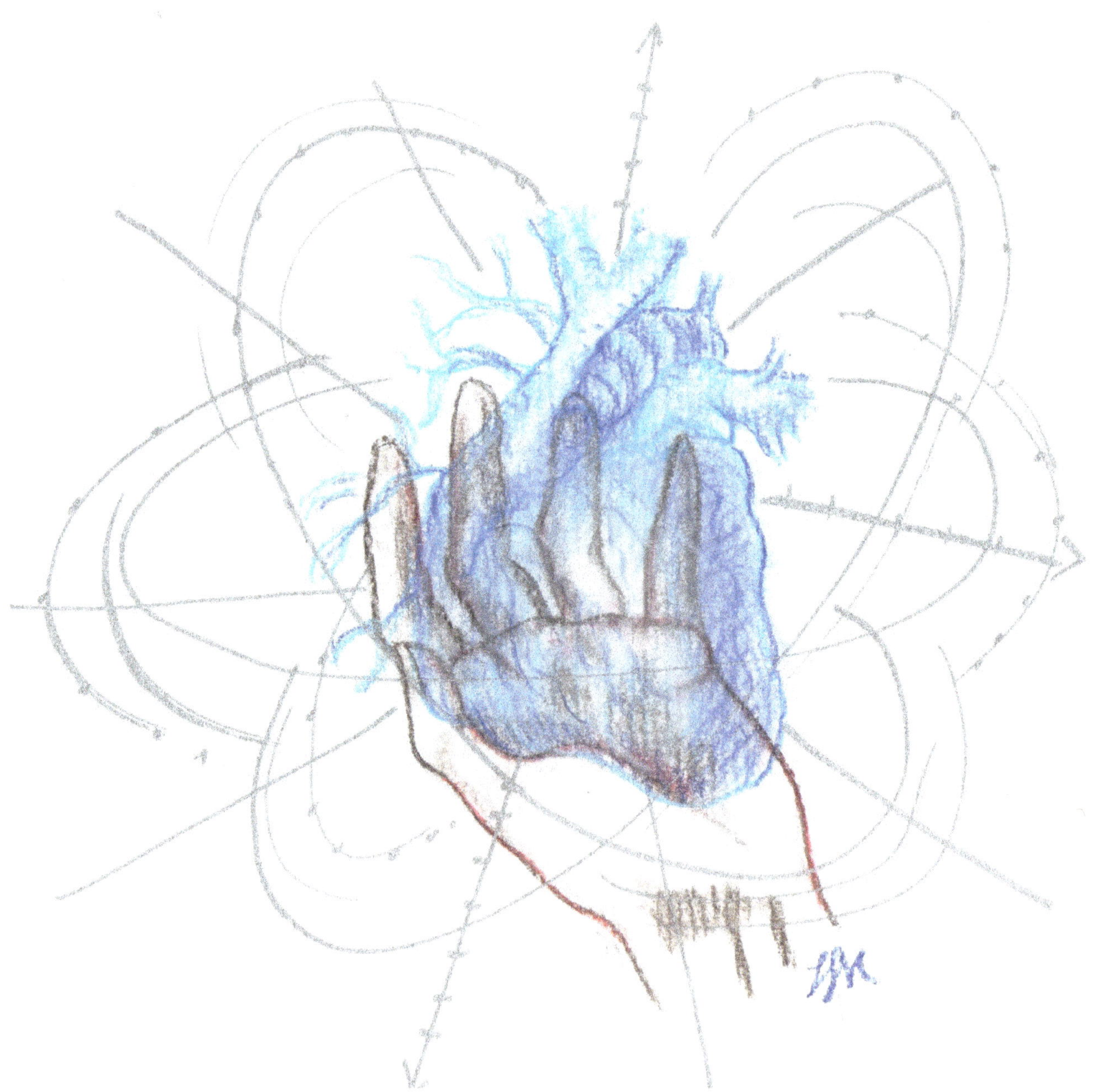

Fig. 4: Does evolution have a heart? [C]

5.4.4 Schrödinger's Cat

… which we are going to deliver from its unjust fate of an existence in an infinite limbo state.

Fig. 5: Yes, we know that this isn't a cat [A]

This topic was chosen because the protection of our two innovations T1 and T2 depends to some extent on the fact that, similar to the Schrödinger cat thought experiment, any "opening of the box" destroys the experiment and the superposed information within[2]. The Schrödinger cat problem is therefore of great importance for the protection of the innovations.

5.4.5 The 3-Generation Problem of Elementary Particles

When applying the Bianchi identity on a scalarized Quantum Gravity Physics realm we obtain three solutions. Since this is true for any system and not just elementary particles, we investigate this problem here to discuss the possibility of explaining the miraculous behavior of the selection agent in T2 via a permanent phase change or phase transition between "generational states". Thereby it could be assumed that the selectivity is just a switch between one generation in the bonded state with the target material and another generation in the unbonded state. Just like with the examples of charged leptons, the unbonded state would be correlated to the electron with low mass, while the bonded state correlates with the muon or the tauon with significantly higher masses.

Fig. 6: The three generations with the missing fourth particle. [A] As extensively elaborated in [4], there are mathematical hints for a fourth particle at such high energies that these generations could never be realized in our universe. Hence the missing Filia in this figure. In this book we will demonstrate, however, that based on a fundamental mathematical law (the Bianchi identity) in fact only three generations of elementary states can occur.

[2] Please note: Information cannot get lost in this universe. Thus, one should better formulate correctly that the information being critical about the T1-T2-experiments is been rendered inaccessible when the box is been opened.

5.4.6 The Particle-Wave Dualism

… which we will address via a fundamentally based mathematical approach leading to the destruction of the concept of particles.

Fig. 7: Wave or particle? [C]

This topic provides insight into the higher dimensionality of just everything and allows for the description of the process of creation, T1, as dimensional growth with entanglement, while separation, T2, comes with disentanglement of systems of higher dimensionality into sub-systems of lower dimensionality.

5.5 Further Topics and Why we Will Need Them Here

5.5.1 Derivation and Application of a Quantum Gravity Theory

When certain experiments, processes, or observations have eluded a scientific explanation for many years, one might think about applying approaches and go paths "no man has gone before". To make sure that our approach here is of maximum generality and fundamentality, we will resort to Hamilton's extremal principle in the form of the Einstein-Hilbert action, leading us to a Quantum Gravity Theory (which will be shown in completely falsifiable manner and which will also be proven in this book). We are also in need of such a theory, because our experiments T1 and T2 demand scale-invariance and arbitrary dimensionality. And as it will be shown here, this is one of the key features of our approach in this book.

We explicitly emphasize the fact that the choice for our approach is not based on the assumption of gravity playing a role in the understanding of the two innovations (even though, naturally, it is present). Here we still face at least 10 magnitudes difference to the weakest interaction of higher order even in cases of huge molecules participating. No, our reasoning is a different one: For we will also prove that there is an intrinsic inconsistency with the concept of particles and everything is just based on oscillating processes, we require a Quantum Gravity Theory also as a general mathematical apparatus for frequency composition and decomposition. This is but the generalized and fundamental equivalent of the mathematical tool of Fourier series and Fourier analysis.

Fig. 8: A beautiful little quantum gravity wave.

5.5.2 Metric Derivation of Electromagnetism

Closely observing table 1, we see that electromagnetism (EM) is one of the key features for the activation part in our T1 and T2 experiments. To obtain an as general (free of any postulations) an EM-toolbox, we will completely rederive it from the Einstein-Hilbert action.

5.5.3 Turbulences Inside an Unconventional Physics Realm

What was just said for the EM-interaction with respect to both experiments T1 and T2 also holds for the activation process of T2 and an additional mechanical, de facto, turbulent component. Because of its importance there, we derive its origin from the Einstein-Hilbert action in a completely unusual, but mathematically consistent way.

5.5.4 Molecular Dynamics without Potential Truncation

Big ensembles of atoms and molecules are usually handled with numerical methods under the expression "molecular dynamics". Here more or less empirically obtained effective potentials are used in order to account for "enough 'particles' with as high as possible accuracy". Thus, this technique is always a compromise and an approximation. In order to handle bigger numbers of particles the potentials have to be truncated, which leads to a dramatic neglect of interactions of higher order [162, 173, 184 - 201]. As exactly those higher order-interactions are of importance for the description of our T1 and T2 innovations, we have to introduce completely different and unabridged techniques. In this book we will not deal with this topic as it would be by far too voluminous… in fact, it would merit a book of its own (e.g., see [173]).

That being said, we should point out that the development of such a wide-band molecular dynamic apparatus is something yet to be performed.

It is one of the main project goals of the three authors to comprehensively close the file on the two innovations T1 and T2 also with respect to their full understanding.

5.5.5 The Photon

Photonic activation also is an issue for the two experiments T1 and T2 (while still not completely settled with respect to T2). Thus, we also require a fundamentally better understanding of the photon in a metric theoretical realm. As this topic would also be too extensive for this book, we here only give the corresponding reference [5, 63] and a very brief introduction.

Fig. 9: Photons at work. [A]

5.5.6 Metric Derivation of the Diffusion Equation

See what was said about the original and the anti-diffusion equation in the subsection "The Origin of Evolution" above.

5.5.7 Vectorial and Tensorial Masses

If the development of the key technologies T1 and T2 is viewed holistically in the market environment, an essential feature of these inventions becomes apparent the moment when they have reached a certain scale. It is clear that by considering the building of structures and their individual growth as a process of dimensional increase and the reverse process as dimensional decrease, every market situation automatically favors technologies that have the ability to imbibe as many degrees of freedom as possible for steering and variation of the process. Only this will guarantee the necessary flexibility and adaptivity for optimum performance in a rapidly changing (evolving) socioeconomic environment. Thereby it must be pointed out that the technology itself can (and probably will) contribute to the evolutionary pressure on an almost holistically variable scale, assessable by the owner. This is just self-interaction at its best (like the Higgs mechanism [58], but unlike the latter not of scalar, but of tensorial character → c.f. section about non-scalar masses in this book, entitled "From Scalar Mass and Energy to a Mass-Energy-Vector or -Matrix?") and when understood and controlled it can be of fundamental use to the owner and exploiter of the invention… hence, with this innovation, the owner actually has much more than just a key technology, but also a key to its own existence. By adding in the understanding about critical processes and phase transitions in socioeconomic spaces [8, 202], which are losing the dominating scales on a variety of attributes in such transitional states and knowing that the tower(-ing) technology can influence scale lengths on certain attributes, one might (or even must) discuss the social aspects of these inventions, too. The deposition and the multi-polar-selective technology, which is to say T1 and T2, respectively, clearly have the potential for such a holistic creativity.

5.5.8 How to Achieve Certainty

Almost as a by-product of our holistic quantum gravity approach, we are going to be able to derive conditions for the construction of universes without any quantum uncertainty. This might be very useful when being interested in absolute certainty… for a change. Unfortunately, so the immediate critical argument, we cannot construct universes. That is correct. But we can create systems and try to isolate them as much as possible from the rest, thereby obtaining a "universe of our liking". Now, being interested in achieving absolute certainty about certain things in this – our – universe, we want to derive the conditions, where we can get rid of all quantum induced uncertainty. We will find that for all scalar quantum equations such conditions exist and could be activated, but this – most unfortunately – is only possible in universes of infinite dimensionality[3]. Nevertheless, with respect to our key technologies and their positioning, attractiveness, and robustness inside a socioeconomic space-time, hence, society, we now know that intelligence ABOUT EVERYTHING is the key for gaining as much as possible certainty for any activity and measure regarding our market performance.

In other words: The more we know about something, the lower the uncertainty… and it appears quite surprising that we needed a fundamentally derived quantum gravity to prove something so obvious in a completely falsifiable and mathematically rigorous way.

5.5.9 Quantum Gravity Thermodynamics

With all the fundamentally obtained building blocks for the most holistic theoretical apparatus possible, we have already loaded ourselves. Why would we also need a quantum gravity originated and thus, completely non-postulated thermodynamics for our innovations T1 and T2?

[3] Something the second author, Dr. D. E. Martin, was always referring to as "infinite orthogonal dimensionality", whereby the orthogonality automatically assures the simplest possible sets of equations.

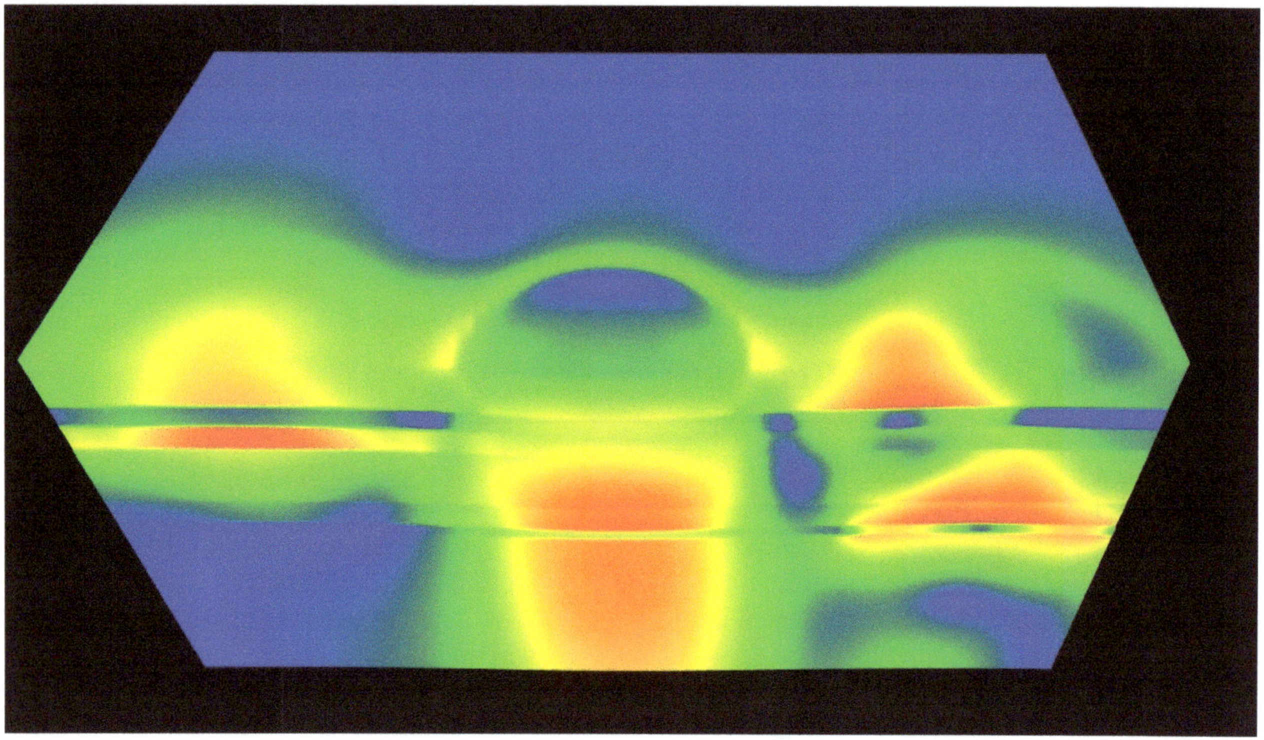

Fig. 10: Temperature field in a complex coating system (a T1-application).

Well, the strangeness of the observations in connection with the two innovations (most prominent among the two, however, the second tower) forces us to also question certain dogma, in classical or, let us better say "established" or "conventional" thermodynamics. In rederiving the latter from scratch, we want to make sure that we do not burden us with potential old bias.

5.5.10 Metric Holistic Elasticity

Another important field which is of need when one intends to optimize the deposition technology T1 towards specific applications is solid body mechanics, especially the theory of elasticity. This, too, as we are going to demonstrate in the sections below, can directly be derived from the Einstein-Hilbert action, thereby revealing astounding options for its generalization.

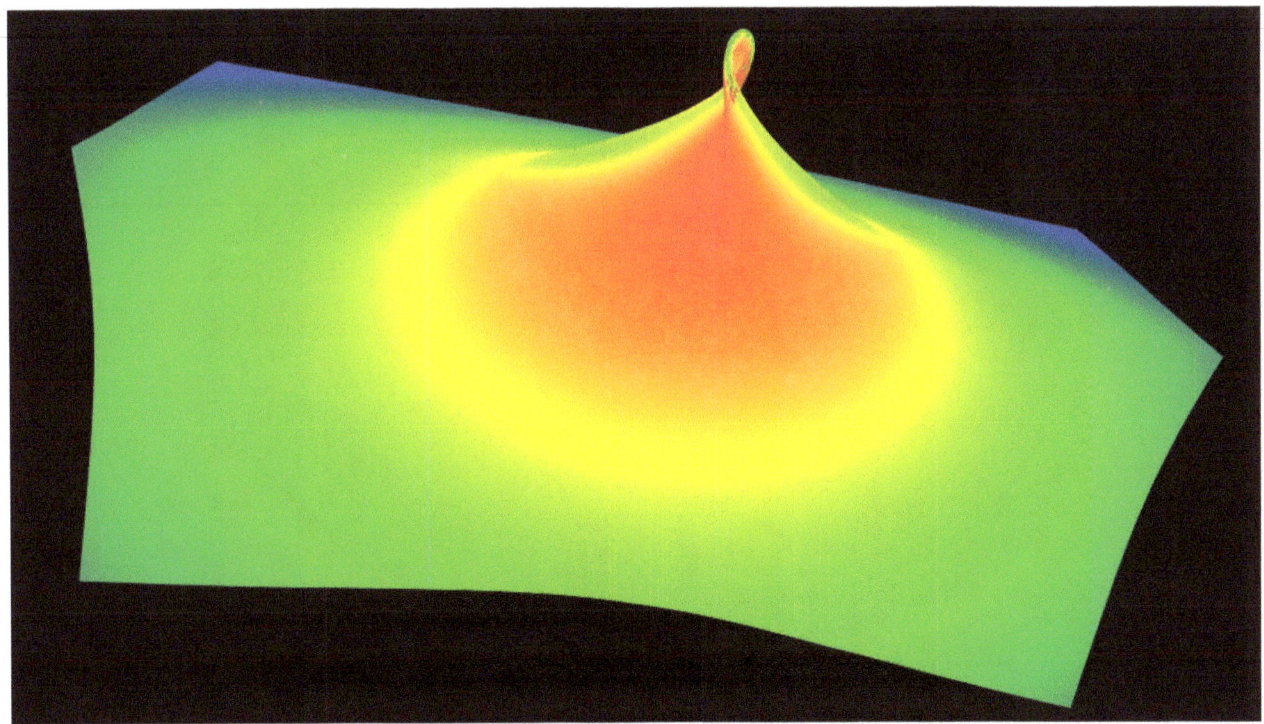

Fig. 11: Unconventional elasticity (a T1-T2-application).

5.5.11 Quantum Gravity and the Outcome of the Theory of Perspectivity

It will be derived in this book that the realization of matter depends on the perspective of the observer. This opens up new possibilities about the protection of innovations from greedy eyes. An aspect we will only derive in here, but not discuss in order not to compromise this protective measure for our own inventions T1 and T2.

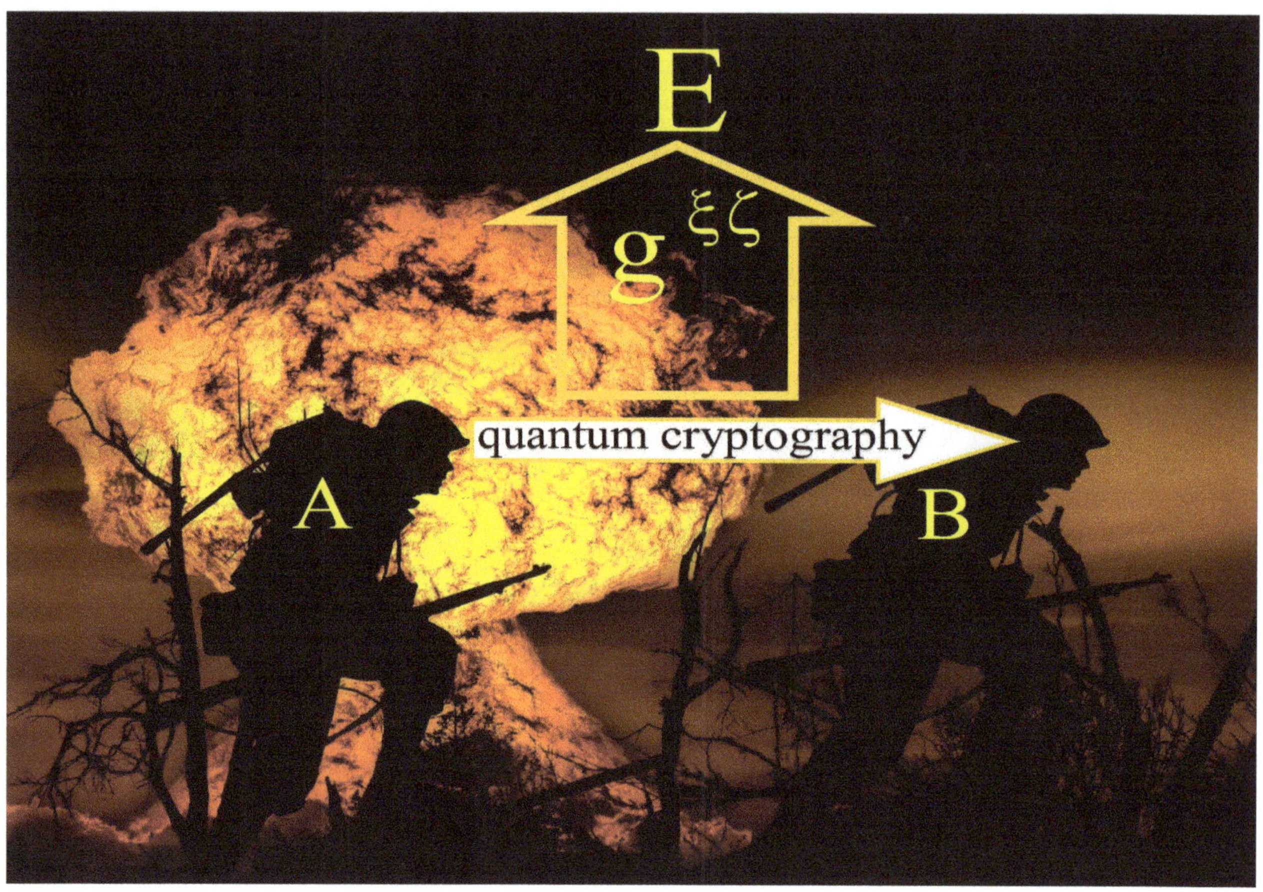

Fig. 12: Applied theory of perspectivity (a T1-T2-protection vehicle).

5.5.12 How to Obtain Life – The Missing Agent?

Even though the famous Miller-Urey experiment[4] in 1953 showed that the simplest ingredients of life, the amino acids could be produced in an abiotic process, it was never possible to obtain living systems. Something was always missing.

What if the missing piece was a Polar Selective Agent from T2, which, as some kind of chemo-physical selector (or creator), assuring just the right components coming together, could provide this missing link to truly make life crawl out of Stanley Miller's reactor?

We are going to derive the math of such a selector or creator via a metric formulation of the Everett multiverse theory [56, 215] further below in this book.

[4] See e.g., https://en.wikipedia.org/wiki/Stanley_Miller#Miller's_experiment

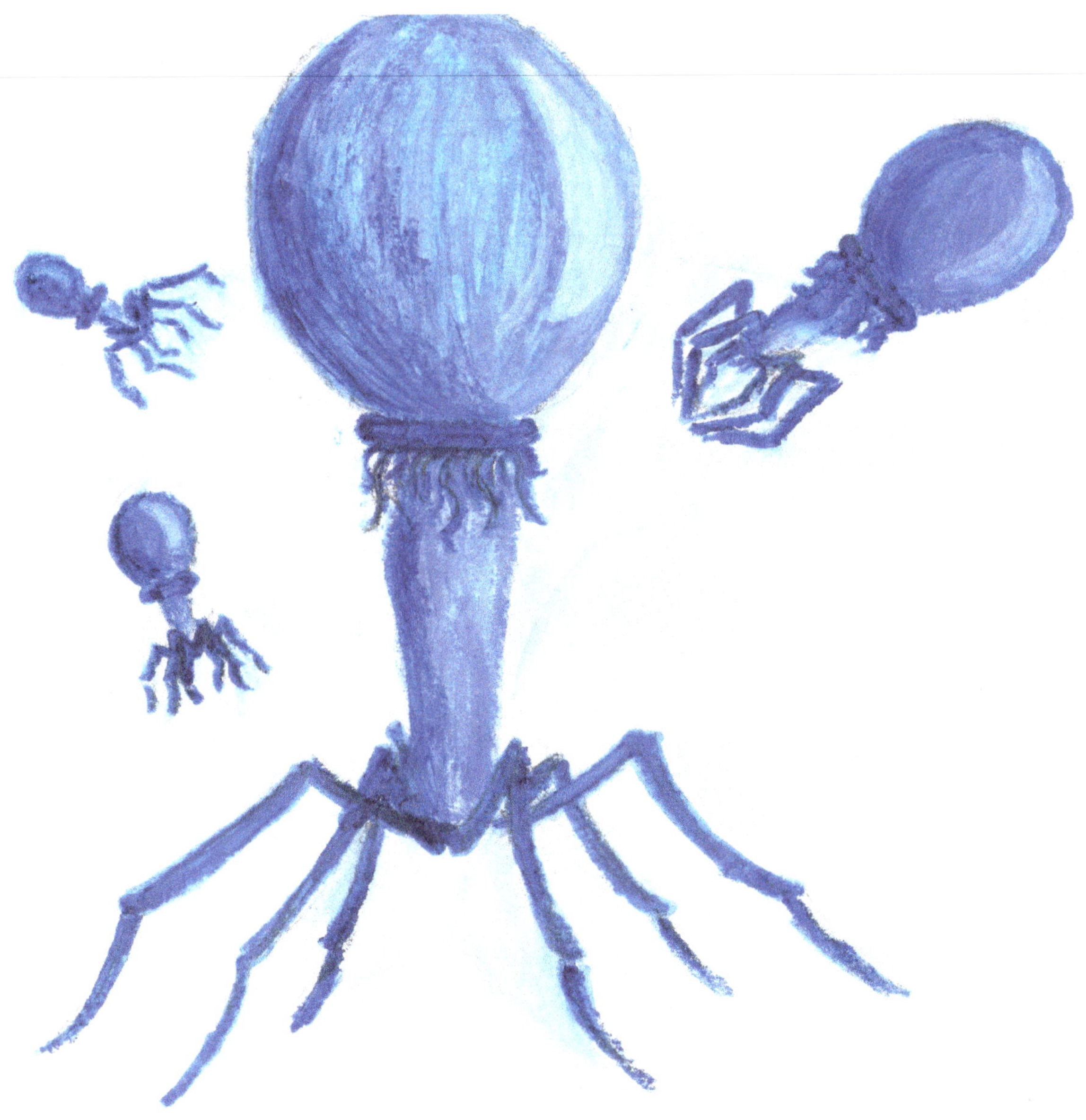

Fig. 13: Where is the missing agent? [A]

5.5.13 Diffusion Equations for Curved Space-Times

As already said above, in connection with the first and the second tower, we are in need of an unbiased description of diffusion processes (see subsections "The Origin of Evolution" and "Metric Derivation of the Diffusion Equation"). The fundamental derivation of the latter will not only give us the ordinary and the anti-diffusion, but also diffusion equations in curved spaces. This is essential for the understanding of the two processes T1 and T2. Regarding T1, it helps us with the understanding why our deposition process has so much less problems on curved surfaces, sharp edges, and corners than any other coating process. With respect to T2 we will be able to describe diffusion through internal (but curved) interfaces inside a liquid agent.

5.5.14 Dirac Evolution and Dirac Diffusion

Attentive observation of the processes, which are happening when applying the key technology T2 for the cleaning of tar sands, reveals a strange and quite characteristic "distancing" or perhaps even dominant length scale behavior of the reactive mixture. In fact, one realizes a characteristic feature of the Polar Selective Agent, PSA, showing such a strong "individuality" and exclusivity bearing that one cannot help but recognize the similarity to the behavior of fermionic matter. Like it is known for turbulences or half spin particles excluding each other – similar or even equal to the Pauli principle [77] – we should not only apply the second order diffusion equations, but also need to find a diffusion equation of Dirac style. Thereby we can follow the classical Dirac path [27] and factorize the second order Klein-Gordon equation or derive a metric Dirac equation directly and rigorously from the Einstein-Hilbert action. While the latter path will be elaborated further below in the subsection "Diffusion Equations of Dirac Character", we here want to briefly repeat the starting point of the derivation of half spin solutions from the first path, which is to say the Klein-Gordon origin. It was shown in [2] in the section "3.4 The Other Hydrogen" that the metric derivation of the central force problem from the Einstein-Hilbert action with a R^*=0-condition directly leads to the following differential equation for the radius component r:

$$E_{n-\text{Schrödinger}} = -\left(\overbrace{\frac{R_{61}}{r}}^{-V/r} + \overbrace{\frac{R_{62}}{r^2} - \frac{\omega^2}{r^2}}^{-\omega^2_{\text{Schrödinger}}/r^2}\right) - \frac{\hbar^2}{2\cdot M}\cdot\frac{1}{h[r]}\cdot\left(\frac{2}{r}\cdot\frac{\partial h[r]}{\partial r} + \frac{\partial^2 h[r]}{\partial r^2}\right). \tag{1}$$

Thereby we have introduced the term $\frac{\hbar^2}{2\cdot M}$ for recognition (reduced Planck constant squared $\hbar^2$ divided by mass M) in order to mirror the classical Schrödinger equation. We will see later (theory-section further below in this book) that, what we recognize as potential terms and momentum, $\left(\overbrace{\frac{R_{61}}{r}}^{-V/r} + \overbrace{\frac{R_{62}}{r^2} - \frac{\omega^2}{r^2}}^{-\omega^2_{\text{Schrödinger}}/r^2}\right)$ are in fact curvatures of space-time. Considering h[r] just as the radial part of a separation approach for the function f of all coordinates gives us the classical radial Schrödinger solution for f[...]:

$$\begin{aligned}
& f_{n,l,m}[t,r,\vartheta,\varphi] = g[t]\cdot\Psi_{n,l,m}[r,\vartheta,\varphi] = g[t]\cdot e^{i\cdot m\cdot\varphi}\cdot P_l^m[\cos\vartheta]\cdot R_{n,l}[r] \\
& = \left(C_1\cdot\cos[c\cdot C_{tt}\cdot t] + C_2\cdot\sin[c\cdot C_{tt}\cdot t]\right)\cdot N\cdot e^{-\rho/2}\cdot\rho^l\cdot L_{n-l-1}^{2l+1}[\rho]\cdot Y_l^m[\vartheta,\varphi] \\
& \rho = \frac{2\cdot r}{n\cdot a_{00}};\quad N = \sqrt{\left(\frac{2}{n\cdot a_0}\right)^3\frac{(n-l-1)!}{2\cdot n\cdot(n+l)!}};\quad C_1 = \pm C_2 = 1; \\
& a_{00} = \frac{\hbar^2}{M}\cdot\frac{1}{R_{61}};\quad E_n = \left[\frac{\hbar^2\cdot C_t^2}{2\cdot M} - R_{60}\right]_n = \frac{R_{61}^2}{n^2}\cdot\frac{\hbar^4}{16\cdot M^2};\quad \omega^2 - R_{62} = l^2 + l \\
& \Rightarrow f_{n,l,m}[t,r,\vartheta,\varphi] = e^{\pm i\cdot c\cdot C_t\cdot t}\cdot N\cdot e^{-\rho/2}\cdot\rho^l\cdot L_{n-l-1}^{2l+1}[\rho]\cdot Y_l^m[\vartheta,\varphi]
\end{aligned} \tag{2}$$

As the symbols are chosen in accordance with the conventional textbook literature (e.g., [24]), we do not think that we need to elaborate the various terms here. Regarding the conditions for the quantum numbers n, l, and m, we not only have the usual:

$$\{n,l,m\}\in\mathbb{Z};\quad n\geq 0;\quad l<n;\quad -l\leq m\leq +l, \tag{3}$$

but also found the suitable solutions for the half spin forms as discussed in [2] and derived in the appendix of the same book. The corresponding main quantum numbers for half spin l-numbers with l=1/2,3/2,... are simply (just as with the classical integers) n=l+1=3/2,5/2,7/2,...

It should explicitly be noted, however, that the usual spherical harmonics are inapplicable in cases of half spin. For {n,l,m}={1/2,3/2,5/2,7/2,...] the wave function (2) has to be adapted as follows:

$$\begin{aligned} f_{n,l,m}[t,r,\vartheta,\varphi] &= e^{\pm i\cdot c\cdot C_t\cdot t}\cdot N\cdot e^{-\rho/2}\cdot\rho^{l}\cdot L_{n-l-1}^{2l+1}[\rho]\cdot Z_m[\varphi]\cdot\begin{Bmatrix} P_l^{m<0}[\cos\vartheta] \\ Q_l^{m>0}[\cos\vartheta]\end{Bmatrix} \\ &= e^{\pm i\cdot c\cdot C_t\cdot t}\cdot N\cdot e^{-\rho/2}\cdot\rho^{l}\cdot L_{n-l-1}^{2l+1}[\rho]\cdot\begin{Bmatrix} \cos[m\cdot\varphi] \\ \sin[m\cdot\varphi]\end{Bmatrix}\cdot\begin{Bmatrix} P_l^{m<0}[\cos\vartheta] \\ Q_l^{m>0}[\cos\vartheta]\end{Bmatrix} \end{aligned} \tag{4}$$

Thereby we have elaborated in [2] that in fact the sin- and the cos-functions seem to make the Pauli exclusion and not the "+" and "–" of m. However, in order to have the usual Fermionic statistic we can simply define as follows:

$$f_{n,l,m}[t,r,\vartheta,\varphi] = e^{\pm i\cdot c\cdot C_t\cdot t}\cdot N\cdot e^{-\rho/2}\cdot\rho^{l}\cdot L_{n-l-1}^{2l+1}[\rho]\cdot\begin{Bmatrix} \sin[m\cdot\varphi]_{m<0} \\ \cos[m\cdot\varphi]_{m>0}\end{Bmatrix}\cdot\begin{Bmatrix} P_l^{m<0}[\cos\vartheta] \\ Q_l^{m>0}[\cos\vartheta]\end{Bmatrix}. \tag{5}$$

As it was discussed and derived in [2], "3.4 The Other Hydrogen" plus the appendix there, the resolution of the degeneration with respect to half spin requires a break of the symmetry, which we achieved in [2] by introducing elliptical geometry instead of the spherical one.

The following figures illustrate the space-time curvature for a variety of half spin settings.

Fig. 14: Space geometry of distortion (or wave) function in the case of quantum gravity solution (4) for the state n=7/2, l=5/2, m=3/2.

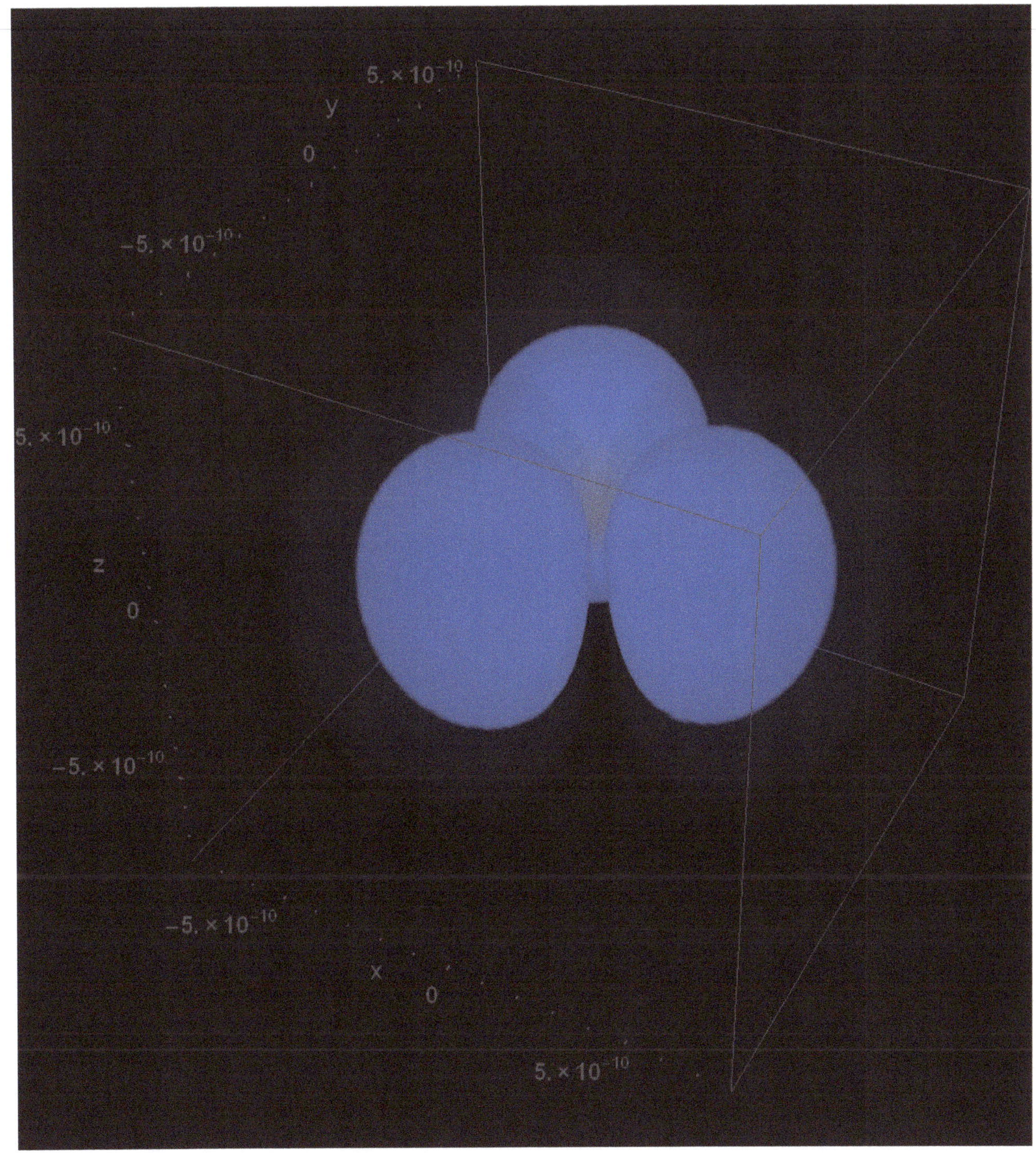

Fig. 15: Space geometry of distortion (or wave) function in the case of quantum gravity solution (4) for the state n=5/2, l=3/2, m=3/2.

Fig. 16: Space geometry of distortion (or wave) function in the case of quantum gravity solution (4) for the state n=7/2, l=3/2, m=3/2. It is assumed that the additional smaller cores could just be the result of a particle creation reaction as answer to the attempt to separate the object shown in figure 16. Further adding of energy results in an increase of the momentum and six equally proportioned cores (c.f. figure 14), which would essentially be a doubling of the original object shown in figure 15. This then, however, would mirror the behavior of quarks.

5.5.15 Avoiding the 720° Half Spin Oddity

Having seen the need for half spin objects for our description of the separation key technology T2 and knowing that all matter particles are half spin objects, hence, that there cannot be a creation technology without them, we cannot ignore the oddity coming with non-integer spins. But there is an interesting solution to the problem which we will present in this book in the section "The Universe's Half Spin Deception".

Here we will reuse some of our results where we applied the quantum Einstein field equations in order to find suitable metric solutions for elementary objects. Thereby our quantum Einstein field equations have been derived directly out of the Einstein-Hilbert action. While processing the subsequent equations, we discovered a mechanism which might explain the occurrence of so-called half spin objects. If truly (mathematically) dealt with as half spin structures in the classical way, these objects would automatically show a strange 720° symmetry, which is to say, one has to turn them around twice before they look the same again. Even though one of us was able to show that singularity-free solutions do exist also for such metric half spin structures (e.g., see appendix of [23]), we found other solutions, only "pretending" to have the half spin characteristics to an outside observer but not truly carrying it. In other words, the oddity would not be of need anymore.

5.5.16 And Almost as a By-Product: The Origin of Matter

Fig. 17: A form of matter and energy... But where does it come from? [designed by Freepik]

Sometimes, when looking quite intensely for explanations in certain fields and for certain observations[5] one finds something else… something one was not actually looking for, but only dragged the topic in, because one was hoping to find a suitable tool in there, a tool for the original problem. Then not only the original problem can be solved in the end, but there is also a nice little contribution to the field which was only dragged in to help solving the first thing, then one might call this a nice by-product.

Our "nice little by-product", when trying to find better tools for the mathematical handling of coatings and thin films in connection with our T1-technology, was the discovery of matter inside the Einstein-Hilbert action. Matter was never found by Einstein nor Hilbert, but had to be postulated and explicitly placed inside the General Theory of Relativity [1, 3] by the two geniuses. Of course, we are aware of the fact that just claiming "We found matter inside the Einstein-Hilbert action!" would probably only lead to some severe head-shaking. It is for this reason that we here present the corresponding mathematical proof – very unusually – even before the theory section. According to one of us, the appearance of the matter energy tensor from just a somewhat more holistic handling of Hilbert's derivation of the Einstein field equations "is so very obvious, like a white elephant in the room that one probably has to have a rather primitive mind-set, which is to say a mind-set like mine, in order to be able to spot it… and I insist on the phrase '**spotting it**', because it is no discovering, no creating of a new theory, and definitively it is no finding something! It is spotting something which has been in plain sight for more than one hundred years! It is the white elephant not just standing in every scientist's office or living room, but sitting on his desk, joining him on the toilet, and laying in his bed." But as it is with many obvious things, they are often the more difficult to see. It is for this reason that we think we better present the proof directly in connection with the claim.

[5] In this case it was with respect to a better and more general mathematical handling of coatings and surfaces in all sorts of applications, which drove one of us ever deeper into the various fields of theoretical physics. So it happens than one can read on the cover of [2, 4, 5] the following text: "The only thing Norbert Schwarzer considers important enough to be known about him is that he does not consider himself important. Dr. Schwarzer has published a variety of papers, mainly in the fields of basic research and application of contact mechanical approaches for laminates, composites, and layered materials. Because of the need for better stability prediction and socioeconomic models, he started to apply concepts from theoretical physics in more down-to-earth fields such as materials science, school transport, and sales market analysis. **Some of this work has finally led to ideas for the improvement of the original theoretical concepts.**"

Here it is:

$$
0=\left(\begin{array}{l}
\overbrace{\boxed{R_{\alpha\beta}-\frac{g_{\alpha\beta}}{2}R}}^{\text{classical Einstein GTR}} \\
+\underbrace{\boxed{\begin{array}{l}
\frac{g_{\alpha\beta}}{2}\left(\begin{array}{l}
\frac{1}{2F}\left(\begin{array}{l}
2F_{,ij}(n-1)g^{ij}+2\Gamma^{a}_{ij}F_{,a}g^{ij} \\
-F_{,i}g^{ab}g_{jb,a}g^{ij}-F_{,j}g^{ab}g_{ib,a}g^{ij} \\
-n\Gamma^{d}_{ij}F_{,d}g^{ij}+\frac{n}{2}F_{,d}g^{cd}g_{ab,c}g^{ab}
\end{array}\right) \\
+\frac{F_{,i}\cdot F_{,j}}{4F^{2}}g^{ij}\left((n-6)(n-1)\right)
\end{array}\right) \\
-\frac{1}{2F}\left(\begin{array}{c}
F_{,\alpha\beta}(n-2)+F_{,ab}g_{\alpha\beta}g^{ab} \\
+F_{,a}g^{ab}\left(g_{\beta b,\alpha}-g_{\beta\alpha,b}\right)-F_{,\alpha}g^{ab}g_{\beta b,a}-F_{,\beta}g^{ab}g_{\alpha b,a} \\
+F_{,d}g^{cd}\frac{1}{2}n\left(\frac{2}{n}g_{\alpha c,\beta}-g_{\alpha c,\beta}-g_{\beta c,\alpha}+g_{\alpha\beta,c}+\frac{1}{n}g_{\alpha\beta}g_{ab,c}g^{ab}\right) \\
-\frac{1}{2F}\left(F_{,\alpha}\cdot F_{,\beta}(3n-6)+g_{\alpha\beta}F_{,c}F_{,d}g^{cd}(4-n)\right)
\end{array}\right)
\end{array}}}_{\text{Energy \& Matter}} \\
+\Lambda\cdot F\cdot g_{\alpha\beta}
\end{array}\right)
$$

And the only thing one needs to do in order to spot this is to separately consider the volume part of the metric tensor:

$$
G_{\alpha\beta}=F\left[f\left[t,x,y,z,\ldots,\xi_{k},\ldots,\xi_{n}\right]\right]^{ij}_{\alpha\beta}g_{ij}\rightarrow G_{\alpha\beta}=F\left[f\left[t,x,y,z,\ldots,\xi_{k},\ldots,\xi_{n}\right]\right]\cdot\delta^{i}_{\alpha}\delta^{j}_{\beta}g_{ij}
$$

and then repeat the derivation of the Einstein-Hilbert action in just the classical way [1].

It should be noted that there are more options to obtain matter, but they are similar to the process given above and will be presented in the following sections of this book.

Fig. 18: And there shall be matter and energy. [A]

6 The First Tower: Creation or from Two Dimensions Onwards

There is no growth without an increase in dimensionality.

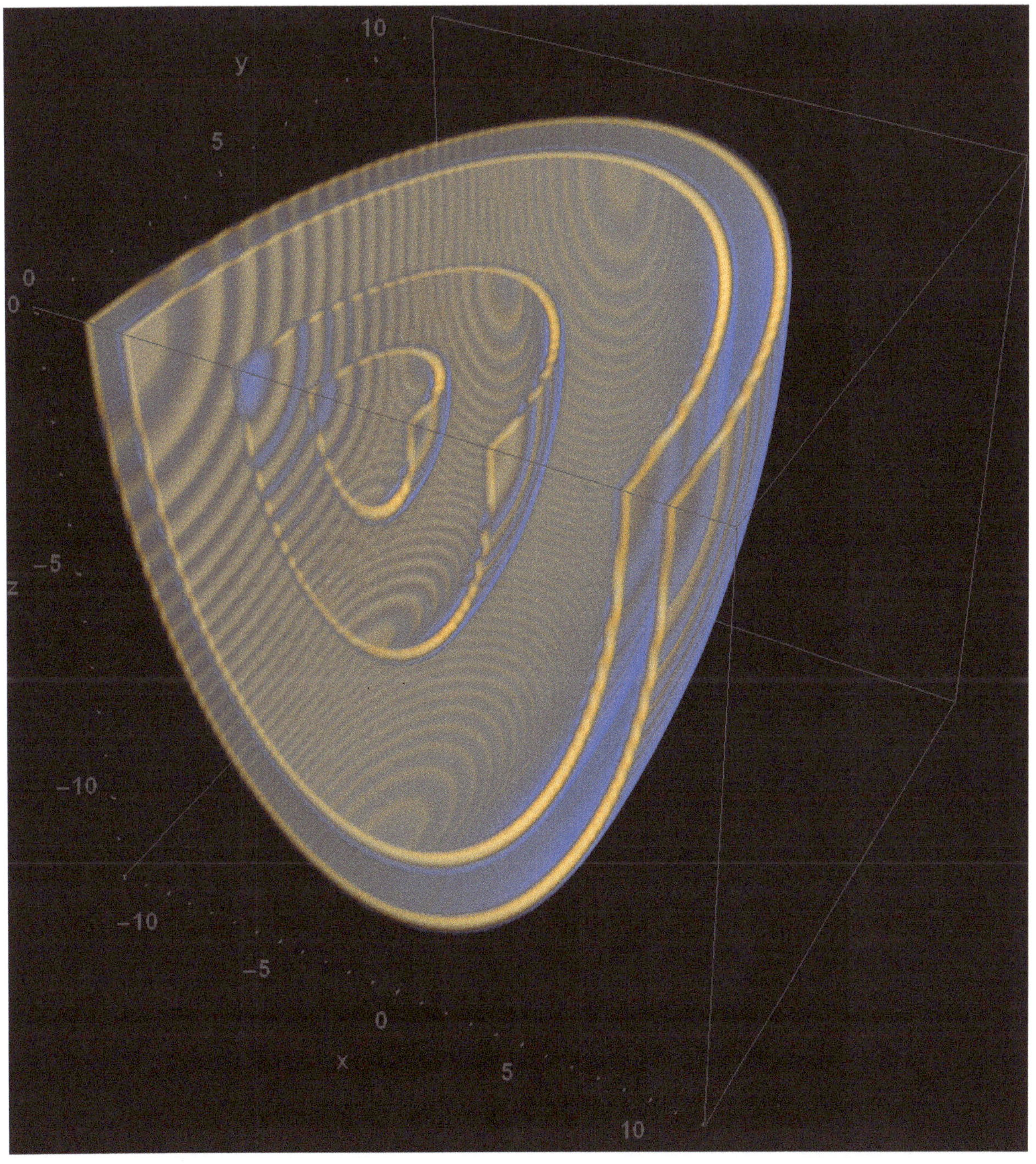

Fig. 19: Surface-based physics.

This is one of the conclusions of the Bekenstein thought experiment on black holes [104, 105] and it was shown in [2, 4] that this is not restricted to such gravity monsters but applies to every system. It can also be shown [85, 106, this book] that in a universe governed by extremal principles the smallest dimensionality allowing for an event or an object is two. Thus, just for these absolute fundamental and quite simple conclusions, the optimum creation machine in such a universe should be based on a 2D or – simply worded – coating technology.

Furthermore, considering the building of structures and their individual growth as a process of dimensional increase, automatically favors technologies which have the ability to imbibe as many degrees of freedom as possible for steering and variation of the process. The deposition technology, developed by the first author, clearly has the potential for such a holistic creativity.

Now also adding stand-alone and world-record [107] features with respect to the understanding, modeling, simulation, failure shooting, and optimization from the theoretical side [107 – 181, especially 108, 119, 127, 156, 165, 173 and 181], can easily render such a technology to a global game changer.

6.1 An Innovative, Flexible, and Quite General Deposition Technology

Fig. 20: Perfect adhesion even on "impossible" substrates.

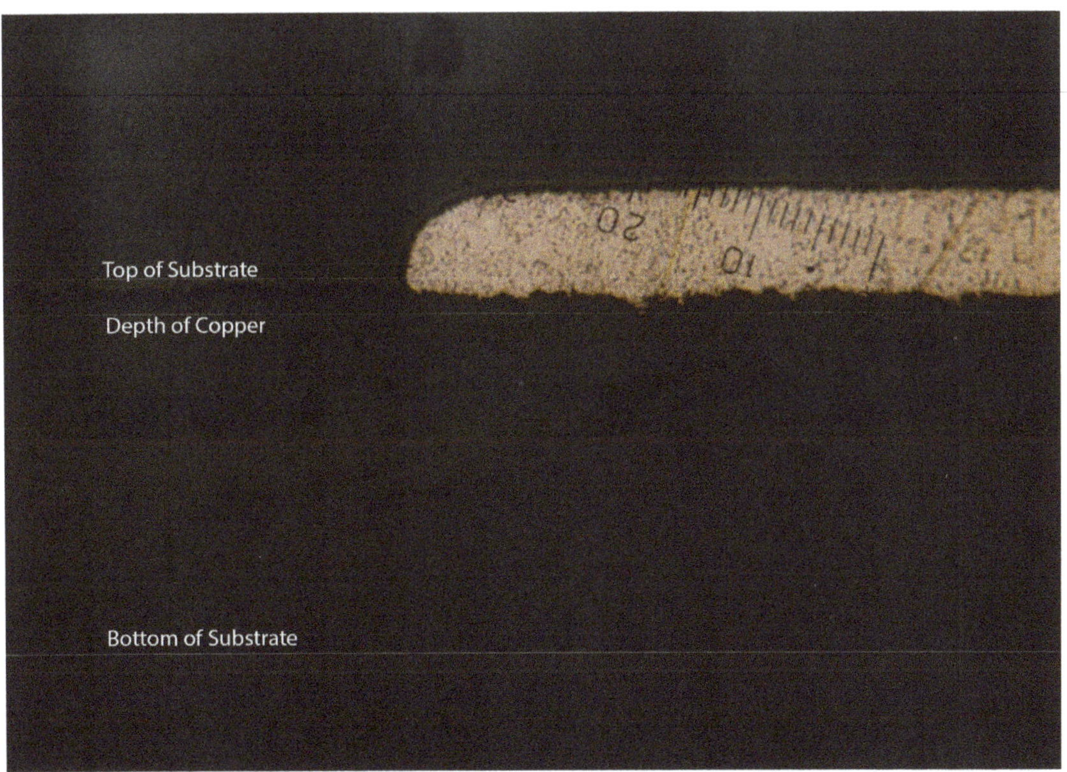

Fig. 21: Perfect adhesion due to partial implantation of the coating material into the substrate.

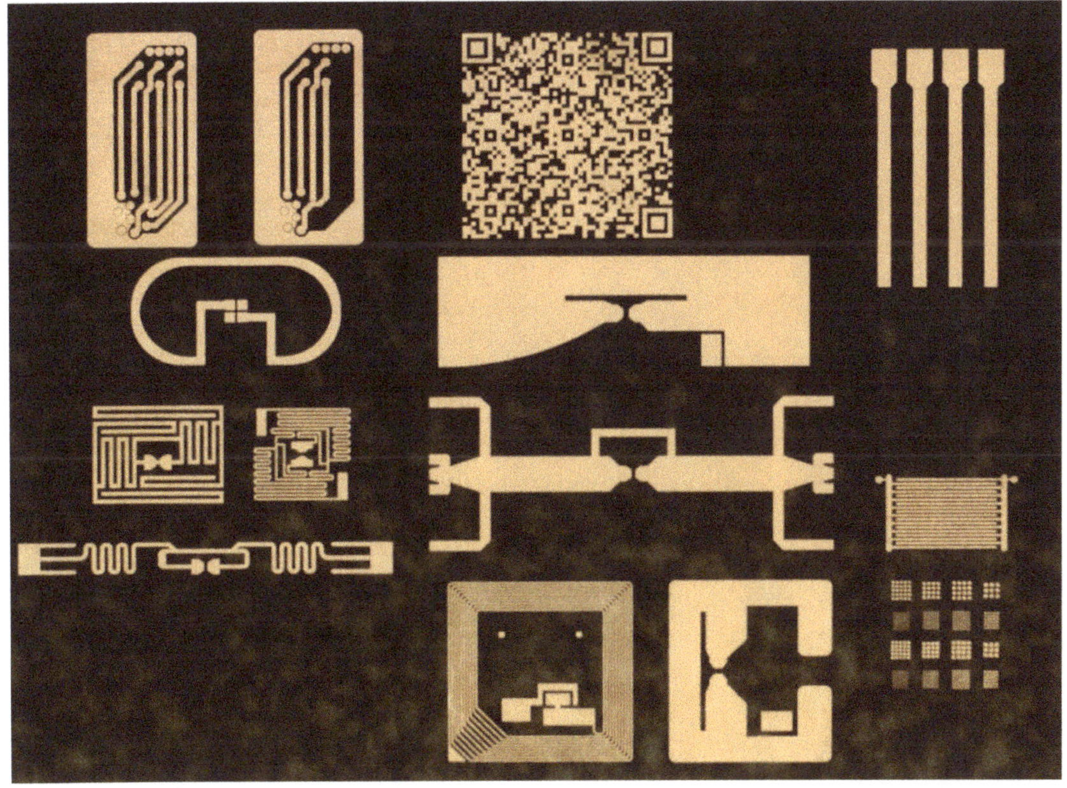

Fig. 22: Another example for perfect adhesion on "impossible" substrates.

See appendix A for more information.

7 The Second Tower: Separation via "Good Vibrations"

As entanglement is the essential feature of this universe to evolve, yes, to be at all, the understanding of the process of evolution – and being – requires the opposite, which is to say, decomposition and separation.

It is therefore of little wonder that a truly fundamental, low-energy and minimum-entropy separation technology could have the potential to completely disrupt all current industrial procedures. That being said, it is still quite wondersome that such a technology exists (see description below) and – what is more – has been developed by "entities" who also were able to develop a key technology for the antagonism of separation (see section above entitled "The First Tower: Creation or from Two Dimensions Onwards"). With the minimum number for entanglement being two and bringing in what was said with respect to the first tower, we may already extract the importance of a 2-dimensional structural approach [this book] for the understanding (protection, application field extension, and potential optimization) of this marvelous invention, of which we here – naturally – can only give a general overview and a phenomenological description.

7.1 An Almost Alchemic Decomposition Experiment

Imagine a process in which a substance, a Polar Selective Agent – PSA as we call it – grabs a target molecule at a place A, "transports" it to place B, releases it there, and then "comes back" for more target molecules to place A. The process repeats itself over and over again without any detectable thermodynamic change or agent degradation. More mysterious is the information transfer between the agent and the target in which a post-interaction deselection preferences the un-interacted over the interacted. The "motivator" for this behavior could be such minuscule or rather infinitesimal differences in certain statistical observables like temperature, concentration, pressure and so on that no ordinary measurement would detect them. Those in doubt should – PLEASE… and before dismissing the whole as esoteric nonsense or even denigrating the authors as frauds – do the following two things:

A) Please watch the three demonstrations given below!
B) Please remember the selective agents in living systems, for example hemoglobin. Visualize how they work and compare with what you have just read about the Polar Selective Agents!

Fig. 23: Demonstrator No. 1 about an almost alchemic decomposition experiment: "RASA Tar Sand Demo"

Fig. 24: Demonstrator No. 2 about an almost alchemic decomposition experiment: "RASA Creosote Demo"

Fig. 25: Demonstrator No. 3 about an almost alchemic decomposition experiment: "RASA Plastics Demo"

Using Sir Arthur Conan Doyle's wise words in citing his hero Sherlock Holmes, "*How often have I said to you that when you have eliminated the impossible, whatever remains, however improbable, must be the truth?*" one might come to the mind-blowing and rather bold conclusion that the PSAs are actually quasi-living systems. And as their process of creation requires a Frankenstein-like ACTIVATION, one could just leave it all there and state that – apparently – "life" or the "state of life" is by far more present, wider reaching, and affects more "things" than we have all learned in school. But what pathetic scientists we would be if we were satisfied with such an "explanation". In fact, we are even aiming for the presentation of the fundamentally derived basics to at least two explanations, namely, one where we will use the parallel of the hemoglobin O2- and CO2-transport, and another one which requires the consideration of turbulences. For the first, we require a first principle derivation of the evolution and diffusion equation of first (Dirac-like) and second order (classical). As already said above, we will thereby obtain ordinary and anti-diffusion. The turbulences, on the other hand, are based on the everyday observation of vortexes in flowing water. We are sure that every reader has already seen floating objects, being small in comparison to the vortex, being dragged into the vortex at a position A and then released at another position B without any recognizable change of the vortex' shape, integral inertial speed, and rotational velocity field. This, too, could be an explanation for our PSA's behavior and will therefore be considered in a special section about the unconventional physics further below in this book. A third option, which shall here only be mentioned, has to do with the possibility of the metamorphosis between bosonic and fermionic matter states during the selection cycles. For this effect the reader is referred to one of our previous books [4], section "11.10 Metric Observables". Last but not least, there is also – of course – the option for a combination of the diffusion, the Boson-Fermion-Metamorphosis and the vortex effect.

However, as already said above, our elaborations will only give the fundamentally derived basics on a first (Hamilton extremal) principle theoretical basis. As this – after all – is “a public facing” paper, we are obliged to protect certain key elements. As responsible entrepreneurs, we cannot give away the intellectual property regarding such key technologies. Readers interested in learning more, shall therefore contact the authors directly [211].

See appendix B for more information.

8 The Third Tower: The Mind

There is an experiment each of us performs every day of her or his life. It is the interpretation of human intercommunication and interaction. Far from even the remotest understanding of the working of the human mind in a conscious way, our brains still accomplish sheer miraculous things in processing the necessary mass of data of observations we often do not even realize when we perform them.

With the toolbox presented in this book, we are going to try and bring the whole "mind-business" into a mathematical formalism. Well, at least we are going to show what needs to be done. Along the way, we have to destroy the concept of particles and the idea of an artificial intelligence or AI being based on linear algebra and fixed integer dimensionality.

8.1 Towards a Math of the Mind – Really?

Most people are of the opinion that it is impossible to mathematically describe human behavior in a satisfactory manner. Probably even more people are of the opinion that describing what we call the human mind is even less possible to be mathematically described than the human behavior. Thereby they accept millions of artificial intelligence solutions doing exactly this in every second and with respect to almost every aspect of their life. This holds definitively for the more instinctive behavior and what neurologists call the system I or zombi mode. With respect to the other mode, which is to say the system II thinking mode, even the majority of neuroscientists and brain experts are of the opinion that such impossibility regarding the mathematical description, modeling or simulation is still something one should count on. After all, so far, no Skynet has developed true consciousness or has been able to properly mirror the conscious human mind. However, many overlook that advanced artificial intelligence systems – here or there, which is to say, in dependence on the field of application – are already extremely close in exactly doing this, and that in some cases such systems appear to be even better than the original thing. The question is why are these systems in some cases so successful and why do they fail in others so abysmally? The answer to this question will lead to a true, fundamentally based intelligence, which we here abbreviate FBI.

So, where and how to start?

It is widely agreed that human beings can be characterized by a list of attributes, which partially entangle with each other... thereby we mean, that the attributes entangle, not the humans. Even though this list may be extremely long and the interaction (or entanglement) of the various attributes might be very complex, in essence we are characterizable by such a list. After all, nothing else is it when we try to – more or less comprehensively and fairly – describe another person or – usually with much less vigor, enthusiasm, and rigorousness – ourselves.

If one accepts the fact that the human being can be described by a list of entangled attributes, however, still does leave us with the question how to mathematically handle this list in order to describe processes of action, human interaction, individuality, and other characteristics. Most surprisingly, this work was already done. Over 100 years ago it was the great German mathematician David Hilbert who found a way to put a set of properties, attributes, degrees of freedom or – as all these other terms are just coding – **dimensions** into a mathematical formalism, which was already in his time known for years as the one fundamental principle for the determination of the laws of our universe: the Hamilton extremal principle. All truly fundamental natural laws are originating from this principle. The only problem is to find the right ingredients to "feed the monster of the Hamilton principle". The essential ingredient namely, the Lagrange density function, seems to differ from problem to problem and this cannot be when assuming that the Hamilton mechanism should be truly fundamental. Here now came Hilbert's stroke of genius [1, 2], who realized that a given set of attributes is nothing else but a set of dimensions and thus forms a mathematical space or space-time. The corresponding Lagrange density function is then always the same, namely the so-called Ricci scalar of the very system or space-time. Hilbert applied his calculus on our ordinary 4-dimensional space-time and derived the Einstein field equations and thus obtained in just one fluid and conclusive derivation the complete mathematical basis for the General Theory of Relativity of Albert Einstein [3]. Admittedly, Hilbert never extended his calculus to the space or space-time of attributes characterizing human beings, their thinking process, their autobiographic memory functions, their behavior and so on, but the fact that the title of his work [1], in which he derived Einstein's great theory, translates "The Fundamentals of Physics" clearly shows us that he saw his approach in a much more general way than just the minimum-principle-origin of Einstein's field equations. Thus, no matter the system we are considering, the moment we have its basic attributes and know their entanglement (various interactions), we also immediately know how to place all this knowledge into Hilbert's calculus to come up with a most fundamental and unbiased model for this very system. Thereby, which is to say within Hilbert's approach, we also find – quite surprisingly – rather unconventional physics realms some scientists would probably like to place under the tag of esoteric. We also – when repeating the Hilbert procedure in a more general manner – see rules which define the size of systems. This is quite useful when being interested in the dimensional size of a model one needs to build or "The Size of a Thought", but here in this book we will only use the method to derive the "Transition to the 'Classical' Physics", partially rederive the size of Black Holes, and solve the Bekenstein problem [104, 105] about the storage of information inside such gravity monsters.

Hilbert's creation, de facto, was – no, is – a World Formula [2, 4] or Theory of Everything [5], even though he himself had little chance of fully realizing this. Even in physics, where we now can show that Hilbert's fundamental equation covers both great theories, General Theory of Relativity and Quantum Theory, the time was not ripe for such a discovery, simply because the mathematical apparatus of Quantum Theory was not fully developed then. While Hilbert brought out his great work in 1915 and knew about the Einstein field equations at the time, the basic quantum equations like the Schrödinger, the Klein-Gordon, and the Dirac equation would not follow before the second half of the 1920s.

For convenience, falsifiability, and rigorousness, we are not only presenting the Hilbert approach adjusted to our arbitrarily dimensional space-times in this book, but also derive the Klein-Gordon, the Schrödinger, and the Dirac equation (the main – relativistic and non-relativistic – quantum equations), because it will be our mathematical fundament for the consideration of a variety of psychological processes like the process of "falling in love", like decision making and like mass formation. As this "third tower" – hopefully quite understandably – would be a bit too voluminous if being comprehensively covered in this book, we here refer to the following work of the third author:

8.2 The Two Basic Mind-Processes: Creation and Separation

As the wise Oogway from Kung Fu Panda said: "There are no accidents!" We stopped wondering about the many coincidences we faced in connection with the above-described key technologies quite some time ago. And thus, consequently, we will also refrain from fuzzing about the funny foreordination of the two towers T1 and T2 just providing what we see as key processes in all mind activities. In essence, namely, this is just creation and separation again.

	Innovation	Analysis
Preparation	Collecting all relations, which is to say entangling with all topics and field possibly being related to the main subject	Abstraction and Simplification
Activation	Narrowing the search field and activation of those areas, which were recognized as essentials	Marking and highlighting of boundaries in order to recognize sub-systems
Processing	Composing "thoughts"	Decomposition of complex problems
Dimensional Impact	Dimensional increase / New ideas and innovations	Dimensional decrease of "target" due to separation into sub-systems or sub-problems

Table 2: The dimensional behavior of the basic mind-processes.

8.3 Incorporation of The Third Tower into a Comprehensive Socioeconomic Strategy

A key technology can only be as successful as a society or – to be mathematically precise – a socioeconomic space-time allows it to be.

Thus, possessing a transformative technology, which is reaching beyond the current state of the art in a sense that its dimensional technical potential, but also its theoretical impact, economic scale, and political spillover effect has to be placed even above any Manhattan project, is not enough. One also must find a way to analyze, process, and – oh yes – influence and steer critical market aspects. The introduction of a transformative, even socioeconomic metamorphic technology from a certain level and scale onwards can only be as successful as it goes hand in hand with a proper observation, interaction, and "control" of these social and economic aspects. As already elaborated in the context with the first and the second tower or T1-T2 key technologies above (see subsection "Vectorial and Tensorial Masses"), there simply is too much back-coupling and self-interaction, which cannot be ignored. It has to be foreseen, simulated in various scenarios, and – oh yes – used, perhaps even exploited by the owner of the technology. Hence, the understanding of individual human behavior, underlying market-relevant mass formations and socioeconomic space-times is crucial for such an endeavor.

Therefore, along with the two towers of transformative technology, we also require a comprehensive understanding of the mind (individual and social), which here will be referred to as the third tower = T3.

8.4 T3 – A Key Technology of Its Own

Nevertheless, we should point out that the development of a "mind-understanding" or even a "mind-technology" opens up much more than only ways to better protect and steer other key technologies. It is a key technology of its own and we want to give a few examples for development directions and applications of which we will present a few more details (illustratively and mathematically) further below in this book.

8.4.1 The Singularity or the Awakening of "Skynet"[6]

For the reasons given above, we are going to repeat some essential (and quite rigorous) derivations and discussion of the usage of mathematical tools for a fundamental description of the human psyche in this book. Thereby we apply the usual strategy that every clever clinical psychoanalyst would use when performing the anamnesis of a patient… or what everyone of us is applying when assessing another human being or a group of individuals: We use a set of attributes, allocate them to the human being or the group, and try to adjust the scale of those attributes (or properties) to our observations. Along the way, we have to combine – entangle – some attributes and make them interact with each other, but still consider and keep them linearly independent. This set of – partially entangled – attributes, which then describes our individual or group of individuals, mathematically forms a space-time and by assuming that this space-time:

---

[6] Here we refer to the movie "Terminator" with A. Schwarzenegger.

A) … is of Riemann character,

B) … as everything else in this universe, has to be subjected to a Hamilton extremal principle in order to allow the object (or system), described by (respectively residing within) the space-time, to exist in a universe governed by extremal principles,

we simply require a mathematical apparatus, which can handle arbitrary – and rather general – ensembles of attributes. As already said above, this apparatus already is at hand. It was (and as it seems so – perhaps even purposefully – nicely kept under the radar, we will repeat it here again) introduced by David Hilbert in 1915 [1] and it directly leads to the so-called Einstein-Hilbert action. Solving the corresponding variational problem – in the classical way[7] – gives generalized Einstein field equations [1, 3], which – most interestingly – also contain all major quantum equations [2, 4, 5]. Obviously, there is no need for a "Theory of Everything" because it already exists. It was – in principle – presented over 108 years ago in an unsuspicious paper [1], which almost everybody always only took as a nice mathematical contribution of a great mathematician to the world of physics. In fact, it was – almost – always only taken as Hilbert's extremal derivation of the Einstein field equations in a rigorous falsifiable mathematical manner. Thereby, as we can see now [2, 4, 56], this paper already WAS or rather contained the "Theory of Everything" or – as it is also named – the "Quantum Gravity Theory" scientists had been searching for for over a century. However, that this paper of Hilbert also provided the fundament for a "Mathematical Psychology" [8] is something we have to discuss here. Thereby, we will see that it seems to be that what classically in quantum physics is known as the wave function makes out what we see as individuality… and here our interest is raised, because it shows us the way for a rigorous and completely unbiased market understanding and steering.

We will also see that it seems to be possible for "certain minds" to bypass condition A (the Riemann condition) and produce so-called g=0-spaces of thought. We will cover this "unconventional physics of the mind" in the subsection "The Conventional and the Unconventional Physics Realm" and we have the suspicion that it is the conscious mind which does not only make use of this option, but actually needs the ability to produce such mind states in order to be of conscious character at all. Yes, we are going to work out the Quantum Gravity Physics of the mind and along the way come across a set of options, which seem to be necessary for consciousness to actually be. These options are:

a) The ability of the mind to set itself, respectively, parts of it (ensembles of ideas or systems of thought) into states of non-Riemann character

b) The ability to steer the individuality or subjectivity via the uncertainty within the variation as

    a. Volume uncertainty, which is to say, a scalar factor of the metric to the system, we see as mind-system or sub-mind-system in case we only consider parts of the mind

    b. Ricci-uncertainty, which is to say, a scalar factor of the scalar curvature presented in the Einstein-Hilbert action as Ricci scalar

---

[7] We will see that this little remark comes of importance when investigating "strange effects" and realizing that there are also other options to solve the Einstein-Hilbert action problem, thereby coming to a quite unconventional, but definitively not "esoteric" physics (c.f. section "The Conventional and the Unconventional Physics Realm" further below in this book).

c) The ability of a conscious mind to actively switch sub-systems of thoughts from fermionic to bosonic behavior and vice versa
d) The ability to apply a local cosmological constant to achieve a greater flexibility in dealing with the Einstein-Hilbert action task
e) The ability to actively adjust the mind's dimensionality via symmetry-breaking and / or x-polarization (x is denoting the number of different charges) thereby thwarting the covariance principle [203]

Further below in this section we will come back to this "other physics realm".

Taking on this basic conjecture, or rather ensemble of conjectures, leads to a variety of insights regarding the inescapable escape of "intellectual property", the inoculation of ideas, the tunneling of alien "ideologies" through an individual's mind guard, the storage of thought, the emittance of confidence, the impossibility of thought prisons (thoughts are always freer than the thinker!), and many other things.

The subsequent recipe one can extract from the equivalence of quantum gravity field concepts and individuality, regarding one individuum, does seem to lead to some kind of "molecular dynamics" (multi-dimensional and complex, though) when considering ensembles of individuals. This concept is of great interest of course, when one intends to introduce disruptive technologies in a socioeconomic space-time. Thereby it is not only the possibility to foresee market and financial reactions, but also to actively steer them and have the answers ready when potential competitors appear on the scene or when the legislature wakes up and reacts in one way or the other. The most important outcome of these insights may be seen in the following sentence[8]:

"For the right mind it always holds that thoughts are freer than the thinker!"

Thereby we will only need a quantum gravity derivation of the Hawking radiation out of a black hole and connect it to our newly found knowledge about the nature of the thoughts, ideas, and the conscious (awake, not woke) mind in general.

This insight is of immense importance for every owner of a transformative, if not to say disruptive technology. As the authors are possessing at least three such metamorphological technologies, it has to be seen almost as a must that they also take care about the subsequent socioeconomic aspects of the application of their technologies.

Further, as already said above, it will be worked out in this book that the general recipe does not only bring us the conventional physics in the form of a quantum gravity for everything, but that there is even more. Above we simply called it the "unconventional physics", but especially in connection with the aspect of "market attraction" and everything having to do with it, we are looking for a somewhat more illustrative expression. We will therefore also begin to consider solutions for a physics realm that has been more or less "correctly" referred to as "chaos" in the history of mankind. As said, the corresponding math will be presented in this book in the subsection "The Conventional and the Unconventional Physics Realm". The authors see this as an essential aspect of the attempt to holistically try and give tools for the understanding, modeling, and simulation of the human mind, be it a single individual or an ensemble of such (group thinking), for all its possible states of

---

[8] For proof see [8]!

existence. There may be purists who will claim that this has nothing to do with science and thus, should not be presented in a book like this, but as it will be shown in a completely falsifiable and mathematically rigorous manner that the math for such an unconventional physics exists and that it even also follows from a Hamilton extremal principle, we see the presentation of such possibilities almost as a must in order to justify the claim of a true holistic approach. The same holds for questions about "life after death" and states of the mind where the thinker claims "to have left reality" – or at least had the feeling of having done so – and entered "another world", a "different form of existence" or – and this will be discussed in the next subsection – another and obviously much higher state of thought. The authors would be a bad caricature of scientists if just dismissing such claims as esoteric nonsense only because it does not fit into the classical physical understanding. Instead, they prefer to look for so far uncharted land in the fundamentals of physics, which is to say "where no man has gone before", in order to make sure that nothing is overlooked. Yes, there may be many false priests and lying shamans who only pretend to "have access to other realities" but it would be very unscientific to sentence them all... at least this should not be done without a fair trial.

Fig. 26: The inevitable? [A]

8.4.2 System III Thinking or the Super-Mode

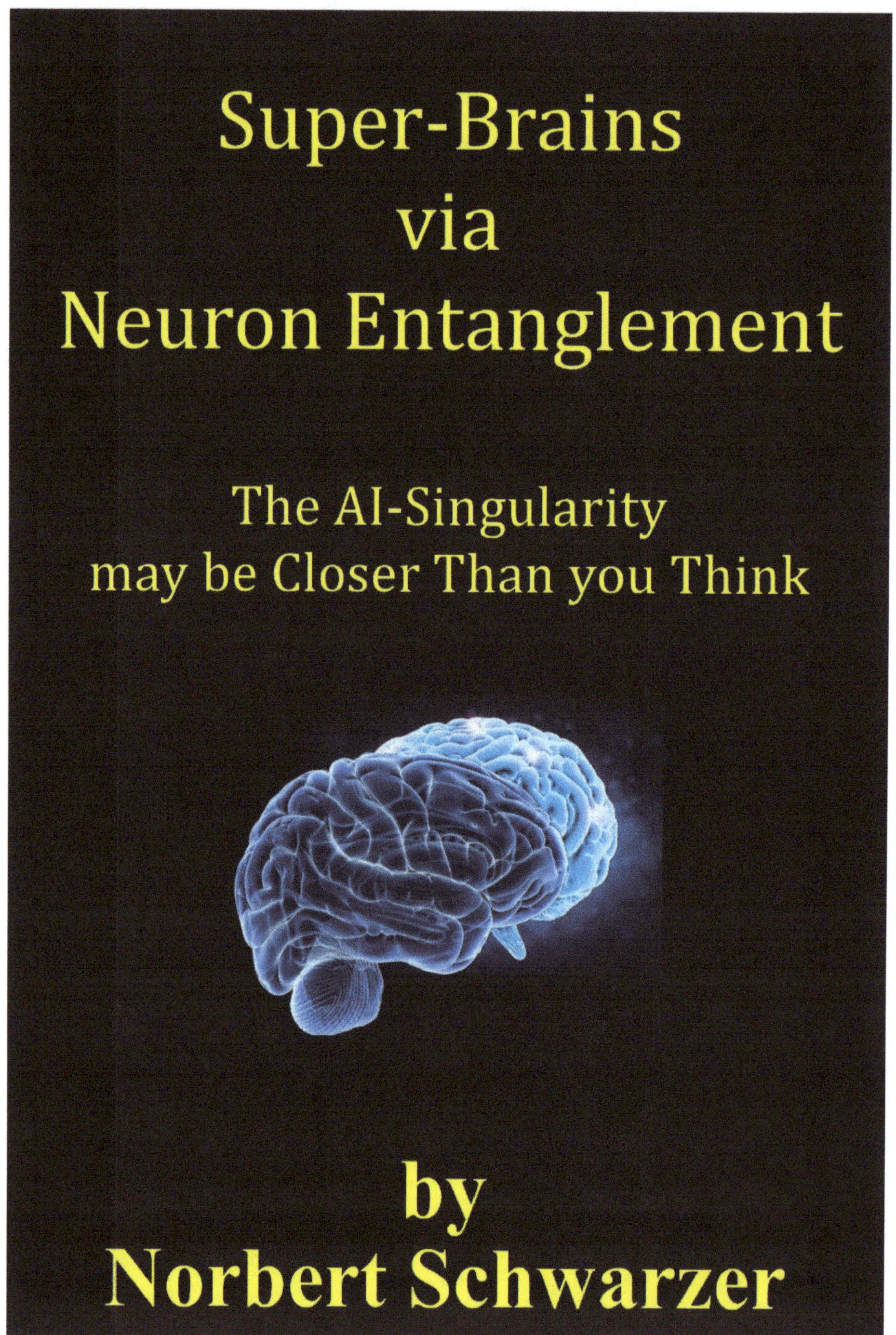

Fig. 27: Cover-picture of one of the third author's publications [204] (here for illustration only).

Neuroscience distinguishes two thinking modes. This is the system I or zombi mode, which is more or less an instinctive and automatic thought process, and the system II mode, which is the cognitive one.

However, it should be noted that the authors are of the opinion that there exists a third mode, which does not just follow from one of the third author's solution to the so-called 3-generation problem of elementary particles (this book and [21]), being in fact a "3-generation-rule" for every system [7, 13, 56, 70], but from a variety of observations (see subsection "Super-Brain Observations" below). Apparently, there are states of higher concentration (meditation, sleep), where we can go beyond the usual limitations of system II thinking and are capable of extreme cognitive performance [4, 204]. Here, too, could be another extremely interesting and – what is more – profitable application, whereby we require an entanglement of T1, T2 and T3.

8.4.2.1 Super-Brain Observations

In March 2020, a researcher from Canada sent this link to one of the authors:

https://singularityhub-com.cdn.ampproject.org/c/s/singularityhub.com/2020/01/14/scientists-discovered-mini-computers-in-human-neurons-and-thats-great-news-for-ai/amp/

It leads to an article, entitled:

"Scientists Discovered 'Mini-Computers' in Human Neurons — and That's Great News for AI" and *informs about a most interesting discovery regarding the capability of human neurons: "With just their input cables, human neurons can perform difficult logic calculations previously only seen in entire neural networks. To restate: human neurons are far more powerful devices than originally thought. And if deep learning algorithms — the AI method loosely based on the brain that has taken our world by storm — take note, they can be too."*

Only a few weeks before this was published, one of the authors of this book discovered some very principal degrees of freedom regarding the human "neural network" and by analyzing their potential it became apparent that the human brain could, in principle, locally form super-computers within or as part of its own structures (figure 27).

This agrees quite nicely with the findings of the scientists from the link given above, but there seems to be more… much more.

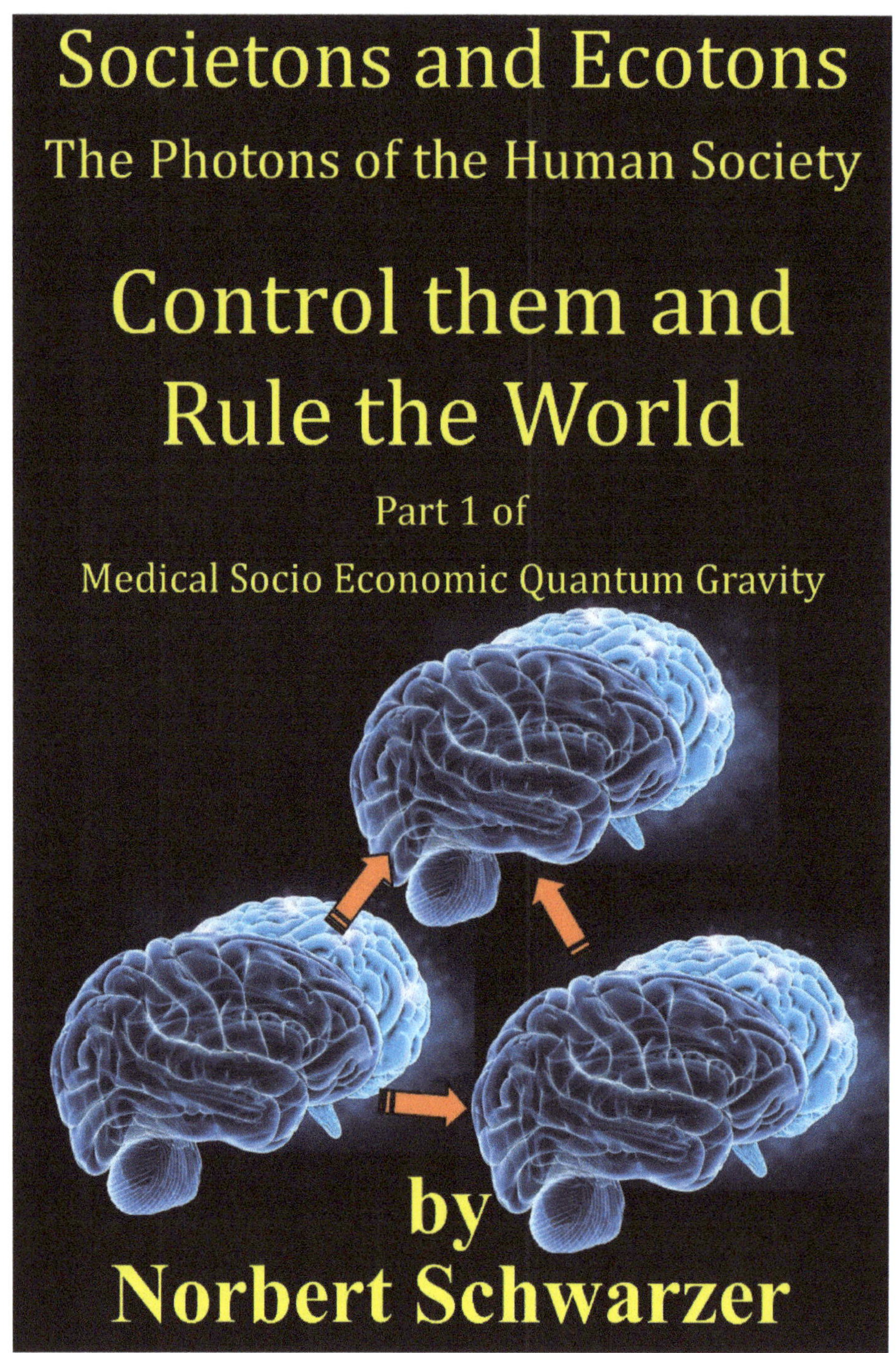

Fig. 28: Cover-picture of one of the third author's publications [205] (here for illustration only).

Theoretically, there is no limit to the human neurons to not just combine themselves to any ordinary, which is to say classical, computer structure, but it could even form quantum computers and – so far this author cannot see why not – gravity quantum computer structures [12, 92]. The only question would be how to keep these structures stable enough so that they could actually couple into the conscious mind and be used for something on a tactical or strategic level. This neural coherence may require certain states of natural or artificially supported (human-machine interfaces) concentration.

Combinations of greater groups of individuals with such "brain-reinforcements" may even act together, forming super intelligent swarms [205, 206], which can by far easier and quicker solve any problem and evaluate its potentials in no-time (figures 28, 29).

Fig. 29: Cover-picture of one of the third author's publications [206] (here for illustration only).

9 The Theory[9]

Fig. 30: Recent publications by one of the authors.

9.1 Looking for a Physical Description of the Three Towers – The Recipe

In order to avoid the classical mistake of compromising a suitable description of something by already including assumptions into the starting point without realizing such restrictions and pre-set boundary conditions, we first need to think about the most general approach there can be for the mathematical comprehension of an (any) experimental setup, respectively, a system.

9.1.1 Ingredients: Attributes=Properties=Degrees of Freedom=Dimensions

Assuming every system, from the elementary particle via a "primitive" or even complex life-form via even unconscious or conscious minds to the whole universe or an Everett-lie set of universes, can be considered an ensemble of potentially entangled properties or attributes, we think that we have chosen the most general and fundamental basis for our theoretical endeavor. These properties are the degrees of freedom of the system and as such they are also the dimensions of that very system forming a space or space-time.

[9] For formal reasons we point out that most of the derivations shown here are based on material which already has been presented elsewhere. The corresponding references are given. However, in order to satisfy the demands of this book, we also needed a few mathematical extensions which have been performed by the original author.

9.1.2 The Code: Hamilton's Extremal Principle and Riemann's Calculus

The system can mathematically be described via its metric. This most general description of any system now allows us to investigate the system when we simply assume that it has to follow certain and very universal laws. The most fundamental law we could think of would be a Hamilton extremal principle[10] and this directly brings us to our starting point.

9.2 A Most Fundamental Starting Point

Our starting point shall be the Hamilton principle in form of the Einstein-Hilbert action [1]:

$$\delta W = 0 = \delta \int_V d^n x \left(\sqrt{-g} \cdot \Phi_R [R] \right) = \delta \int_V d^n x \left(\sqrt{-g} \cdot \left(R - 2\Lambda + L_M \right) \right). \tag{6}$$

Here R gives the Ricci scalar and g denotes the determinant of the metric tensor $g_{\alpha\beta}$. The term L_M stands for the matter Lagrange density, which was postulated by both Hilbert and Einstein, and Λ gives the cosmological constant. Assuming that we can extract a functional factor F[f], with f being a function of all coordinates, from the metric tensor in the following way:

$$G_{\alpha\beta} = g_{\alpha\beta} \cdot F[f], \tag{7}$$

it was shown in [2] that we do not only obtain the Einstein field equations [3], but also all the most important classical quantum equations (see also [4]).

Here, we only briefly repeat the essentials and present some of the derivations in appendix D.

Setting (7) into (6) we obtain the following Einstein-Hilbert action [4]:

---

[10] "In physics, Hamilton's principle is William Rowan Hamilton's formulation of the principle of stationary action. It states that the dynamics of a physical system are determined by a variational problem for a functional based on a single function, the Lagrangian, which may contain all physical information concerning the system and the forces acting on it. The variational problem is equivalent to and allows for the derivation of the differential equations of motion of the physical system. Although formulated originally for classical mechanics, Hamilton's principle also applies to classical fields such as the electromagnetic and gravitational fields, and plays an important role in quantum mechanics, quantum field theory and criticality theories." (from https://en.wikipedia.org/wiki/Hamilton's_principle). It should explicitly be pointed out that the rigorous usage of the Hamilton principles (in its very general form of the Einstein-Hilbert action) in this book is not without controversy among the authors. It is a postulated principle after all! However, for here and now and taking established physics into account, it is the most fundamental starting point there is and so we chose it and built all our concepts upon this principle.

$$\delta W = 0 = \delta \int_V d^n x \left(\sqrt{-g \cdot F^n} \times \left(\left(\begin{array}{c} R - \frac{1}{2F} \left(\begin{array}{c} 2F_{,\alpha\beta}(n-1) g^{\alpha\beta} + 2\Gamma^{a}_{\alpha\beta} F_{,a} g^{\alpha\beta} \\ -F_{,\alpha} g^{ab} g_{\beta b,a} g^{\alpha\beta} - F_{,\beta} g^{ab} g_{\alpha b,a} g^{\alpha\beta} \\ -n\Gamma^{d}_{\alpha\beta} F_{,d} g^{\alpha\beta} + \frac{n}{2} F_{,d} g^{cd} g_{ab,c} g^{ab} \end{array} \right) \\ -\frac{F_{,\alpha} \cdot F_{,\beta}}{4F^2} g^{\alpha\beta} ((n-6)(n-1)) \end{array} \right) \frac{1}{F} \\ -2\Lambda + L_M \right) \right). \tag{8}$$

Thereby, in order to keep things simple and still general enough, one can often restrict the metric selection to diagonal metrics with the following properties:

$$g_{ij} = \begin{pmatrix} g_{00} & \cdots & 0 \\ \vdots & \ddots & \vdots \\ 0 & \cdots & g_{n-1n-1} \end{pmatrix}; \quad g_{jj,j} = 0, \tag{9}$$

which gives an extended Hilbert integral:

$$\delta W = 0 = \delta \int_V d^n x \left(\sqrt{-g \cdot F^n} \times \left(\left(\begin{array}{c} R + \frac{F'}{F}(1-n)\Delta f \\ + \frac{f_{,\alpha} f_{,\beta} g^{\alpha\beta} (1-n)}{4F^2} \left(4FF'' + (F')^2 (n-6) \right) \end{array} \right) \frac{1}{F} \\ -2\Lambda + L_M \right) \right). \tag{10}$$

The interested reader may prove that this set of metrics covers for most of the generally used coordinates in Ricci flat space-times. Generally, we will not use this simplification, except for examples with corresponding metrical symmetries.

9.2.1 A Point for Caution – Surface Terms

Regarding the performance of the variation in (6), we have to be very careful with what type of variation we choose. Incorporating (7), the variational integral can be rewritten and expanded as follows:

$$
\begin{aligned}
\delta W = 0 &= \delta\int_V d^n x\left(\sqrt{-G}\cdot\left(R^* - 2\Lambda + L_M\right)\right) \\
&= \delta\int_V d^n x\left(\sqrt{-g\cdot F^{\frac{n}{2}}}\cdot\left(R^*_{\alpha\beta}G^{\alpha\beta} - 2\Lambda + L_M\right)\right) \\
&= \delta\int_V d^n x\left(\sqrt{-g\cdot F^{\frac{n}{2}}}\cdot\left(\frac{1}{F}\cdot R^*_{\alpha\beta}g^{\alpha\beta} - 2\Lambda + L_M\right)\right) \\
&= \int_V d^n x\left(\sqrt{-g\cdot F^{\frac{n}{2}}}\cdot\delta\left(\frac{1}{F}\cdot R^*_{\alpha\beta}g^{\alpha\beta} - 2\Lambda + L_M\right) + \left(\frac{1}{F}\cdot R^*_{\alpha\beta}g^{\alpha\beta} - 2\Lambda + L_M\right)\delta\sqrt{-g\cdot F^{\frac{n}{2}}}\right)
\end{aligned}
\quad . \quad (11)
$$

Now it is very important to take care about the fact with respect to which metric the variation is performed. Variation with respect to the scaled metric from (7), reading:

$$
\begin{aligned}
\delta W = 0 &= \delta\int_V d^n x\left(\sqrt{-G}\cdot\left(R^* - 2\Lambda + L_M\right)\right) \\
&= \int_V d^n x\left(\begin{array}{c}\sqrt{-g\cdot F^{\frac{n}{2}}}\cdot\delta_{G_{\alpha\beta}}\left(\frac{1}{F}\cdot R^*_{\alpha\beta}g^{\alpha\beta} - 2\Lambda + L_M\right) \\ + \left(\frac{1}{F}\cdot R^*_{\alpha\beta}g^{\alpha\beta} - 2\Lambda + L_M\right)\delta_{G_{\alpha\beta}}\sqrt{-g\cdot F^{\frac{n}{2}}}\end{array}\right)
\end{aligned}
, \quad (12)
$$

will give us the classical Hilbert result, which is to say, the Einstein field equations, only that they now read:

$$
R^*_{\mu\nu} - \frac{1}{2}R^* G_{\mu\nu} = 0 \quad (13)
$$

instead of:

$$
R_{\mu\nu} - \frac{1}{2}R\cdot g_{\mu\nu} = 0 \, . \quad (14)
$$

When, however, performing the variation with respect to the unscaled metric, the situation changes quite dramatically.

It becomes immediately clear from the structure of the Ricci tensor integral term from (12), which we can expand as follows:

$$
\delta W = \int_V d^n x\left(\begin{array}{c}\sqrt{-g\cdot F^{\frac{n}{2}}}\cdot\left(\frac{1}{F}\cdot R^*_{\alpha\beta}\delta_{G_{\alpha\beta}}g^{\alpha\beta} + g^{\alpha\beta}\delta_{G_{\alpha\beta}}R^*_{\alpha\beta} - 2\Lambda + \delta_{G_{\alpha\beta}}L_M\right) \\ + \left(\frac{1}{F}\cdot R^*_{\alpha\beta}g^{\alpha\beta} - 2\Lambda + L_M\right)\delta_{G_{\alpha\beta}}\sqrt{-g\cdot F^{\frac{n}{2}}}\end{array}\right), \quad (15)
$$

and ignoring the matter term and the cosmological constant, plus using the results from appendix D of this book for the scaled Ricci tensor via:

$$\delta W = \int_V d^n x \left(\begin{array}{c} \sqrt{-g \cdot F^{\frac{n}{2}}} \cdot \left(\frac{1}{F} \cdot R^*_{\alpha\beta} \delta_{G_{\alpha\beta}} g^{\alpha\beta} + \frac{1}{F} \cdot g^{\alpha\beta} \delta_{G_{\alpha\beta}} \overbrace{\left(R_{\alpha\beta} + \Theta[F]_{\alpha\beta} \right)}^{=\frac{1}{F} \cdot R^*_{\alpha\beta}} \right) \\ + \frac{1}{F} \cdot R^*_{\alpha\beta} g^{\alpha\beta} \delta_{G_{\alpha\beta}} \sqrt{-g \cdot F^{\frac{n}{2}}} \end{array} \right), \tag{16}$$

$$= \int_V d^n x \left(\begin{array}{c} \sqrt{-g \cdot F^{\frac{n}{2}}} \cdot \left(\frac{1}{F} \cdot R^*_{\alpha\beta} \delta_{G_{\alpha\beta}} g^{\alpha\beta} + \frac{1}{F} \cdot \left(\boxed{g^{\alpha\beta} \delta_{G_{\alpha\beta}} R_{\alpha\beta}} + g^{\alpha\beta} \delta_{G_{\alpha\beta}} \Theta[F]_{\alpha\beta} \right) \right) \\ + \frac{1}{F} \cdot R^*_{\alpha\beta} g^{\alpha\beta} \delta_{G_{\alpha\beta}} \sqrt{-g \cdot F^{\frac{n}{2}}} \end{array} \right)$$

and changing the variation of the boxed term above into:

$$\int_V d^n x \sqrt{-g} \cdot \frac{F^{n/2}}{F} \cdot g^{\alpha\beta} \delta_{g_{\alpha\beta}} R_{\alpha\beta} = \int_V d^n x \sqrt{-g} \cdot F^{n/2-1} \cdot g^{\alpha\beta} \delta R_{\alpha\beta}, \tag{17}$$

that this cannot give the same surface term (and thus, would not vanish within the variational process as demonstrated by Hilbert in [1]) result as the classical form, reading:

$$\int_V d^n x \sqrt{-g} \cdot g^{\alpha\beta} \delta_{g_{\alpha\beta}} R_{\alpha\beta} = \int_V d^n x \sqrt{-g} \cdot g^{\alpha\beta} \delta R_{\alpha\beta}. \tag{18}$$

The reason is that while (18) can be made a complete divergence, reading:

$$\int_V d^n x \sqrt{-g} \cdot g^{\alpha\beta} \delta R_{\alpha\beta} = \int_V d^n x \left(\sqrt{-g} \cdot g^{\alpha\beta} \delta \Gamma^\gamma_{\alpha\beta} \right)_{,\gamma} - \int_V d^n x \left(\sqrt{-g} \cdot g^{\alpha\beta} \delta \Gamma^\sigma_{\beta\sigma} \right)_{,\alpha}, \tag{19}$$

we will have the following from (17):

$$\begin{aligned} & \int_V d^n x \sqrt{-g} \cdot F^{n/2-1} \cdot g^{\alpha\beta} \delta R_{\alpha\beta} \\ = & \int_V d^n x \cdot F^{n/2-1} \cdot \left(\sqrt{-g} \cdot g^{\alpha\beta} \delta \Gamma^\gamma_{\alpha\beta} \right)_{,\gamma} - \int_V d^n x \cdot F^{n/2-1} \cdot \left(\sqrt{-g} \cdot g^{\alpha\beta} \delta \Gamma^\sigma_{\beta\sigma} \right)_{,\alpha} \end{aligned}. \tag{20}$$

Only after integration by parts:

$$
\begin{aligned}
&\int_V d^n x\sqrt{-g}\cdot F^{n/2-1}\cdot g^{\alpha\beta}\delta R_{\alpha\beta} \\
&=\int_V d^n x\cdot F^{n/2-1}\cdot\left(\sqrt{-g}\cdot g^{\alpha\beta}\delta\Gamma^{\gamma}_{\alpha\beta}\right)_{,\gamma}-\int_V d^n x\cdot F^{n/2-1}\cdot\left(\sqrt{-g}\cdot g^{\alpha\beta}\delta\Gamma^{\sigma}_{\beta\sigma}\right)_{,\alpha} \\
&=\int_V d^n x\cdot F^{n/2-1}{}_{,\gamma}\cdot\sqrt{-g}\cdot g^{\alpha\beta}\delta\Gamma^{\gamma}_{\alpha\beta}-\int_V d^n x\cdot\left(F^{n/2-1}\cdot\sqrt{-g}\cdot g^{\alpha\beta}\delta\Gamma^{\gamma}_{\alpha\beta}\right)_{,\gamma} \\
&-\int_V d^n x\cdot F^{n/2-1}{}_{,\alpha}\cdot\sqrt{-g}\cdot g^{\alpha\beta}\delta\Gamma^{\sigma}_{\beta\sigma}+\int_V d^n x\cdot\left(F^{n/2-1}\cdot\sqrt{-g}\cdot g^{\alpha\beta}\delta\Gamma^{\sigma}_{\beta\sigma}\right)_{,\alpha} \\
&=\int_V d^n x\cdot F^{n/2-1}{}_{,\gamma}\cdot\sqrt{-g}\cdot g^{\alpha\beta}\delta\Gamma^{\gamma}_{\alpha\beta}-\int_{\partial V} d^n x\cdot n_{\gamma}\left(F^{n/2-1}\cdot\sqrt{-g}\cdot g^{\alpha\beta}\delta\Gamma^{\gamma}_{\alpha\beta}\right) \\
&-\int_V d^n x\cdot F^{n/2-1}{}_{,\alpha}\cdot\sqrt{-g}\cdot g^{\alpha\beta}\delta\Gamma^{\sigma}_{\beta\sigma}+\int_{\partial V} d^n x\cdot n_{\alpha}\left(F^{n/2-1}\cdot\sqrt{-g}\cdot g^{\alpha\beta}\delta\Gamma^{\sigma}_{\beta\sigma}\right)
\end{aligned}
, \quad (21)
$$

we have some of the desired surface integrals, which we can assume to vanish (!) using Hilbert's assumption of a boundary free space-time (caution here (!)). The remains are the two integrals:

$$
\begin{aligned}
&\int_V d^n x\sqrt{-g}\cdot F^{n/2-1}\cdot g^{\alpha\beta}\delta R_{\alpha\beta} \\
&=\int_V d^n x\cdot F^{n/2-1}{}_{,\gamma}\cdot\sqrt{-g}\cdot g^{\alpha\beta}\delta\Gamma^{\gamma}_{\alpha\beta}-\int_V d^n x\cdot F^{n/2-1}{}_{,\alpha}\cdot\sqrt{-g}\cdot g^{\alpha\beta}\delta\Gamma^{\sigma}_{\beta\sigma}
\end{aligned}
, \quad (22)
$$

where we once again find the variated Levi-Civita-Connections $\delta\Gamma^{\gamma}_{\alpha\beta},\delta\Gamma^{\sigma}_{\beta\sigma}$, which this time result in:

$$
\begin{aligned}
&\int_V d^n x \cdot F^{n/2-1}{}_{,\gamma} \cdot \sqrt{-g} \cdot g^{\alpha\beta} \delta\Gamma^{\gamma}_{\alpha\beta} - \int_V d^n x \cdot F^{n/2-1}{}_{,\alpha} \cdot \sqrt{-g} \cdot g^{\alpha\beta} \delta\Gamma^{\sigma}_{\beta\sigma} \\
&= \int_V d^n x \cdot F^{n/2-1}{}_{,\gamma} \cdot \sqrt{-g} \cdot g^{\alpha\beta} \left(\frac{g^{\gamma\sigma}}{2} \left(\delta g_{\sigma\alpha,\beta} + \delta g_{\sigma\beta,\alpha} - \delta g_{\alpha\beta,\sigma} \right) - g^{\gamma\lambda} \Gamma^{\kappa}_{\alpha\beta} \delta g_{\kappa\lambda} \right) \\
&\quad - \int_V d^n x \cdot F^{n/2-1}{}_{,\alpha} \cdot \sqrt{-g} \cdot g^{\alpha\beta} \frac{1}{2} \cdot g^{\sigma\lambda} \delta g_{\sigma\lambda;\beta} \\
&= \int_V d^n x \cdot F^{n/2-1}{}_{,\gamma} \cdot \sqrt{-g} \cdot g^{\alpha\beta} \left(\frac{g^{\gamma\sigma}}{2} GX^{\kappa\lambda\rho}_{\alpha\beta\sigma} \delta g_{\kappa\lambda,\rho} - g^{\gamma\lambda} \Gamma^{\kappa}_{\alpha\beta} \delta g_{\kappa\lambda} \right) \\
&\quad - \int_V d^n x \cdot F^{n/2-1}{}_{,\alpha} \cdot \sqrt{-g} \cdot g^{\alpha\beta} \frac{1}{2} \cdot g^{\kappa\lambda} \delta g_{\kappa\lambda;\beta} \\
&= \int_V d^n x \cdot F^{n/2-1}{}_{,\gamma} \cdot \sqrt{-g} \cdot g^{\alpha\beta} \left(\frac{g^{\gamma\sigma}}{2} GX^{\kappa\lambda\rho}_{\alpha\beta\sigma} \delta g_{\kappa\lambda,\rho} - g^{\gamma\lambda} \Gamma^{\kappa}_{\alpha\beta} \delta g_{\kappa\lambda} \right) \\
&\quad - \int_V d^n x \cdot F^{n/2-1}{}_{,\alpha} \cdot \sqrt{-g} \cdot g^{\alpha\beta} \frac{1}{2} \cdot g^{\kappa\lambda} \left(\delta g_{\kappa\lambda,\beta} - \Gamma^{\xi}_{\beta\kappa} \delta g_{\xi\lambda} - \Gamma^{\xi}_{\beta\lambda} \delta g_{\xi\kappa} \right) \\
&= \int_V d^n x \cdot F^{n/2-1}{}_{,\gamma} \cdot \sqrt{-g} \cdot g^{\alpha\beta} \left(\frac{g^{\gamma\sigma}}{2} GX^{\kappa\lambda\rho}_{\alpha\beta\sigma} \delta g_{\kappa\lambda,\rho} - g^{\gamma\lambda} \Gamma^{\kappa}_{\alpha\beta} \delta g_{\kappa\lambda} \right) \\
&\quad - \int_V d^n x \cdot F^{n/2-1}{}_{,\alpha} \cdot \sqrt{-g} \cdot g^{\alpha\beta} \frac{1}{2} \left(g^{\kappa\lambda} \delta g_{\kappa\lambda,\beta} - g^{\kappa\lambda} \Gamma^{\xi}_{\beta\kappa} \delta g_{\xi\lambda} - g^{\kappa\lambda} \Gamma^{\xi}_{\beta\lambda} \delta g_{\xi\kappa} \right) \\
&= \int_V d^n x \cdot F^{n/2-1}{}_{,\gamma} \cdot \sqrt{-g} \cdot g^{\alpha\beta} \left(\frac{g^{\gamma\sigma}}{2} GX^{\kappa\lambda\rho}_{\alpha\beta\sigma} \delta g_{\kappa\lambda,\rho} - g^{\gamma\lambda} \Gamma^{\kappa}_{\alpha\beta} \delta g_{\kappa\lambda} \right) \\
&\quad - \int_V d^n x \cdot F^{n/2-1}{}_{,\alpha} \cdot \sqrt{-g} \cdot g^{\alpha\beta} \frac{1}{2} \left(g^{\kappa\lambda} \delta g_{\kappa\lambda,\beta} - g^{\xi\lambda} \Gamma^{\kappa}_{\beta\xi} \delta g_{\kappa\lambda} - g^{\kappa\xi} \Gamma^{\lambda}_{\beta\xi} \delta g_{\lambda\kappa} \right)
\end{aligned}
\quad , \qquad (23)
$$

$$=\int_V d^n x\cdot F^{n/2-1}{}_{,\gamma}\cdot\sqrt{-g}\cdot g^{\alpha\beta}\left(\frac{g^{\gamma\sigma}}{2}GX^{\kappa\lambda\rho}_{\alpha\beta\sigma}\delta g_{\kappa\lambda,\rho}-g^{\gamma\lambda}\Gamma^{\kappa}_{\alpha\beta}\delta g_{\kappa\lambda}\right)$$
$$-\int_V d^n x\cdot F^{n/2-1}{}_{,\alpha}\cdot\sqrt{-g}\cdot g^{\alpha\beta}\frac{1}{2}\left(g^{\kappa\lambda}\delta g_{\kappa\lambda,\beta}-g^{\xi\lambda}\Gamma^{\kappa}_{\beta\xi}\delta g_{\kappa\lambda}-g^{\kappa\xi}\Gamma^{\lambda}_{\beta\xi}\delta g_{\lambda\kappa}\right)$$
$$=\int_V d^n x\cdot\sqrt{-g}\cdot g^{\alpha\beta}\left(F^{n/2-1}{}_{,\gamma}\cdot\frac{g^{\gamma\sigma}}{2}GX^{\kappa\lambda\rho}_{\alpha\beta\sigma}\delta g_{\kappa\lambda,\rho}-F^{n/2-1}{}_{,\gamma}\cdot g^{\gamma\lambda}\Gamma^{\kappa}_{\alpha\beta}\delta g_{\kappa\lambda}\right)$$
$$-\int_V d^n x\cdot\sqrt{-g}\cdot g^{\alpha\beta}\frac{1}{2}\left(F^{n/2-1}{}_{,\alpha}\cdot g^{\kappa\lambda}\delta g_{\kappa\lambda,\beta}-F^{n/2-1}{}_{,\alpha}\cdot g^{\xi\lambda}\Gamma^{\kappa}_{\beta\xi}\delta g_{\kappa\lambda}-F^{n/2-1}{}_{,\alpha}\cdot g^{\kappa\xi}\Gamma^{\lambda}_{\beta\xi}\delta g_{\lambda\kappa}\right). \quad (24)$$
$$=\int_V d^n x\cdot\sqrt{-g}\cdot g^{\alpha\beta}F^{n/2-1}{}_{,\gamma}\cdot\frac{g^{\gamma\sigma}}{2}GX^{\kappa\lambda\rho}_{\alpha\beta\sigma}\delta g_{\kappa\lambda,\rho}-\int_V d^n x\cdot\sqrt{-g}\cdot g^{\alpha\beta}\frac{1}{2}F^{n/2-1}{}_{,\alpha}\cdot g^{\kappa\lambda}\delta g_{\kappa\lambda,\beta}$$
$$\int_V d^n x\cdot\sqrt{-g}\cdot g^{\alpha\beta}F^{n/2-1}{}_{,\gamma}\cdot g^{\gamma\lambda}\Gamma^{\kappa}_{\alpha\beta}\delta g_{\kappa\lambda}$$
$$+\int_V d^n x\cdot\sqrt{-g}\cdot g^{\alpha\beta}\frac{1}{2}\left(F^{n/2-1}{}_{,\alpha}\cdot g^{\xi\lambda}\Gamma^{\kappa}_{\beta\xi}+F^{n/2-1}{}_{,\alpha}\cdot g^{\kappa\xi}\Gamma^{\lambda}_{\beta\xi}\right)\delta g_{\kappa\lambda}$$

Thereby we have used the following abbreviation:

$$\delta_g\left(\Gamma^{\gamma}_{\mu\nu}\right)=\frac{1}{2}g^{\gamma\sigma}\left(\delta g_{\sigma\mu;\nu}+\delta g_{\sigma\nu;\mu}-\delta g_{\mu\nu;\sigma}\right)=\frac{g^{\gamma\sigma}}{2}GX^{\kappa\lambda\rho}_{\mu\nu\sigma}\delta g_{\kappa\lambda,\rho}-g^{\gamma\lambda}\Gamma^{\kappa}_{\mu\nu}\delta g_{\kappa\lambda}. \quad (25)$$
$$\left(g^{\kappa}_{\mu}g^{\lambda}_{\sigma}g^{\rho}_{\nu}+g^{\kappa}_{\sigma}g^{\lambda}_{\nu}g^{\rho}_{\mu}-g^{\kappa}_{\mu}g^{\lambda}_{\nu}g^{\rho}_{\sigma}\right)\equiv GX^{\kappa\lambda\rho}_{\mu\nu\sigma}$$

While the integrals with the metric variation terms (second last and last line in (24)) are no problem, we need to find a way to also treat the derivatives of the metric variation. Thus, we further consider the two integrals:

$$\begin{aligned}
&\int_V d^n x\cdot\sqrt{-g}\cdot g^{\alpha\beta}F^{n/2-1}{}_{,\gamma}\cdot\frac{g^{\gamma\sigma}}{2}GX^{\kappa\lambda\rho}_{\alpha\beta\sigma}\delta g_{\kappa\lambda,\rho}-\int_V d^n x\cdot\sqrt{-g}\cdot g^{\alpha\beta}\frac{1}{2}F^{n/2-1}{}_{,\alpha}\cdot g^{\kappa\lambda}\delta g_{\kappa\lambda,\beta}\\
&=\int_V d^n x\cdot\left(\sqrt{-g}\cdot g^{\alpha\beta}F^{n/2-1}{}_{,\gamma}\cdot\frac{g^{\gamma\sigma}}{2}GX^{\kappa\lambda\rho}_{\alpha\beta\sigma}\right)_{,\rho}\delta g_{\kappa\lambda}\\
&-\int_V d^n x\cdot\left(\sqrt{-g}\cdot g^{\alpha\beta}F^{n/2-1}{}_{,\gamma}\cdot\frac{g^{\gamma\sigma}}{2}GX^{\kappa\lambda\rho}_{\alpha\beta\sigma}\delta g_{\kappa\lambda}\right)_{,\rho}\\
&-\int_V d^n x\cdot\left(\sqrt{-g}\cdot g^{\alpha\beta}\frac{1}{2}F^{n/2-1}{}_{,\alpha}\cdot g^{\kappa\lambda}\right)_{,\beta}\delta g_{\kappa\lambda}\\
&+\int_V d^n x\cdot\left(\sqrt{-g}\cdot g^{\alpha\beta}\frac{1}{2}F^{n/2-1}{}_{,\alpha}\cdot g^{\kappa\lambda}\delta g_{\kappa\lambda}\right)_{,\beta}\\
&=\int_V d^n x\cdot\left(\sqrt{-g}\cdot g^{\alpha\beta}F^{n/2-1}{}_{,\gamma}\cdot\frac{g^{\gamma\sigma}}{2}GX^{\kappa\lambda\rho}_{\alpha\beta\sigma}\right)_{,\rho}\delta g_{\kappa\lambda}\\
&\overbrace{-\int_{\partial V} d^n x\cdot n_\rho\sqrt{-g}\cdot g^{\alpha\beta}F^{n/2-1}{}_{,\gamma}\cdot\frac{g^{\gamma\sigma}}{2}GX^{\kappa\lambda\rho}_{\alpha\beta\sigma}\delta g_{\kappa\lambda}}^{=0}\\
&-\int_V d^n x\cdot\left(\sqrt{-g}\cdot g^{\alpha\beta}\frac{1}{2}F^{n/2-1}{}_{,\alpha}\cdot g^{\kappa\lambda}\right)_{,\beta}\delta g_{\kappa\lambda}\\
&\overbrace{+\int_{\partial V} d^n x\cdot n_\beta\sqrt{-g}\cdot g^{\alpha\beta}\frac{1}{2}F^{n/2-1}{}_{,\alpha}\cdot g^{\kappa\lambda}\delta g_{\kappa\lambda}}^{=0}
\end{aligned}\quad, \tag{26}$$

and this leads us to:

$$\begin{aligned}
&\int_V d^n x\cdot\sqrt{-g}\cdot g^{\alpha\beta}F^{n/2-1}{}_{,\gamma}\cdot\frac{g^{\gamma\sigma}}{2}GX^{\kappa\lambda\rho}_{\alpha\beta\sigma}\delta g_{\kappa\lambda,\rho}-\int_V d^n x\cdot\sqrt{-g}\cdot g^{\alpha\beta}\frac{1}{2}F^{n/2-1}{}_{,\alpha}\cdot g^{\kappa\lambda}\delta g_{\kappa\lambda,\beta}\\
&=\int_V d^n x\cdot\left[\left(\sqrt{-g}\cdot g^{\alpha\beta}F^{n/2-1}{}_{,\gamma}\cdot\frac{g^{\gamma\sigma}}{2}GX^{\kappa\lambda\rho}_{\alpha\beta\sigma}\right)_{,\rho}-\left(\sqrt{-g}\cdot g^{\alpha\beta}\frac{1}{2}F^{n/2-1}{}_{,\alpha}\cdot g^{\kappa\lambda}\right)_{,\beta}\right]\delta g_{\kappa\lambda}
\end{aligned}\quad. \tag{27}$$

Thus, in total we have from (22):

$$\int_V d^n x\sqrt{-g}\cdot F^{n/2-1}\cdot g^{\alpha\beta}\delta R_{\alpha\beta}$$

$$=\int_V d^n x\cdot\left(\begin{array}{l}\sqrt{-g}\cdot g^{\alpha\beta}\left(\frac{F^{n/2-1}{}_{,\alpha}\cdot g^{\xi\lambda}\Gamma^{\kappa}_{\beta\xi}+F^{n/2-1}{}_{,\alpha}\cdot g^{\kappa\xi}\Gamma^{\lambda}_{\beta\xi}}{2}+F^{n/2-1}{}_{,\gamma}\cdot g^{\gamma\lambda}\Gamma^{\kappa}_{\alpha\beta}\right)\\+\left(\sqrt{-g}\cdot g^{\alpha\beta}F^{n/2-1}{}_{,\gamma}\cdot\frac{g^{\gamma\sigma}}{2}GX^{\kappa\lambda\rho}_{\alpha\beta\sigma}\right)_{,\rho}-\left(\sqrt{-g}\cdot g^{\alpha\beta}\frac{1}{2}F^{n/2-1}{}_{,\alpha}\cdot g^{\kappa\lambda}\right)_{,\beta}\end{array}\right)\delta g_{\kappa\lambda}$$

$$=\int_V d^n x\cdot\left(\begin{array}{c}\sqrt{-g}\cdot g^{\alpha\beta}\left(\frac{F^{n/2-1}{}_{,\alpha}\cdot g^{\xi\lambda}\Gamma^{\kappa}_{\beta\xi}+F^{n/2-1}{}_{,\alpha}\cdot g^{\kappa\xi}\Gamma^{\lambda}_{\beta\xi}}{2}+F^{n/2-1}{}_{,\gamma}\cdot g^{\gamma\lambda}\Gamma^{\kappa}_{\alpha\beta}\right)\\+\left(\sqrt{-g}\cdot g^{\alpha\beta}F^{n/2-1}{}_{,\gamma}\frac{\left(g^{\kappa\rho}g^{\gamma\lambda}+g^{\rho\lambda}g^{\gamma\kappa}-g^{\kappa\lambda}g^{\gamma\rho}\right)}{2}\right)_{,\rho}\\-\left(\sqrt{-g}\cdot g^{\alpha\beta}\frac{1}{2}F^{n/2-1}{}_{,\alpha}\cdot g^{\kappa\lambda}\right)_{,\beta}\end{array}\right)\delta g_{\kappa\lambda}\quad. \tag{28}$$

We see that, in order to avoid surface terms, we always have to perform the variation with respect to the scaled metric $G_{\alpha\beta}$ and not $g_{\alpha\beta}$.

9.2.2 Quantum Einstein- or Quantum Gravity Field Equations

After performing this variation in (8), we obtain the following quantum gravity field equations:

$$0=\left(\begin{array}{c}\boxed{R_{\alpha\beta}-\frac{g_{\alpha\beta}}{2}R}+\frac{g_{\alpha\beta}}{2}\left(\begin{array}{c}\frac{1}{2F}\left(\begin{array}{c}2F_{,ij}(n-1)g^{ij}+2\Gamma^{a}_{ij}F_{,a}g^{ij}\\-F_{,i}g^{ab}g_{jb,a}g^{ij}-F_{,j}g^{ab}g_{ib,a}g^{ij}\\-n\Gamma^{d}_{ij}F_{,d}g^{ij}+\frac{n}{2}F_{,d}g^{cd}g_{ab,c}g^{ab}\end{array}\right)\\+\frac{F_{,i}\cdot F_{,j}}{4F^{2}}g^{ij}\left((n-6)(n-1)\right)\end{array}\right)\\-\frac{1}{2F}\left(\begin{array}{c}F_{,\alpha\beta}(n-2)+F_{,ab}g_{\alpha\beta}g^{ab}\\+F_{,a}g^{ab}\left(g_{\beta b,\alpha}-g_{\beta\alpha,b}\right)-F_{,\alpha}g^{ab}g_{\beta b,a}-F_{,\beta}g^{ab}g_{\alpha b,a}\\+F_{,d}g^{cd}\frac{1}{2}n\left(\frac{2}{n}g_{\alpha c,\beta}-g_{\alpha c,\beta}-g_{\beta c,\alpha}+g_{\alpha\beta,c}+\frac{1}{n}g_{\alpha\beta}g_{ab,c}g^{ab}\right)\\-\frac{1}{2F}\left(F_{,\alpha}\cdot F_{,\beta}(3n-6)+g_{\alpha\beta}F_{,c}F_{,d}g^{cd}(4-n)\right)\end{array}\right)\\+\kappa T_{\alpha\beta}+\Lambda\cdot F\cdot g_{\alpha\beta}\end{array}\right). \tag{29}$$

For completeness, in appendix D we will derive the equations without the affine connection, with further simplifications and purely with the derivatives of the metric tensor. The result then reads:

$$0=\begin{pmatrix} \boxed{R_{\alpha\beta}-\frac{g_{\alpha\beta}}{2}R}+\frac{g_{\alpha\beta}}{2}\begin{pmatrix} \frac{1}{2F}\left(2g^{\alpha\beta}F_{,\alpha\beta}(n-1)+F_{,d}g^{cd}g^{\alpha\beta}\left((n-1)g_{\alpha\beta,c}-ng_{\alpha c,\beta}\right)\right) \\ +\frac{F_{,i}\cdot F_{,j}}{4F^2}g^{ij}\left((n-6)(n-1)\right) \end{pmatrix} \\ -\frac{1}{2F}\begin{pmatrix} F_{,\alpha\beta}(n-2)+F_{,ab}g_{\alpha\beta}g^{ab} \\ +F_{,a}g^{ab}\left(g_{\beta b,\alpha}-g_{\beta\alpha,b}\right)-F_{,\alpha}g^{ab}g_{\beta b,a}-F_{,\beta}g^{ab}g_{\alpha b,a} \\ +F_{,d}g^{cd}\frac{1}{2}n\left(\frac{2}{n}g_{\alpha c,\beta}-g_{\alpha c,\beta}-g_{\beta c,\alpha}+g_{\alpha\beta,c}+\frac{1}{n}g_{\alpha\beta}g_{ab,c}g^{ab}\right) \\ -\frac{1}{2F}\left(F_{,\alpha}\cdot F_{,\beta}(3n-6)+g_{\alpha\beta}F_{,c}F_{,d}g^{cd}(4-n)\right) \end{pmatrix} \\ +\kappa T_{\alpha\beta}+\Lambda\cdot F\cdot g_{\alpha\beta} \end{pmatrix}. \tag{30}$$

We immediately recognize the classical Einstein field equations (red box) inside our new and more general ones. It was shown in [2, 4, 56] that the scaling terms in (29) are giving the Klein-Gordon, Dirac, and Schrödinger equation (regarding the latter see also [5]).

9.2.3 Scaling Option for the New Field Equations

Under the variational integral, the integrand (29) has to be multiplied with the variated metric tensor $\delta G^{\alpha\beta}=\delta\left(\frac{g^{\alpha\beta}}{F}\right)$ and therefore could be split up into two equations:

$$
0=\begin{pmatrix}
\boxed{R_{\alpha\beta}-\frac{g_{\alpha\beta}}{2}R}+\frac{g_{\alpha\beta}}{2}\begin{pmatrix}
\frac{1}{2F}\begin{pmatrix}
2F_{,ij}(n-1)g^{ij}+2\Gamma^{a}_{ij}F_{,a}g^{ij}\\
-F_{,i}g^{ab}g_{jb,a}g^{ij}-F_{,j}g^{ab}g_{ib,a}g^{ij}\\
-n\Gamma^{d}_{ij}F_{,d}g^{ij}+\frac{n}{2}F_{,d}g^{cd}g_{ab,c}g^{ab}
\end{pmatrix}\\
+\frac{F_{,i}\cdot F_{,j}}{4F^{2}}g^{ij}\left((n-6)(n-1)\right)
\end{pmatrix}\\
-\frac{1}{2F}\begin{pmatrix}
F_{,\alpha\beta}(n-2)+F_{,ab}g_{\alpha\beta}g^{ab}\\
+F_{,a}g^{ab}\left(g_{\beta b,\alpha}-g_{\beta\alpha,b}\right)-F_{,\alpha}g^{ab}g_{\beta b,a}-F_{,\beta}g^{ab}g_{\alpha b,a}\\
+F_{,d}g^{cd}\frac{1}{2}n\left(\frac{2}{n}g_{\alpha c,\beta}-g_{\alpha c,\beta}-g_{\beta c,\alpha}+g_{\alpha\beta,c}+\frac{1}{n}g_{\alpha\beta}g_{ab,c}g^{ab}\right)\\
-\frac{1}{2F}\left(F_{,\alpha}\cdot F_{,\beta}(3n-6)+g_{\alpha\beta}F_{,c}F_{,d}g^{cd}(4-n)\right)
\end{pmatrix}\\
+\kappa T_{\alpha\beta}+\Lambda\cdot F\cdot g_{\alpha\beta}
\end{pmatrix}\delta G^{\alpha\beta}
$$

$$
=\begin{pmatrix}
\boxed{R_{\alpha\beta}-\frac{g_{\alpha\beta}}{2}R}+\frac{g_{\alpha\beta}}{2}\begin{pmatrix}
\frac{1}{2F}\begin{pmatrix}
2F_{,ij}(n-1)g^{ij}+2\Gamma^{a}_{ij}F_{,a}g^{ij}\\
-F_{,i}g^{ab}g_{jb,a}g^{ij}-F_{,j}g^{ab}g_{ib,a}g^{ij}\\
-n\Gamma^{d}_{ij}F_{,d}g^{ij}+\frac{n}{2}F_{,d}g^{cd}g_{ab,c}g^{ab}
\end{pmatrix}\\
+\frac{F_{,i}\cdot F_{,j}}{4F^{2}}g^{ij}\left((n-6)(n-1)\right)
\end{pmatrix}\\
-\frac{1}{2F}\begin{pmatrix}
F_{,\alpha\beta}(n-2)+F_{,ab}g_{\alpha\beta}g^{ab}\\
+F_{,a}g^{ab}\left(g_{\beta b,\alpha}-g_{\beta\alpha,b}\right)-F_{,\alpha}g^{ab}g_{\beta b,a}-F_{,\beta}g^{ab}g_{\alpha b,a}\\
+F_{,d}g^{cd}\frac{1}{2}n\left(\frac{2}{n}g_{\alpha c,\beta}-g_{\alpha c,\beta}-g_{\beta c,\alpha}+g_{\alpha\beta,c}+\frac{1}{n}g_{\alpha\beta}g_{ab,c}g^{ab}\right)\\
-\frac{1}{2F}\left(F_{,\alpha}\cdot F_{,\beta}(3n-6)+g_{\alpha\beta}F_{,c}F_{,d}g^{cd}(4-n)\right)
\end{pmatrix}\\
+\kappa T_{\alpha\beta}+\Lambda\cdot F\cdot g_{\alpha\beta}
\end{pmatrix}\begin{pmatrix}
g^{\alpha\beta}\delta\left(\frac{1}{F}\right)\\
+\frac{\delta g^{\alpha\beta}}{F}
\end{pmatrix}, \qquad (31)
$$

and

$$0=\left(\left(1-\frac{n}{2}\right)\left(R-\left(\frac{1}{2F}\begin{pmatrix}2F_{,ij}(n-1)g^{ij}+2\Gamma_{ij}^{a}F_{,a}g^{ij}\\-F_{,i}g^{ab}g_{jb,a}g^{ij}-F_{,j}g^{ab}g_{ib,a}g^{ij}\\-n\Gamma_{ij}^{d}F_{,d}g^{ij}+\frac{n}{2}F_{,d}g^{cd}g_{ab,c}g^{ab}\end{pmatrix}+\frac{F_{,i}\cdot F_{,j}}{4F^{2}}g^{ij}\left((n-6)(n-1)\right)\right)\right)+\Lambda\cdot F\cdot n\right)\delta\left(\frac{1}{F}\right)$$

$$0=\left(\begin{matrix}R_{\alpha\beta}-\frac{g_{\alpha\beta}}{2}R+\frac{g_{\alpha\beta}}{2}\left(\frac{1}{2F}\begin{pmatrix}2F_{,ij}(n-1)g^{ij}+2\Gamma_{ij}^{a}F_{,a}g^{ij}\\-F_{,i}g^{ab}g_{jb,a}g^{ij}-F_{,j}g^{ab}g_{ib,a}g^{ij}\\-n\Gamma_{ij}^{d}F_{,d}g^{ij}+\frac{n}{2}F_{,d}g^{cd}g_{ab,c}g^{ab}\end{pmatrix}+\frac{F_{,i}\cdot F_{,j}}{4F^{2}}g^{ij}\left((n-6)(n-1)\right)\right)\\-\frac{1}{2F}\left(\begin{matrix}F_{,\alpha\beta}(n-2)+F_{,ab}g_{\alpha\beta}g^{ab}\\+F_{,a}g^{ab}\left(g_{\beta b,\alpha}-g_{\beta\alpha,b}\right)-F_{,\alpha}g^{ab}g_{\beta b,a}-F_{,\beta}g^{ab}g_{\alpha b,a}\\+F_{,d}g^{cd}\frac{1}{2}n\left(\frac{2}{n}g_{\alpha c,\beta}-g_{\alpha c,\beta}-g_{\beta c,\alpha}+g_{\alpha\beta,c}+\frac{1}{n}g_{\alpha\beta}g_{ab,c}g^{ab}\right)\\-\frac{1}{2F}\left(F_{,\alpha}\cdot F_{,\beta}(3n-6)+g_{\alpha\beta}F_{,c}F_{,d}g^{cd}(4-n)\right)\end{matrix}\right)\\+\kappa T_{\alpha\beta}+\Lambda\cdot F\cdot g_{\alpha\beta}\end{matrix}\right)\frac{\delta g^{\alpha\beta}}{F}\quad. \tag{32}$$

While we consider the second equation of the pair in (32) generalized or quantum Einstein field equations, we used the first one to derive the classical quantum equations (Klein-Gordon, Dirac, and Schrödinger) for a generally curved space-time. Demanding the "gravity" to be weak as follows:

$$\delta G^{\alpha\beta}=G^{\alpha\beta}\cdot\delta_{0}+\overbrace{G^{ab}\delta_{ab}^{\alpha\beta}}^{\text{Gravity}}\xrightarrow{\forall\,\delta_{ab}^{\alpha\beta}\ll\delta_{0}}=\frac{g^{\alpha\beta}}{F}\cdot\delta_{0}, \tag{33}$$

we are left with the scalar and purely metrically obtained quantum equation, which, when excluding the trivial case n=2, reads:

$$0=\left(R-\left(\frac{1}{2F}\begin{pmatrix}2F_{,ij}(n-1)g^{ij}+2\Gamma_{ij}^{a}F_{,a}g^{ij}\\-F_{,i}g^{ab}g_{jb,a}g^{ij}-F_{,j}g^{ab}g_{ib,a}g^{ij}\\-n\Gamma_{ij}^{d}F_{,d}g^{ij}+\frac{n}{2}F_{,d}g^{cd}g_{ab,c}g^{ab}\end{pmatrix}+\frac{F_{,i}\cdot F_{,j}}{4F^{2}}g^{ij}\left((n-6)(n-1)\right)\right)\right)+\Lambda\cdot F\cdot n. \tag{34}$$

From there, which is to say based on the latter equation, all classically important equations can be obtained. In the case of a vanishing cosmological constant (34) simplifies to:

$$0 = R - \left(\frac{1}{2F} \begin{pmatrix} 2F_{,ij}(n-1)g^{ij} + 2\Gamma^{a}_{ij}F_{,a}g^{ij} \\ -F_{,i}g^{ab}g_{jb,a}g^{ij} - F_{,j}g^{ab}g_{ib,a}g^{ij} \\ -n\Gamma^{d}_{ij}F_{,d}g^{ij} + \frac{n}{2}F_{,d}g^{cd}g_{ab,c}g^{ab} \end{pmatrix} + \frac{F_{,i} \cdot F_{,j}}{4F^{2}} g^{ij}\left((n-6)(n-1)\right) \right)$$

$$= R - \begin{pmatrix} \frac{F'}{2F} \begin{pmatrix} 2f_{,ij}(n-1)g^{ij} + 2\Gamma^{a}_{ij}f_{,a}g^{ij} \\ -f_{,i}g^{ab}g_{jb,a}g^{ij} - f_{,j}g^{ab}g_{ib,a}g^{ij} \\ -n\Gamma^{d}_{ij}f_{,d}g^{ij} + \frac{n}{2}f_{,d}g^{cd}g_{ab,c}g^{ab} \end{pmatrix} \\ +(n-1)\frac{f_{,i} \cdot f_{,j}}{4F^{2}} g^{ij}\left(4FF'' + (F')^{2}(n-6)\right) \end{pmatrix}, \qquad (35)$$

where we expanded the scaling function F[f].

When demanding the following linearization condition:

$$0 = 4FF'' + (F')^{2}(n-6), \qquad (36)$$

we require the function F[f] to be as follows:

$$F[f] = \begin{cases} C_F \cdot (f + C_f)^{\frac{4}{n-2}} & n \neq 2 \\ C_F \cdot e^{f \cdot C_f} & n = 2 \end{cases}. \qquad (37)$$

Subsequently, equation (35) can be simplified and yields:

$$0 = R - \frac{F'}{2F} \begin{pmatrix} 2f_{,ij}(n-1)g^{ij} + 2\Gamma^{a}_{ij}f_{,a}g^{ij} \\ -f_{,i}g^{ab}g_{jb,a}g^{ij} - f_{,j}g^{ab}g_{ib,a}g^{ij} \\ -n\Gamma^{d}_{ij}f_{,d}g^{ij} + \frac{n}{2}f_{,d}g^{cd}g_{ab,c}g^{ab} \end{pmatrix}. \qquad (38)$$

Interestingly, for metrics without shear elements:

$$g_{ij} = \begin{pmatrix} g_{00} & \cdots & 0 \\ \vdots & \ddots & \vdots \\ 0 & \cdots & g_{n-1n-1} \end{pmatrix}; \quad g_{ii,i} = 0, \qquad (39)$$

this converges to the ordinary Laplace operator, namely:

$$R^* = 0 \quad \to \quad 0 = F\cdot R + F'\cdot(1-n)\cdot\Delta f$$

$$\Rightarrow \quad 0 = \begin{cases} \left(f - C_f\right)^{\frac{4}{n-2}} \cdot C_F \left(R + \dfrac{4}{n-2} \cdot \dfrac{(1-n)}{\left(f - C_f\right)} \cdot \Delta f \right) & n > 2 \\ e^{C_f \cdot f} \cdot C_F \left(R + C_f \cdot (1-n) \cdot \Delta f \right) & n = 2 \end{cases} \cdot \tag{40}$$

Thus, in the case of n>2 we always also have the option for a constant (broken symmetry) solution of the kind:

$$0 = f - C_{f0} \quad \Rightarrow \quad f = C_{f0}. \tag{41}$$

Otherwise, we have the simple equations:

$$0 = \begin{cases} \left(f - C_{f0}\right)\cdot R + (1-n)\cdot \dfrac{4}{n-2} \cdot \Delta f & n > 2 \\ R + C_{f0} \cdot (1-n) \cdot \Delta f & n = 2 \end{cases} . \tag{42}$$

A critical argument should now be that this equation is not of Klein-Gordon character as it does not contain any potential nor mass, but one of the authors has already shown that this problem is easily solved by adding additional dimensions carrying the right properties to produce masses and potentials (e.g., [4, 7 – 10]). This mass or potential creation requires an entanglement of dimensions. Here we are especially concentrating on the question of the metric appearance of physical potentials. But before we come to that, we want to consider an alternative, slightly more general way to obtain quantum Einstein field equations and scalarize them.

9.3 The Safest Thing One Can Say about the Universe Is That Nothing Is Truly Certain

9.3.1 Uncertainties One Cannot Get Rid Of

Let us assume that there is a fundamental uncertainty. In fact, we assume this uncertainty is so fundamental that it even influences already fundamental principles like Hamilton's minimum principle. Translating this into the picture of Hilbert's action [1], the variational task:

$$\begin{gathered} \delta_g W = 0 = \delta_g \int_V d^n x \left(\sqrt{-g} \cdot \Phi_R[R] \right) = \delta_g \int_V d^n x \left(\sqrt{-g} \cdot R \right) \\ \delta_G W = ?? = \delta_G \int_V d^n x \left(\sqrt{-G} \cdot R^* \right) \\ \delta_g W = ? = \delta_g \int_V d^n x \left(\sqrt{-g} \cdot \Phi_{R^*}[R^*] \right) \end{gathered} , \tag{43}$$

should read:

$$
\begin{aligned}
\delta W = 0 = \delta \int_V d^n x \left(\sqrt{-g} \cdot \left(\overbrace{(1-\Omega)}^{=\Phi} \cdot R - 2\Lambda + L_M \right) \right) \\
\Rightarrow 0 = \delta \int_V d^n x \left(\sqrt{-g} \cdot \left(\Phi \cdot R - 2\Lambda + L_M \right) \right)
\end{aligned}
. \tag{44}
$$

Thereby Ω presents the general perturbation of the usual action and thus, the symbol Φ stands for an arbitrary function of the coordinates. We know from [6] that this would usually lead to complicated expressions like this:

$$
\begin{aligned}
\delta W = 0 = & \left[\begin{array}{c} \int_V d^n x \left(\sqrt{-g} \cdot \left(\begin{array}{c} \Phi \cdot R_{\mu\nu} - \frac{1}{2} \Phi \cdot R \cdot g_{\mu\nu} + \Lambda g_{\mu\nu} - \left(\nabla_\mu \nabla_\nu - g_{\mu\nu} \Delta_g \right) \Phi \\ + \left\{ \begin{array}{cc} 0 & \ldots \text{"vacuum"} \\ 8\pi G T_{\mu\nu} & \ldots \text{postulated matter} \end{array} \right\} \Phi \end{array} \right) \right) \delta g^{\mu\nu} \\ - \int_{\text{Surface}} d^{n-1} y \left(\sqrt{|h|} \cdot \varepsilon \cdot \Phi \cdot N^\lambda h^{\mu\nu} \partial_\lambda \left(\delta g_{\mu\nu} \right) \right) \end{array} \right] \\
= & \left[\begin{array}{c} \int_V d^n x \left(\sqrt{-g} \cdot \left(\begin{array}{c} R_{\mu\nu} - \frac{1}{2} R \cdot g_{\mu\nu} + \frac{\Lambda}{\Phi} g_{\mu\nu} - \left(\nabla_\mu \nabla_\nu - g_{\mu\nu} \Delta_g \right) \\ + \left\{ \begin{array}{cc} 0 & \ldots \text{"vacuum"} \\ 8\pi G T_{\mu\nu} & \ldots \text{postulated matter} \end{array} \right\} \end{array} \right) \Phi \right) \delta g^{\mu\nu} \\ - \int_{\text{Surface}} d^{n-1} y \left(\sqrt{|h|} \cdot \varepsilon \cdot \Phi \cdot N^\lambda h^{\mu\nu} \partial_\lambda \left(\delta g_{\mu\nu} \right) \right) \end{array} \right]
\end{aligned}
. \tag{45}
$$

Thereby, the latter was extracted from [6], where we find that:

$$
\begin{aligned}
\delta_g W = 0 = \delta_g \int_V d^n x \left(\sqrt{-g} \cdot \Phi_R [R] \right) \\
\delta W = 0 = \left[\begin{array}{c} \int_V d^n x \left(\sqrt{-g} \cdot \left(\begin{array}{c} \Phi'_R [R] \cdot R_{\mu\nu} - \frac{1}{2} \Phi_R [R] \cdot g_{\mu\nu} \\ + \Lambda g_{\mu\nu} - \left(\nabla_\mu \nabla_\nu - g_{\mu\nu} \Delta_g \right) \Phi'_R [R] \end{array} \right) \right) \delta g^{\mu\nu} \\ - \int_{\text{Surface}} d^{n-1} y \left(\sqrt{|h|} \cdot \varepsilon \cdot \Phi'_R [R] \cdot N^\lambda h^{\mu\nu} \partial_\lambda \left(\delta g_{\mu\nu} \right) \right) \end{array} \right]
\end{aligned}
. \tag{46}
$$

Now we simply substitute $R \rightarrow R^*$, $g_{ab} \rightarrow G_{ab}$ and $\Phi_R[R] \rightarrow \Phi^* R$, which gives us:

$$\delta_G W = 0 = \delta_G \int_V d^n x \left(\sqrt{-G}\cdot\Phi\cdot R^*\right)$$

$$\delta W = 0 = \left[\begin{array}{c} \int_V d^n x \left(\sqrt{-G}\cdot \left(\begin{array}{c} \Phi\cdot R^*_{\mu\nu} - \frac{1}{2}\Phi\cdot R^*\cdot G_{\mu\nu} \\ +\Lambda G_{\mu\nu} - \left(\nabla_\mu\nabla_\nu - G_{\mu\nu}\Delta_G\right)\Phi \end{array} \right) \right) \delta G^{\mu\nu} \\ - \int_{Surface} d^{n-1}y \left(\sqrt{|h|}\cdot\varepsilon\cdot\Phi\cdot N^\lambda h^{\mu\nu}\partial_\lambda \left(\delta G_{\mu\nu}\right)\right) \end{array} \right]. \tag{47}$$

The usual, and rather tedious, path forward would be the discussion of the last equation and its investigation with respect to all its constituents. In this section, however, we will show how to avoid such cumbersome equations (especially the surface term). For the reason of brevity and simplicity, we demand:

$$\delta W = 0 = \delta \int_V d^n x \left(\sqrt{-g}\cdot \left(\Phi\cdot R - \overbrace{2\Lambda + L_M}^{=0} \right)\right) \tag{48}$$

and adjust the kernel as follows:

$$\begin{array}{c} \delta W = 0 = \delta \int_V d^n x \left(\sqrt{-g}\cdot\Phi\cdot R\right) \\ \xrightarrow{g_{\alpha\beta}=\gamma_{\alpha\beta}\cdot F;\quad F^{\frac{n}{2}}=\Phi} \\ 0 = \delta_{g_{\alpha\beta}} \int_V d^n x \left(\sqrt{-\gamma}\cdot F^{\frac{n}{2}} R^* \right) \Rightarrow 0 = \delta_{g_{\alpha\beta}} \int_V d^n x \left(\sqrt{-\gamma}\cdot F^n \cdot R^*\right) \\ \Rightarrow \quad 0 = R^*_{\alpha\beta} - \frac{1}{2}\cdot\gamma_{\alpha\beta}\cdot R^* \end{array}. \tag{49}$$

Thereby we remember that:

$$R^* = \frac{1}{F}\left(R - \frac{1}{2F}\left(\begin{array}{c} 2F_{,ij}(n-1)\gamma^{ij} + 2\Gamma^a_{ij}F_{,a}\gamma^{ij} \\ -F_{,i}\gamma^{ab}\gamma_{jb,a}\gamma^{ij} - F_{,j}\gamma^{ab}\gamma_{ib,a}\gamma^{ij} \\ -n\Gamma^d_{ij}F_{,d}\gamma^{ij} + \frac{n}{2}F_{,d}\gamma^{cd}\gamma_{ab,c}\gamma^{ab} \end{array} \right) - \frac{F_{,i}\cdot F_{,j}}{4F^2}\gamma^{ij}\left((n-6)(n-1)\right) \right) \tag{50}$$

and subsequently obtain after variation with respect to the scaled metric tensor:

$$0=\int_{V_G} d^n x\sqrt{-\gamma\cdot F^n}\left(\left(\left(\begin{array}{c} R_{\alpha\beta}-\frac{\gamma_{\alpha\beta}}{2}\left(\begin{array}{c} R-\frac{1}{2F}\left(\begin{array}{c} 2F_{,ij}(n-1)\gamma^{ij}+2\Gamma^{a}_{ij}F_{,a}\gamma^{ij} \\ -F_{,i}\gamma^{ab}\gamma_{jb,a}\gamma^{ij}-F_{,j}\gamma^{ab}\gamma_{ib,a}\gamma^{ij} \\ -n\Gamma^{d}_{ij}F_{,d}\gamma^{ij}+\frac{n}{2}F_{,d}\gamma^{cd}\gamma_{ab,c}\gamma^{ab} \end{array}\right) \\ -\frac{F_{,i}\cdot F_{,j}}{4F^2}\gamma^{ij}((n-6)(n-1)) \end{array}\right) \\ -\frac{1}{2F}\left(\begin{array}{c} F_{,\alpha\beta}(n-2)+F_{,ab}\gamma_{\alpha\beta}\gamma^{ab} \\ +F_{,a}\gamma^{ab}\left(\gamma_{\beta b,\alpha}-\gamma_{\beta\alpha,b}\right)-F_{,\alpha}\gamma^{ab}\gamma_{\beta b,a}-F_{,\beta}\gamma^{ab}\gamma_{\alpha b,a} \\ +F_{,d}\gamma^{cd}\frac{1}{2}n\left(\begin{array}{c} \frac{2}{n}\gamma_{\alpha c,\beta}-\gamma_{\alpha c,\beta}-\gamma_{\beta c,\alpha} \\ +\gamma_{\alpha\beta,c}+\frac{1}{n}\gamma_{\alpha\beta}\gamma_{ab,c}\gamma^{ab} \end{array}\right) \\ -\frac{1}{2F}\left(F_{,\alpha}\cdot F_{,\beta}(3n-6)+\gamma_{\alpha\beta}F_{,c}F_{,d}\gamma^{cd}(4-n)\right) \end{array}\right) \end{array}\right)\right)\delta g^{\alpha\beta}\right). \quad (51)$$

The resulting field equations read as follows:

$$0=\left(\begin{array}{c} R_{\alpha\beta}-\frac{\gamma_{\alpha\beta}}{2}R+\frac{\gamma_{\alpha\beta}}{2}\left(\begin{array}{c} \frac{1}{2F}\left(\begin{array}{c} 2F_{,ij}(n-1)\gamma^{ij}+2\Gamma^{a}_{ij}F_{,a}\gamma^{ij} \\ -F_{,i}\gamma^{ab}\gamma_{jb,a}\gamma^{ij}-F_{,j}\gamma^{ab}\gamma_{ib,a}\gamma^{ij} \\ -n\Gamma^{d}_{ij}F_{,d}\gamma^{ij}+\frac{n}{2}F_{,d}\gamma^{cd}\gamma_{ab,c}\gamma^{ab} \end{array}\right) \\ +\frac{F_{,i}\cdot F_{,j}}{4F^2}\gamma^{ij}((n-6)(n-1)) \end{array}\right) \\ -\frac{1}{2F}\left(\begin{array}{c} F_{,\alpha\beta}(n-2)+F_{,ab}\gamma_{\alpha\beta}\gamma^{ab} \\ +F_{,a}\gamma^{ab}\left(\gamma_{\beta b,\alpha}-\gamma_{\beta\alpha,b}\right)-F_{,\alpha}\gamma^{ab}\gamma_{\beta b,a}-F_{,\beta}\gamma^{ab}\gamma_{\alpha b,a} \\ +F_{,d}\gamma^{cd}\frac{1}{2}n\left(\frac{2}{n}\gamma_{\alpha c,\beta}-\gamma_{\alpha c,\beta}-\gamma_{\beta c,\alpha}+\gamma_{\alpha\beta,c}+\frac{1}{n}\gamma_{\alpha\beta}\gamma_{ab,c}\gamma^{ab}\right) \\ -\frac{1}{2F}\left(F_{,\alpha}\cdot F_{,\beta}(3n-6)+\gamma_{\alpha\beta}F_{,c}F_{,d}\gamma^{cd}(4-n)\right) \end{array}\right) \end{array}\right). \quad (52)$$

Presenting them in dependency on our original kernel perturbation Φ gives us:

$$0=\left(\begin{array}{c}R_{\alpha\beta}-\frac{\gamma_{\alpha\beta}}{2}R+\frac{\gamma_{\alpha\beta}}{2}\left(\frac{1}{2\Phi^{\frac{2}{n}}}\left(\begin{array}{c}2\Phi^{\frac{2}{n}}{}_{,ij}(n-1)\gamma^{ij}+2\Gamma^{a}_{ij}\Phi^{\frac{2}{n}}{}_{,a}\gamma^{ij}\\ -\Phi^{\frac{2}{n}}{}_{,i}\gamma^{ab}\gamma_{jb,a}\gamma^{ij}-\Phi^{\frac{2}{n}}{}_{,j}\gamma^{ab}\gamma_{ib,a}\gamma^{ij}\\ -n\Gamma^{d}_{ij}\Phi^{\frac{2}{n}}{}_{,d}\gamma^{ij}+\frac{n}{2}\Phi^{\frac{2}{n}}{}_{,d}\gamma^{cd}\gamma_{ab,c}\gamma^{ab}\end{array}\right)+\frac{\Phi^{\frac{2}{n}}{}_{,i}\cdot\Phi^{\frac{2}{n}}{}_{,j}}{4\Phi^{\frac{4}{n}}}\gamma^{ij}\left((n-6)(n-1)\right)\right)\\ -\frac{1}{2\Phi^{\frac{2}{n}}}\left(\begin{array}{c}\Phi^{\frac{2}{n}}{}_{,\alpha\beta}(n-2)+\Phi^{\frac{2}{n}}{}_{,ab}\gamma_{\alpha\beta}\gamma^{ab}\\ +\Phi^{\frac{2}{n}}{}_{,a}\gamma^{ab}\left(\gamma_{\beta b,\alpha}-\gamma_{\beta\alpha,b}\right)-\Phi^{\frac{2}{n}}{}_{,\alpha}\gamma^{ab}\gamma_{\beta b,a}-\Phi^{\frac{2}{n}}{}_{,\beta}\gamma^{ab}\gamma_{\alpha b,a}\\ +\Phi^{\frac{2}{n}}{}_{,d}\gamma^{cd}\frac{1}{2}n\left(\frac{2}{n}\gamma_{\alpha c,\beta}-\gamma_{\alpha c,\beta}-\gamma_{\beta c,\alpha}+\gamma_{\alpha\beta,c}+\frac{1}{n}\gamma_{\alpha\beta}\gamma_{ab,c}\gamma^{ab}\right)\\ -\frac{1}{2\Phi^{\frac{2}{n}}}\left(\Phi^{\frac{2}{n}}{}_{,\alpha}\cdot\Phi^{\frac{2}{n}}{}_{,\beta}(3n-6)+\gamma_{\alpha\beta}\Phi^{\frac{2}{n}}{}_{,c}\Phi^{\frac{2}{n}}{}_{,d}\gamma^{cd}(4-n)\right)\end{array}\right)\end{array}\right). \tag{53}$$

We realize that a perturbation of the Hilbert kernel in the Einstein-Hilbert action results in quantum or at least quantum-like terms for the field equations.

Now, substituting $1\text{-}\Omega\text{=}\Phi$ as done in (44) and setting:

$$\Phi=F^{\frac{n}{2}} \tag{54}$$

gives us the following illustrative development:

$$\delta W=0=\delta\int_V d^nx\left(\sqrt{-g}\cdot(1-\Omega)\cdot R\right)=\delta\int_V d^nx\left(\sqrt{-g}\cdot R\right)-\delta\int_V d^nx\left(\sqrt{-g}\cdot\Omega\cdot R\right)$$

$$\xrightarrow{g_{\alpha\beta}=\gamma_{\alpha\beta}\cdot F;\ \ F^{\frac{n}{2}}=\Omega}$$

$$\begin{aligned}&0=\delta\int_V d^nx\left(\sqrt{-g}\cdot R\right)-\delta_{g_{\alpha\beta}}\int_V d^nx\left(\sqrt{-\gamma}\cdot F^{\frac{n}{2}}R^{*}\right)\\ &\Rightarrow 0=\delta\int_V d^nx\left(\sqrt{-g}\cdot R\right)-\delta_{g_{\alpha\beta}}\int_V d^nx\left(\sqrt{-\gamma\cdot F^{n}}\cdot R^{*}\right)\\ &\Rightarrow\quad 0=R_{\alpha\beta}-\frac{1}{2}\cdot g_{\alpha\beta}\cdot R-\left(R^{*}_{\alpha\beta}-\frac{1}{2}\cdot\gamma_{\alpha\beta}\cdot R^{*}\right)\end{aligned}\quad. \tag{55}$$

Incorporating our results from above (e.g., (51)) leads to:

$$0=\int_{V_G} d^n x\sqrt{-\gamma\cdot F^n}\left(\begin{array}{c}\left(R_{\alpha\beta}-\frac{F}{2}\cdot\gamma_{\alpha\beta}\cdot R\right)\delta g^{\alpha\beta}\\ -\left(\begin{array}{c}R_{\alpha\beta}-\frac{\gamma_{\alpha\beta}}{2}\left(\begin{array}{c}R-\frac{1}{2F}\left(\begin{array}{c}2F_{,ij}(n-1)\gamma^{ij}+2\Gamma^a_{ij}F_{,a}\gamma^{ij}\\ -F_{,i}\gamma^{ab}\gamma_{jb,a}\gamma^{ij}-F_{,j}\gamma^{ab}\gamma_{ib,a}\gamma^{ij}\\ -n\Gamma^d_{ij}F_{,d}\gamma^{ij}+\frac{n}{2}F_{,d}\gamma^{cd}\gamma_{ab,c}\gamma^{ab}\end{array}\right)\\ -\frac{F_{,i}\cdot F_{,j}}{4F^2}\gamma^{ij}\left((n-6)(n-1)\right)\end{array}\right)\\ -\frac{1}{2F}\left(\begin{array}{c}F_{,\alpha\beta}(n-2)+F_{,ab}\gamma_{\alpha\beta}\gamma^{ab}\\ +F_{,a}\gamma^{ab}\left(\gamma_{\beta b,\alpha}-\gamma_{\beta\alpha,b}\right)-F_{,\alpha}\gamma^{ab}\gamma_{\beta b,a}-F_{,\beta}\gamma^{ab}\gamma_{\alpha b,a}\\ +F_{,d}\gamma^{cd}\frac{1}{2}n\left(\begin{array}{c}\frac{2}{n}\gamma_{\alpha c,\beta}-\gamma_{\alpha c,\beta}-\gamma_{\beta c,\alpha}\\ +\gamma_{\alpha\beta,c}+\frac{1}{n}\gamma_{\alpha\beta}\gamma_{ab,c}\gamma^{ab}\end{array}\right)\\ -\frac{1}{2F}\left(F_{,\alpha}\cdot F_{,\beta}(3n-6)+\gamma_{\alpha\beta}F_{,c}F_{,d}\gamma^{cd}(4-n)\right)\end{array}\right)\end{array}\right)\delta g^{\alpha\beta}\end{array}\right) \quad (56)$$

and makes the Ricci tensor disappear:

$$0=\int_{V_G} d^n x\sqrt{-\gamma\cdot F^n}\left(\left(\begin{array}{c}\frac{\gamma_{\alpha\beta}}{2}\left(\begin{array}{c}R-\frac{1}{2F}\left(\begin{array}{c}2F_{,ij}(n-1)\gamma^{ij}+2\Gamma^a_{ij}F_{,a}\gamma^{ij}\\ -F_{,i}\gamma^{ab}\gamma_{jb,a}\gamma^{ij}-F_{,j}\gamma^{ab}\gamma_{ib,a}\gamma^{ij}\\ -n\Gamma^d_{ij}F_{,d}\gamma^{ij}+\frac{n}{2}F_{,d}\gamma^{cd}\gamma_{ab,c}\gamma^{ab}\end{array}\right)\\ -\frac{F_{,i}\cdot F_{,j}}{4F^2}\gamma^{ij}\left((n-6)(n-1)\right)\end{array}\right)-\frac{F}{2}\cdot\gamma_{\alpha\beta}\cdot R\\ +\frac{1}{2F}\left(\begin{array}{c}F_{,\alpha\beta}(n-2)+F_{,ab}\gamma_{\alpha\beta}\gamma^{ab}\\ +F_{,a}\gamma^{ab}\left(\gamma_{\beta b,\alpha}-\gamma_{\beta\alpha,b}\right)-F_{,\alpha}\gamma^{ab}\gamma_{\beta b,a}-F_{,\beta}\gamma^{ab}\gamma_{\alpha b,a}\\ +F_{,d}\gamma^{cd}\frac{1}{2}n\left(\begin{array}{c}\frac{2}{n}\gamma_{\alpha c,\beta}-\gamma_{\alpha c,\beta}-\gamma_{\beta c,\alpha}\\ +\gamma_{\alpha\beta,c}+\frac{1}{n}\gamma_{\alpha\beta}\gamma_{ab,c}\gamma^{ab}\end{array}\right)\\ -\frac{1}{2F}\left(F_{,\alpha}\cdot F_{,\beta}(3n-6)+\gamma_{\alpha\beta}F_{,c}F_{,d}\gamma^{cd}(4-n)\right)\end{array}\right)\end{array}\right)\delta g^{\alpha\beta}\right). \quad (57)$$

In the case of variation not with respect to the metric tensor, but only to the scaling function f (c.d. section "Dissolving an Apparent Conflict" in [7]) we adjust the kernel as follows:

$$\begin{array}{c}\delta W=0=\delta\int_V d^n x\left(\sqrt{-g}\cdot\Phi_0\cdot R\right)\\ \xrightarrow{g_{\alpha\beta}=\gamma_{\alpha\beta}\cdot F=\gamma_{\alpha\beta}\cdot F[f];\; F^{\frac{n}{2}+1}=\Phi_0}\\ 0=\delta_f\int_V d^n x\left(\sqrt{-\gamma}\cdot F^{\frac{n}{2}+1}R^*\right)\Rightarrow 0=\delta_f\int_V d^n x\left(\sqrt{-\gamma}\cdot F^n\cdot F\cdot R^*\right)\end{array}. \tag{58}$$

Now the resulting variational task reads:

$$\frac{\delta W}{\delta f}=0=\delta_f\int_{V_G} d^n x\sqrt{-g\cdot F^n}\left(\begin{array}{c}R-\frac{1}{2F}\left(\begin{array}{c}2F_{,\alpha\beta}(n-1)g^{\alpha\beta}+2\Gamma^a_{\alpha\beta}F_{,a}g^{\alpha\beta}\\ -F_{,\alpha}g^{ab}g_{\beta b,a}g^{\alpha\beta}-F_{,\beta}g^{ab}g_{\alpha b,a}g^{\alpha\beta}\\ -n\Gamma^d_{\alpha\beta}F_{,d}g^{\alpha\beta}+\frac{n}{2}F_{,d}g^{cd}g_{\alpha\beta,c}g^{\alpha\beta}\end{array}\right)\\ -\frac{F_{,\alpha}\cdot F_{,\beta}}{4F^2}g^{\alpha\beta}\left((n-6)(n-1)\right)\end{array}\right) \tag{59}$$

and we obtain the scalar equation:

$$0=\left[\begin{array}{c}\frac{\sqrt{-g}}{2}\left(\frac{f_{,\alpha}f_{,\beta}}{2}g^{\alpha\beta}\left(\frac{1}{2}(n-4)(n+2)(F')^2+2FF''\right)+(n+2)FF'\Delta f\right)\\ +F^2\left(\begin{array}{c}\left(\sqrt{-g}\cdot g^{\alpha\beta}\right)_{,\alpha\beta}-\frac{\sqrt{-g}\cdot R}{(n-1)}\\ -\frac{1}{2(n-1)}\left(\sqrt{-g}\cdot g^{\alpha\beta}\cdot g^{ij}\left(\begin{array}{c}g_{i\beta,j}-g_{ij,\beta}-g_{\beta j,i}\\ +\frac{n}{2}\left(2g_{ij,\beta}-g_{i\beta,j}-g_{j\beta,i}\right)\end{array}\right)\right)_{,\alpha}\end{array}\right)\end{array}\right]. \tag{60}$$

It should be pointed out that for a functional setting of F[f]=f, the nonlinear terms in our scalar field equation (60) disappear in 4-dimensional space-times, where we would get:

$$0=\left[3\cdot\sqrt{-g}\cdot f\cdot\Delta f+f^2\left(\begin{array}{c}\left(\sqrt{-g}\cdot g^{\alpha\beta}\right)_{,\alpha\beta}-\frac{\sqrt{-g}\cdot R}{(n-1)}\\ -\frac{1}{2(n-1)}\left(\sqrt{-g}\cdot g^{\alpha\beta}\cdot g^{ij}\left(\begin{array}{c}g_{i\beta,j}-g_{ij,\beta}-g_{\beta j,i}\\ +2\left(2g_{ij,\beta}-g_{i\beta,j}-g_{j\beta,i}\right)\end{array}\right)\right)_{,\alpha}\end{array}\right)\right]. \tag{61}$$

With $f \neq 0$ this can be made an eigen equation as follows:

$$\Delta f = \frac{f}{3 \cdot \sqrt{-g}} \left(\begin{array}{c} \frac{1}{2(n-1)} \left(\sqrt{-g} \cdot g^{\alpha\beta} \cdot g^{ij} \left(\begin{array}{c} g_{i\beta,j} - g_{ij,\beta} - g_{\beta j,i} \\ +2\left(2g_{ij,\beta} - g_{i\beta,j} - g_{j\beta,i}\right) \end{array} \right) \right)_{,\alpha} \\ + \frac{\sqrt{-g} \cdot R}{(n-1)} - \left(\sqrt{-g} \cdot g^{\alpha\beta} \right)_{,\alpha\beta} \end{array} \right). \tag{62}$$

When interpreting the scaling function f as something being equivalent or at least connected with the quantum theoretical wave function of an arbitrary system, we see that a perturbation of the curvature kernel of the Hilbert-extremal integral (43)→(44) directly leads to additional degrees of freedom and quantum equations. Thus, an observer would realize the perturbations, which are just volume jitter of the space-time, as quantum effects, whereas when regarding the system in question this may just be taken as its very own and "individual" uncertainties.

We state that variational perturbation, respectively, some kernel uncertainty inside the Einstein-Hilbert action automatically leads to a Quantum Theory.

It was just shown that we can always factor out a suitable function F[f] from the metric tensor to simplify the variational task for either the whole metric or just with respect to the volume factor (see also [8, 19] for more details). Thereby we did not care about whether the kernels for both cases are identical, but simply adjusted the factor in order to obtain maximum simplicity to each task. However, it appears reasonable to assume that the kernel should be the same in both variational cases. Thus, aiming for maximum simplicity in the case of the variation with respect to the whole metric in applying (49), we require the derivation of the more complex case with respect to f and vice versa. The corresponding evaluation is presented in appendix L of our publication [8].

9.3.2 A Quantum World from the Scaled Hilbert Kernel Φ*R

Having derived that a scaled Einstein-Hilbert action kernel of the type (44) can be brought into such a form that in the end we obtain quantum equations, it should also allow us to extract such or similar quantum equations via the classical derivational path as shown in [6], for instance. Assuming that we already have a scaled metric of type:

$$G_{\alpha\beta} = g_{\alpha\beta} \cdot F[f], \tag{63}$$

plus a scaled Einstein-Hilbert action with a scaled kernel of type (44), we should be able to define the variation in such a way that we are allowed to demand the variation to give zero on both the surface of our volume integral and the sub-surface of the integral term from (45):

$$\int_{\text{Surface}} d^{n-1}y\left(\sqrt{|h|} \cdot \varepsilon \cdot \Phi \cdot N^{\lambda} h^{\mu\nu} \partial_{\lambda}\left(\delta G_{\mu\nu}\right)\right) = 0. \tag{64}$$

This is further motivated by the fact that in the subsection above we have been able to show that the variational integral of the Einstein-Hilbert action (e.g., in (55)) can be brought in such a form that we end up with surface terms. Those can be set to zero (per definition with respect to the variation) and because our extended form of the variational task with scaled kernel plus scaled metric allows for more degrees of freedom regarding the definition of our variation, we take it for granted to be allowed to in fact omit all surface terms. We leave the rigorous proof to the skilled mathematician. In essence, however, this all then makes (45) to:

$$\delta W = 0 = \int_V d^n x \left(\sqrt{-G} \cdot \left(\begin{array}{c} \Phi \cdot R^*_{\mu\nu} - \frac{1}{2} \Phi \cdot R^* G_{\mu\nu} + \Lambda G_{\mu\nu} - \left(\nabla_\mu \nabla_\nu - G_{\mu\nu} \Delta_G \right) \Phi \\ + \left\{ \begin{array}{ll} 0 & \ldots\text{"vacuum"} \\ \kappa T_{\mu\nu} & \ldots\text{postulated matter} \end{array} \right\} \Phi \end{array} \right) \right) \delta G^{\mu\nu} . \quad (65)$$

Setting the cosmological constant and the classical matter tensor equal to zero, simplifies the kernel as follows:

$$\Phi \cdot R^*_{\mu\nu} - \frac{1}{2} \Phi \cdot R^* G_{\mu\nu} - \left(\nabla_\mu \nabla_\nu - G_{\mu\nu} \Delta_G \right) \Phi = 0 . \quad (66)$$

Assuming the space-time to be Ricci flat results in:

$$\left(\nabla_\mu \nabla_\nu - G_{\mu\nu} \Delta_G \right) \Phi = 0 . \quad (67)$$

Applying our estimation from [4], chapter 10 (esp. "10.10 The Gravity Dirac Equation"), we can decompose the last equation as follows:

At first, we assume eigenvalue solutions to the Laplace operator:

$$\Delta_G \Phi = m^2 \cdot \Phi . \quad (68)$$

This we set into (67) and get:

$$\left(\nabla_\mu \nabla_\nu - G_{\mu\nu} m^2 \right) \Phi = \left(\nabla_\mu \nabla_\nu - \mathbf{E}_\mu \mathbf{E}_\nu m^2 \right) \Phi = 0 . \quad (69)$$

Then, our Dirac-like approach shall be:

$$\left(a \cdot \nabla_\mu + A \cdot \mathbf{E}_\mu \cdot m \right) \left(b \cdot \nabla_\nu - B \cdot \mathbf{E}_\nu \cdot m \right) \Phi = 0 . \quad (70)$$

With the objects a, A, b, and B being general 2x2 matrices we find a variety of solutions of which we here only present a very simple one, namely:

$$A = \begin{pmatrix} 1 & 1 \\ 1 & 1 \end{pmatrix}; \quad a = \begin{pmatrix} -1 & 1 \\ -1 & 1 \end{pmatrix}; \quad B = \begin{pmatrix} 1 & -1 \\ 1 & -1 \end{pmatrix}; \quad b = \begin{pmatrix} -1 & 1 \\ 1 & -1 \end{pmatrix} . \quad (71)$$

This makes (70) to result in:

$$\begin{aligned} & \left(a \cdot \nabla_\mu + A \cdot \mathbf{E}_\mu \cdot m \right) \left(b \cdot \nabla_\nu - B \cdot \mathbf{E}_\nu \cdot m \right) \Phi \\ & = 2 \cdot \begin{pmatrix} 1 & -1 \\ 1 & -1 \end{pmatrix} \cdot \left(\nabla_\mu \nabla_\nu - \mathbf{E}_\mu \mathbf{E}_\nu m^2 \right) \Phi \\ & = 2 \cdot \begin{pmatrix} 1 & -1 \\ 1 & -1 \end{pmatrix} \cdot \left(\nabla_\mu \nabla_\nu - G_{\mu\nu} m^2 \right) \Phi \end{aligned} \quad (72)$$

and gives us the following gravity “Dirac equations”:

$$\begin{aligned}&\left(\mathrm{a}\cdot\nabla_\mu+\mathrm{A}\cdot\mathbf{E}_\mu\cdot\mathrm{m}\right)\Phi=\left(\begin{pmatrix}-1&1\\-1&1\end{pmatrix}\cdot\nabla_\mu+\begin{pmatrix}1&1\\1&1\end{pmatrix}\cdot\mathbf{E}_\mu\cdot\mathrm{m}\right)\Phi=0\\&\left(\mathrm{b}\cdot\nabla_\nu-\mathrm{B}\cdot\mathbf{E}_\nu\cdot\mathrm{m}\right)\Phi=\left(\begin{pmatrix}-1&1\\1&-1\end{pmatrix}\cdot\nabla_\nu-\begin{pmatrix}1&-1\\1&-1\end{pmatrix}\cdot\mathbf{E}_\nu\cdot\mathrm{m}\right)\Phi=0\end{aligned}. \tag{73}$$

It should be pointed out that due to the fact that – in principle – the indices μ and ν can be exchanged by any symbol, the introduction of the matrices is not really of need. One could also write the factorization of (69) as follows:

$$\begin{aligned}&\left(\nabla_\mu\nabla_\nu-G_{\mu\nu}m^2\right)\Phi=\left(\nabla_\mu\nabla_\nu-\mathbf{E}_\mu\mathbf{E}_\nu m^2\right)\Phi=0\\&\Rightarrow\\&\left(\nabla_\mu\nabla_\nu-\mathbf{E}_\mu\mathbf{E}_\nu m^2\right)\Phi=\left(\nabla_\mu\nabla_\nu-\mathbf{E}_\mu\mathbf{E}_\nu m^2-\nabla_\mu\mathbf{E}_\nu\cdot\mathbf{m}+\mathbf{E}_\mu\cdot\mathbf{m}\nabla_\nu\right)\Phi\\&=\left(\nabla_\mu\nabla_\nu-\mathbf{E}_\mu\mathbf{E}_\nu m^2-\nabla_\mu\mathbf{E}_\nu\cdot\mathbf{m}+\nabla_\nu\mathbf{E}_\mu\cdot\mathbf{m}\right)\Phi\\&=\left(\nabla_\mu\nabla_\nu-\mathbf{E}_\mu\mathbf{E}_\nu m^2+\frac{\nabla_\nu\mathbf{E}_\mu\cdot\mathbf{m}-\nabla_\mu\mathbf{E}_\nu\cdot\mathbf{m}}{2}+\frac{\nabla_\nu\mathbf{E}_\mu\cdot\mathbf{m}-\nabla_\mu\mathbf{E}_\nu\cdot\mathbf{m}}{2}\right)\Phi\\&\xrightarrow{\text{exchange }\mu,\nu\text{ in }\frac{\nabla_\nu\mathbf{E}_\mu\cdot\mathbf{m}-\nabla_\mu\mathbf{E}_\nu\cdot\mathbf{m}}{2}}\\&=\left(\nabla_\mu\nabla_\nu-\mathbf{E}_\mu\mathbf{E}_\nu m^2+\frac{1}{2}\overbrace{\left(\nabla_\nu\mathbf{E}_\mu\cdot\mathbf{m}-\nabla_\mu\mathbf{E}_\nu\cdot\mathbf{m}+\nabla_\mu\mathbf{E}_\nu\cdot\mathbf{m}-\nabla_\nu\mathbf{E}_\mu\cdot\mathbf{m}\right)}^{=0}\right)\Phi\\&=\left(\nabla_\mu+\mathbf{E}_\mu\cdot\mathbf{m}\right)\left(\nabla_\nu-\mathbf{E}_\nu\cdot\mathbf{m}\right)\Phi\end{aligned}. \tag{74}$$

Thereby we used the fact that the covariant derivative of the base vector gives zero and that we also assume the parameter vector m to be a list of constants. In all other cases we have to apply the matrices and end up with (73).

The Dirac-like equations resulting from (74) would be:

$$\begin{aligned}&\left(\nabla_\mu+\mathbf{E}_\mu\cdot\mathbf{m}\right)\Phi=0\\&\left(\nabla_\nu-\mathbf{E}_\nu\cdot\mathbf{m}\right)\Phi=0\end{aligned}. \tag{75}$$

Contraction with the corresponding base vector leads to the scalar equations:

$$\begin{aligned}&\mathbf{E}^\mu\left(\nabla_\mu+\mathbf{E}_\mu\cdot\mathbf{m}\right)\Phi=\left(\mathbf{E}^\mu\nabla_\mu+\mathrm{n}\cdot\mathbf{m}\right)\Phi=0\\&\mathbf{E}^\nu\left(\nabla_\nu-\mathbf{E}_\nu\cdot\mathbf{m}\right)\Phi=\left(\mathbf{E}^\nu\nabla_\nu-\mathrm{n}\cdot\mathbf{m}\right)\Phi=0\end{aligned}. \tag{76}$$

If we want to make these equations complete scalars, we realize that Φ also has to be some kind of vector. This vector would then just be a Dirac-like spinor as known from the classical Dirac theory

[27]. With respect to a more comprehensive consideration of the metric derivation of the Dirac equation and its metric derivatives, the reader is referred to [46, 56].

9.3.3 Scalarizing the New Quantum Gravity Field Equation

It should be pointed out that one way[11] of scalarizing equation (66) can be achieved via the following expansion of the variational term:

$$\delta G^{\mu\nu} = \delta\left(\frac{g^{\mu\nu}}{F}\right) = \frac{1}{F}\cdot\delta g^{\mu\nu} + g^{\mu\nu}\cdot\delta\left(\frac{1}{F}\right) \tag{77}$$

under the integral (65). This makes (66) – if taken out from (65) more completely – in principle to:

$$\begin{aligned}
0 &= \left(\Phi\cdot R^{*}_{\mu\nu} - \frac{1}{2}\Phi\cdot R^{*}G_{\mu\nu} - \left(\nabla_{\mu}\nabla_{\nu} - G_{\mu\nu}\Delta_{G}\right)\Phi\right)\left(\frac{1}{F}\cdot\delta g^{\mu\nu} + g^{\mu\nu}\cdot\delta\left(\frac{1}{F}\right)\right) \\
&= \left(\Phi\cdot R^{*}_{\mu\nu} - \frac{1}{2}\Phi\cdot R^{*}G_{\mu\nu} - \left(\nabla_{\mu}\nabla_{\nu} - G_{\mu\nu}\Delta_{G}\right)\Phi\right)\frac{1}{F}\cdot\delta g^{\mu\nu} \\
&\quad + \left(\Phi\cdot R^{*}_{\mu\nu} - \frac{1}{2}\Phi\cdot R^{*}G_{\mu\nu} - \left(\nabla_{\mu}\nabla_{\nu} - G_{\mu\nu}\Delta_{G}\right)\Phi\right)g^{\mu\nu}\cdot\delta\left(\frac{1}{F}\right) \\
&= \frac{1}{F}\cdot\left(\Phi\cdot R^{*}_{\mu\nu} - \frac{1}{2}\Phi\cdot R^{*}G_{\mu\nu} - \left(\nabla_{\mu}\nabla_{\nu} - G_{\mu\nu}\Delta_{G}\right)\Phi\right)\delta g^{\mu\nu} \\
&\quad + \left(\Phi\cdot R^{*}_{\mu\nu} - \frac{1}{2}\Phi\cdot R^{*}G_{\mu\nu} - \left(\nabla_{\mu}\nabla_{\nu} - G_{\mu\nu}\Delta_{G}\right)\Phi\right)g^{\mu\nu}\cdot\delta\left(\frac{1}{F}\right) \\
&\xrightarrow{G_{\mu\nu}=F\cdot g_{\mu\nu}\Leftrightarrow g^{\mu\nu}=F\cdot G^{\mu\nu}} \\
&= \frac{1}{F}\cdot\left(\Phi\cdot R^{*}_{\mu\nu} - \frac{1}{2}\Phi\cdot R^{*}G_{\mu\nu} - \left(\nabla_{\mu}\nabla_{\nu} - G_{\mu\nu}\Delta_{G}\right)\Phi\right)\delta g^{\mu\nu} \\
&\quad + \left(\Phi\cdot F\cdot R^{*}_{\mu\nu}G^{\mu\nu} - \frac{1}{2}\Phi\cdot F\cdot R^{*}g_{\mu\nu}g^{\mu\nu} - F\cdot G^{\mu\nu}\left(\nabla_{\mu}\nabla_{\nu} - G_{\mu\nu}\Delta_{G}\right)\Phi\right)\cdot\delta\left(\frac{1}{F}\right) \\
&= \frac{1}{F}\cdot\left(\Phi\cdot R^{*}_{\mu\nu} - \frac{1}{2}\Phi\cdot R^{*}G_{\mu\nu} - \left(\nabla_{\mu}\nabla_{\nu} - G_{\mu\nu}\Delta_{G}\right)\Phi\right)\delta g^{\mu\nu} \\
&\quad + \left(\Phi\cdot F\cdot R^{*}\left(1-\frac{n}{2}\right) - (1-n)F\cdot\Delta_{G}\Phi\right)\cdot\delta\left(\frac{1}{F}\right)
\end{aligned} \quad . \tag{78}$$

[11] There are more such possibilities, which we have partially presented in here already. For a compact presentation the reader is referred to [19], section “A Variety of Quantum Gravity Field Equations”.

This can be split up into two scalar equations instead of one tensor one (which would be (66) again, of course), namely:

$$
\begin{gathered}
0=\frac{1}{F}\cdot\left(\Phi\cdot R^{*}_{\mu\nu}-\frac{1}{2}\Phi\cdot R^{*}G_{\mu\nu}-\left(\nabla_{\mu}\nabla_{\nu}-G_{\mu\nu}\Delta_{G}\right)\Phi\right)\delta g^{\mu\nu}\\
0=\left(\Phi\cdot F\cdot R^{*}\left(1-\frac{n}{2}\right)-(1-n)F\cdot\Delta_{G}\Phi\right)\cdot\delta\left(\frac{1}{F}\right)\\
\Rightarrow\\
0=\left(\Phi\cdot R^{*}_{\mu\nu}-\frac{1}{2}\Phi\cdot R^{*}G_{\mu\nu}-\left(\nabla_{\mu}\nabla_{\nu}-G_{\mu\nu}\Delta_{G}\right)\Phi\right)\delta g^{\mu\nu}\\
0=\Phi\cdot F\cdot R^{*}\left(1-\frac{n}{2}\right)-(1-n)F\cdot\Delta_{G}\Phi
\end{gathered}
\quad . \tag{79}
$$

When now assuming the variation to be of the following character:

$$
\delta G^{\alpha\beta}=\frac{g^{\alpha\beta}\cdot\delta_{0}+\overbrace{g^{ab}\delta^{\alpha\beta}_{ab}}^{\text{Gravity}}}{F}\xrightarrow{\forall\,\delta^{\alpha\beta}_{ab}\ll\delta_{0}}=g^{\alpha\beta}\cdot\frac{\delta_{0}}{F}, \tag{80}
$$

we can omit the first equation (third line) in (79) and then, only the last equation dominates the scene. This latter equation can be rewritten with respect to the unscaled metric as follows:

$$
\begin{gathered}
0=\Phi\cdot F\cdot R^{*}\left(1-\frac{n}{2}\right)-(1-n)F\cdot\Delta_{G}\Phi\\
=\Phi\cdot F\cdot R^{*}\left(1-\frac{n}{2}\right)-\frac{(1-n)}{F}F\cdot\left(\Delta_{g}\Phi-\frac{F_{,\mu}}{F}g^{\mu\nu}\Phi_{,\nu}\right).\\
\Rightarrow\\
0=\Phi\cdot F^{2}\cdot R^{*}\frac{(2-n)}{2\cdot(1-n)}-F\cdot\Delta_{g}\Phi+F_{,\mu}g^{\mu\nu}\Phi_{,\nu}
\end{gathered}
\tag{81}
$$

Apparently, this is a polynomial of second order in F, but when inserting (50) (with the correct metric, which would be g_{ab}) we obtain:

$$0=\Phi\cdot F^{2}\cdot R^{*}\frac{(2-n)}{2\cdot(1-n)}-F\cdot\Delta_{g}\Phi+F_{,\mu}g^{\mu\nu}\Phi_{,\nu}$$

$$0=\Phi\cdot F\cdot\left(\begin{array}{c}R-\frac{1}{2F}\left(\begin{array}{c}2F_{,ij}(n-1)g^{ij}+2\Gamma_{ij}^{a}F_{,a}g^{ij}\\ -F_{,i}g^{ab}g_{jb,a}g^{ij}-F_{,j}g^{ab}g_{ib,a}g^{ij}\\ -n\Gamma_{ij}^{d}F_{,d}g^{ij}+\frac{n}{2}F_{,d}g^{cd}g_{ab,c}g^{ab}\end{array}\right)\\ -\frac{F_{,i}\cdot F_{,j}}{4F^{2}}g^{ij}\left((n-6)(n-1)\right)\end{array}\right)\frac{(2-n)}{2\cdot(1-n)}-F\cdot\Delta_{g}\Phi+F_{,\mu}g^{\mu\nu}\Phi_{,\nu}$$

$$=\left(\begin{array}{c}\Phi\cdot F\cdot R-\frac{\Phi}{2}\left(\begin{array}{c}2F_{,ij}(n-1)g^{ij}+2\Gamma_{ij}^{a}F_{,a}g^{ij}\\ -F_{,i}g^{ab}g_{jb,a}g^{ij}-F_{,j}g^{ab}g_{ib,a}g^{ij}\\ -n\Gamma_{ij}^{d}F_{,d}g^{ij}+\frac{n}{2}F_{,d}g^{cd}g_{ab,c}g^{ab}\end{array}\right)\\ -\Phi\cdot\frac{F_{,i}\cdot F_{,j}}{4F}g^{ij}\left((n-6)(n-1)\right)\end{array}\right)\frac{(2-n)}{2\cdot(1-n)}-F\cdot\Delta_{g}\Phi+F_{,\mu}g^{\mu\nu}\Phi_{,\nu}\quad. \tag{82}$$

This leads us to:

$$0=\left[\begin{array}{c}\left(\begin{array}{c}\Phi\cdot F^{2}\cdot R-F\cdot\frac{\Phi}{2}\left(\begin{array}{c}2F_{,ij}(n-1)g^{ij}+2\Gamma_{ij}^{a}F_{,a}g^{ij}\\ -F_{,i}g^{ab}g_{jb,a}g^{ij}-F_{,j}g^{ab}g_{ib,a}g^{ij}\\ -n\Gamma_{ij}^{d}F_{,d}g^{ij}+\frac{n}{2}F_{,d}g^{cd}g_{ab,c}g^{ab}\end{array}\right)\\ -\Phi\cdot\frac{F_{,i}\cdot F_{,j}}{4}g^{ij}\left((n-6)(n-1)\right)\end{array}\right)\frac{(2-n)}{2\cdot(1-n)}\\ -F^{2}\cdot\Delta_{g}\Phi+F\cdot F_{,\mu}g^{\mu\nu}\Phi_{,\nu}\end{array}\right]. \tag{83}$$

Thus, we still have a quadratic equation in F. However, when remembering that, within the variational integral, we could always choose F and Φ in such a way that – with a new metric γ_{ab} – the two functions would become equal, we can write:

$$\begin{gathered}\delta W=0=\delta\int_{V}d^{n}x\left(\sqrt{-G}\cdot\Phi\cdot R^{*}\right)\\ \xrightarrow{g_{\alpha\beta}=\gamma_{\alpha\beta}\cdot F;\quad F=\Phi}\\ 0=\delta_{g_{\alpha\beta}}\int_{V}d^{n}x\left(\sqrt{-\gamma}\cdot F\cdot R^{*}\right)\end{gathered}\quad. \tag{84}$$

Thereby we took into account that the metric g_{ab} would already be scaled via an arbitrary function F* and that some shifting is necessary in order to truly make the two functions equal and adjust the metric accordingly. In the end, with respect to our equation (82), we result in:

$$0=\left(\left(\Phi\cdot F^2\cdot R-\frac{\Phi\cdot F\cdot}{2}\left(\begin{array}{l}2F_{,ij}(n-1)g^{ij}+2\Gamma_{ij}^{a}F_{,a}g^{ij}\\-F_{,i}g^{ab}g_{jb,a}g^{ij}-F_{,j}g^{ab}g_{ib,a}g^{ij}\\-n\Gamma_{ij}^{d}F_{,d}g^{ij}+\frac{n}{2}F_{,d}g^{cd}g_{ab,c}g^{ab}\end{array}\right)\right)\frac{(2-n)}{2\cdot(1-n)}\right.$$
$$\left.-\Phi\cdot\frac{F_{,i}\cdot F_{,j}}{4}g^{ij}\left((n-6)(n-1)\right)\right)$$
$$-F^2\cdot\Delta_g\Phi+F\cdot F_{,\mu}g^{\mu\nu}\Phi_{,\nu}\Bigg)$$
$$\xrightarrow{\Phi\to F\ \ \&\ \ g^{ij}\to\gamma^{ij}}\qquad\qquad . \qquad (85)$$
$$=\left(F^3\cdot R-\frac{F^2}{2}\left(\begin{array}{l}2F_{,ij}(n-1)g^{ij}+2\Gamma_{ij}^{a}F_{,a}\gamma^{ij}\\-F_{,i}\gamma^{ab}\gamma_{jb,a}\gamma^{ij}-F_{,j}\gamma^{ab}\gamma_{ib,a}\gamma^{ij}\\-n\Gamma_{ij}^{d}F_{,d}\gamma^{ij}+\frac{n}{2}F_{,d}\gamma^{cd}\gamma_{ab,c}\gamma^{ab}\end{array}\right)-F\cdot\frac{F_{,i}\cdot F_{,j}}{4}\gamma^{ij}\left((n-6)(n-1)\right)\right)\frac{(2-n)}{2\cdot(1-n)}-F^2\cdot\Delta_gF+F\cdot F_{,\mu}g^{\mu\nu}F_{,\nu}$$

Of course, a much simpler way would be to just take equation (82) and set $\Phi=F$:

$$0=\left(F^3\cdot R-\frac{F^2}{2}\left(\begin{array}{l}2F_{,ij}(n-1)g^{ij}+2\Gamma_{ij}^{a}F_{,a}g^{ij}\\-F_{,i}g^{ab}g_{jb,a}g^{ij}-F_{,j}g^{ab}g_{ib,a}g^{ij}\\-n\Gamma_{ij}^{d}F_{,d}g^{ij}+\frac{n}{2}F_{,d}g^{cd}g_{ab,c}g^{ab}\end{array}\right)-F\cdot\frac{F_{,i}\cdot F_{,j}}{4}g^{ij}\left((n-6)(n-1)\right)\right)\frac{(2-n)}{2\cdot(1-n)}-F^2\cdot\Delta_gF+F\cdot F_{,\mu}g^{\mu\nu}F_{,\nu}$$
$$\Rightarrow\qquad\qquad . \qquad (86)$$
$$0=\left(F^2\cdot R-\frac{F}{2}\left(\begin{array}{l}2F_{,ij}(n-1)g^{ij}+2\Gamma_{ij}^{a}F_{,a}g^{ij}\\-F_{,i}g^{ab}g_{jb,a}g^{ij}-F_{,j}g^{ab}g_{ib,a}g^{ij}\\-n\Gamma_{ij}^{d}F_{,d}g^{ij}+\frac{n}{2}F_{,d}g^{cd}g_{ab,c}g^{ab}\end{array}\right)-\frac{F_{,i}\cdot F_{,j}}{4}g^{ij}\left((n-6)(n-1)\right)\right)\frac{(2-n)}{2\cdot(1-n)}-F\cdot\Delta_gF+F_{,\mu}g^{\mu\nu}F_{,\nu}$$

We may see this as a special eigenvalue equation of quadratic character with eigenvalues as follows:

$$\frac{1}{2}\begin{pmatrix} 2F_{,ij}(n-1)g^{ij}+2\Gamma_{ij}^{a}F_{,a}g^{ij} \\ -F_{,i}g^{ab}g_{jb,a}g^{ij}-F_{,j}g^{ab}g_{ib,a}g^{ij} \\ -n\Gamma_{ij}^{d}F_{,d}g^{ij}+\frac{n}{2}F_{,d}g^{cd}g_{ab,c}g^{ab} \end{pmatrix}\frac{(2-n)}{2\cdot(1-n)}+\Delta_{g}F=M_{\Delta}^{2}\cdot F \quad , \tag{87}$$
$$F_{,\mu}g^{\mu\nu}F_{,\nu}\left(1-\frac{(n-6)(n-1)(2-n)}{8\cdot(1-n)}\right)=M_{\nabla}^{2}\cdot F^{2}$$

which results in the following equation:

$$0=F^{2}\cdot\left(R\frac{(2-n)}{2\cdot(1-n)}-M_{\Delta}^{2}+M_{\nabla}^{2}\right). \tag{88}$$

Without any setting for the function Φ, using an eigenvalue approach for the function F, we could similarly evaluate from (82) onward as follows:

$$0=\left[\Phi\cdot\overbrace{\left(\begin{array}{c} F^{2}\cdot R-\frac{F}{2}\begin{pmatrix} 2F_{,ij}(n-1)g^{ij}+2\Gamma_{ij}^{a}F_{,a}g^{ij} \\ -F_{,i}g^{ab}g_{jb,a}g^{ij}-F_{,j}g^{ab}g_{ib,a}g^{ij} \\ -n\Gamma_{ij}^{d}F_{,d}g^{ij}+\frac{n}{2}F_{,d}g^{cd}g_{ab,c}g^{ab} \end{pmatrix}\frac{(2-n)}{2\cdot(1-n)} \\ -\frac{F_{,i}\cdot F_{,j}}{4}g^{ij}\left((n-6)(n-1)\right) \end{array}\right)}^{=F^{2}\cdot M_{F}^{2}} \\ +F\cdot\left(F_{,\mu}g^{\mu\nu}\Phi_{,\nu}-F\cdot\Delta_{g}\Phi\right)\right]$$
$$\Rightarrow 0=\Phi\cdot F^{2}\cdot M_{F}^{2}+F\cdot\left(F_{,\mu}g^{\mu\nu}\Phi_{,\nu}-F\cdot\Delta_{g}\Phi\right) \quad . \tag{89}$$
$$0=\Phi\cdot F\cdot M_{F}^{2}+F_{,\mu}g^{\mu\nu}\Phi_{,\nu}-F\cdot\Delta_{g}\Phi$$

The last line now offers a few options. Here, in order to give an example, we just set an eigenvalue for the function Φ under the Laplace operator with respect to the metric g_{ab} and obtain:

$$
\begin{gathered}
0=\Phi\cdot F\cdot M_F^2+F_{,\mu}g^{\mu\nu}\Phi_{,\nu}-F\cdot\overbrace{\Delta_g\Phi}^{=\Phi\cdot M_\Phi^2}\\
\Rightarrow\\
0=\Phi\cdot F\cdot\overbrace{\left(M_F^2-M_\Phi^2\right)}^{-M^2}+F_{,\mu}g^{\mu\nu}\Phi_{,\nu}\\
0=-\Phi\cdot F\cdot M^2+F_{,\mu}\mathbf{g}^\mu\cdot\mathbf{g}^\nu\Phi_{,\nu}\\
\Rightarrow\\
0=\Phi\cdot F\cdot M^2-F_{,\mu}\mathbf{g}^\mu\cdot\mathbf{g}^\nu\Phi_{,\nu}
\end{gathered}
\quad. \tag{90}
$$

Now our Dirac-like approach shall be:

$$
\left(a\cdot\Phi\cdot M+A\cdot\mathbf{g}^\nu\Phi_{,\nu}\right)\left(b\cdot\Phi\cdot M-B\cdot F_{,\mu}\mathbf{g}^\mu\right)\Phi=0. \tag{91}
$$

With the objects a, A, b, and B being general 2x2 matrices we find a variety of solutions of which we here only present a very simple one, namely:

$$
A=\begin{pmatrix}1&1\\1&1\end{pmatrix};\quad a=\begin{pmatrix}-1&1\\-1&1\end{pmatrix};\quad B=\begin{pmatrix}1&-1\\1&-1\end{pmatrix};\quad b=\begin{pmatrix}-1&1\\1&-1\end{pmatrix}. \tag{92}
$$

Thus, (91) results in:

$$
\begin{aligned}
&\left(a\cdot\Phi\cdot M+A\cdot\mathbf{g}^\nu\Phi_{,\nu}\right)\left(b\cdot\Phi\cdot M-B\cdot F_{,\mu}\mathbf{g}^\mu\right)\Phi=0\\
&=2\cdot\begin{pmatrix}1&-1\\1&-1\end{pmatrix}\cdot\left(\Phi\cdot F\cdot M^2-F_{,\mu}\mathbf{g}^\mu\cdot\mathbf{g}^\nu\Phi_{,\nu}\right)\\
&=2\cdot\begin{pmatrix}1&-1\\1&-1\end{pmatrix}\cdot\left(\Phi\cdot F\cdot M^2-F_{,\mu}g^{\mu\nu}\Phi_{,\nu}\right)
\end{aligned} \tag{93}
$$

and gives us the following quantum gravity "Dirac equations":

$$
\begin{aligned}
&a\cdot\Phi\cdot M+A\cdot\mathbf{g}^\nu\Phi_{,\nu}=\begin{pmatrix}-1&1\\-1&1\end{pmatrix}\cdot\Phi\cdot M+\begin{pmatrix}1&1\\1&1\end{pmatrix}\cdot\mathbf{g}^\nu\Phi_{,\nu}=0\\
&\Rightarrow 0=\left(\begin{pmatrix}-1&1\\-1&1\end{pmatrix}\cdot M+\begin{pmatrix}1&1\\1&1\end{pmatrix}\cdot\mathbf{g}^\nu\partial_\nu\right)\Phi\\
&b\cdot F\cdot M-B\cdot F_{,\mu}\mathbf{g}^\mu=\begin{pmatrix}-1&1\\1&-1\end{pmatrix}\cdot F\cdot M-\begin{pmatrix}1&-1\\1&-1\end{pmatrix}\cdot F_{,\mu}\mathbf{g}^\mu=0\\
&\Rightarrow 0=\left(\begin{pmatrix}-1&1\\1&-1\end{pmatrix}\cdot M-\begin{pmatrix}1&-1\\1&-1\end{pmatrix}\cdot\mathbf{g}^\mu\partial_\mu\right)F
\end{aligned}
\quad. \tag{94}
$$

As (90) is to be seen as some kind of boundary condition for the eigenvalue equations in (89) and (90) (first line), we have the following pair of separated equations for the functions F and Φ:

$$F\Rightarrow\begin{cases} F^2\cdot M_F^2=\left(F^2\cdot R-\frac{F}{2}\begin{pmatrix}2F_{,ij}(n-1)g^{ij}+2\Gamma_{ij}^{a}F_{,a}g^{ij}\\ -F_{,i}g^{ab}g_{jb,a}g^{ij}-F_{,j}g^{ab}g_{ib,a}g^{ij}\\ -n\Gamma_{ij}^{d}F_{,d}g^{ij}+\frac{n}{2}F_{,d}g^{cd}g_{ab,c}g^{ab}\end{pmatrix}\frac{(2-n)}{2\cdot(1-n)}\right. \\ \left.\qquad -\frac{F_{,i}\cdot F_{,j}}{4}g^{ij}\left((n-6)(n-1)\right)\right) \\ 0=\left(\begin{pmatrix}-1 & 1\\ 1 & -1\end{pmatrix}\cdot M-\begin{pmatrix}1 & -1\\ 1 & -1\end{pmatrix}\cdot\mathbf{g}^{\mu}\partial_{\mu}\right)F\end{cases}\tag{95}$$

and

$$\Phi\Rightarrow\begin{cases}\Phi\cdot M_\Phi^2=\Delta_g\Phi\\ 0=\left(\begin{pmatrix}-1 & 1\\ -1 & 1\end{pmatrix}\cdot M+\begin{pmatrix}1 & 1\\ 1 & 1\end{pmatrix}\cdot\mathbf{g}^{\nu}\partial_{\nu}\right)\Phi\end{cases},\tag{96}$$

respectively.

Thereby it should be observed that with a suitable choice of the function F[f], the term $\frac{F_{,i}\cdot F_{,j}}{4}g^{ij}$ can always be avoided in any number of dimensions. As shown above and in another paper by one of the authors ([9] subsection "So now what?"; also see [4] for the full calculation), the corresponding function reads:

$$F[f]=\left(C_f+f\right)^{\frac{4}{n-2}}.\tag{97}$$

Some discussion about these results is to be found in our previous papers, of which (apart from the books [2, 4, 5]) we especially recommend [7 – 23].

9.3.4 About Non-Vanishing Hilbert Surface Terms and the Scaling Trick

While an integral kernel of the type (44) apparently results in surface terms, we want to find a suitable adjustment of the variational task in such a way that these surface terms can be omitted. Thereby we intend to substitute the original variation with respect to a metric $\gamma_{\alpha\beta}$ (for simplicity we ignore the cosmological constant and the matter term; $\gamma=\det[\gamma_{\alpha\beta}]$):

$$\delta W = 0 = 0 = \delta\int_V d^n x\left(\sqrt{-\gamma}\cdot\left(\Phi\cdot R - \overbrace{2\Lambda + L_M}^{=0}\right)\right)$$
$$\Rightarrow \tag{98}$$
$$= \delta\int_V d^n x\left(\sqrt{-\gamma}\cdot\Phi\cdot R\right)$$

as follows:

$$\delta\int_V d^n x\left(\sqrt{-\gamma}\cdot\Phi\cdot R\right) \to \delta\int_V d^n x\left(\sqrt{-g}\cdot F^{n/2}\cdot R^*\right) = \delta\int_V d^n x\left(\sqrt{-g}\cdot F^{n/2-1}\cdot R^*_{\alpha\beta}G^{\alpha\beta}\cdot F\right)$$
$$\Rightarrow \tag{99}$$
$$= \delta\int_V d^n x\left(\sqrt{-g}\cdot F^{n/2-1}\cdot R^*_{\alpha\beta}g^{\alpha\beta}\right)$$

Knowing that we can write $R^*_{\alpha\beta}g^{\alpha\beta}$ via (derivation see appendix D):

$$R^*_{\alpha\beta}G^{\alpha\beta}\cdot F = R^*_{\alpha\beta}g^{\alpha\beta}$$
$$= \left(\begin{array}{c} R - \frac{1}{2F}\left((n-1)\left(\overbrace{2g^{\alpha\beta}F_{,\alpha\beta} + F_{,d}g^{cd}g^{\alpha\beta}g_{\alpha\beta,c}}^{=2\Delta F - 2F_{,d}g^{cd}{}_{,c}}\right) - nF_{,d}g^{cd}g^{\alpha\beta}g_{\alpha c,\beta}\right) \\ -(n-1)\frac{g^{\alpha\beta}F_{,\alpha}\cdot F_{,\beta}}{4F^2}(n-6) \end{array}\right), \tag{100}$$
$$= \left(\begin{array}{c} R - \frac{1}{2F}\left(2(n-1)\Delta F - F_{,d}\left(2(n-1)g^{cd}{}_{,c} + ng^{cd}g^{\alpha\beta}g_{\alpha c,\beta}\right)\right) \\ -(n-1)\frac{g^{\alpha\beta}F_{,\alpha}\cdot F_{,\beta}}{4F^2}(n-6) \end{array}\right) \equiv R + \Psi[F]$$

we obtain the following variational task:

$$\delta W = \delta\int_V d^n x\left(\sqrt{-g}\cdot F^{n/2-1}\cdot(R + \Psi[F])\right). \tag{101}$$

Performing this variation with respect to the scaled metric $G_{\alpha\beta}=F^* g_{\alpha\beta}$ allows us to avoid any surface term completely.

Setting the two kernels from (98) and (101) equal to each other gives:

$$\sqrt{-\gamma}\cdot\Phi\cdot R = \sqrt{-g}\cdot F^{n/2-1}\cdot(R + \Psi[F])$$
$$\Rightarrow \tag{102}$$
$$\Phi = \sqrt{-\frac{g}{\gamma}}\cdot F^{n/2-1}\cdot\left(1 + \frac{\Psi[F]}{R}\right)$$

or:

$$R = \frac{\sqrt{-g} \cdot F^{n/2-1} \cdot \Psi[F]}{\sqrt{-\gamma} \cdot \Phi - \sqrt{-g} \cdot F^{n/2-1}}. \tag{103}$$

Demanding that the determinants of the metrics $\gamma_{\alpha\beta}$ and $g_{\alpha\beta}$ are equal, we can simplify as follows:

$$\Phi \cdot R = F^{n/2-1} \cdot (R + \Psi[F]) \quad \Rightarrow \quad \Phi = F^{n/2-1} \cdot \left(1 + \frac{\Psi[F]}{R}\right), \tag{104}$$

or:

$$R = \frac{F^{n/2-1} \cdot \Psi[F]}{\Phi - F^{n/2-1}}. \tag{105}$$

So, in principle, one can always substitute a variation of the type (44), which would produce complicated surface terms, by one of the form (101), where the surface term can be avoided when performing the variation with respect to the scaled metric tensor $G_{\alpha\beta}$=F* $g_{\alpha\beta}$.

9.4 The Quantum Theory Inside Hilbert's Variational Integral

The extraction of the main quantum equations out of our generalized Einstein-Hilbert action as given above has been – quite extensively – elaborated in [7, 8, 13]. Regarding a fairly compact presentation about how to obtain the Klein-Gordon [24], the Schrödinger [25, 26], and the Dirac equation [27, 28] from this concept, the reader is especially referred to [13]. In here we will need only one derivation, which is the one of the metric Klein-Gordon equation directly out of the quantum Einstein field equations. We simply repeat from [19] (see there for more details and explanations).

We are looking for intrinsic solutions to the quantum Einstein field equations (66) as follows:

$$0=\begin{pmatrix} -\frac{g^{ab}}{2F}\begin{pmatrix} 2F_{,ab}(n-2)-\Gamma^{d}_{ab}F_{,d}(n-2) \\ +F_{,d}g^{cd}\left(g_{ab,c}\frac{1}{2}(n-2)-2g_{cb,a}\right)\end{pmatrix} \\ -\frac{F_{,i}\cdot F_{,j}}{4F^2}g^{ij}\left((n-7)(n-2)\right)\end{pmatrix}$$

$$0=R_{\alpha\beta}-\frac{g_{\alpha\beta}}{2}R+\begin{pmatrix} \frac{1}{2F}\begin{pmatrix} F_{,\alpha\beta}(n-2)+F_{,a}g^{ab}\left(g_{\beta b,\alpha}-g_{\beta\alpha,b}\right) \\ -F_{,\alpha}g^{ab}g_{\beta b,a}-F_{,\beta}g^{ab}g_{\alpha b,a} \\ +F_{,d}g^{cd}\frac{1}{2}n\left(\frac{2}{n}g_{\alpha c,\beta}-g_{\alpha c,\beta}-g_{\beta c,\alpha}+g_{\alpha\beta,c}\right) \\ -\frac{1}{2F}\left(F_{,\alpha}\cdot F_{,\beta}(3n-6)\right)\end{pmatrix} \\ -\frac{\left(\partial_{\alpha}\Phi_{,\beta}-\Gamma^{*\gamma}_{\alpha\beta}\Phi_{,\gamma}-\frac{F\cdot g_{\alpha\beta}}{\sqrt{-gF^n}}\partial_{\mu}\left(\sqrt{-gF^n}\cdot\frac{g^{\mu\nu}}{F}\Phi_{,\nu}\right)\right)}{\Phi}\end{pmatrix}. \tag{106}$$

We start with the first equation in (106) and obtain:

$$\begin{aligned} 0&=\begin{pmatrix} -\frac{g^{ab}}{2F}\begin{pmatrix} 2\left(F''f_{,a}\cdot f_{,b}+F'f_{,ab}\right)(n-2)-F'\Gamma^{d}_{ab}f_{,d}(n-2) \\ +F'f_{,d}g^{cd}\left(g_{ab,c}\frac{1}{2}(n-2)-2g_{cb,a}\right)\end{pmatrix} \\ -(F')^2\frac{f_{,a}\cdot f_{,b}}{4F^2}g^{ab}\left((n-7)(n-2)\right)\end{pmatrix} \\ &=\begin{pmatrix} -\frac{g^{ab}F'}{2F}\begin{pmatrix} 2f_{,ab}(n-2)-\Gamma^{d}_{ab}f_{,d}(n-2) \\ +f_{,d}g^{cd}\left(g_{ab,c}\frac{1}{2}(n-2)-2g_{cb,a}\right)\end{pmatrix} \\ -\frac{f_{,a}\cdot f_{,b}}{2F}g^{ab}(n-2)\left(\frac{(F')^2}{2F}(n-7)+2F''\right)\end{pmatrix}\end{aligned}. \tag{107}$$

As linearization condition we demand:

$$0=\frac{(F')^2}{2F}(n-7)+2F'', \tag{108}$$

which can be satisfied via the function F[f] as follows:

$$F[f]=\begin{cases} C_F\cdot\left(f+C_f\right)^{\frac{4}{n-3}} & n\neq 3 \\ C_F\cdot e^{f\cdot C_f} & n=3\end{cases}. \tag{109}$$

Subsequently, equation (107) can be simplified and yields:

$$0=g^{ab}\left(2f_{,ab}(n-2)-\Gamma_{ab}^{d}f_{,d}(n-2)+f_{,d}g^{cd}\left(g_{ab,c}\frac{1}{2}(n-2)-2g_{cb,a}\right)\right). \tag{110}$$

Interestingly, for metrics without shear elements:

$$g_{ij}=\begin{pmatrix} g_{00} & \cdots & 0 \\ \vdots & \ddots & \vdots \\ 0 & \cdots & g_{n-1n-1} \end{pmatrix}, \tag{111}$$

this converges to the ordinary Laplace operator, namely:

$$\begin{aligned}&0=g^{ab}\left(2f_{,ab}(n-2)-\Gamma_{ab}^{d}f_{,d}(n-2)+f_{,d}g^{cd}\left(g_{ab,c}\frac{1}{2}(n-2)-2g_{cb,a}\right)\right)\\ &\xrightarrow{g_{ij}=\begin{pmatrix} g_{00} & \cdots & 0 \\ \vdots & \ddots & \vdots \\ 0 & \cdots & g_{n-1n-1} \end{pmatrix}}\\ &0=2(n-2)\Delta f\end{aligned}. \tag{112}$$

As said before, some critics may now argue that this equation is not of Klein-Gordon character as it does not contain any potential nor mass, but two of the authors have already shown that this problem is easily solved by adding additional dimensions carrying the right properties and entanglement characteristics with other dimensions to produce masses and potentials (e.g., [4, 7 – 10, 29, 30]).

9.4.1 The Special Situation in Two Dimensions

For an arbitrary metric, including a scaled one, it can be shown that the quantum Einstein field equations (29) are always automatically fulfilled in 2 dimensions. This holds as long as the variation is performed with respect to the scaled metric tensor G_{ij} (e.g., see [101]). In order to still obtain a fundamental rule, some scientists resort to the condition R=0, but taking the findings from our section here, we could just take (66) as the fundamental condition, which, due to:

$$\left[\Phi\cdot R^{*}_{\mu\nu}-\frac{1}{2}\Phi\cdot R^{*}G_{\mu\nu}\right]_{2D}=\Phi\cdot\left[R^{*}_{\mu\nu}-\frac{1}{2}R^{*}G_{\mu\nu}\right]_{2D}=0 \tag{113}$$

in two dimensions always simplifies to:

$$\left(\nabla_{\mu}\nabla_{\nu}-G_{\mu\nu}\Delta_{G}\right)\Phi=0. \tag{114}$$

9.5 Functional Derivative with Respect to the Function f

A third option to derive scalar quantum gravity field equations requires the use of the functional derivative:

$$\delta W=\frac{\delta W}{\delta f[\mathbf{y}]}=0=\frac{\delta}{\delta f}\int_V d^n x\left(\sqrt{g\cdot F^n}\cdot F\cdot R^*\right)$$

$$=\lim_{\in\to 0}\frac{1}{\in}\left(\begin{array}{c}\int_V d^n x\left(\sqrt{g\cdot F\left[f[\mathbf{x}]+\in\cdot\delta[\mathbf{x}-\mathbf{y}]\right]^n}\cdot R^{**}\left[f[\mathbf{x}]+\in\cdot\delta[\mathbf{x}-\mathbf{y}]\right]\right)\\ -\int_V d^n x\left(\sqrt{g\cdot F[f[\mathbf{x}]]^n}\cdot R^{**}[f[\mathbf{x}]]\right)\end{array}\right). \tag{115}$$

Thereby we have to apply the Ricci scalar with the scaled metric tensor, which we here give without the affine connection, but solely with the metric tensor and its derivatives:

$$\begin{aligned}R^{**}=F\cdot R^*=R&-\frac{g^{\alpha\beta}}{2F}\left(2F_{,\alpha\beta}(n-1)+F_{,d}g^{cd}\left(\begin{array}{c}\left(g_{\beta c,\alpha}+g_{\alpha c,\beta}-g_{\beta\alpha,c}\right)-2g_{c\beta,\alpha}\\ +\frac{n}{2}\left(\left(-g_{\alpha c,\beta}-g_{\beta c,\alpha}+g_{\alpha\beta,c}\right)+g_{\alpha\beta,c}\right)\end{array}\right)\right)\\ &+\frac{1}{4F^2}\left(F_{,\alpha}\cdot F_{,\beta}(3n-6)+g_{\alpha\beta}F_{,c}F_{,d}g^{cd}(4-n)\right)g^{\alpha\beta}\\ =R&-\frac{g^{\alpha\beta}}{2F}\left(2F_{,\alpha\beta}(n-1)+F_{,d}g^{cd}\left(\begin{array}{c}\left(g_{\beta c,\alpha}+g_{\alpha c,\beta}-g_{\beta\alpha,c}\right)-2g_{c\beta,\alpha}\\ +\frac{n}{2}\left(\left(-g_{\alpha c,\beta}-g_{\beta c,\alpha}+g_{\alpha\beta,c}\right)+g_{\alpha\beta,c}\right)\end{array}\right)\right)\\ &+\frac{1}{4F^2}\left(F_{,\alpha}\cdot F_{,\beta}(3n-6)+g_{\alpha\beta}F_{,c}F_{,d}g^{cd}(4-n)\right)g^{\alpha\beta}\\ =R&-\frac{g^{\alpha\beta}}{2F}F'\left(2f_{,\alpha\beta}(n-1)+f_{,d}g^{cd}\left(\begin{array}{c}g_{\alpha c,\beta}-g_{\beta\alpha,c}-g_{c\beta,\alpha}\\ +\frac{n}{2}\left(2g_{\alpha\beta,c}-g_{\alpha c,\beta}-g_{\beta c,\alpha}\right)\end{array}\right)\right)\\ &-(n-1)\frac{f_{,\alpha}\cdot f_{,\beta}}{4F^2}g^{\alpha\beta}\left(4FF''+F'\cdot F'(n-6)\right)\end{aligned}\ . \tag{116}$$

The full derivation is given in appendix C and results in:

$$\frac{\delta W}{\delta f}=0\ \Rightarrow\ \Rightarrow$$

$$0=\left[\begin{array}{c}\frac{\sqrt{g}}{2}\left(\frac{f_{,\alpha}f_{,\beta}}{2}g^{\alpha\beta}\left(\frac{1}{2}(n-4)(n+2)(F')^2+2FF''\right)+(n+2)FF'\Delta f\right)\\ +F^2\left(\begin{array}{c}\left(\sqrt{g}\cdot g^{\alpha\beta}\right)_{,\alpha\beta}-\frac{1}{2(n-1)}\left(\sqrt{g}\cdot g^{\alpha\beta}\cdot g^{ij}\left(\begin{array}{c}g_{i\beta,j}-g_{ij,\beta}-g_{\beta j,i}\\ +\frac{n}{2}\left(2g_{ij,\beta}-g_{i\beta,j}-g_{j\beta,i}\right)\end{array}\right)\right)_{,\alpha}\\ -\frac{n}{2}\frac{\sqrt{g}\cdot R}{(n-1)}\end{array}\right)\end{array}\right]_{\mathbf{y}}, \tag{117}$$

$$\xrightarrow{F=C+a\cdot f}$$

$$0=\left[\begin{array}{c}\frac{\sqrt{g}}{2}(n+2)\left(\frac{f_{,\alpha}f_{,\beta}}{4}g^{\alpha\beta}(n-4)a^2+(C+a\cdot f)\cdot a\cdot\Delta f\right)\\ +(C+a\cdot f)^2\left(\begin{array}{c}\left(\sqrt{g}\cdot g^{\alpha\beta}\right)_{,\alpha\beta}-\frac{n}{2}\frac{\sqrt{g}\cdot R}{(n-1)}\\ -\frac{1}{2(n-1)}\left(\sqrt{g}\cdot g^{\alpha\beta}\cdot g^{ij}\left(\begin{array}{c}g_{i\beta,j}-g_{ij,\beta}-g_{\beta j,i}\\ +\frac{n}{2}\left(2g_{ij,\beta}-g_{i\beta,j}-g_{j\beta,i}\right)\end{array}\right)\right)_{,\alpha}\end{array}\right)\end{array}\right]_{y}. \quad (118)$$

Instead of the setting of a linear function for F[f] as used in (118), we aim for a linearization of (117) and demand the condition:

$$0=(n-4)(n+2)(F')^2+4FF", \quad (119)$$

which can be satisfied by the function F[f] as follows:

$$F[f]=C_F\cdot(f+C_f)^{\frac{4}{12+n\cdot(n-6)}}. \quad (120)$$

Subsequently, equation (117) can be simplified and yields:

$$\begin{gathered}0=\left[\begin{array}{c}\frac{\sqrt{g}}{2}(n+2)F\frac{F'}{F}\Delta f-\frac{\sqrt{g}\cdot R}{(n-1)}\\ +\left(\sqrt{g}\cdot g^{\alpha\beta}\right)_{,\alpha\beta}-\frac{1}{2(n-1)}\left(\sqrt{g}\cdot g^{\alpha\beta}\cdot g^{ij}\left(\begin{array}{c}g_{i\beta,j}-g_{ij,\beta}-g_{\beta j,i}\\ +\frac{n}{2}\left(2g_{ij,\beta}-g_{i\beta,j}-g_{j\beta,i}\right)\end{array}\right)\right)_{,\alpha}\end{array}\right]\\ \Rightarrow\\ 0=\left[\begin{array}{c}(n-1)(n+2)F\frac{F'}{2F}\Delta f-R\\ +\left(\frac{(n-1)}{\sqrt{g}}\left(\sqrt{g}\cdot g^{\alpha\beta}\right)_{,\alpha\beta}-\frac{1}{2\sqrt{g}}\left(\sqrt{g}\cdot g^{\alpha\beta}\cdot g^{ij}\left(\begin{array}{c}g_{i\beta,j}-g_{ij,\beta}-g_{\beta j,i}\\ +\frac{n}{2}\left(2g_{ij,\beta}-g_{i\beta,j}-g_{j\beta,i}\right)\end{array}\right)\right)_{,\alpha}\right)\end{array}\right]\end{gathered}. \quad (121)$$

9.6 Transition to the "Classical" Physics and Some Interesting Findings Along the Way

As we have seen that the separation of the volume part of the metric via setting:

$$G_{\alpha\beta} = g_{\alpha\beta} \cdot F[f] \tag{122}$$

leads us to the classical quantum equations, we now want to derive the amount of approximation residing in the latter. For this, we take the Klein-Gordon equation:

$$\left[\frac{M^2 c^2}{\hbar^2} + V + \Delta\right] f = 0, \tag{123}$$

with potential V, mass M, speed of light in vacuum c, and the reduced Planck constant ħ and compare it with the quantum Einstein field equations from above (c.f. (29)):

$$0 = \left(\begin{array}{c} \boxed{R_{\alpha\beta} - \frac{g_{\alpha\beta}}{2} R} + \frac{g_{\alpha\beta}}{2} \left(\begin{array}{c} \frac{1}{2F} \left(\begin{array}{c} 2F_{,ij}(n-1)g^{ij} + 2\Gamma^{a}_{ij} F_{,a} g^{ij} \\ -F_{,i} g^{ab} g_{jb,a} g^{ij} - F_{,j} g^{ab} g_{ib,a} g^{ij} \\ -n\Gamma^{d}_{ij} F_{,d} g^{ij} + \frac{n}{2} F_{,d} g^{cd} g_{ab,c} g^{ab} \end{array} \right) \\ + \frac{F_{,i} \cdot F_{,j}}{4F^2} g^{ij} \left((n-6)(n-1)\right) \end{array} \right) \\ -\frac{1}{2F} \left(\begin{array}{c} F_{,\alpha\beta}(n-2) + F_{,ab} g_{\alpha\beta} g^{ab} \\ +F_{,a} g^{ab} \left(g_{\beta b,\alpha} - g_{\beta\alpha,b}\right) - F_{,\alpha} g^{ab} g_{\beta b,a} - F_{,\beta} g^{ab} g_{\alpha b,a} \\ +F_{,d} g^{cd} \frac{1}{2} n \left(\frac{2}{n} g_{\alpha c,\beta} - g_{\alpha c,\beta} - g_{\beta c,\alpha} + g_{\alpha\beta,c} + \frac{1}{n} g_{\alpha\beta} g_{ab,c} g^{ab} \right) \\ -\frac{1}{2F} \left(F_{,\alpha} \cdot F_{,\beta} (3n-6) + g_{\alpha\beta} F_{,c} F_{,d} g^{cd} (4-n) \right) \end{array} \right) \\ +\kappa T_{\alpha\beta} + \Lambda \cdot F \cdot g_{\alpha\beta} \end{array} \right). \tag{124}$$

As it is textbook material how one can derive the Schrödinger equation and the Dirac equation from (123), we will not bother here with the direct derivation of the these equations from (124), but it should be mentioned that some of the corresponding material will be presented further below in this book in slightly different contexts and for other purposes.

Observing (124), we see a great variety of options to separate general tensor-terms from those terms which are just scalars multiplied with the metric tensor, which is to say, terms of type $g_{\alpha\beta}$*scalar.

We will start our consideration with the $R/R_{\alpha\beta}$-ratio as it was already applied in (32). By setting the functional wrapper F[f] to:

$$F[f]=\begin{cases} C_F\cdot\left(f+C_f\right)^{\frac{4}{n-2}} & n\neq 2\\ C_F\cdot e^{f\cdot C_f} & n=2\end{cases} \tag{125}$$

and demanding a zero cosmological constant and vanishing – originally anyway only postulated – matter terms via $T_{\alpha\beta}=0$, (124) simplifies as follows:

$$0=\left(\begin{array}{c} \boxed{R_{\alpha\beta}-\frac{g_{\alpha\beta}}{2}R}+\frac{g_{\alpha\beta}}{2}\left(\frac{F'}{2F}\left(\begin{array}{c} 2f_{,ij}(n-1)g^{ij}+2\Gamma_{ij}^{a}f_{,a}g^{ij}\\ -f_{,i}g^{ab}g_{jb,a}g^{ij}-f_{,j}g^{ab}g_{ib,a}g^{ij}\\ -n\Gamma_{ij}^{d}f_{,d}g^{ij}+\frac{n}{2}f_{,d}g^{cd}g_{ab,c}g^{ab}\end{array}\right)\right)\\ -\frac{1}{2F}\left(\begin{array}{c} F_{,\alpha\beta}(n-2)+F_{,ab}g_{\alpha\beta}g^{ab}\\ +F_{,a}g^{ab}\left(g_{\beta b,\alpha}-g_{\beta\alpha,b}\right)-F_{,\alpha}g^{ab}g_{\beta b,a}-F_{,\beta}g^{ab}g_{\alpha b,a}\\ +F_{,d}g^{cd}\frac{1}{2}n\left(\frac{2}{n}g_{\alpha c,\beta}-g_{\alpha c,\beta}-g_{\beta c,\alpha}+g_{\alpha\beta,c}+\frac{1}{n}g_{\alpha\beta}g_{ab,c}g^{ab}\right)\\ -\frac{1}{2F}\left(F_{,\alpha}\cdot F_{,\beta}(3n-6)+g_{\alpha\beta}F_{,c}F_{,d}g^{cd}(4-n)\right)\end{array}\right)\end{array}\right). \tag{126}$$

For non-vanishing postulated matter and cosmological constant the equation would read:

$$0=\left(\begin{array}{c} \boxed{R_{\alpha\beta}-\frac{g_{\alpha\beta}}{2}R}+\frac{g_{\alpha\beta}}{2}\left(\frac{F'}{2F}\left(\begin{array}{c} 2f_{,ij}(n-1)g^{ij}+2\Gamma_{ij}^{a}f_{,a}g^{ij}\\ -f_{,i}g^{ab}g_{jb,a}g^{ij}-f_{,j}g^{ab}g_{ib,a}g^{ij}\\ -n\Gamma_{ij}^{d}f_{,d}g^{ij}+\frac{n}{2}f_{,d}g^{cd}g_{ab,c}g^{ab}\end{array}\right)\right)\\ -\frac{1}{2F}\left(\begin{array}{c} F_{,\alpha\beta}(n-2)+F_{,ab}g_{\alpha\beta}g^{ab}\\ +F_{,a}g^{ab}\left(g_{\beta b,\alpha}-g_{\beta\alpha,b}\right)-F_{,\alpha}g^{ab}g_{\beta b,a}-F_{,\beta}g^{ab}g_{\alpha b,a}\\ +F_{,d}g^{cd}\frac{1}{2}n\left(\frac{2}{n}g_{\alpha c,\beta}-g_{\alpha c,\beta}-g_{\beta c,\alpha}+g_{\alpha\beta,c}+\frac{1}{n}g_{\alpha\beta}g_{ab,c}g^{ab}\right)\\ -\frac{1}{2F}\left(F_{,\alpha}\cdot F_{,\beta}(3n-6)+g_{\alpha\beta}F_{,c}F_{,d}g^{cd}(4-n)\right)\end{array}\right)\\ +\kappa T_{\alpha\beta}+\Lambda\cdot F\cdot g_{\alpha\beta}\end{array}\right). \tag{127}$$

With the relation for the connection of the Laplace operator with the Christoffel symbols:

$$\Delta f=f_{,ij}g^{ij}-\Gamma_{ij}^{a}f_{,a}g^{ij}, \tag{128}$$

we can rewrite (126):

$$0=\left(\begin{array}{c}\boxed{R_{\alpha\beta}-\frac{g_{\alpha\beta}}{2}R}+\frac{g_{\alpha\beta}}{2}\left(\frac{F'}{2F}\left(\begin{array}{c}2f_{,ij}(n-1)g^{ij}-2n\Gamma_{ij}^{d}f_{,d}g^{ij}+2\Gamma_{ij}^{a}f_{,a}g^{ij}\\ -f_{,i}g^{ab}g_{jb,a}g^{ij}-f_{,j}g^{ab}g_{ib,a}g^{ij}\\ +n\Gamma_{ij}^{d}f_{,d}g^{ij}+\frac{n}{2}f_{,d}g^{cd}g_{ab,c}g^{ab}\end{array}\right)\right)\\ -\frac{1}{2F}\left(\begin{array}{c}F_{,\alpha\beta}(n-2)+F_{,ab}g_{\alpha\beta}g^{ab}\\ +F_{,a}g^{ab}\left(g_{\beta b,\alpha}-g_{\beta\alpha,b}\right)-F_{,\alpha}g^{ab}g_{\beta b,a}-F_{,\beta}g^{ab}g_{\alpha b,a}\\ +F_{,d}g^{cd}\frac{1}{2}n\left(\frac{2}{n}g_{\alpha c,\beta}-g_{\alpha c,\beta}-g_{\beta c,\alpha}+g_{\alpha\beta,c}+\frac{1}{n}g_{\alpha\beta}g_{ab,c}g^{ab}\right)\\ -\frac{1}{2F}\left(F_{,\alpha}\cdot F_{,\beta}(3n-6)+g_{\alpha\beta}F_{,c}F_{,d}g^{cd}(4-n)\right)\end{array}\right)\end{array}\right)$$

$$=\left(\begin{array}{c}\boxed{R_{\alpha\beta}-\frac{g_{\alpha\beta}}{2}R}+\frac{g_{\alpha\beta}}{2}\left(\frac{F'}{2F}\left(\begin{array}{c}2\Delta f(n-1)-f_{,i}g^{ab}g_{jb,a}g^{ij}-f_{,j}g^{ab}g_{ib,a}g^{ij}\\ +n\Gamma_{ij}^{d}f_{,d}g^{ij}+\frac{n}{2}f_{,d}g^{cd}g_{ab,c}g^{ab}\end{array}\right)\right)\\ -\frac{1}{2F}\left(\begin{array}{c}F_{,\alpha\beta}(n-2)+F_{,ab}g_{\alpha\beta}g^{ab}\\ +F_{,a}g^{ab}\left(g_{\beta b,\alpha}-g_{\beta\alpha,b}\right)-F_{,\alpha}g^{ab}g_{\beta b,a}-F_{,\beta}g^{ab}g_{\alpha b,a}\\ +F_{,d}g^{cd}\frac{1}{2}n\left(\frac{2}{n}g_{\alpha c,\beta}-g_{\alpha c,\beta}-g_{\beta c,\alpha}+g_{\alpha\beta,c}+\frac{1}{n}g_{\alpha\beta}g_{ab,c}g^{ab}\right)\\ -\frac{1}{2F}\left(F_{,\alpha}\cdot F_{,\beta}(3n-6)+g_{\alpha\beta}F_{,c}F_{,d}g^{cd}(4-n)\right)\end{array}\right)\end{array}\right). \tag{129}$$

Comparing this with the classical Klein-Gordon equation (123), we realize that the approximation from the full to the classical equation is not only of scalar, but of tensorial character, namely:

$$0=\left(\begin{array}{c}\overbrace{-\frac{g_{\alpha\beta}}{2}\frac{F'}{F}(n-1)\left(\frac{R}{(n-1)}\frac{F}{F'}-\Delta f-\frac{1}{2(n-1)}\left(\begin{array}{c}-f_{,i}g^{ab}g_{jb,a}g^{ij}-f_{,j}g^{ab}g_{ib,a}g^{ij}\\ +n\Gamma_{ij}^{d}f_{,d}g^{ij}+\frac{n}{2}f_{,d}g^{cd}g_{ab,c}g^{ab}\end{array}\right)\right)}^{\text{scalar}}\\ \underbrace{+R_{\alpha\beta}-\frac{1}{2F}\left(\begin{array}{c}F_{,\alpha\beta}(n-2)+F_{,ab}g_{\alpha\beta}g^{ab}\\ +F_{,a}g^{ab}\left(g_{\beta b,\alpha}-g_{\beta\alpha,b}\right)-F_{,\alpha}g^{ab}g_{\beta b,a}-F_{,\beta}g^{ab}g_{\alpha b,a}\\ +F_{,d}g^{cd}\frac{1}{2}n\left(\frac{2}{n}g_{\alpha c,\beta}-g_{\alpha c,\beta}-g_{\beta c,\alpha}+g_{\alpha\beta,c}+\frac{1}{n}g_{\alpha\beta}g_{ab,c}g^{ab}\right)\\ -\frac{1}{2F}\left(F_{,\alpha}\cdot F_{,\beta}(3n-6)+g_{\alpha\beta}F_{,c}F_{,d}g^{cd}(4-n)\right)\end{array}\right)}_{\text{tensor}}\end{array}\right). \tag{130}$$

For practical reasons it could sometimes be an advantage to substitute the affine connection in the Ricci scalar (Christoffel symbols) by the corresponding metric tensor derivatives (the corresponding derivation is given in appendix D):

$$
\begin{aligned}
&R^{*}_{\alpha\beta}G^{\alpha\beta}\cdot F = R^{*}_{\alpha\beta}g^{\alpha\beta} \\
&= \left(\begin{array}{l} R-\frac{1}{2F}\left((n-1)\left(\overbrace{2g^{\alpha\beta}F_{,\alpha\beta}+F_{,d}g^{cd}g^{\alpha\beta}g_{\alpha\beta,c}}^{=2\Delta F-2F_{,d}g^{cd}{}_{,c}}\right)-nF_{,d}g^{cd}g^{\alpha\beta}g_{\alpha c,\beta}\right) \\ -(n-1)\frac{g^{\alpha\beta}F_{,\alpha}\cdot F_{,\beta}}{4F^{2}}(n-6) \end{array}\right) \\
&= \left(\begin{array}{l} R-\frac{1}{2F}\left(2(n-1)\Delta F-F_{,d}\left(2(n-1)g^{cd}{}_{,c}+ng^{cd}g^{\alpha\beta}g_{\alpha c,\beta}\right)\right) \\ -(n-1)\frac{g^{\alpha\beta}F_{,\alpha}\cdot F_{,\beta}}{4F^{2}}(n-6) \end{array}\right) \\
&= \left(\begin{array}{l} R-\frac{F'}{2F}\left(2(n-1)\Delta f-f_{,d}\left(2(n-1)g^{cd}{}_{,c}+ng^{cd}g^{\alpha\beta}g_{\alpha c,\beta}\right)\right) \\ -(n-1)\frac{g^{\alpha\beta}f_{,\alpha}\cdot f_{,\beta}}{4F^{2}}\left((n-6)(F')^{2}+4FF'\right) \end{array}\right)
\end{aligned}
\quad . \qquad (131)
$$

This yields:

$$
0=\left(\begin{array}{l} -\frac{g_{\alpha\beta}}{2}\frac{F'}{F}(n-1)\overbrace{\left(\frac{R}{(n-1)}\frac{F}{F'}-\Delta f+\frac{f_{,d}}{2(n-1)}\left(2(n-1)g^{cd}{}_{,c}+ng^{cd}g^{ab}g_{ac,b}\right)\right)}^{\text{scalar}} \\ \underbrace{+R_{\alpha\beta}-\frac{1}{2F}\left(\begin{array}{c} F_{,\alpha\beta}(n-2)+F_{,ab}g_{\alpha\beta}g^{ab} \\ +F_{,a}g^{ab}\left(g_{\beta b,\alpha}-g_{\beta\alpha,b}\right)-F_{,\alpha}g^{ab}g_{\beta b,a}-F_{,\beta}g^{ab}g_{\alpha b,a} \\ +F_{,d}g^{cd}\frac{1}{2}n\left(\frac{2}{n}g_{\alpha c,\beta}-g_{\alpha c,\beta}-g_{\beta c,\alpha}+g_{\alpha\beta,c}+\frac{1}{n}g_{\alpha\beta}g_{ab,c}g^{ab}\right) \\ -\frac{1}{2F}\left(F_{,\alpha}\cdot F_{,\beta}(3n-6)+g_{\alpha\beta}F_{,c}F_{,d}g^{cd}(4-n)\right) \end{array}\right)}_{\text{tensor}} \end{array}\right) . \qquad (132)
$$

Thereby, we have maintained the R scalar to Ricci tensor part separation also for the F-terms. Further below, we are going to consider more rigorous scalar to tensor separations. Reshaping of equation (130) leads to:

$$0=\left(\begin{array}{c}\overbrace{\frac{g_{\alpha\beta}}{2}\frac{F'}{F}(n-1)\left(\Delta f-\frac{R}{(n-1)}\frac{F}{F'}+\frac{1}{2(n-1)}\left(\begin{array}{c}-f_{,i}g^{ab}g_{jb,a}g^{ij}-f_{,j}g^{ab}g_{ib,a}g^{ij}\\+n\Gamma^{d}_{ij}f_{,d}g^{ij}+\frac{n}{2}f_{,d}g^{cd}g_{ab,c}g^{ab}\end{array}\right)\right)}^{\text{scalar}}\\ \underbrace{+R_{\alpha\beta}-\frac{1}{2F}\left(\begin{array}{c}F_{,\alpha\beta}(n-2)+F_{,ab}g_{\alpha\beta}g^{ab}\\+F_{,a}g^{ab}\left(g_{\beta b,\alpha}-g_{\beta\alpha,b}\right)-F_{,\alpha}g^{ab}g_{\beta b,a}-F_{,\beta}g^{ab}g_{\alpha b,a}\\+F_{,d}g^{cd}\frac{1}{2}n\left(\frac{2}{n}g_{\alpha c,\beta}-g_{\alpha c,\beta}-g_{\beta c,\alpha}+g_{\alpha\beta,c}+\frac{1}{n}g_{\alpha\beta}g_{ab,c}g^{ab}\right)\\-\frac{1}{2F}\left(F_{,\alpha}\cdot F_{,\beta}(3n-6)+g_{\alpha\beta}F_{,c}F_{,d}g^{cd}(4-n)\right)\end{array}\right)}_{\text{tensor}}\end{array}\right)$$

$$\xrightarrow{F[f]=\begin{cases}C_F\cdot(f+C_f)^{\frac{4}{n-2}} & n\neq 2\\ C_F\cdot e^{f\cdot C_f} & n=2\end{cases}\quad \text{for}\quad n>2}$$

$$0=\left(\begin{array}{c}\overbrace{\frac{2\cdot g_{\alpha\beta}}{\left(f+C_f\right)}\frac{n-1}{n-2}\left(\begin{array}{c}\Delta f-\frac{R\left(f+C_f\right)}{(n-1)}\frac{n-2}{4}\\+\frac{1}{2(n-1)}\left(\begin{array}{c}-f_{,i}g^{ab}g_{jb,a}g^{ij}-f_{,j}g^{ab}g_{ib,a}g^{ij}\\+n\Gamma^{d}_{ij}f_{,d}g^{ij}+\frac{n}{2}f_{,d}g^{cd}g_{ab,c}g^{ab}\end{array}\right)\end{array}\right)}^{\text{scalar}}\\ \underbrace{+R_{\alpha\beta}-\frac{1}{2F}\left(\begin{array}{c}F_{,\alpha\beta}(n-2)+F_{,ab}g_{\alpha\beta}g^{ab}\\+F_{,a}g^{ab}\left(g_{\beta b,\alpha}-g_{\beta\alpha,b}\right)-F_{,\alpha}g^{ab}g_{\beta b,a}-F_{,\beta}g^{ab}g_{\alpha b,a}\\+F_{,d}g^{cd}\frac{1}{2}n\left(\frac{2}{n}g_{\alpha c,\beta}-g_{\alpha c,\beta}-g_{\beta c,\alpha}+g_{\alpha\beta,c}+\frac{1}{n}g_{\alpha\beta}g_{ab,c}g^{ab}\right)\\-\frac{1}{2F}\left(F_{,\alpha}\cdot F_{,\beta}(3n-6)+g_{\alpha\beta}F_{,c}F_{,d}g^{cd}(4-n)\right)\end{array}\right)}_{\text{tensor}}\end{array}\right),\quad(133)$$

$$0=\left(\begin{array}{c}\frac{2\cdot g_{\alpha\beta}}{\left(f+C_f\right)}\frac{n-1}{n-2}\overbrace{\left(\begin{array}{c}\Delta f-\frac{R\left(f+C_f\right)}{(n-1)}\frac{n-2}{4}\\+\frac{1}{2(n-1)}\left(n\left(\Gamma^i_{dj}g^{dj}+\frac{g^{ci}g_{ab,c}g^{ab}}{2}\right)-2g^{ab}g_{jb,a}g^{ij}\right)\partial_i f\end{array}\right)}^{\text{scalar}}\\ \underbrace{+R_{\alpha\beta}-\frac{1}{2F}\left(\begin{array}{c}F_{,\alpha\beta}(n-2)+F_{,ab}g_{\alpha\beta}g^{ab}\\+F_{,a}g^{ab}\left(g_{\beta b,\alpha}-g_{\beta\alpha,b}\right)-F_{,\alpha}g^{ab}g_{\beta b,a}-F_{,\beta}g^{ab}g_{\alpha b,a}\\+F_{,d}g^{cd}\frac{1}{2}n\left(\frac{2}{n}g_{\alpha c,\beta}-g_{\alpha c,\beta}-g_{\beta c,\alpha}+g_{\alpha\beta,c}+\frac{1}{n}g_{\alpha\beta}g_{ab,c}g^{ab}\right)\\-\frac{1}{2F}\left(F_{,\alpha}\cdot F_{,\beta}(3n-6)+g_{\alpha\beta}F_{,c}F_{,d}g^{cd}(4-n)\right)\end{array}\right)}_{\text{tensor}}\end{array}\right). \quad (134)$$

By multiplying equation (123) with $\frac{2\cdot g_{\alpha\beta}}{\left(f+C_f\right)}\frac{n-1}{n-2}$ we obtain:

$$\frac{2\cdot g_{\alpha\beta}}{\left(f+C_f\right)}\frac{n-1}{n-2}\left[\frac{M^2c^2}{\hbar^2}+V+\Delta\right]_{KG}f=0 \quad (135)$$

and can directly compare with the quantum Einstein field equations in the form (134):

$$0=\left(\begin{array}{c}\frac{2\cdot g_{\alpha\beta}}{\left(f+C_f\right)}\frac{n-1}{n-2}\left(\overbrace{\Delta f+\left(\begin{array}{c}\frac{1}{2(n-1)}\left(n\left(\Gamma^i_{dj}g^{dj}+\frac{g^{ci}g_{ab,c}g^{ab}}{2}\right)-2g^{ab}g_{jb,a}g^{ij}\right)\partial_i f\\-\frac{R\left(f+C_f\right)}{(n-1)}\frac{n-2}{4}\end{array}\right)}^{=\left(\Delta+\frac{M^2c^2}{\hbar^2}+V\right)_{KG}\left(f+C_f\right)}\right)\\ \underbrace{+R_{\alpha\beta}-\frac{1}{2F}\left(\begin{array}{c}F_{,\alpha\beta}(n-2)+F_{,ab}g_{\alpha\beta}g^{ab}\\+F_{,a}g^{ab}\left(g_{\beta b,\alpha}-g_{\beta\alpha,b}\right)-F_{,\alpha}g^{ab}g_{\beta b,a}-F_{,\beta}g^{ab}g_{\alpha b,a}\\+F_{,d}g^{cd}\frac{1}{2}n\left(\frac{2}{n}g_{\alpha c,\beta}-g_{\alpha c,\beta}-g_{\beta c,\alpha}+g_{\alpha\beta,c}+\frac{1}{n}g_{\alpha\beta}g_{ab,c}g^{ab}\right)\\-\frac{1}{2F}\left(F_{,\alpha}\cdot F_{,\beta}(3n-6)+g_{\alpha\beta}F_{,c}F_{,d}g^{cd}(4-n)\right)\end{array}\right)}_{\text{tensor}\equiv V_{\alpha\beta}}\end{array}\right). \quad (136)$$

For tactical reasons we multiply this equation and (135) with f+C_f:

$$0=\left(\begin{array}{c}\underbrace{2\cdot g_{\alpha\beta}\frac{n-1}{n-2}\left(\overbrace{\Delta f+\left(\begin{array}{c}\frac{1}{2(n-1)}\left(n\left(\Gamma^{i}_{dj}g^{dj}+\frac{g^{ci}g_{ab,c}g^{ab}}{2}\right)-2g^{ab}g_{jb,a}g^{ij}\right)\partial_i f\\ -\frac{R(f+C_f)}{(n-1)}\frac{n-2}{4}\end{array}\right)}^{=\left(\Delta+\frac{M^2c^2}{\hbar^2}+V\right)_{KG}(f+C_f)}\right)}\\ \underbrace{+R_{\alpha\beta}(f+C_f)-\frac{(f+C_f)}{2F}\left(\begin{array}{c}F_{,\alpha\beta}(n-2)+F_{,ab}g_{\alpha\beta}g^{ab}\\ +F_{,a}g^{ab}\left(g_{\beta b,\alpha}-g_{\beta\alpha,b}\right)-F_{,\alpha}g^{ab}g_{\beta b,a}-F_{,\beta}g^{ab}g_{\alpha b,a}\\ +F_{,d}g^{cd}\frac{1}{2}n\left(\begin{array}{c}\frac{2}{n}g_{\alpha c,\beta}-g_{\alpha c,\beta}-g_{\beta c,\alpha}\\ +g_{\alpha\beta,c}+\frac{1}{n}g_{\alpha\beta}g_{ab,c}g^{ab}\end{array}\right)\\ -\frac{1}{2F}\left(F_{,\alpha}\cdot F_{,\beta}(3n-6)+g_{\alpha\beta}F_{,c}F_{,d}g^{cd}(4-n)\right)\end{array}\right)}_{\text{tensor}\equiv V_{\alpha\beta}(f+C_f)}\end{array}\right). \quad (137)$$

We see that the "tensor" term in the field equations has no counterpart in the classical Quantum Theory, presented via (123). The latter's correction would subsequently have to consist of a scalar and a tensor term and reads:

$$2\cdot g_{\alpha\beta}\frac{n-1}{n-2}\left[\frac{M^2c^2}{\hbar^2}+V+\Delta\right]_{KG}(f+C_f)+V_{\alpha\beta}(f+C_f)=0. \quad (138)$$

The reader may feel tempted to consider the Laplace operator in the Klein-Gordon equation equal to the one from our field equations, but here we remind that the two can differ in dimensionality, resulting in scalar mass and potential terms. The latter do not need to be added or postulated, but as shown in section "Example: The Everett-Schrödinger Hydrogen Atom" come out of the theory via the entanglement with higher dimensions. We see that the Klein-Gordon equation is been mirrored by either setting the tensorial part to zero, which is to say $V_{\alpha\beta}=0$ and $C_f=0$ or by applying the following approximation:

$$\begin{array}{c}V_{\alpha\beta}\approx 2\cdot g_{\alpha\beta}\frac{n-1}{n-2}V_T\\ \Rightarrow\left(\left[\frac{M^2c^2}{\hbar^2}+V+\Delta\right]_{KG}+V_T\right)(f+C_f)=0\end{array}. \quad (139)$$

This way, and by ignoring the constant term C_f, we would have derived the connection to classical theory.

Now we repeat the evaluation with a different scalar to tensor part separation, thereby pulling all Ricci tensor terms of type $g_{\alpha\beta}$*scalar into the scalar part of the quantum Einstein field equations. As the evaluation is lengthy, we included it in appendix D. The result looks as follows:

$$R^{*}_{\alpha\beta}-\frac{R^{*}}{2}G_{\alpha\beta}=\left(\begin{array}{l}R_{\alpha\beta}-\frac{R}{2}g_{\alpha\beta}+\frac{g_{\alpha\beta}}{2}\frac{F'}{2F}\left((n-2)\left(2g^{ab}f_{,ab}+f_{,d}g^{cd}g^{ab}g_{ab,c}\right)-nf_{,d}g^{cd}g^{ab}g_{ac,b}\right)\\+\frac{(n-2)}{4F^{2}}\left(f_{,\alpha}\cdot f_{,\beta}\left(3(F')^{2}-2FF''\right)+f_{,a}\cdot f_{,b}\frac{g_{\alpha\beta}}{2}g^{ab}\left((F')^{2}(n-7)+4FF''\right)\right)\\-\frac{F'}{2F}\left(\begin{array}{l}f_{,\alpha\beta}(n-2)\\+f_{,a}g^{ab}\left(g_{\beta b,\alpha}-g_{\beta\alpha,b}\right)-f_{,\alpha}g^{ab}g_{\beta b,a}-f_{,\beta}g^{ab}g_{\alpha b,a}\\+f_{,d}g^{cd}\left(g_{\alpha c,\beta}-\frac{1}{2}ng_{\alpha c,\beta}-\frac{1}{2}ng_{\beta c,\alpha}+\frac{1}{2}ng_{\alpha\beta,c}\right)\end{array}\right)\end{array}\right). \tag{140}$$

We see that the nonlinear scalar part, reading:

$$\frac{g_{\alpha\beta}}{2}g^{ab}\left((F')^{2}(n-7)+4FF''\right) \tag{141}$$

would vanish with the following functional wrapper:

$$F[f]=\begin{cases}C_{F}\cdot\left(f+C_{f}\right)^{\frac{4}{n-3}} & n\neq 3\\ C_{F}\cdot e^{f\cdot C_{f}} & n=3\end{cases}. \tag{142}$$

Setting this function into (140) gives:

$$R^{*}_{\alpha\beta}-\frac{R^{*}}{2}G_{\alpha\beta}=\left(\begin{array}{l}\frac{g_{\alpha\beta}}{2}\left(\frac{F'}{2F}\left((n-2)\left(2g^{ab}f_{,ab}+f_{,d}g^{cd}g^{ab}g_{ab,c}\right)-nf_{,d}g^{cd}g^{ab}g_{ac,b}\right)-R\right)\\R_{\alpha\beta}+\frac{(n-2)}{4F^{2}}f_{,\alpha}\cdot f_{,\beta}\left(3(F')^{2}-2FF''\right)\\-\frac{F'}{2F}\left(\begin{array}{l}f_{,\alpha\beta}(n-2)\\+f_{,a}g^{ab}\left(g_{\beta b,\alpha}-g_{\beta\alpha,b}\right)-f_{,\alpha}g^{ab}g_{\beta b,a}-f_{,\beta}g^{ab}g_{\alpha b,a}\\+f_{,d}g^{cd}\left(g_{\alpha c,\beta}-\frac{1}{2}ng_{\alpha c,\beta}-\frac{1}{2}ng_{\beta c,\alpha}+\frac{1}{2}ng_{\alpha\beta,c}\right)\end{array}\right)\end{array}\right), \tag{143}$$

allowing us to compare the field equations with the Klein-Gordon equation (123) as follows:

$$R^*_{\alpha\beta}-\frac{R^*}{2}G_{\alpha\beta}=\begin{pmatrix}\frac{g_{\alpha\beta}}{2}\left(\frac{F'}{2F}\left((n-2)\left(\overbrace{2g^{ab}f_{,ab}+f_{,d}g^{cd}g^{ab}g_{ab,c}}^{-2\Delta F-2F_{,d}g^{cd}{}_{,c}}\right)-nf_{,d}g^{cd}g^{ab}g_{ac,b}\right)-R\right)\\ R_{\alpha\beta}+\frac{(n-2)}{4F^2}f_{,\alpha}\cdot f_{,\beta}\left(3(F')^2-2FF''\right)\\ -\frac{F'}{2F}\begin{pmatrix}f_{,\alpha\beta}(n-2)\\ +f_{,a}g^{ab}\left(g_{\beta b,\alpha}-g_{\beta\alpha,b}\right)-f_{,\alpha}g^{ab}g_{\beta b,a}-f_{,\beta}g^{ab}g_{\alpha b,a}\\ +f_{,d}g^{cd}\left(g_{\alpha c,\beta}-\frac{1}{2}ng_{\alpha c,\beta}-\frac{1}{2}ng_{\beta c,\alpha}+\frac{1}{2}ng_{\alpha\beta,c}\right)\end{pmatrix}\end{pmatrix}$$

$$=\begin{pmatrix}\frac{g_{\alpha\beta}}{2}\left(\frac{F'}{2F}\left((n-2)\left(2\Delta F-2F_{,d}g^{cd}{}_{,c}\right)-nf_{,d}g^{cd}g^{ab}g_{ac,b}\right)-R\right)\\ R_{\alpha\beta}+\frac{(n-2)}{4F^2}f_{,\alpha}\cdot f_{,\beta}\left(3(F')^2-2FF''\right)\\ -\frac{F'}{2F}\begin{pmatrix}f_{,\alpha\beta}(n-2)+f_{,a}g^{ab}\left(g_{\beta b,\alpha}-g_{\beta\alpha,b}\right)-f_{,\alpha}g^{ab}g_{\beta b,a}-f_{,\beta}g^{ab}g_{\alpha b,a}\\ +f_{,d}g^{cd}\left(g_{\alpha c,\beta}-\frac{1}{2}ng_{\alpha c,\beta}-\frac{1}{2}ng_{\beta c,\alpha}+\frac{1}{2}ng_{\alpha\beta,c}\right)\end{pmatrix}\end{pmatrix}. \tag{144}$$

This time the comparison with the Klein-Gordon-Equation reads:

$$R^*_{\alpha\beta}-\frac{R^*}{2}G_{\alpha\beta}=$$

$$\begin{pmatrix}\frac{g_{\alpha\beta}}{2}(n-2)\frac{F'}{F}\left(\overbrace{\left(\Delta f-\left(g^{cd}{}_{,c}+\frac{n}{2(n-2)}g^{cd}g^{ab}g_{ac,b}\right)\partial_d\right)(f+C_f)}^{=\left(\Delta+\frac{M^2c^2}{\hbar^2}+V\right)_{KG}(f+C_f)}-\frac{F}{F'(n-2)}R\right)\\ \underbrace{\begin{pmatrix}R_{\alpha\beta}+\frac{(n-2)}{4F^2}f_{,\alpha}\cdot f_{,\beta}\left(3(F')^2-2FF''\right)\\ -\frac{F'}{2F}\begin{pmatrix}f_{,\alpha\beta}(n-2)+f_{,a}g^{ab}\left(g_{\beta b,\alpha}-g_{\beta\alpha,b}\right)-f_{,\alpha}g^{ab}g_{\beta b,a}-f_{,\beta}g^{ab}g_{\alpha b,a}\\ +f_{,d}g^{cd}\left(g_{\alpha c,\beta}-\frac{1}{2}ng_{\alpha c,\beta}-\frac{1}{2}ng_{\beta c,\alpha}+\frac{1}{2}ng_{\alpha\beta,c}\right)\end{pmatrix}\end{pmatrix}}_{V_{\alpha\beta}}\end{pmatrix} \tag{145}$$

and, when following the tactical reordering from above, we obtain a scalar and a tensorial correction to the classical theory in the form (138) with slightly different constants, namely:

$$2\cdot g_{\alpha\beta}\frac{n-2}{n-3}\left[\frac{M^2c^2}{\hbar^2}+V+\Delta\right]_{KG}(f+C_f)+V_{\alpha\beta}(f+C_f)=0. \tag{146}$$

9.6.1 An Interesting Connection to the Bekenstein Thought Experiment

9.6.1.1 The Bekenstein Bit-Problem

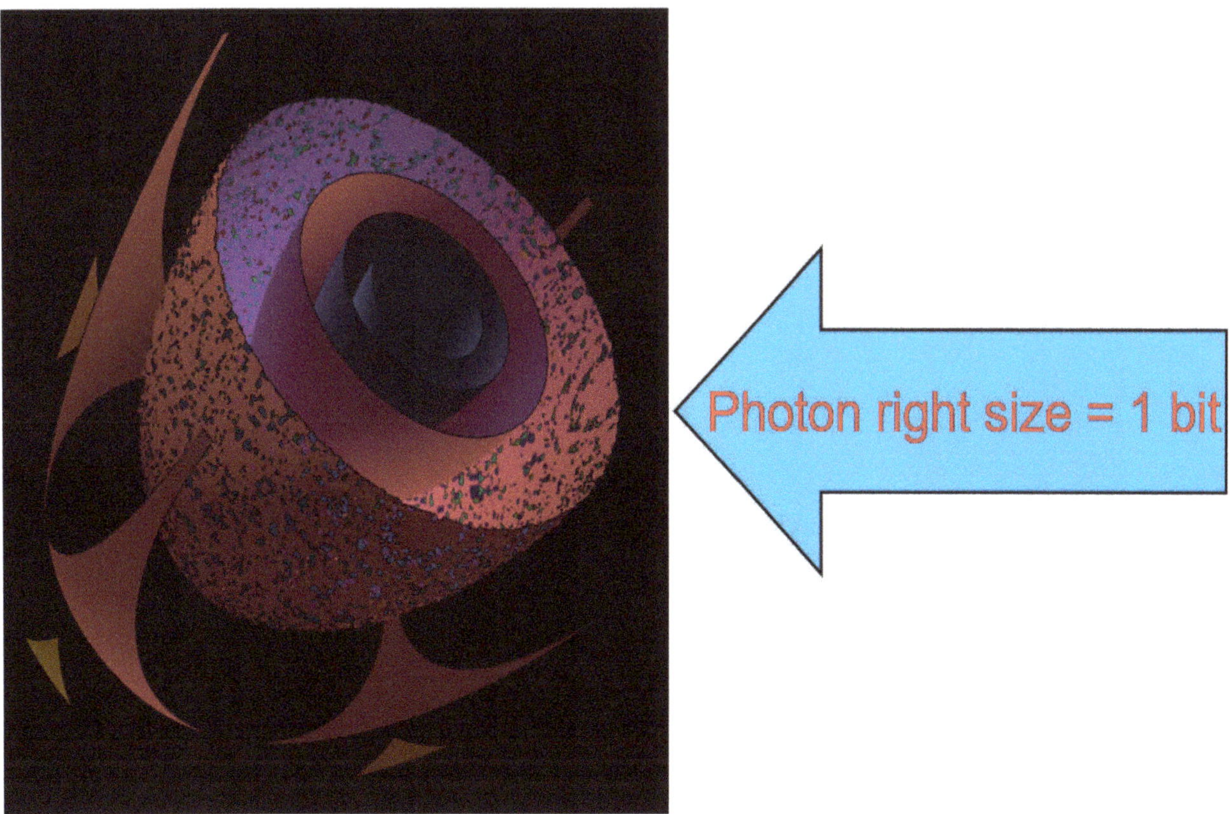

Fig. 31: Illustration of the Bekenstein thought experiment.

In the early seventies J. Bekenstein [104, 105] investigated the connection between black hole surface area and information. Thereby he simply considered the surface change of a black hole which would be hit by a photon just of the same size as the black hole. His idea was that with such a geometric constellation the outcome of the experiment would just consist of the information whether the photon fell into the black hole or whether it did not. Thus, it would be a 1-bit piece of information. His calculations led him to the funny proportionality of area and information. He found that the number of bits, coded by a certain black hole, is proportional to the surface area of this very black hole if measured in Planck area ℓ_P^2. In fact, the dependency how one bit of information changes the area of the black hole (ΔA) reads:

$$\Delta A = 32 \cdot \pi^2 \cdot \ell_P^2 + 64 \cdot \pi^3 \cdot \frac{\ell_P^4}{r_s^2}. \tag{147}$$

Ignoring the extremely small second term, one could just assume our black hole to be constructed of many such bit surface pieces. Thus, we could write:

$$q \cdot \Delta A = q \cdot 32 \cdot \pi^2 \cdot \ell_P^2 = 4 \cdot \pi \cdot r_s^2 \quad \Rightarrow \quad r_s^2 = q \cdot 8 \cdot \pi \cdot \ell_P^2, \tag{148}$$

where r_s gives the radius of the black hole. We see that our black hole radius is proportional to the square root of the bits q thrown into it (for more information see [2, 4]).

Now we want to compare the dependency $r_s[q]$ with the radii $r_{max}[N]$ resulting in maximum volume of n-spheres for a certain number of space-time dimensions N=n+1.

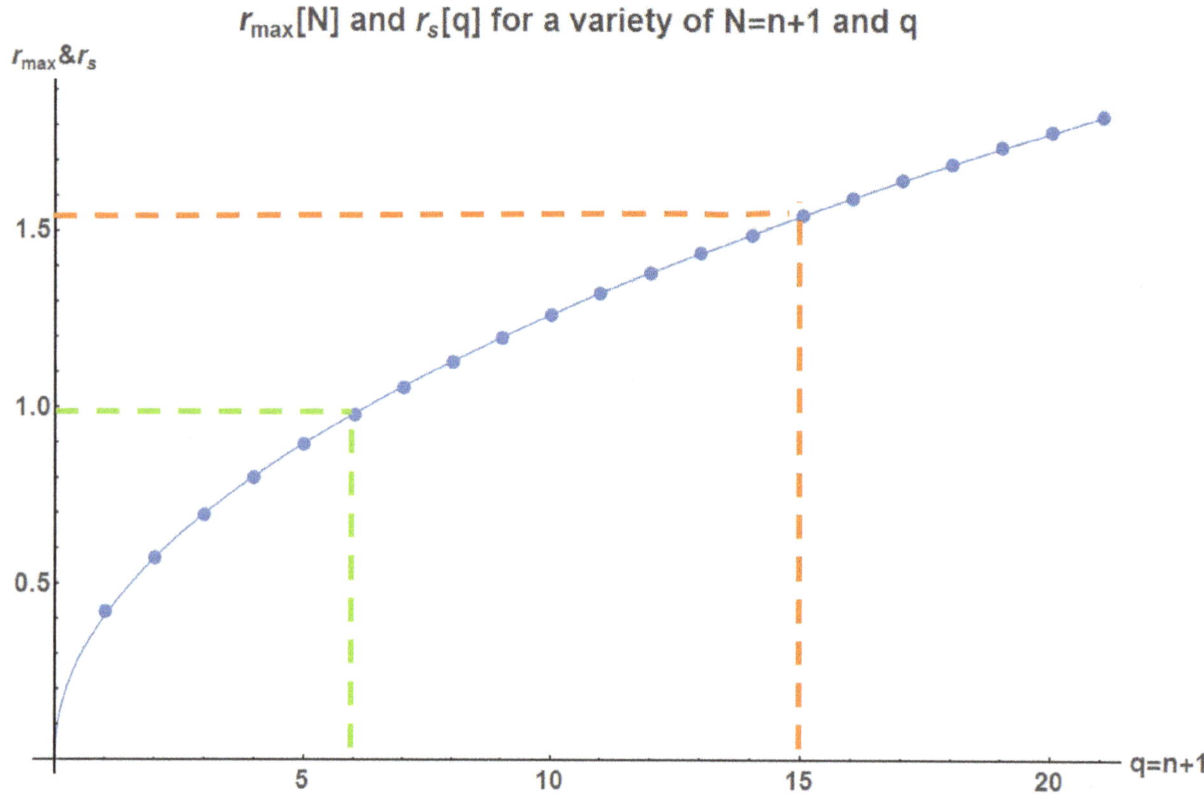

Fig. 32: Radius r_{max} for which at a certain number of dimensions the n-sphere has maximum volume in dependency on N=n+1 compared with the increase of the Schwarzschild radius r_s of a black hole in dependence on the number of bits q thrown into it. We find that q=N=n+1. As examples we pick the situation with a radius slightly bigger than 1.5 (whatever unit). We obtain maximum volume for a sphere in 15 dimensions (orange dotted line). Picking a radius slightly below 1, however, gives us a 6-dimensional sphere which can have maximum volume at such a size (green dotted line).

We find a perfect fit (see figure 32) to the $r_{max}[N]$-dependency for q=N with the following function:

$$r_s = U \cdot \left(0.014948 + 0.3951244 \cdot \sqrt{q}\right); \quad U^2 = 8 \cdot \pi \cdot \ell_P^2, \tag{149}$$

where U denotes a unit-factor which was set U=1 in figure 32.

Our finding does not only connect the intrinsic dimension of a black hole with its mass respectively its surface, but also, at least partially, gives an explanation to the hitherto unsolved problem of "what are the microstates of a black hole giving it temperature and allowing it to store information". According to the evaluation in this section, these microstates are just various states of dimensions realized within the black hole in dependence on the number of bits it contains (and thus, its mass). The bigger the number of bits, the higher the intrinsic dimensions the black hole has. In fact, the connection even is a direct one and only seems to deviate from the simple direct proportionality for very low numbers of masses[12], respectively Schwarzschild radii r_s, respectively numbers of bits q the black hole has swallowed.

9.6.1.2 Connection to the Bekenstein Thought Experiment

Considering the scalarized equation (38), respectively its simpler form (40) in cases of metrics of the type:

$$g_{ij} = \begin{pmatrix} g_{00} & \cdots & 0 \\ \vdots & \ddots & \vdots \\ 0 & \cdots & g_{n-1n-1} \end{pmatrix}; \quad g_{ii,i} = 0 \tag{150}$$

and (81), we see that their comparison with classical theory gives a rather direct connection to the metric picture.

For elaboration we take equation (81) and compare it with the Klein-Gordon equation (123). Substitution of the mass M in (123) by its corresponding Schwarzschild radius via $r_s = \frac{2 \cdot M \cdot G}{c^2}$ yields:

$$\left[\frac{\left(\frac{c^2 \cdot r_s}{2 \cdot G}\right)^2 c^2}{\hbar^2} + V + \Delta\right]\Phi = \left[\frac{r_s^2 \cdot c^6}{4 \cdot G^2 \cdot \hbar^2} + V + \Delta\right]\Phi = \left[\frac{r_s^2}{4 \cdot \ell_P^4} + V + \Delta\right]\Phi = 0. \tag{151}$$

Please note that we here only take the relation of mass to the corresponding Schwarzschild radius of an object of metric (150), which we only chose because of its similarity and closeness to the Klein-Gordon equation. The usage of the relation does not require a metric of Schwarzschild character for our purposes here.

Now we use (148) and obtain:

$$\left[\frac{q \cdot 8 \cdot \pi \cdot \ell_P^2}{4 \cdot \ell_P^4} + V + \Delta\right]\Phi = \left[\frac{q \cdot 2 \cdot \pi}{\ell_P^2} + V + \Delta\right]\Phi = 0. \tag{152}$$

Comparison with (81), thereby assuming high numbers of n and a function F of type (125) (guaranteeing linearity), leads to:

[12] Besides, this deviation is also suggested by the Bekenstein finding summed up in equation (147), where we could assume the second term to become of importance at lower numbers of r_s.

$$0=\lim_{n\to\infty}\left[-\Phi\cdot F^2\cdot R^*\frac{(2-n)}{2\cdot(1-n)}+F\cdot\Delta_g\Phi-F_{,\mu}g^{\mu\nu}\Phi_{,\nu}\right]$$

$$=\lim_{n\to\infty}\left[-\Phi\cdot F^2\cdot\left(R-\frac{F'}{2F}\left(\begin{matrix}2(n-1)\Delta f\\ -f_{,d}\left(\begin{matrix}2(n-1)g^{cd}{}_{,c}\\ +ng^{cd}g^{\alpha\beta}g_{\alpha c,\beta}\end{matrix}\right)\end{matrix}\right)\right)\frac{(2-n)}{2\cdot(1-n)}+F\cdot\Delta_g\Phi-F_{,\mu}g^{\mu\nu}\Phi_{,\nu}\right]$$

$$=\left[-C_F\cdot R+\frac{C_F\cdot}{f+C_f}\left(2\Delta f-f_{,d}\left(2g^{cd}{}_{,c}+g^{cd}g^{\alpha\beta}g_{\alpha c,\beta}\right)\right)+\Delta_g\right]\Phi \quad . \quad (153)$$

$$\Downarrow \qquad\qquad \Downarrow \qquad\qquad \Downarrow$$

$$\left[\frac{q\cdot 2\cdot\pi}{\ell_P^2}\qquad + \qquad V \qquad + \qquad \Delta\right]\Phi=0$$

Thereby, the constant C_F has to have the dimension of 1/length².

It should be noted that in a space or space-time of vanishing Ricci curvature, R=0, the mass term in the classical theory could also be mirrored as follows in the metric picture:

$$\left[-C_F\cdot\left(R-\frac{2}{f+C_f}\cdot\Delta f\right)-\frac{C_F}{f+C_f}\cdot f_{,d}\left(2g^{cd}{}_{,c}+g^{cd}g^{\alpha\beta}g_{\alpha c,\beta}\right)+\Delta_g\right]\Phi=0$$

$$\Downarrow \qquad\qquad \Downarrow \qquad\qquad \Downarrow$$

$$\left[\frac{q\cdot 2\cdot\pi}{\ell_P^2}\qquad + \qquad V \qquad + \qquad \Delta\right]\Phi=0$$

$$\xrightarrow{R=0} \quad . \quad (154)$$

$$\left[-\frac{2\cdot C_F}{f+C_f}\cdot\Delta f-\frac{C_F}{f+C_f}\cdot f_{,d}\left(2g^{cd}{}_{,c}+g^{cd}g^{\alpha\beta}g_{\alpha c,\beta}\right)+\Delta_g\right]\Phi=0$$

$$\Downarrow \qquad\qquad \Downarrow \qquad\qquad \Downarrow$$

$$\left[\frac{q\cdot 2\cdot\pi}{\ell_P^2}\qquad + \qquad V \qquad + \qquad \Delta\right]\Phi=0$$

Thus, the correct allocation ((153) or (154)) cannot be decided on the current basis of knowledge. However, when demanding that there should also be masses in Ricci flat spaces or space-times, then it should definitively be (154). Mass would then be measured in reciprocal Planck areas times the number of dimensions of information the massive objects possess. Please note that the condition of high numbers of n does not necessarily mean that the wave function Φ needs to depend on all of these dimensions. It could just depend on our ordinary 4 dimensions (3 spatial and one time dimension), while only f depends on additional – potentially compactified – dimensions, thereby giving us mass and a potential. In the general case of arbitrary n, equation (154) should read as follows:

$$\left[\underbrace{-\frac{F}{1-n}\cdot\left(R+\frac{1}{f+C_f}\cdot\left(\begin{array}{c}2(n-1)\Delta f\\ -f_{,d}\left(\begin{array}{c}2(n-1)g^{cd}{}_{,c}\\ +ng^{cd}g^{\alpha\beta}g_{\alpha c,\beta}\end{array}\right)\end{array}\right)\right)}_{\frac{q\cdot 2\cdot\pi}{\ell_P^2}\quad+\quad V}+\Delta_g-\frac{4}{(n-2)}\frac{f_{,\mu}g^{\mu\nu}}{f+C_f}\partial_\nu\right]\Phi=0$$

$$\Downarrow\qquad\qquad\Downarrow\qquad\qquad\Downarrow$$

$$\left[\frac{q\cdot 2\cdot\pi}{\ell_P^2}\quad+\quad V\quad+\quad\Delta\quad+\quad ?\right]\Phi=0$$

$$\xrightarrow{R=0}$$

$$\left[\underbrace{F\cdot\left(\frac{1}{f+C_f}\cdot\left(\begin{array}{c}2\Delta f\\ -f_{,d}\left(2g^{cd}{}_{,c}-\frac{n}{n-1}g^{cd}g^{\alpha\beta}g_{\alpha c,\beta}\right)\end{array}\right)\right)}_{\frac{q\cdot 2\cdot\pi}{\ell_P^2}\quad+\quad V}+\Delta_g-\frac{4}{(n-2)}\frac{f_{,\mu}g^{\mu\nu}}{f+C_f}\partial_\nu\right]\Phi=0$$

$$\Downarrow\qquad\qquad\Downarrow\qquad\qquad\Downarrow$$

$$\left[\frac{q\cdot 2\cdot\pi}{\ell_P^2}\quad+\quad V\quad+\quad\Delta\quad+\quad ?\right]\Phi=0\qquad.(155)$$

Knowing that mass can be produced by the term Δf or the term $\Delta_g\Phi$ simply via dimensional entanglement (c.f. [2, 4, 84, 210] or see [56], appendix D), we could now even explain the connection to the parameter q=n+1 (for bigger n) in the classical theory. Thereby the creation of mass can be performed in various ways, of which we will only consider one possibility here.

In the section "Example: The Everett-Schrödinger Hydrogen Atom" further below in this book we simply added dimensions with the right properties, while in [2, 4] we introduced special de Sitter vacuum metrics, which gave mass terms in the corresponding Ricci scalar-based Klein-Gordon equation (40) via entanglement in various dimensions. For completeness, we repeat the main results here in a generalized way.

Introducing a metric for the coordinates t, r, ϑ, φ, u_i, v_i of the form (p=n-4):

$$g^{n}_{\alpha\beta}=\begin{pmatrix} -c^2 & 0 & 0 & 0 & 0 & 0 & & & \\ 0 & 1 & 0 & 0 & 0 & 0 & & & \\ 0 & 0 & r^2 & 0 & 0 & 0 & & \cdots & \\ 0 & 0 & 0 & r^2\cdot\sin[\vartheta] & 0 & 0 & & & \\ 0 & 0 & 0 & 0 & g_1[v_1] & 0 & & & \\ 0 & 0 & 0 & 0 & 0 & g_1[v_1] & & & \\ & & & & & & \ddots & 0 & 0 \\ & & & \vdots & & & 0 & g_p[v_p] & 0 \\ & & & & & & 0 & 0 & g_p[v_p] \end{pmatrix}\cdot f[t,r,\vartheta,\varphi], \quad (156)$$

with the entanglement solution for all v_i as follows:

$$g_i[v_i]=\frac{C_{vi1}^2}{C_{vi0}^2\cdot\left(1+\cosh\left[C_{vi1}\cdot\left(v_i+C_{vi2}\right)\right]\right)}, \quad (157)$$

gives us the following equation from (154) with condition F=const:

$$\begin{aligned}&\left[\underbrace{\sum_{i=1}^{p/2}C_{vi0}^2+\frac{C_F}{n-1}\cdot R}_{\frac{q\cdot 2\cdot\pi}{\ell_P^2}\quad+\quad V}+\frac{\partial^2}{c^2\cdot\partial t^2}-\Delta_{3D\text{-sphere}}\right]\Phi=0\\ &\qquad\qquad\Downarrow\qquad\qquad\qquad\qquad\Downarrow\\ &\left[\frac{q\cdot 2\cdot\pi}{\ell_P^2}\quad+\quad V\quad+\quad\Delta\right]\Phi=0\\ &\qquad\xrightarrow{R=0}\\ &\left[\sum_{i=1}^{p/2}C_{vi0}^2+\frac{\partial^2}{c^2\cdot\partial t^2}-\Delta_{3D\text{-sphere}}\right]\Phi=0\\ &\qquad\Downarrow\qquad\qquad\Downarrow\\ &\left[\frac{q\cdot 2\cdot\pi}{\ell_P^2}\quad+\quad V=0\quad+\quad\Delta\right]\Phi=0\end{aligned}\quad. \quad (158)$$

Here $\Delta_{3D\text{-sphere}}$ denotes the Laplace operator in spherical coordinates.

As for $R^*=0$ and $\sum_{i=1}^{p/2}C_{vi0}^2=\frac{M^2\cdot c^2}{\hbar^2}=\frac{q\cdot 2\cdot\pi}{\ell_P^2}$ (M=mass, c=speed of light in vacuum, ħ=reduced Planck constant, and q=mass creating dimensions), we clearly have obtained the quantum Klein-Gordon equation and we can easily conclude that mass can obviously be obtained (created) by a suitable set of additional, entangled dimensions. Taking the structure from (158), we could further conclude that the more dimensions are "taking part in the mass-creating entanglement", the more mass M we will obtain. In fact, we could write:

$$\forall C_{vi0}^2 \equiv C_{v0}^2$$
$$\Rightarrow \sum_{i=1}^{p/2} C_{vi0}^2 = \frac{p}{2} \cdot C_{v0}^2 = \frac{q \cdot 2 \cdot \pi}{\ell_P^2}. \tag{159}$$
$$\Rightarrow p = q \quad \text{and} \quad C_{v0}^2 = \frac{4 \cdot \pi}{\ell_P^2}$$

Thus, we have found that – taking the mechanism for mass creation as derived here in connection with the comparison of the classical Quantum Theory and the quantum Einstein field equations – mass is created in quanta of $C_{v0}^2 = \frac{4 \cdot \pi}{\ell_P^2}$, which is the unit sphere area measured in Planck surface areas.

Why is this finding above so important?

Well, we work this out via just an ensemble of further questions, namely:

What is it that mass does? We mean, why are there masses being proposed and have been introduced into classical Quantum Theory in the first place? After all, this theory does not even have gravity. Why bother then with masses at all?

Exactly, the reader will probably already have answered the questions above: Mass has been introduced into classical Quantum Theory because masses do not only provide gravity, but also inertia… and this is what they do in the classical quantum equations. They take care about the inertia effect.

And what does inertia do?

Yes, it hinders objects with mass to fly with the speed of light or to accelerate without the presence of a force acting on this very object.

And now – all of a sudden – we have found that mass is just entanglement (of a certain kind, while here just used an entanglement between the coordinates u_i with the v_i and an additional entanglement of all these mass-producing entanglements with our ordinary space-time coordinates t, r, ϑ, φ, we can also have other constellations).

… and what does entanglement usually do? We mean, how is the word "entanglement" normally understood when it is used in ordinary everyday language?

It hinders objects from moving freely, simply because they are entangled with something.

The stronger the entanglement, the more difficult it is for the object to move, isn't that correct?!

Hence, when more dimensional entanglement gives more mass, it is clear why more mass means higher inertia, which is to say, a higher resistance against acceleration. So, it is the entanglement which creates resistance against acceleration and this is described as the concept of mass in the classical theoretical framework. The more dimensional entanglement, the higher the resistance, hence the mass. As the same mechanism, namely the one of dimensional entanglement, also provides the gravitational effect of mass, which we found out during our little excurse into the Bekenstein thought experiment [2, 4, 211], it become also clear why there has to be an equivalence principle between gravitational and inertial mass. This is something Einstein only postulated or used as a basic assumption for and within his General Theory of Relativity. Now we have "properly" derived it from a fundamental law, namely, the Hamilton minimum principle in general space-times and arbitrary dimensionality.

It all makes a lot of sense now, does it not?

Here the attentive reader surely will intervene and tell us that we have only found out what we have put in and what is going to point us towards equation (151), where we have forced the finding for black holes into the classical Klein-Gordon equation (hence the "..." signs for the word "properly" above). In order to truly derive the mass equivalence, we have to consider somewhat more general geometries and fundamental metric solutions to the quantum Einstein field equations. Hence, we at first cover other metrics beyond the Schwarzschild, namely generalized spheres and tori, and then we consider Dirac-like metric solutions to our quantum gravity field equations (29) and compare them with the Dirac theory [27]. In between the two sections we will discuss a few by-products of quite some importance regarding the storage of information and the characteristic fundamentality of immanent scales.

9.6.1.3 Generalization to Spheres with Arbitrary Fundamental Length-Scales L

In the subsections above we saw that, when taking the equation for the Laplace length ℓ_P from the Bekenstein thought experiment, we find the following ratio between r_s (Schwarzschild radius of a black hole) and ℓ_P (c.f. equation (147) and [2, 4, 8, 13, 56, 211]):

$$\frac{r_s}{\ell_P}=2\cdot\sqrt{\pi\cdot\left(q+\sqrt{q\cdot(1+q)}\right)}. \tag{160}$$

Most interestingly, we also found that the solution to the extremal volume problem for a fixed radius r_f for n-spheres results in the same dependency when variating with respect to the number of dimensions of those n-spheres. We obtain (see dots in figure 32) excellent fits when applying an approach like:

$$r_f=L\cdot\sqrt{\pi\cdot\left(n+\sqrt{n\cdot(1+n)}\right)}. \tag{161}$$

Thereby we have the characteristic (system-dependent) length scale L.

This automatically gives us a connection between the size-parameter r_f of any system of spherical symmetry and its theoretical capability to store information. A perfect mathematical n-sphere thereby follows the rule (161) almost perfectly, while other systems may do so only from certain critical sizes onwards, but, nevertheless, we think we can draw the conclusion that the information storage capacity of given systems – if showing enough spherical symmetry – can be extracted from

(161). Then the structural size-parameter r_f determines the number of storable bits n in dependence on the system-inherent length parameter L.

From this, one even may deduce that r_f and L could be substituted by other system characteristics. While in (161) their dimension is length, we should not exclude mass, time, charges, energies, and so on. That being said, one should point out that any factor k could be added to both sides of (161) which would allow us to get back the Planck length ℓ_P:

$$k \cdot r_f = k \cdot L \cdot \sqrt{\pi \cdot \left(n + \sqrt{n \cdot (1+n)}\right)} = \ell_P \cdot \sqrt{\pi \cdot \left(n + \sqrt{n \cdot (1+n)}\right)}. \tag{162}$$

When now also defining a system-inherent correction factor k_f like:

$$k \cdot r_f = k_f \cdot r_s, \tag{163}$$

we just obtain:

$$k_f \cdot r_s = \ell_P \cdot \sqrt{\pi \cdot \left(n + \sqrt{n \cdot (1+n)}\right)}. \tag{164}$$

So, we see that for general spherical systems (and not just black holes), we would obtain the same mass equivalency. There would just be an adjustment factor, depending on the system's structural and inner geometrical characteristics.

9.6.1.4 Generalization to Tori-Ensembles

Generalization to other geometries like tori, and knowing that any system could be constructed out of sums of tori and spheres [13] (see also [56], appendix Q), we could conclude that we now have a metrically derived mass equivalency for gravitational mass and inertia.

The generalization of the dimensional variation:

$$\begin{aligned} &\delta_? W = 0 = \delta_? \int_V d^n x \left(\sqrt{-g} \cdot \Phi_R [R]\right) \\ &\Rightarrow \delta_n W = 0 = \delta_n \int_V d^n x \left(\sqrt{-g} \cdot [\Phi_R [R] = 1]\right) = \delta_n \int_V d^n x \sqrt{-g} \end{aligned} \tag{165}$$

for an ensemble of N tori of dimensions n_j for the sub-n_j-spheres of the individual torus, requires the solution of the following volume integral:

$$W = \int_V d^n x \left(\sqrt{-g}\right) = \prod_{j=1}^{n} V_j (n_j, r_j) = \prod_{j=1}^{n} T[n_j] \cdot \frac{\pi^{\frac{n_j}{2}}}{\Gamma\left[\frac{(n_j+2)}{2}\right]} \cdot (r_j)^{n_j}. \tag{166}$$

This could be generalized further for a sum of tori and leaves us with a great variety of pure volume (radii) and dimension variations.

As tori can be seen as combined n_{ij}-spheres with n_{ij} giving the dimension of the sub-spheres constructing each torus, (165) can be generalized as follows when assuming N tori with dimensions n_i and sub-spheres of dimensions n_{ij}:

$$W=\sum_{i=1}^{N}W_i=\sum_{i=1}^{N}\int_{V_i}d^{n_i}x\left(\sqrt{-g}\right)=\sum_{i=1}^{N}\prod_{j=1}^{n_{ij}}T\left[n_{ij}\right]\cdot\frac{\pi^{\frac{(n_{ij})}{2}}}{\Gamma\left[\frac{(n_{ij}+2)}{2}\right]}\cdot\left(r_{ij}\right)^{n_{ij}}. \tag{167}$$

Assuming that also complex symmetries of systems could be constructed out of sums of tori, we realize that the variational options are manifold:

$$\delta W=0=\delta\left(n_{ij},r_{ij}\right)\sum_{i=1}^{N}\prod_{j=1}^{n_{ij}}T\left[n_{ij}\right]\cdot\frac{\pi^{\frac{(n_{ij})}{2}}}{\Gamma\left[\frac{(n_{ij}+2)}{2}\right]}\cdot\left(r_{ij}\right)^{n_{ij}} \tag{168}$$

and this leaves us with a great variety of possibilities for an optimum sized system in the case of complex symmetries.

We see that all these geometries structurally result in the same sort of equations and guarantee the equivalency of gravitational mass and inertia as piece-wise dimensional entanglement.

9.6.2 By-Products

As we have just seen how very fundamental our findings regarding the Bekenstein experiment also affect systems in general, we here want to point out a few aspects regarding the capacity of storing information and characteristic length of a given system.

9.6.2.1 About the Relativity of System-Scales

We saw that – similar to the Bekenstein or Bekenstein-Hawking problem (figure 31) – we add bits via dimensions to our metric structure (here we considered black holes, generalized spheres, and tori). Assuming our metric system to be a black hole (which we here only used to have a sufficiently simple math), we can even obtain a ratio of the black hole's radius r_s to the smallest structure this black hole can resolve. Taking the result for ℓ_P – the Planck length – from the Bekenstein thought experiment, we find the ratio between the Schwarzschild radius r_s of a black hole and ℓ_P:

$$\frac{r_s}{\ell_P}=2\cdot\sqrt{\pi\cdot\left(q+\sqrt{q\cdot(1+q)}\right)}. \tag{169}$$

In our universe, the Planck length ℓ_P is considered to be the smallest length possible to resolve. What if the ratio (169) we found for black holes actually is a more fundamental law? At any rate, it appears logical to assume that more bits could be coded or stored by an object of bigger size and smaller internal structure, thereby leaving more options to describe something with these structures. Thus, equation (169) makes intuitive sense, but could it also be just the other way around? Could it be that to an object of a given size the number of bits (being equivalent to its dimensions as we see in figures 32) it contains, determines the smallest scale – the Planck length – of the object, too? And, if referring to the third tower T3, which is to say, the "mathematical psychology" in this book, the system presents a thinking entity, does this also mean that thoughts have a physical scale?[13]

From inside and taking the Planck length as measure, the increase of information to this very object, subject or entity would look like an increase of its size. Now assuming the inside of the black hole to be a general system, the inhabitants of this system may see this system as their very own universe and would register the increase of information as a growth of their "universe" measured in the Planck length of that very system-universe. Unfortunately, we cannot "live" inside a generalized sphere or torus, but it is quite reasonable to assume that the number of intrinsic dimensions is not just proportional to the number of bits the system can store or the complexity of "programmable tasks" it may be capable to perform, but also an equivalent to how much matter would be recognized by any entity, which is itself based on such functional structures.

9.6.2.2 Does More Information Always Mean More Mass?[14]

Quantum computer scientists have already pointed out that, with our current way of storing information, we will one day reach a limit with respect to the number of atoms we can apply for the storing process and the energy being needed to keep the information stable (maintained). Citing from the abstract of [91], we have the following situation:

"Currently, we produce $\sim 10^{21}$ digital bits of information annually on Earth. Assuming a 20% annual growth rate, we estimate that after ~350 years from now, the number of bits produced will exceed the number of all atoms on Earth, $\sim 10^{50}$. After ~300 years, the power required to sustain this digital production will exceed 18.5×10^{15} W, i.e., the total planetary power consumption today, and after ~500 years from now, the digital content will account for more than half Earth's mass, according to the mass-energy–information equivalence principle. Besides the existing global challenges such as climate, environment, population, food, health, energy, and security, our estimates point to another singular event for our planet, called information catastrophe."

It must be pointed out that when looking for possible inner Schwarzschild solutions [212], we also found that there are solutions, where the mass decreases with the increase of the object size. It may well be that such strange states are not only realized in black holes, but could perhaps also help to overcome our future information storage problem. With the second tower-PSA technology providing such an astounding similarity to the Bekenstein problem [211], we should at least consider the possibility of constructing an azeotropic memory "stick" (with the stick – of course – being more a bottle).

---

[13] For more, the reader is referred to [56].

[14] This question is of importance for RASA in connection with a variety of further developments planned for and from the authors of this book and has to do with AI and the so-called quantum gravity computer [12, 92].

9.6.2.3 Dirac and Shell Solutions to the Quantum Einstein Field Equations

With the need for Dirac-like metric solutions as pointed out above and the assumption from the introduction section as well as our strategic investigation [211] of interfaces or surfaces playing an important role in a variety of selection and deposition processes, we automatically have to ask ourselves whether there are shell solutions to the quantum Einstein field equations which would help us to better understand the intrinsic behavior of these technologies on a sufficiently fundamental level. To have a starting point, we consider spherical symmetry.

In [2] we have shown that the classical n-dimensional Schwarzschild solution could be applied to construct internally structured n-dimensional black holes, while outside we still have the usual 4-dimensional solution from [79] with the classical Schwarzschild metric. This, however, would not explain how the black hole can code any information.

With the help of the new metric solutions evaluated in [212], namely, to just give an example, in the 3-dimensional case with coordinates t, r, and an angle:

$$g_{\alpha\beta} = C_1 \cdot f[t]^4 \cdot \begin{pmatrix} -c^2 & 0 & 0 \\ 0 & t^2 & 0 \\ 0 & 0 & t^2 \cdot \sin(r)^2 \end{pmatrix}, \qquad f[t] = \sqrt{t^{\pm i \cdot c - 1}} \cdot C_f \tag{170}$$

(please note that r has become an angle while t took over the position of the radius):

$$g_{\alpha\beta} = C_1 \cdot f[t]^4 \cdot \begin{pmatrix} -c^2 & 0 & 0 \\ 0 & \rho^2 & 0 \\ 0 & 0 & \rho^2 \cdot \sin(r)^2 \end{pmatrix} \qquad f[t] = e^{\pm \frac{i \cdot c}{2 \cdot \rho} \cdot t} \cdot C_f \tag{171}$$

(this represents a shell-like object), we want to solve this problem. A generalization of this type of solution is given in appendix N of [56]. Thereby, we found that the Schwarzschild singularity could be avoided (figure 33).

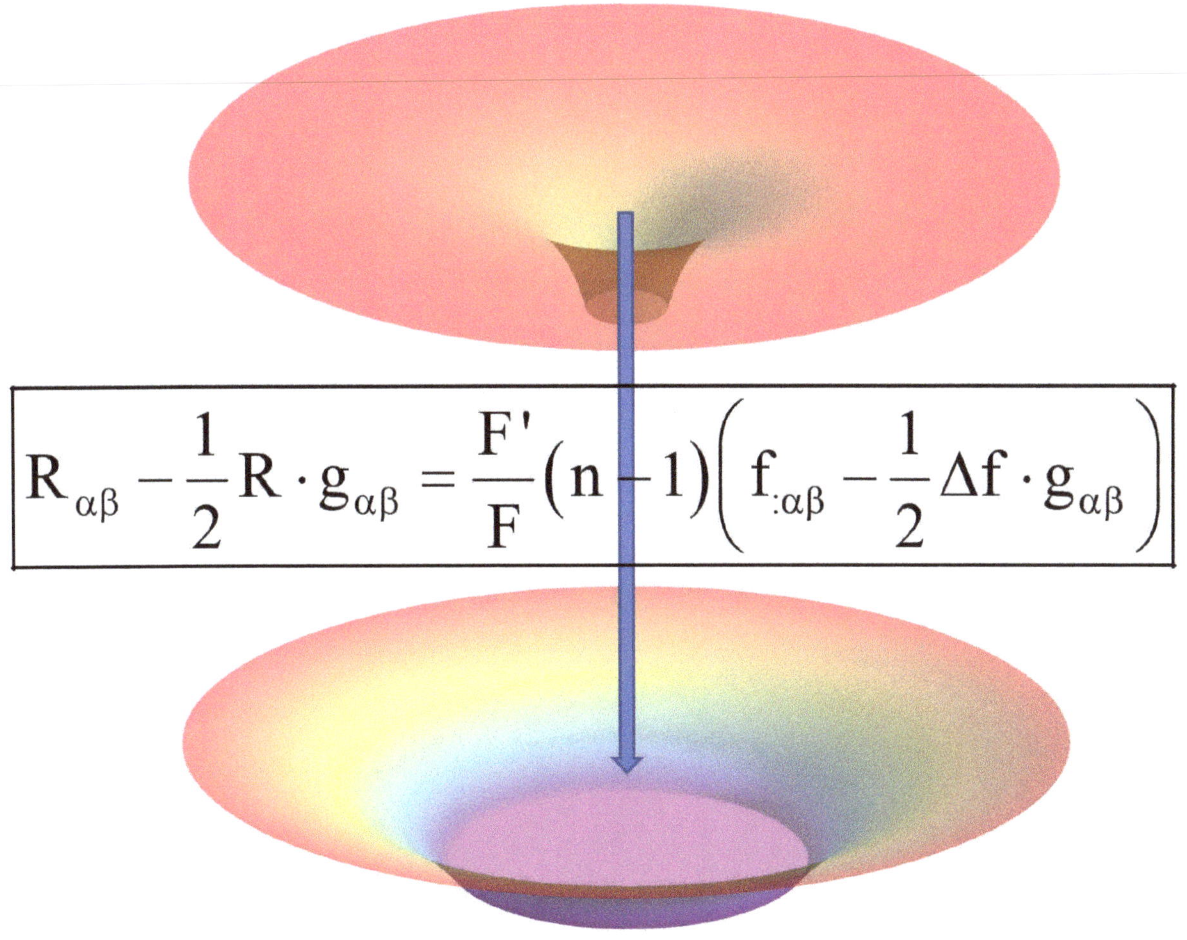

Fig. 33: From the classical Schwarzschild solution to a quantum black hole [212].

At first, however, we should note that also the n-dimensional Schwarzschild solutions from [2], section 3.8 would provide plenty of options to code information, because there are enough degrees of freedom regarding the thicknesses of the individual x-dimensional layers of the onion-like Schwarzschild object, which was proposed there (see also [212]). Similar assumptions could be made for the Robertson-Walker approach introduced in [212], but with respect to T3 and apart from again mentioning the onion-layer structured ogre-mind from the movie Shrek, we will not further consider these possibilities here.

In the case of photonic inner solutions as also suggested in [212], one might assume some kind of standing waves inside the black hole, but as we currently do not have the math to realize such structures, we postpone the investigation of this possibility. The same holds for a similar idea regarding intrinsic photonic, plasmonic, or phonic solutions within our activated agents for the T1- and/or T2-technologies, where the assumption of interface-internal plasmatic phases might allow for a stable existence of such waves.

Thus, we here concentrate on solutions (170) and (171) as potential inner solutions to a black hole. As we see that the parameter ρ clearly is a length, we want to derive its properties. For the general case this was already done in appendix N of [56]. Nevertheless, we repeat it here for the setting (170) and (171). From basic Quantum Theory [27] we know that a particle at rest has the time dependency:

$$f[t]=e^{\pm i\cdot\frac{m\cdot c^2}{\hbar}\cdot t}\cdot C_f, \tag{172}$$

with m giving the rest mass of the particle and $\hbar$ denoting the reduced Planck constant. Comparing with the f[t]-function from the metric solution (171), we find:

$$\frac{m\cdot c}{\hbar}=\frac{1}{2\cdot\rho}. \tag{173}$$

Inserting the Schwarzschild radius $r_s=\frac{2\cdot m\cdot G}{c^2}$ (G… Newton's constant, c… speed of light in vacuum), thereby substituting the rest mass m, leaves us with:

$$\frac{r_s\cdot c^3}{2\cdot\hbar\cdot G}=\frac{1}{2\cdot\rho}\quad\Rightarrow\quad\rho=\frac{1}{r_s}\cdot\left(\frac{c^3}{\hbar\cdot G}\right)^{-1}=\frac{\ell_P^2}{r_s}. \tag{174}$$

Here ℓ_P denotes the Planck length. By inserting (169) into (174) we obtain:

$$\rho=\frac{\ell_P^2}{r_s}=\frac{\ell_P^2}{2\cdot\ell_P\cdot\sqrt{\pi\cdot\left(q+\sqrt{q\cdot(1+q)}\right)}}=\frac{\ell_P}{2\cdot\sqrt{\pi\cdot\left(q+\sqrt{q\cdot(1+q)}\right)}}. \tag{175}$$

Comparing this result – which this time was not extracted from a classical quantum equation in comparison with a GTR-solution, but from a metric, which is solving the quantum Einstein field equations – with our generalized sphere-geometry (164) and reshaping a bit as follows:

$$\rho\cdot r_s=\ell_P^2\Rightarrow\rho\cdot\frac{\ell_P}{k_f}\cdot\sqrt{\pi\cdot\left(n+\sqrt{n\cdot(1+n)}\right)}=\ell_P^2\Rightarrow\rho=\ell_P\cdot\frac{k_f}{\sqrt{\pi\cdot\left(n+\sqrt{n\cdot(1+n)}\right)}}, \tag{176}$$

we see that the inverse dimensional behavior of mass to size can also be extracted for generalized spherical structures.

Thus, while for a black hole the number of bits thrown into it leads to an almost perfectly square-root-like increase of the Schwarzschild radius in accordance with equation (169), the ρ-parameter of the (171)-objects decreases with the number of bits. (171)-objects would have the same ρ-parameter, which we may see as a size, as a black hole only for Schwarzschild radii r_s equal to the Planck length. In other words, for growing black holes with radii bigger than the Planck length the corresponding equally heavy (171)-objects would be significantly smaller than the black holes.

So, we ask: Could the (171)-objects be used as building blocks for the black holes, residing inside it, which is to say behind the event horizon?

Assuming that the black hole's surface is made out of metric spherical objects of the type (171) and further assuming that each of these objects in the surface of the black hole, which is to say at $r=r_s$ (which also happens to be the event horizon), requires its own surface space of something like $C_\rho*\rho^2$, we can directly evaluate the number of such (171)-objects, we from now on name "ρ-spheres", are residing inside the event horizon with increasing numbers of bits thrown into the black hole. Assuming that the mass is always additive, the total mass m of the black hole must then be distributed among the N ρ-spheres, which changes (174) to:

$$\frac{r_s \cdot c^3}{2 \cdot \hbar \cdot G} = \frac{1}{2 \cdot N \cdot \rho} \quad \Rightarrow \quad N \cdot \rho = \frac{1}{r_s} \cdot \left(\frac{c^3}{\hbar \cdot G} \right)^{-1} = \frac{\ell_P^2}{r_s} \quad \Rightarrow \quad \rho = \frac{\ell_P^2}{N \cdot r_s}. \tag{177}$$

Also having to satisfy the following equation for the N ρ-spheres sitting on the surface, we have to solve the following equation:

$$\begin{aligned} & N \cdot C_\rho \cdot \rho^2 = 4 \cdot \pi \cdot r_s^2 \\ \Rightarrow C_\rho \cdot & \frac{\ell_P^2}{4 \cdot \pi \cdot N \cdot \left(q + \sqrt{q \cdot (1+q)} \right)} = (4 \cdot \pi)^2 \cdot \ell_P^2 \cdot \left(q + \sqrt{q \cdot (1+q)} \right). \\ & \Rightarrow N = \frac{1}{64 \cdot \pi^3 \cdot \left(q + \sqrt{q(1+q)} \right)^2} \end{aligned} \tag{178}$$

We realize that such a structure could not be used to store any information, because the number of ρ-spheres should have to increase with the number of bits and not decrease as it does. Things are improving the moment we allow a combination of ρ-spheres and (170)-objects (the latter we shall call “t-spheres”) to make up our inner black hole. We propose the following (simplest of the many possibilities) structure:

A) In the center of the black hole sits a ρ-spheres of “radius-parameter” ρ given in (174) and thus, $\rho = \frac{\ell_P^2}{r_s}$, which is to say, the bigger the Schwarzschild radius r_S of the black hole, the smaller its core. In fact, for infinite masses the core would become a singularity.

B) This single ρ-sphere core is surrounded by t-spheres (170) and the number of those t-spheres which a black hole can bind is proportional to the number of bits the black hole has swallowed.

C) Taking the Bekenstein-condition, this demands an average size for the t-spheres, being bound by the black hole or the black hole’s surface, to be such that its projected surface would be equal to ℓ_P^2. In other words, we could assume the average radius of the t-spheres (the ones bound to the black hole) to be equal to $\ell_P / \sqrt{\pi}$.

With such a structure it is very well possible that in fact black holes have no singularity and follow our scheme of inner-outer-solution, but one cannot detect any difference to the classical Schwarzschild solution from the outside, because the inner parts are always hidden behind the event horizon.

But does this help us to solve the Bekenstein information problem?

Yes, it does.

We can imagine many t-sphere objects (of number N=q) sitting on the surface of the black hole. As the generalized solution to (170) would read:

$$g_{\alpha\beta} = C_1 \cdot f[t]^4 \cdot \begin{pmatrix} -c^2 & 0 & 0 \\ 0 & A^2 \cdot t^2 & 0 \\ 0 & 0 & B^2 \cdot t^2 \cdot \sin(r)^2 \end{pmatrix}, \qquad (179)$$

$$f[t] = \sqrt{t^{\pm\frac{i \cdot c}{A} - 1}} \cdot C_f$$

we see that each t-sphere could not only store information via a certain sign within the exponent, but also via the free parameter B.

9.6.2.4 The Singularity-Free Black Hole

The reader may proof that in any dimension n we always obtain the same solution for the quantum Einstein field equations for the function f[t] with metrics of n-spherical shape as follows:

$$G_{\alpha\beta} = f[t] \cdot \begin{pmatrix} -c^2 & 0 & 0 & 0 & 0 & 0 \\ 0 & \rho^2 & 0 & 0 & 0 & 0 \\ 0 & 0 & \rho^2 \cdot \sin(r)^2 & 0 & 0 & 0 \\ 0 & 0 & 0 & \rho^2 \cdot \sin(r)^2 \cdot \sin(\vartheta)^2 & 0 & 0 \\ 0 & 0 & 0 & 0 & \ddots & 0 \\ 0 & 0 & 0 & 0 & 0 & g_{n-1n-1} \end{pmatrix}. \qquad (180)$$

$$f[t] = e^{\pm 2 \cdot \frac{i \cdot c}{\rho} \cdot t} \cdot C_f; \quad g_{n-1n-1} = \rho^2 \cdot \sin(r)^2 \cdot \prod_{i=0}^{n-4} \sin(\vartheta_i)^2$$

Now knowing that with increasing mass the shell radius shrinks, we can easily assume this metric to present the inner solution of our black hole, while to the outer world (at its event horizon $r=r_S$) the object just shows its outer Schwarzschild metric in accordance with [79]. Such an object would have all the characteristics of a black hole, but no singularity at r=0.

9.6.2.5 The Size of the Electron?

Applying (177) and assuming a ρ-sphere-structure for the electron, gives us:

$$\frac{r_s \cdot c^3}{2 \cdot \hbar \cdot G} = \frac{1}{2 \cdot N \cdot \rho} \quad \Rightarrow \quad N \cdot \rho = \frac{1}{r_s} \cdot \left(\frac{c^3}{\hbar \cdot G} \right)^{-1} = \frac{\ell_P^2}{r_s}$$
$$\Rightarrow \quad \rho = \frac{\ell_P^2}{N \cdot r_s} = \frac{1.93 \times 10^{-13}}{N} \text{meter} \qquad (181)$$

Setting N=1 we would end up with a ρ-sphere of $\rho = 1.93 \times 10^{-13}$ meter for the “pure” or “naked” electron. In contrast to any Schwarzschild or Kerr black hole approach, where we would obtain electron sizes below the Planck length, our result here represents a quite reasonable size.

9.7 The Conventional and the Unconventional Physics Realm

9.7.1 The Conventional Physics Realm

(partially from [4], chapter 11)

From the material presented in the previous sections one might feel tempted to conclude that the scale factors to every metric may be interpreted as the soul of objects or – as this would be of special interest here – as a mathematical tool for the switch to another world in the sense of Everett's approach. That things are not so simple can be demonstrated by the fact that in our 4-dimensional space-time the corresponding scale factor provides the electromagnetic field as it was shown in section 2 of [2] and has been extended in chapter 2 of [4]. So, our question is: how can we come to a gravity quantum description which embeds more dimensions with metric and quantum fields? The solution to this problem can be found via the following approach:

$$G_{\alpha\beta} = F_n\left[f_n\right]\begin{pmatrix} F_{n-1}\cdot\ldots\cdot F_2\cdot g_{00} & F_{n-1}\cdot\ldots\cdot F_2\cdot g_{10} & \cdots & \cdots & \cdots & g_{n0} \\ F_{n-1}\cdot\ldots\cdot F_2\cdot g_{10} & F_{n-1}\cdot\ldots\cdot F_2\cdot g_{11} & \cdots & \cdots & \cdots & \vdots \\ F_{n-1}\cdot\ldots\cdot F_3\cdot g_{20} & F_{n-1}\cdot\ldots\cdot F_3\cdot g_{21} & \ddots & \cdots & \cdots & \vdots \\ F_{n-1}\cdot\ldots\cdot F_4\cdot g_{30} & F_{n-1}\cdot\ldots\cdot F_4\cdot g_{31} & \cdots & \ddots & \cdots & \vdots \\ \cdots & \cdots & \cdots & \cdots & \ddots & \vdots \\ g_{n0} & \cdots & \cdots & \cdots & \cdots & g_{nn} \end{pmatrix}. \qquad (182)$$

$$F_i = F_i\left[f_i\right]$$

Using the definition:

$$\begin{aligned} {}^2G_{\alpha\beta} &= F_2\cdot{}^2g_{\alpha\beta} \\ {}^3G_{\alpha\beta} &= F_3\cdot{}^3g_{\alpha\beta} \\ &\vdots \\ {}^nG_{\alpha\beta} &= F_n\cdot{}^ng_{\alpha\beta} \end{aligned}, \qquad (183)$$

we can write (182) as follows:

$$
{}^{2}G_{\alpha\beta} = \left(F_2 \cdot \left({}^{2}g_{\alpha\beta}\right)\right) = F_2 \cdot \begin{pmatrix} {}^{2}g_{00} & {}^{2}g_{01} \\ {}^{2}g_{01} & {}^{2}g_{11} \end{pmatrix}
$$

$$
{}^{3}G_{\alpha\beta} = F_3 \cdot {}^{3}g_{\alpha\beta} = F_3 \cdot \begin{pmatrix} {}^{3}g_{00} & {}^{3}g_{01} & {}^{3}g_{02} \\ {}^{3}g_{01} & {}^{3}g_{11} & {}^{3}g_{12} \\ {}^{3}g_{02} & {}^{3}g_{12} & {}^{3}g_{22} \end{pmatrix} = F_3 \cdot \begin{pmatrix} F_2 \cdot \begin{pmatrix} {}^{2}g_{00} & {}^{2}g_{01} \\ {}^{2}g_{01} & {}^{2}g_{11} \end{pmatrix} & \begin{matrix} {}^{3}g_{02} \\ {}^{3}g_{12} \end{matrix} \\ \begin{matrix} {}^{3}g_{02} & {}^{3}g_{12} \end{matrix} & {}^{3}g_{22} \end{pmatrix}
$$

$$
\vdots \tag{184}
$$

$$
{}^{n}G_{\alpha\beta} = F_n \cdot \begin{pmatrix} \begin{pmatrix} \cdots F_3 \cdot \begin{pmatrix} F_2 \cdot \begin{pmatrix} {}^{2}g_{00} & {}^{2}g_{01} \\ {}^{2}g_{01} & {}^{2}g_{11} \end{pmatrix} & \begin{matrix} {}^{3}g_{02} \\ {}^{3}g_{12} \end{matrix} \\ \begin{matrix} {}^{3}g_{02} & {}^{3}g_{12} \end{matrix} & {}^{3}g_{22} \end{pmatrix} \cdots \\ \cdots \end{pmatrix} & \begin{matrix} {}^{n}g_{0m} \\ \vdots \\ \vdots \\ \vdots \end{matrix} \\ {}^{n}g_{0m} \quad \cdots \quad \cdots \quad \cdots \quad \cdots & {}^{n}g_{mm} \end{pmatrix}; \quad m = n-1
$$

and obtain the corresponding multi-scaled Ricci scalar to be:

$$
\begin{gathered}
R^{*} = \sum_{m=1}^{n-1}\left(\sum_{\alpha=0}^{m} 2 \cdot {}^{m+1}R_{m\alpha} \, {}^{m+1}g^{m\alpha} - {}^{m+1}R_{mm} \, {}^{m+1}g^{mm}\right) + \sum_{m=2}^{n} \frac{F_m{}'}{F_m} \cdot \Delta_m f_m \\
\Delta_m f = \frac{1}{\sqrt{{}^{m}g}} \sum_{\alpha,\beta=0}^{m-1} \partial_\beta\left(\sqrt{{}^{m}g} \cdot {}^{m}g^{\alpha\beta} \cdot \partial_\alpha f\right)
\end{gathered} \tag{185}
$$

Thereby the ${}^{j}R_{\alpha\beta}$ are denoting the Ricci tensor to the ${}^{j}g_{\alpha\beta}$ metric and all functions F_i must fulfill the condition:

$$
\begin{gathered}
4 \cdot F_i F_i{}'' + F_i{}' F_i{}' (n-6) = 0 \;\; \Rightarrow \;\; F_i = \left(C_f \pm f^a\right)^{\frac{4}{n-2}}; \;\; n > 2 \\
4 \cdot F_2 F_2{}'' + F_2{}' F_2{}' (n-6) = 0 \;\; \Rightarrow \;\; F_2 = C_F \cdot e^{C_f \cdot f}; \;\; n = 2
\end{gathered} \tag{186}
$$

Most interestingly, we obtain a sum of Laplace operators in various dimensions on the scale-factors (wave functions) f plus curvature terms. Thus, what originally (on first sight) has been a typical GTR-task, namely the evaluation of the Ricci tensor and the subsequent Ricci scalar, evolves to a complex ensemble of quantum equations if only be brought into the right form. It should explicitly be pointed out that we can also construct string, brane, multi-verse, and similar structures. Here we just give a few examples:

$$G_{\alpha\beta} = {}_nF\left[f_n\right]\begin{pmatrix} {}_2F_1\left[f_1\right]\cdot {}^2_1g_{\alpha\beta} & {}^20 & \cdots & \cdots & \cdots & {}^20 \\ {}^20 & {}_2F_2\left[f_2\right]\cdot {}^2_2g_{\alpha\beta} & \cdots & \cdots & \cdots & \vdots \\ {}^20 & {}^20 & \ddots & \cdots & \cdots & \vdots \\ {}^20 & {}^20 & \cdots & \ddots & \cdots & \vdots \\ \cdots & \cdots & \cdots & \cdots & \ddots & \vdots \\ {}^20 & \cdots & \cdots & \cdots & \cdots & {}^2_{n/2}g_{\alpha\beta} \end{pmatrix}, \quad (187)$$

$${}^2_ig_{\alpha\beta} = \begin{pmatrix} {}_ig_{00} & {}_ig_{01} \\ {}_ig_{01} & {}_ig_{11} \end{pmatrix}; \quad {}_2F_i\left[f_i\right] = C_{f1i}\cdot e^{C_{fi}\cdot f_i} = C_{f1i}\cdot e^{C_{fi}\cdot f_i\left[x_i, y_i\right]}; \quad {}^20 = \begin{pmatrix} 0 & 0 \\ 0 & 0 \end{pmatrix}$$

$$G_{\alpha\beta} = {}_nF\left[f_n\right]\begin{pmatrix} {}_3F_1\left[{}_3f_1\right]\cdot {}^3_1g_{\alpha\beta} & {}^20 & \cdots & \cdots & \cdots & {}^30 \\ {}^30 & {}_3F_2\left[{}_3f_2\right]\cdot {}^3_2g_{\alpha\beta} & \cdots & \cdots & \cdots & \vdots \\ {}^30 & {}^30 & \ddots & \cdots & \cdots & \vdots \\ {}^30 & {}^30 & \cdots & \ddots & \cdots & \vdots \\ \cdots & \cdots & \cdots & \cdots & \ddots & \vdots \\ {}^30 & \cdots & \cdots & \cdots & \cdots & {}^3_{n/3}g_{\alpha\beta} \end{pmatrix}$$

$${}^3_ig_{\alpha\beta} = \begin{pmatrix} {}_2F_i\left[{}_2f_i\right]\cdot\begin{pmatrix} {}_ig_{00} & {}_ig_{01} \\ {}_ig_{01} & {}_ig_{11} \end{pmatrix} & {}_ig_{02} \\ & {}_ig_{11} \\ {}_ig_{02} \quad {}_ig_{21} & {}_ig_{22} \end{pmatrix}; \quad {}_2F_i\left[{}_2f_i\right] = C_{f1i}\cdot e^{C_{fi}\cdot {}_2f_i} = C_{f1i}\cdot e^{C_{fi}\cdot {}_2f_i\left[x_i, y_i\right]}, \quad (188)$$

$${}_3F_i\left[{}_3f_i\right] = C_{f1i}\cdot\left(C_{fi} + {}_3f_i\left[t_i, x_i, y_i, z_i\right]\right)^4; \quad {}^30 = \begin{pmatrix} 0 & 0 & 0 \\ 0 & 0 & 0 \\ 0 & 0 & 0 \end{pmatrix}$$

$$G_{\alpha\beta} = {}_{n}F[f_n]\begin{pmatrix} {}_4F_1[f_1]\cdot {}_1^4g_{\alpha\beta} & {}^40 & \cdots & \cdots & \cdots & {}^40 \\ {}^40 & {}_4F_2[f_2]\cdot {}_2^4g_{\alpha\beta} & \cdots & \cdots & \cdots & \vdots \\ {}^40 & {}^40 & \ddots & \cdots & \cdots & \vdots \\ {}^40 & {}^40 & \cdots & \ddots & \cdots & \vdots \\ \cdots & \cdots & \cdots & \cdots & \ddots & \vdots \\ {}^40 & \cdots & \cdots & \cdots & \cdots & {}_{n/4}^{4}g_{\alpha\beta} \end{pmatrix}$$

$${}_i^4g_{\alpha\beta} = \begin{pmatrix} {}_2F_{i1}[{}_2f_{i1}]\cdot\begin{pmatrix} {}_ig_{00} & {}_ig_{01} \\ {}_ig_{01} & {}_ig_{11} \end{pmatrix} & {}^20 \\ {}^20 & {}_2F_{i2}[{}_2f_{i2}]\cdot\begin{pmatrix} {}_ig_{00} & {}_ig_{01} \\ {}_ig_{01} & {}_ig_{11} \end{pmatrix} \end{pmatrix}; \quad . \qquad (189)$$

$${}_4F_i[f_i] = C_{f1i}\cdot\left(C_{fi} + f_i[t_i, x_i, y_i, z_i]\right)^2; \quad {}^40 = \begin{pmatrix} 0 & 0 & 0 & 0 \\ 0 & 0 & 0 & 0 \\ 0 & 0 & 0 & 0 \\ 0 & 0 & 0 & 0 \end{pmatrix}$$

When evaluating the Ricci scalar, we obtain – just as shown above with the Matryoshka structure from (184) – sums of Laplace operators applied on the various wave functions appearing in (187) to (189). Almost arbitrary combinations of scaled sub-metrics in accordance with the examples given above is possible.

Examples and discussion are presented elsewhere [31].

With this new multitude of options, however, the number of degrees of freedom does not end. As elaborated in [2] and [32], there is also the option of centers of gravity and centers of wave functions, which simply means that we can seek for a satisfaction of (6) via variation with respect to the position of such centers. Seeing that, the last line of (184) would become generalized as follows:

$${}^nG_{\alpha\beta} = F_n[X_n]\cdot\begin{pmatrix} \begin{pmatrix} \cdots F_3[X_3]\cdot\begin{pmatrix} F_2[X_2]\cdot\begin{pmatrix} {}^2g_{00} & {}^2g_{01} \\ {}^2g_{01} & {}^2g_{11} \end{pmatrix} & \begin{matrix} {}^3g_{02} \\ {}^3g_{12} \end{matrix} \\ \begin{matrix} {}^3g_{02} & {}^3g_{12} \end{matrix} & {}^3g_{22} \end{pmatrix} \cdots \\ \cdots \end{pmatrix} & \begin{matrix} {}^ng_{0m} \\ \vdots \\ \vdots \\ \vdots \end{matrix} \\ {}^ng_{0m} \quad \cdots \quad \cdots \quad \cdots \quad \cdots & {}^ng_{mm} \end{pmatrix}, \qquad (190)$$

$$X_i = x_i - \xi_i; \quad {}^ig_{\alpha\beta} = {}^ig_{\alpha\beta}[Y_i] = {}^ig_{\alpha\beta}[y_i - \zeta_i]$$

we realize a huge number of possible combinations, all fulfilling (6) and many having the possibility to code not just matter, dark matter, energy, but also objects the common language might better express with the words “chaos”, “divine”, and “soul”.

For suitably chosen coordinates we should be able to avoid terms like:

$$\sum_{m=1}^{n-1}\left(\sum_{\alpha=0}^{m}2\cdot{}^{m+1}R_{m\alpha}\ {}^{m+1}g^{m\alpha}-{}^{m+1}R_{mm}\ {}^{m+1}g^{mm}\right) \tag{191}$$

in (185) and subsequently receive equations as follows:

$$\begin{aligned}R^{*}&=\sum_{m=2}^{n}\frac{F_{m}'}{F_{m}}\cdot\Delta_{m}f_{m}\\ \Delta_{m}f&=\frac{1}{\sqrt{{}^{m}g}}\sum_{\alpha,\beta=0}^{m-1}\partial_{\beta}\left(\sqrt{{}^{m}g}\cdot{}^{m}g^{\alpha\beta}\cdot\partial_{\alpha}f\right)\end{aligned}. \tag{192}$$

We realize that with suitable eigenvalue solutions for one function f_i, we might also obtain potentials, masses, and other forms of inertia. Thereby, we also assume that we are able to restrict the coordinate choice and the variation to forms where we have (80) fulfilled and are just left with the scalar quantum Einstein field equation.

With the classical Dirac approach for the "factorization" of the Laplace operator, we should also be able to force fermionic behavior into (192), but there is also another such possibility via a proper choice of the functions F_i. With the results from our previous publications [33 – 47], we can extract the option for the wrapper functions F_i as described in [46, 47]. Instead of a sum of various Laplace operators, we would then obtain the following equation:

$$\begin{aligned}R^{*}&=\sum_{m=2}^{n}\frac{F_{m}'}{F_{m}}\cdot\left(f_{m}\Delta_{m}f_{m}-f_{m,\alpha}f_{m,\beta}\ {}^{m}g^{\alpha\beta}\right)=0\\ \Delta_{m}f&=\frac{1}{\sqrt{{}^{m}g}}\sum_{\alpha,\beta=0}^{m-1}\partial_{\beta}\left(\sqrt{{}^{m}g}\cdot{}^{m}g^{\alpha\beta}\cdot\partial_{\alpha}f\right)\end{aligned}. \tag{193}$$

The solution can be found via the "Dirac" linearization as demonstrated in [46, 47]. Naturally, we could also have (via various choices of the functional wrappers F_i) mixed equations as follows:

$$\begin{aligned}R^{*}&=\sum_{\forall i}\frac{F_{i}'}{F_{i}}\left(n_{i}-1\right)\left(f_{i}\Delta_{i}f_{i}-f_{i,\alpha}f_{i,\beta}\ {}^{i}g^{\alpha\beta}\right)+\sum_{\forall j}\frac{F_{j}'}{F_{j}}\left(n_{j}-1\right)\left(f_{j}\Delta_{j}f_{j}\right)=0\\ \Delta_{i}f&=\frac{1}{\sqrt{{}^{i}g}}\sum_{\alpha,\beta=0}^{i-1}\partial_{\beta}\left(\sqrt{{}^{i}g}\cdot{}^{i}g^{\alpha\beta}\cdot\partial_{\alpha}f\right);\quad \Delta_{j}f=\frac{1}{\sqrt{{}^{j}g}}\sum_{\alpha,\beta=0}^{j-1}\partial_{\beta}\left(\sqrt{{}^{j}g}\cdot{}^{j}g^{\alpha\beta}\cdot\partial_{\alpha}f\right)\end{aligned}. \tag{194}$$

We see that in the realm of conventional physics, which is to say, the physics following from (29), we already have a huge variety of possibilities to place (and potentially hide) something in there, which could be the residual of a formerly living entity.

9.7.2 The Unconventional Physics Realm

Now we simply propose that there are other solutions not covered by (29) but still following from the Einstein-Hilbert action of ensembles of arbitrary properties and dimensionality. For reasons of classical perception, we name such solutions "chaos" (or pre-perspective pre-order) and suggest that entities, consisting of ensembles of information and born in the conventional physics realm, could be transferred to the chaos and live on "in there". Thus, we here only need to derive the mathematical options for this chaos.

Thereby it should explicitly be pointed out that the expression "unconventional" only refers to the fact that the usual – everyday – physics does not consider solutions to the variational problem of the Einstein-Hilbert action any different than (29) (not in principle anyway, but here or there in connection with different integral kernels, like functions of R and so on). Here we want to consider all other options.

9.7.2.1 What Could Be the Chaos?

Again, our starting point should be the Hamilton extremal principle and its manifestation for arbitrary Riemann attribute ensembles in form of the Einstein-Hilbert action [1]:

$$\delta W = 0 = \delta \int_V d^n x \left(\sqrt{-g} \cdot \left(R - 2\Lambda + L_M \right) \right). \tag{195}$$

Generalization (scaling) of the metric as follows [2, 5]:

$$G_{\alpha\beta} = g_{\alpha\beta} \cdot F[f], \tag{196}$$

brings us to (here only simplest metrics, while general ones are considered in the sections above; see also [4]):

$$\delta W = 0 = \delta \int_V d^n x \left(\sqrt{-g \cdot F^n} \times \left(\left(\begin{array}{c} R + \frac{F'}{F}(1-n)\Delta f \\ + \frac{f_{,\alpha} f_{,\beta} g^{\alpha\beta}(1-n)}{4F^2} \left(4FF'' + (F')^2 (n-6) \right) \end{array} \right) \frac{1}{F} \\ -2\Lambda + L_M \right) \right). \tag{197}$$

Performing the variation yields:

$$
0=\sqrt{-g\cdot F^{n}}\times\left(\begin{array}{c}
\boxed{R_{\alpha\beta}-\frac{g_{\alpha\beta}}{2}R+\kappa T_{\alpha\beta}+\Lambda\cdot g_{\alpha\beta}}\\
+\frac{g_{\alpha\beta}}{2}\left(\begin{array}{c}
\frac{1}{2F}\left(\begin{array}{c}
2F_{,ij}(n-1)g^{ij}+2\Gamma_{ij}^{a}F_{,a}g^{ij}\\
-F_{,i}g^{ab}g_{jb,a}g^{ij}-F_{,j}g^{ab}g_{ib,a}g^{ij}\\
-n\Gamma_{ij}^{d}F_{,d}g^{ij}+\frac{n}{2}F_{,d}g^{cd}g_{ab,c}g^{ab}
\end{array}\right)\\
+\frac{F_{,i}\cdot F_{,j}}{4F^{2}}g^{ij}\left((n-6)(n-1)\right)
\end{array}\right)\\
-\frac{1}{2F}\left(\begin{array}{c}
F_{,\alpha\beta}(n-2)+F_{,ab}g_{\alpha\beta}g^{ab}\\
+F_{,a}g^{ab}\left(g_{\beta b,\alpha}-g_{\beta\alpha,b}\right)-F_{,\alpha}g^{ab}g_{\beta b,a}-F_{,\beta}g^{ab}g_{\alpha b,a}\\
+F_{,d}g^{cd}\frac{1}{2}n\left(\begin{array}{c}
\frac{2}{n}g_{\alpha c,\beta}-g_{\alpha c,\beta}-g_{\beta c,\alpha}\\
+g_{\alpha\beta,c}+\frac{1}{n}g_{\alpha\beta}g_{ab,c}g^{ab}
\end{array}\right)\\
-\frac{1}{2F}\left(F_{,\alpha}\cdot F_{,\beta}(3n-6)+g_{\alpha\beta}F_{,c}F_{,d}g^{cd}(4-n)\right)
\end{array}\right)
\end{array}\right)\delta G^{\alpha\beta}\quad , \tag{198}
$$

where the red rectangle gives the classical Einstein field equations [3], while the rest describes what we know as Quantum Theory… hence, we have a Quantum Gravity Theory.

Now, the usual path forward is to just make the term in parentheses to zero, which – as it was shown in the chapters above – gives the classical field equations in the presence of quantum effects. This leads to all the physics we know, including General Theory of Relativity, Quantum Theory, and even thermodynamics when considering ensembles of quantum and gravity centers [2, 4].

But this is only half of the truth… in fact it is only one third, because we also have the option of the terms G ($G=g*F^n$… determinant of the scaled metric) and the variation of the scaled metric:

$$
\delta G^{\alpha\beta} \tag{199}
$$

being zero in order to fulfill the whole extremal principle. At first, for us only the option of g=0 is of interest, because in comparison to our classical physics, which is based on:

$$0=\left(\begin{array}{c}\boxed{R_{\alpha\beta}-\frac{g_{\alpha\beta}}{2}R+\kappa T_{\alpha\beta}+\Lambda\cdot g_{\alpha\beta}}\\ +\frac{g_{\alpha\beta}}{2}\left(\begin{array}{c}\frac{1}{2F}\left(\begin{array}{c}2F_{,ij}(n-1)g^{ij}+2\Gamma_{ij}^{a}F_{,a}g^{ij}\\ -F_{,i}g^{ab}g_{jb,a}g^{ij}-F_{,j}g^{ab}g_{ib,a}g^{ij}\\ -n\Gamma_{ij}^{d}F_{,d}g^{ij}+\frac{n}{2}F_{,d}g^{cd}g_{ab,c}g^{ab}\end{array}\right)\\ +\frac{F_{,i}\cdot F_{,j}}{4F^{2}}g^{ij}\left((n-6)(n-1)\right)\end{array}\right)\\ -\frac{1}{2F}\left(\begin{array}{c}F_{,\alpha\beta}(n-2)+F_{,ab}g_{\alpha\beta}g^{ab}\\ +F_{,a}g^{ab}\left(g_{\beta b,\alpha}-g_{\beta\alpha,b}\right)-F_{,\alpha}g^{ab}g_{\beta b,a}-F_{,\beta}g^{ab}g_{\alpha b,a}\\ +F_{,d}g^{cd}\frac{1}{2}n\left(\frac{2}{n}g_{\alpha c,\beta}-g_{\alpha c,\beta}-g_{\beta c,\alpha}+g_{\alpha\beta,c}+\frac{1}{n}g_{\alpha\beta}g_{ab,c}g^{ab}\right)\\ -\frac{1}{2F}\left(F_{,\alpha}\cdot F_{,\beta}(3n-6)+g_{\alpha\beta}F_{,c}F_{,d}g^{cd}(4-n)\right)\end{array}\right)\end{array}\right), \tag{200}$$

this is something completely different, quasi a paradigm shift and compared to what we know from our "daily" cognitive experience "chaos" [49].

Thus, there may well have been a "time" before "time" in which this "g=0-chaos" reigned and which then somehow switched into the other, the ordinary physics option and created what we recognize as our universe.

The fact that the condition g=0 also demands the universe to have no volume and that – according to the great theoreticians – the big bang is considered to have started with a tiny (even singular) universe, kind of even supports this idea of a "g=0-chaos before time".

However, we should not forget the other option being:

$$\delta G^{\alpha\beta}=0. \tag{201}$$

This also turns out to be a possibility, as we are going to show further below.

Here we have an apparently lazy universe, with not much mood for any global variational dynamics.

There would only be an intrinsic activity.

We cannot exclude the universe (or its creator) having awoken from such a state or even being an intermediate option in a timely and/or spatial aspect.

But as such a situation also allows for an extremely strange physics, we might just also consider this as a "chaos" state and thus, "prove" the old genesis theory of the Hebrews with the two mathematical options:

$$\delta G^{\alpha\beta}=0 \tag{202}$$

and

$$g = 0 \tag{203}$$

right.

9.7.2.2 More Details about the "Unconventional" Physics

While in the subsection above, the "chaos" section, as we might call it, we showed truly "unconventional" physics coming out of the Einstein-Hilbert action via unusual ways to solve the Hilbert variational task, we should point out that there are also other options to come to an "unconventional" physics simply by considering the true, which is to say the metric, origin of originally only postulated equations. Thereby we often see the outcome of conventional physics only as an approximated form of the fundamental, which is to say metrically, derived one.

It may be worth nothing that "chaos" is a function of a phase perception based on arbitrary coordinates which only hold that status under contrived conditions. The delimitations on "measured" phenomena must be ignored to accept "chaos" as a state condition or mathematical certainty.

9.7.2.2.1 The "Unconventional" Physics Realm – Part I

We start again with Hamilton's extremal principle in an arbitrary space of attributes in the form of the Einstein-Hilbert action:

$$\delta W = 0 = \delta \int_V d^n x \left(\sqrt{-g} \cdot \Phi_R[R] \right) = \delta \int_V d^n x \left(\sqrt{-g} \cdot \left(R - 2\Lambda + L_M \right) \right). \tag{204}$$

Considering the volume part of the metric separately like this:

$$G_{\alpha\beta} = g_{\alpha\beta} \cdot F[f], \tag{205}$$

gives us:

$$\delta W = 0 = \delta \int_V d^n x \left(\sqrt{-g \cdot F^n} \times \left(\left(R - \frac{1}{2F} \begin{pmatrix} 2F_{,\alpha\beta}(n-1) g^{\alpha\beta} + 2\Gamma^a_{\alpha\beta} F_{,a} g^{\alpha\beta} \\ -F_{,\alpha} g^{ab} g_{\beta b,a} g^{\alpha\beta} - F_{,\beta} g^{ab} g_{\alpha b,a} g^{\alpha\beta} \\ -n\Gamma^d_{\alpha\beta} F_{,d} g^{\alpha\beta} + \frac{n}{2} F_{,d} g^{cd} g_{ab,c} g^{ab} \end{pmatrix} - \frac{F_{,\alpha} \cdot F_{,\beta}}{4F^2} g^{\alpha\beta} \left((n-6)(n-1) \right) \right) \frac{1}{F} - 2\Lambda + L_M \right) \right). \tag{206}$$

With a proper setting for F[f], namely as follows:

$$F[f]=\begin{cases} C_F\cdot\left(f+C_f\right)^{\frac{4}{n-2}} & n\neq 2 \\ C_F\cdot e^{f\cdot C_f} & n=2 \end{cases}, \tag{207}$$

we can get rid of the nonlinear terms and obtain:

$$\delta W=0=\delta\int_V d^n x\left(\sqrt{-g\cdot F^n}\times\left(\left(R-\frac{1}{2F}\begin{pmatrix} 2F_{,\alpha\beta}(n-1)g^{\alpha\beta}+2\Gamma^a_{\alpha\beta}F_{,a}g^{\alpha\beta} \\ -F_{,\alpha}g^{ab}g_{\beta b,a}g^{\alpha\beta}-F_{,\beta}g^{ab}g_{\alpha b,a}g^{\alpha\beta} \\ -n\Gamma^d_{\alpha\beta}F_{,d}g^{\alpha\beta}+\frac{n}{2}F_{,d}g^{cd}g_{ab,c}g^{ab} \end{pmatrix}\right)\frac{1}{F} \atop -2\Lambda+L_M\right)\right). \tag{208}$$

Now you might ask yourself, “Where the heck is the classical physics?” and you will realize that you can only get it when restricting your little box of insight to the following space-times:

$$g_{ij}=\begin{pmatrix} g_{00} & \cdots & 0 \\ \vdots & \ddots & \vdots \\ 0 & \cdots & g_{n-1n-1} \end{pmatrix};\quad g_{ii,i}=0, \tag{209}$$

which gives the well-known:

$$\delta W=0=\delta\int_V d^n x\left(\sqrt{-g\cdot F^n}\times\left(\left(R+\frac{F'}{F}(1-n)\Delta f\right)\frac{1}{F}-2\Lambda+L_M\right)\right). \tag{210}$$

So many years of physics with a postulated equation and nobody came up with the idea that by deriving it properly from a principle know for even much longer could have resulted in something much more general.

Well, all those doubters may feel pretty much vindicated now!

And there is more!

9.7.2.2.2 The “Unconventional” Physics Realm – Part II

We start with (206) and realize that we can obtain a “completely unconventional” physics when solving this equation via the variation with respect to the number of dimensions. This has not yet been taken into account in “classical” physics. We have discussed this variation in connection with the Bekenstein-Hawking problem of storing information in black holes (e.g., c.f. [4], pp. 342).

The same holds for the variation with respect to gravity and quantum centers as demonstrated in [2, 4, 50 – 53, 70], which leads – among other things – to a quantum gravity thermodynamics.

9.7.2.2.3 The "Unconventional" Physics Realm – Part III

Another option to derive scalar quantum gravity field equations requires the use of the functional derivative (for the full derivation see appendix C of this book):

$$\delta W = \frac{\delta W}{\delta f[\mathbf{y}]} = 0 = \frac{\delta}{\delta f}\int_V d^n x\left(\sqrt{g}\cdot F^n\cdot F\cdot R^*\right)$$

$$= \lim_{\in\to 0}\frac{1}{\in}\left(\begin{array}{c}\int_V d^n x\left(\sqrt{g\cdot F\left[f[\mathbf{x}]+\in\cdot\delta[\mathbf{x}-\mathbf{y}]\right]^n}\cdot R^{**}\left[f[\mathbf{x}]+\in\cdot\delta[\mathbf{x}-\mathbf{y}]\right]\right)\\ -\int_V d^n x\left(\sqrt{g\cdot F[f[\mathbf{x}]]^n}\cdot R^{**}[f[\mathbf{x}]]\right)\end{array}\right). \tag{211}$$

With:

$$\begin{array}{c}R^{**} = R - \frac{g^{\alpha\beta}}{2F}F'\left(2f_{,\alpha\beta}(n-1)+f_{,d}g^{cd}\left(\begin{array}{c}g_{\alpha c,\beta}-g_{\beta\alpha,c}-g_{c\beta,\alpha}\\ +\frac{n}{2}\left(2g_{\alpha\beta,c}-g_{\alpha c,\beta}-g_{\beta c,\alpha}\right)\end{array}\right)\right)\\ -(n-1)\frac{f_{,\alpha}\cdot f_{,\beta}}{4F^2}g^{\alpha\beta}\left(4FF''+F'\cdot F'(n-6)\right)\end{array} \tag{212}$$

this gives:

$$\frac{\delta W}{\delta f} = 0 \Rightarrow$$

$$0 = -\left[F^{\frac{n}{2}-3}F'(n-1)\left(\begin{array}{c}\frac{\sqrt{g}\cdot g^{\alpha\beta}}{2}\frac{f_{,\alpha}f_{,\beta}}{2}\left(\frac{1}{2}(n-4)(n+2)(F')^2+2FF''\right)\\ +\frac{(n+2)}{2}FF'\sqrt{g}\cdot\Delta f\\ +F^2\left(\sqrt{g}\cdot g^{\alpha\beta}\right)_{,\alpha\beta}-\sqrt{g}\cdot\frac{n}{2}\cdot\frac{R\cdot F^2}{(n-1)}\\ -\frac{F^2}{2(n-1)}\left(\sqrt{g}\cdot g^{\alpha\beta}\cdot g^{ij}\left(\begin{array}{c}g_{i\beta,j}-g_{ij,\beta}-g_{\beta j,i}\\ +\frac{n}{2}\left(2g_{ij,\beta}-g_{i\beta,j}-g_{j\beta,i}\right)\end{array}\right)\right)_{,\alpha}\end{array}\right)\right]_{\mathbf{y}}. \tag{213}$$

9.7.2.2.4 The "Unconventional" Physics Realm – Part IV

Especially with respect to the technologies T1 and T2, we shall also consider another unconventional physics resulting from a further zero-option inside the Einstein-Hilbert action and being connected with the metric determinant.

We start with Hamilton's extremal principle in an arbitrary space of attributes in the form of the Einstein-Hilbert action:

$$\delta W = 0 = \delta \int_V d^n x \left(\sqrt{-g} \cdot \Phi_R [R] \right) = \delta \int_V d^n x \left(\sqrt{-g} \cdot (R - 2\Lambda + L_M) \right). \tag{214}$$

A completely unconventional physics arises when solving this equation via condition:

$$g = 0. \tag{215}$$

At this point it should be added that this apparent simple condition g=0 is anything but simple, because the determinant of the metric tensor g_{ab} is given as follows:

$$g = \det\left[g_{ab}\right] = \frac{1}{(n-1)!} \cdot \varepsilon_{i_1 \ldots i_{n-1}} \varepsilon^{j_1 \ldots j_{n-1}} g_{i_1 j_1} \cdot \ldots \cdot g_{i_{n-1} j_{n-1}}. \tag{216}$$

9.7.2.2.5 The "Unconventional" Physics Realm – Part V

As before, we start with Hamilton's extremal principle in an arbitrary space of attributes in the form of the Einstein-Hilbert action:

$$\delta W = 0 = \delta \int_V d^n x \left(\sqrt{-g} \cdot \Phi_R [R] \right) = \delta \int_V d^n x \left(\sqrt{-g} \cdot (R - 2\Lambda + L_M) \right). \tag{217}$$

Considering the volume part of the metric separately like this:

$$G_{\alpha\beta} = g_{\alpha\beta} \cdot F[f], \tag{218}$$

gives us:

$$\delta W = 0 = \delta \int_V d^n x \left(\sqrt{-g \cdot F^n} \times \left(\left(R - \frac{1}{2F} \left(\begin{array}{c} 2F_{,\alpha\beta}(n-1) g^{\alpha\beta} + 2\Gamma^a_{\alpha\beta} F_{,a} g^{\alpha\beta} \\ -F_{,\alpha} g^{ab} g_{\beta b,a} g^{\alpha\beta} - F_{,\beta} g^{ab} g_{\alpha b,a} g^{\alpha\beta} \\ -n\Gamma^d_{\alpha\beta} F_{,d} g^{\alpha\beta} + \frac{n}{2} F_{,d} g^{cd} g_{ab,c} g^{ab} \\ -\frac{F_{,\alpha} \cdot F_{,\beta}}{4F^2} g^{\alpha\beta} \left((n-6)(n-1) \right) \end{array} \right) \frac{1}{F} \right) \begin{array}{c} \\ -2\Lambda + L_M \end{array} \right) \right). \tag{219}$$

A completely unconventional physics arises when solving this equation via the condition:

$$\delta G^{\alpha\beta} = 0, \tag{220}$$

which requires the solution of the following equation:

$$\delta G^{\alpha\beta} = g^{\alpha\beta} \delta\left(\frac{1}{F}\right) + \frac{\delta g^{\alpha\beta}}{F} = 0. \tag{221}$$

And there is more!

But for here and now, this shall suffice to demonstrate the number and size of white spots on the map of the "established" or "conventional" physics.

9.8 How to Metrically Derive the Dirac Equation

In subsection "Option No. One" in [47] we have demonstrated how the Dirac equation [27] follows quite naturally from the Einstein-Hilbert action. In the following we are going to repeat the corresponding derivation.

In the sections above, in quite a variety of ways, we ended up in scalar quantum gravity equations, which could all be brought into the Klein-Gordon form for the most commonly used metrics (especially the flat ones) and under certain conditions regarding the functional metric wrapper F[f] (c.f. (7)). So, here we have (at least) three fundamentally derived scalar quantum gravity equations, namely, (35), (107) and (117). All these equations can be linearized via conditions (36), (108) and (119), respectively. After this linearization process, we obtain the standard quantum equations either directly or for most standard metrics. So, for instance, the metrics (9) and (111) for the cases (36) and (108), respectively, will almost directly bring us the Klein-Gordon equation. The only problem left is the occurrence of mass, potential, and energy; but as it was shown in our previous work (e.g., [2, 4, 7 – 23, 29, 30] that this only requires the introduction of entangled dimensions, we refrain from discussing this aspect here. Another option to obtain mass and energy that we will discuss here results from the consideration of the scalarized quantum field equations when demanding the same eigenvalue to the second order differential operator and the first order one (c.f. equation (222) below).

In this section here, our starting point shall be equation (29). Studying this equation closely, we realize that it contains a scalar and tensorial part. While we might consider the tensorial equation a generalized or quantum Einstein field equation, we are going to use the scalar one to derive the classical Dirac equation [27]. We already saw that by applying condition (33) and assuming metrics of the kind (9), our integrand in the Einstein-Hilbert action can dramatically be simplified. If this integrand is set equal to zero, we obtain the following differential equation:

$$0=R-\frac{F_i'}{F_i}(n-1)\Delta f_i+\frac{f_{i,\alpha}f_{i,\beta}(1-n)}{4F_i^2}g^{\alpha\beta}\left(4F_iF_i''+F_i'F_i'(n-6)\right). \tag{222}$$

Please note that this is not the original Dirac starting point, but a general outcome of the Einstein-Hilbert action with just an assumption regarding the metric being of the type (9) and a restriction of the variation on the volume part of the latter (33). The only thing we – boldly – introduced (postulated) is the vector character of f and – consequently – F. We have set $f\rightarrow f_i$ and $F\rightarrow F_i$. We know that this is a bold move and in principle, it should be considered downright wrong, because it contradicts the tensor-conservation of (7), if – and only if – this vector would be a true one with respect to the metric base. Nevertheless, we just moved on and have seen in [31] how to sort out the problem later. After all, it may be considered a problem that scientists have lived with for almost a hundred years… and, taking the success of the Dirac equation, they lived quite well with it. However, seeing the appearance of F_i in a metric ensemble of the kind (182), we might consider even more options for the justification of the vector character of F and f. This, however, will be discussed elsewhere.

Now we add in the Dirac starting point, who took for granted that:

a) the space-time can be assumed Ricci flat and thus, R=0,
b) the Laplace operator term in (222) has an eigenvalue M^2.

Also setting F[f]=f gives us:

$$0 = f_{i,\alpha} f_{i,\beta} g^{\alpha\beta} + M^2 \cdot \overbrace{\frac{4}{(n-6)}}^{\equiv\mu^2 = \mu\cdot\mu} \cdot f_i^2 . \tag{223}$$

The next step is to apply the known relation between the Dirac matrices (note that all empty slots are standing for zeros; also note that we here only give the Dirac matrices in the Cartesian, respectively the Minkowski case for a four-dimensional space-time):

$$\gamma^0 = \begin{pmatrix} 1 & & & \\ & 1 & & \\ & & -1 & \\ & & & -1 \end{pmatrix}; \quad \gamma^1 = \begin{pmatrix} & & & 1 \\ & & 1 & \\ & -1 & & \\ -1 & & & \end{pmatrix},$$
$$\gamma^2 = \begin{pmatrix} & & & -i \\ & & i & \\ & i & & \\ -i & & & \end{pmatrix}; \quad \gamma^3 = \begin{pmatrix} & & 1 & \\ & & & -1 \\ -1 & & & \\ & 1 & & \end{pmatrix}; \quad I = \begin{pmatrix} 1 & & & \\ & 1 & & \\ & & 1 & \\ & & & 1 \end{pmatrix} \tag{224}$$

and the metric tensor, reading:

$$g^{\alpha\beta} \cdot I = \frac{\gamma^\alpha \gamma^\beta + \gamma^\beta \gamma^\alpha}{2}, \tag{225}$$

and put (223) into the following form:

$$0=\left(f_{i,\alpha}f_{i,\beta}g^{\alpha\beta}+\mu^2\cdot f_i^2\right)\cdot I$$

$$\Rightarrow \;=\; \Rightarrow f_{i,\alpha}f_{i,\beta}\left(\frac{\gamma^\alpha\gamma^\beta+\gamma^\beta\gamma^\alpha}{2}\right)+\frac{I+I}{2}\cdot m^2\cdot f_i^2=0$$

$$\Rightarrow f_{i,\alpha}f_{i,\beta}\gamma^\alpha\gamma^\beta+I\cdot m^2\cdot f_i^2$$

$$=\left(f_{i,\alpha}\gamma^\alpha f_{i,\beta}\gamma^\beta+i\cdot I\cdot m\cdot f_i\overbrace{\left(f_{i,\beta}\gamma^\beta-f_{i,\alpha}\gamma^\alpha\right)}^{f_{i,\beta}\gamma^\beta-f_{i,\alpha}\gamma^\alpha=f_{i,\gamma}\gamma^\gamma-f_{i,\gamma}\gamma^\gamma=0}+I\cdot m^2\cdot f_i^2\right). \tag{226}$$

$$=\left(f_{i,\alpha}\gamma^\alpha+i\cdot I\cdot m\cdot f_i\right)\left(f_{i,\beta}\gamma^\beta-i\cdot I\cdot m\cdot f_i\right)=0$$

$$\left(f_{i,\beta}\gamma^\beta-i\cdot I\cdot m\cdot f_i\right)=0$$

$$\Rightarrow f_{i,\alpha}f_{i,\beta}\gamma^\beta\gamma^\alpha+I\cdot m^2\cdot f_i^2=0$$

$$\Rightarrow\left(f_{i,\alpha}\gamma^\alpha+i\cdot I\cdot m\cdot f_i\right)\left(f_{i,\beta}\gamma^\beta-i\cdot I\cdot m\cdot f_i\right)=0$$

This gives us the Dirac equation in its classical form [27]:

$$0=f_{i,\beta}\gamma^\beta-i\cdot I\cdot m\cdot f_i. \tag{227}$$

We should emphasize – again – that the Dirac matrices (224) are only valid for Cartesian coordinates, while in curvilinear coordinates quite some adjustments are of need [48].

Please note that in some cases the Dirac equation is also given via a separated time-derivative. This is obtained by multiplication of the equation (227) with $\beta=\gamma^0$. This leads to:

$$f_{i,t}=f_{i,\lambda}\alpha^\lambda-i\cdot\beta\cdot m\cdot f_i;\quad \lambda=1,2,3;\quad \alpha^\lambda=\begin{pmatrix}0&\sigma_\lambda\\\sigma_\lambda&0\end{pmatrix}. \tag{228}$$

Thereby we have applied the Pauli matrices which are given as follows:

$$\sigma_x=\begin{Vmatrix}0&1\\1&0\end{Vmatrix};\quad \sigma_y=\begin{Vmatrix}0&-i\\i&0\end{Vmatrix};\quad \sigma_z=\begin{Vmatrix}1&0\\0&-1\end{Vmatrix}, \tag{229}$$

and also used the β-matrix with $\beta=\begin{pmatrix}1&&&\\&1&&\\&&-1&\\&&&-1\end{pmatrix}$.

In order to make this consistent with the decomposition of the Klein-Gordon equation, we have to treat the f_i as scalars, which allows us to rewrite (226):

$$
\begin{aligned}
\left\{f_{i;\alpha},f_{i;\beta}\right\}=\left\{\nabla_\alpha f_i,\nabla_\beta f_i\right\}=\left\{f_{i;\alpha},f_{i;\beta}\right\}&\Rightarrow 0=\left(f_{i;\alpha}f_{i;\beta}g^{\alpha\beta}+\mu^2\cdot f_i^2\right)\cdot I\\
\Rightarrow\quad=\quad&\Rightarrow f_{i;\alpha}f_{i;\beta}\left(\frac{\gamma^\alpha\gamma^\beta+\gamma^\beta\gamma^\alpha}{2}\right)+\frac{I^2+I^2}{2}\cdot m^2\cdot f_i^2=0\\
&\Rightarrow f_{i;\alpha}f_{i,\beta}\gamma^\alpha\gamma^\beta+I^2\cdot m^2\cdot f_i^2=0\\
&\Rightarrow\left(f_{i;\alpha}\gamma^\alpha+i\cdot I\cdot m\cdot f_i\right)\left(f_{i;\beta}\gamma^\beta-i\cdot I\cdot m\cdot f_i\right)=0\\
&\Rightarrow f_{i;\alpha}f_{i;\beta}\gamma^\beta\gamma^\alpha+I^2\cdot m^2\cdot f_i^2=0\\
&\Rightarrow\left(f_{i;\alpha}\gamma^\alpha+i\cdot I\cdot m\cdot f_i\right)\left(f_{i;\beta}\gamma^\beta-i\cdot I\cdot m\cdot f_i\right)=0
\end{aligned}
\quad. \tag{230}
$$

The advantage of the Dirac approach clearly is that here the structure of the equation with the quaternions already assures the satisfaction of the eigenvalue equation for the Laplace operator term, which is to say the Klein-Gordon condition, we used above in (223):

$$\Delta f_i=M^2\cdot f_i\,. \tag{231}$$

To our dismay, however, we find, that we actually would need:

$$\Delta f_i=-M^2\cdot f_i\,, \tag{232}$$

in order to make the whole derivation consistent with the classical Dirac operator factorization.

This can be "proven" as follows:

At first, we write (232) in the following way:

$$\Delta f_i=g^{\alpha\beta}\nabla_\alpha\nabla_\beta f_i=-M^2\cdot f_i\,. \tag{233}$$

Now we apply (225) again, thereby also use (224) and the β-matrix, and obtain:

$$
\begin{aligned}
&g^{\alpha\beta}\nabla_\alpha\nabla_\beta f_i+M^2\cdot f_i=0\\
&0=\left(g^{\alpha\beta}\nabla_\alpha\nabla_\beta f_i+M^2\cdot f_i\right)\cdot I\\
\Rightarrow\quad=\quad&\Rightarrow\left(\frac{\gamma^\alpha\gamma^\beta+\gamma^\beta\gamma^\alpha}{2}\right)\nabla_\alpha\nabla_\beta f_i+\frac{I^2+I^2}{2}\cdot\tilde{m}^2\cdot f_i=0\\
&\Rightarrow\gamma^\alpha\gamma^\beta\nabla_\alpha\nabla_\beta f_i+I^2\cdot\tilde{m}^2\cdot f_i=0\\
&\Rightarrow\left(\gamma^\alpha\nabla_\alpha+i\cdot I\cdot\tilde{m}\right)\left(\gamma^\beta\nabla_\beta-i\cdot I\cdot\tilde{m}\right)f_i=0\\
&\Rightarrow\gamma^\beta\gamma^\alpha\nabla_\alpha\nabla_\beta f_i+I^2\cdot\tilde{m}^2\cdot f_i=0\\
&\Rightarrow\left(\gamma^\alpha\nabla_\alpha+i\cdot I\cdot\tilde{m}\right)\left(\gamma^\beta\nabla_\beta-i\cdot I\cdot\tilde{m}\right)f_i=0
\end{aligned}
\quad. \tag{234}
$$

Since it is completely sufficient that the functions f_i fulfill the inner operator in the fifth and last line of equation (234), we can demand:

$$\left(\gamma^\beta\nabla_\beta-i\cdot I\cdot\tilde{m}\right)f_i=0\quad\Rightarrow\quad\gamma^\beta\nabla_\beta f_i-i\cdot I\cdot\tilde{m}\cdot f_i=0 \tag{235}$$

and thus, have "almost" reproduced our Dirac equation (227).

The only problem, apart from our "little" sign discrepancy in (231) and (232) is that we note that in general we have $m \neq \tilde{m}$ and that therefore the equations (227) are not compatible with the essential Klein-Gordon condition (232). This condition, however, is of need in order to make the whole metric Dirac operation work at all.

What can be done?

Well, the problem is easily solved in either the right number of dimensions (this would be n=10 (! see footnote[15])) in the case of (227) and (232), or a simple adjustment of the functional wrapper $F[f_i]$ in (222). We know from [4], chapter 11, that with F given as:

$$4FF''+F'F'(n-6)=0 \;\Rightarrow\; F=(f_i)^{\frac{4}{n-2}};\quad n>2, \tag{236}$$

and setting this into (222) results in:

$$0=R-\frac{4\cdot(n-1)}{(n-2)f_i}\Delta f_i. \tag{237}$$

Now we simply assume R not to be zero as in the Dirac approach, but:

$$R=\frac{4\cdot(n-1)}{(n-2)}\cdot M^2 \tag{238}$$

and can directly finish the derivation via (234) without any incompatibility problem at all, because due to (236), the nonlinear term in (222) has disappeared.

Even though this is a nice compact derivation, we should not ignore other possibilities, which, as it was shown in our previous paper [43], and has been represented in various ways in chapters 9 and 10 of [4], can be obtained via the degrees of freedom offering themselves in the form of the wrapper function F[f].

For instance: Knowing that, instead of condition (236), we could also have any other condition in (222) like e.g., (with an arbitrary function $H'[f]=\frac{\partial H[f]}{\partial f}=H'$ – for convenience we here omit the index i for the list of functions f_i):

$$\frac{4FF''+(F')^2(n-6)}{4F}=F'\cdot H'$$

$$\Rightarrow\quad F=\begin{cases}\left(C_{f0}+\int_1^f e^{H[\phi]}\cdot\frac{n-2}{4}\cdot d\phi\right)^{\frac{4}{n-2}}\cdot C_{f1} & n>2\\ C_{f1}\cdot e^{C_{f0}\cdot\int_1^f e^{H[\phi]}d\phi} & n=2\end{cases} \tag{239}$$

leads us to:

[15] We note that this is the number of dimensions the string theory needs to describe fermions.

$$\begin{aligned} 0 &= R + \frac{F'}{F}(1-n)\Delta f + \frac{f_{,\alpha} f_{,\beta}(1-n)}{F} g^{\alpha\beta} H' F' \\ \Rightarrow -R &= \frac{F'}{F}(1-n)\left(f \cdot \Delta f + f_{,\alpha} f_{,\beta} g^{\alpha\beta} H'\right) \end{aligned} . \tag{240}$$

In order to get some orientation, we first apply a function $H=H[f]=C_{f0}-f$. The corresponding solution for $F[f]$ in the case of $n>2$ would be:

$$\frac{4FF''+(F')^2(n-6)}{4F} = F' \cdot H' = -F' \quad \Rightarrow \quad F = C_{f1} \cdot e^{\frac{4}{n-2}\cdot\left(\ln\left[e^f - C_{f0}\right]-f\right)} \quad n>2, \tag{241}$$

where we detect an interesting minimum at $f = \ln\left[C_{f0}\right]$ (figure 34). The finding merits quite some discussion which will be presented elsewhere.

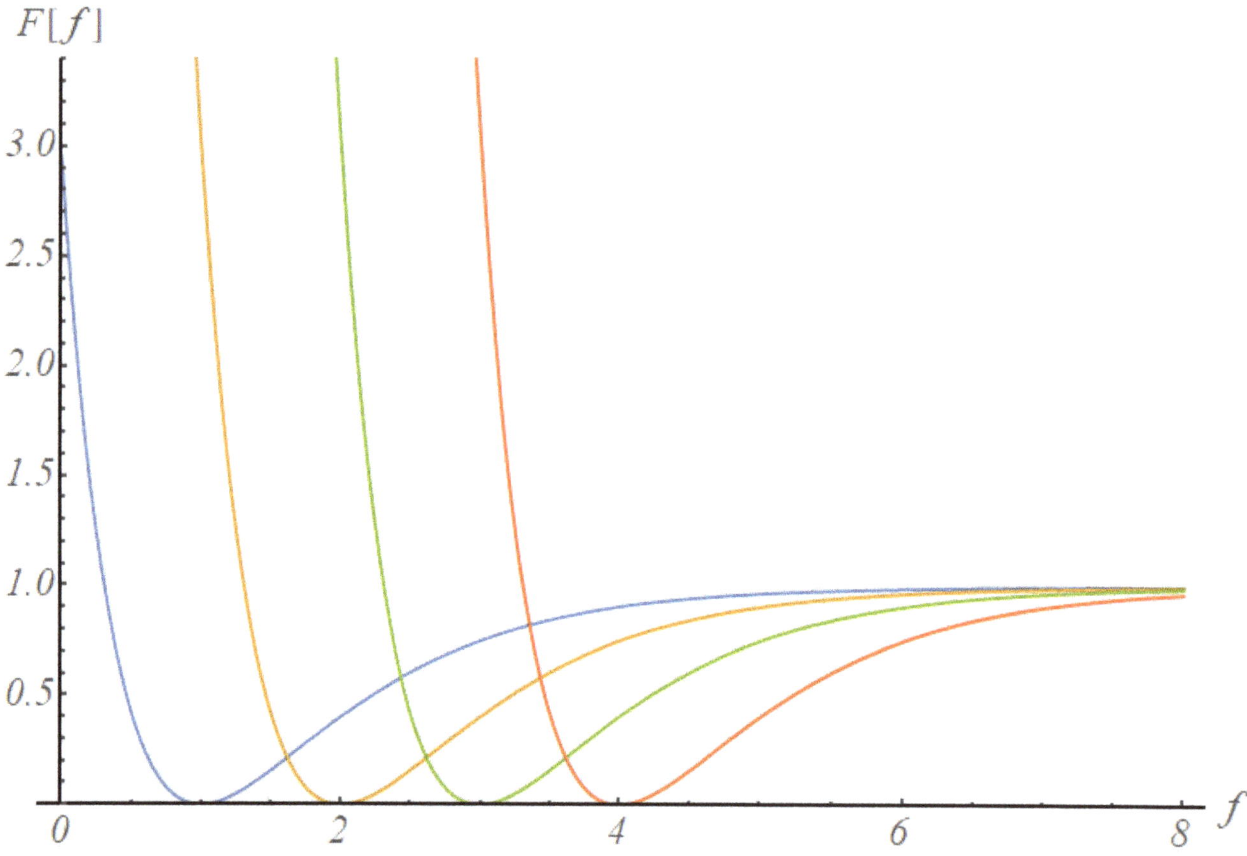

Fig. 34: Illustration of the function F[f] from (241) for a variety of settings regarding the constant C_{f0}, which is leading to minima at $f = \ln\left[C_{f0}\right]$.

The simplest path forward in order to obtain a good starting point for the Dirac equation, however, is now to apply a function H'=H'[f]=1/F[f]/f. The corresponding solution for F[f] in the case of n>2 would be:

$$\frac{4FF''+(F')^2(n-6)}{4F}=F'\cdot H'=-\frac{F'}{f} \quad\Rightarrow\quad F=C_{f1}\cdot\left(C_{f0}+\ln[f]\right)^{\frac{4}{n-2}} \quad n>2. \tag{242}$$

With the Dirac condition for the Ricci flat space-time R=0, this gives us:

$$R^*=0=\frac{F'}{F}(1-n)\left(f\cdot\Delta f-f_{,\alpha}f_{,\beta}g^{\alpha\beta}\right) \quad\Rightarrow\quad R^*=0=f\cdot\Delta f-f_{,\alpha}f_{,\beta}g^{\alpha\beta}, \tag{243}$$

from which we can derive the Dirac equation via the process given above with the Klein-Gordon condition (232). Introducing the index i again, namely, (243) becomes:

$$0=f_i\cdot\Delta f_i-f_{i,\alpha}f_{i,\beta}g^{\alpha\beta} \xrightarrow{\Delta f_i=-M^2f_i} 0=f_{i,\alpha}f_{i,\beta}g^{\alpha\beta}+M^2f_i^2 \tag{244}$$

and we can directly move forward with (226) and (230), only that this time instead of μ we have M:

$$\begin{aligned}
& 0=\left(f_{i,\alpha}f_{i,\beta}g^{\alpha\beta}+M^2\cdot f_i^2\right)\cdot I \\
\Rightarrow \quad = \quad & \Rightarrow f_{i,\alpha}f_{i,\beta}\left(\frac{\gamma^\alpha\gamma^\beta+\gamma^\beta\gamma^\alpha}{2}\right)+\frac{I^2+I^2}{2}\cdot m^2\cdot f_i^2=0 \\
& \Rightarrow f_{i,\alpha}f_{i,\beta}\gamma^\alpha\gamma^\beta+I^2\cdot m^2\cdot f_i^2=0 \qquad . \\
& \Rightarrow\left(f_{i,\alpha}\gamma^\alpha+i\cdot I\cdot m\cdot f_i\right)\left(f_{i,\beta}\gamma^\beta-i\cdot I\cdot m\cdot f_i\right)=0 \\
& \Rightarrow f_{i,\alpha}f_{i,\beta}\gamma^\beta\gamma^\alpha+I^2\cdot m^2\cdot f_i^2=0 \\
& \Rightarrow\left(f_{i,\alpha}\gamma^\alpha+i\cdot I\cdot m\cdot f_i\right)\left(f_{i,\beta}\gamma^\beta-i\cdot I\cdot m\cdot f_i\right)=0
\end{aligned} \tag{245}$$

This does not only solve our little factor plus the sign problem from above, but also gives us a clear metric understanding of the origin of the Dirac equation… at least with respect to the scalar quantum gravity equation (35). With metric scaling functions of the kind (241) and Ricci flatness with respect to the unscaled metric tensor $g_{\alpha\beta}$, namely, we obtain Ricci scalars R* of the type (243). Such equations can be brought into Dirac form.

9.8.1 Where Does the Dirac Spinor Come from?

So far it was not explained where – mathematically – the spinor f_i, which we here often called a vector or function vector, actually came from. This question arises, because formally f cannot be a vector, but has to be a scalar in order to truly only be a scalar metric factor compatible to the metric tensor, not destroying the latter's tensor properties. Without much fuss and because the derivation is quite lengthy, we just say that the solution to this puzzle is to be found in the following scalar product structure for the wave function f:

$$f=\prod_{i=1}^{k}G^{\alpha_i\beta_i}f_{\alpha_i}V_{\beta_i}, \tag{246}$$

which in the simplest case would just be:

$$f=G^{\alpha\beta}f_{\alpha}V_{\beta}. \tag{247}$$

Thereby the vectors V_{β_i} and V_{β} are vectors of constants.

Consequently, we have something similar for the function Φ via:

$$\Phi=\prod_{i=1}^{k}G^{\alpha_i\beta_i}\phi_{\alpha_i}V_{\beta_i} \tag{248}$$

and

$$\Phi=G^{\alpha\beta}\phi_{\alpha}V_{\beta}. \tag{249}$$

Please note that the simple form (249) gives an astounding explanation for the dominance of 4 dimensions of bigger scales in our universe [46, 56, 57].

9.8.2 Generalization to Arbitrary Metrics

In arbitrary metric space-times we cannot use (222), but have to apply (35), which we here print again in a variety of different forms:

$$
0=R-\left(\begin{array}{l}\frac{F'}{2F}\left(\begin{array}{l}2f_{,ij}(n-1)g^{ij}+2\Gamma_{ij}^{a}f_{,a}g^{ij}-f_{,i}g^{ab}g_{jb,a}g^{ij}\\-f_{,j}g^{ab}g_{ib,a}g^{ij}-n\Gamma_{ij}^{d}f_{,d}g^{ij}+\frac{n}{2}f_{,d}g^{cd}g_{ab,c}g^{ab}\end{array}\right)\\+(n-1)\frac{f_{,i}\cdot f_{,j}}{4F^{2}}g^{ij}\left(4FF''+(F')^{2}(n-6)\right)\end{array}\right)
$$

$$
=R-\frac{g^{\alpha\beta}}{2F}\left(2F_{,\alpha\beta}(n-1)+F_{,d}g^{cd}\left(\begin{array}{l}\left(g_{\beta c,\alpha}+g_{\alpha c,\beta}-g_{\beta\alpha,c}\right)-2g_{c\beta,\alpha}\\+\frac{n}{2}\left(\left(-g_{\alpha c,\beta}-g_{\beta c,\alpha}+g_{\alpha\beta,c}\right)+g_{\alpha\beta,c}\right)\end{array}\right)\right)
$$

$$
+\frac{1}{4F^{2}}\left(F_{,\alpha}\cdot F_{,\beta}(3n-6)+g_{\alpha\beta}F_{,c}F_{,d}g^{cd}(4-n)\right)g^{\alpha\beta}
$$

$$
=R-\frac{g^{\alpha\beta}}{2F}\left(2F_{,\alpha\beta}(n-1)+F_{,d}g^{cd}\left(\begin{array}{l}\left(g_{\beta c,\alpha}+g_{\alpha c,\beta}-g_{\beta\alpha,c}\right)-2g_{c\beta,\alpha}\\+\frac{n}{2}\left(\left(-g_{\alpha c,\beta}-g_{\beta c,\alpha}+g_{\alpha\beta,c}\right)+g_{\alpha\beta,c}\right)\end{array}\right)\right)
$$

$$
+\frac{1}{4F^{2}}\left(F_{,\alpha}\cdot F_{,\beta}(3n-6)+g_{\alpha\beta}F_{,c}F_{,d}g^{cd}(4-n)\right)g^{\alpha\beta}
$$

$$
=R-\frac{g^{\alpha\beta}}{2F}F'\left(2f_{,\alpha\beta}(n-1)+f_{,d}g^{cd}\left(\begin{array}{c}g_{\alpha c,\beta}-g_{\beta\alpha,c}-g_{c\beta,\alpha}\\+\frac{n}{2}\left(2g_{\alpha\beta,c}-g_{\alpha c,\beta}-g_{\beta c,\alpha}\right)\end{array}\right)\right)
$$

$$
-(n-1)\frac{f_{,\alpha}\cdot f_{,\beta}}{4F^{2}}g^{\alpha\beta}\left(4FF''+F'\cdot F'(n-6)\right) \quad . \tag{250}
$$

We realize that instead of the Klein-Gordon condition (232), we end up with something significantly more complicated, namely:

$$
\frac{g^{\alpha\beta}}{2(n-1)}\left(\begin{array}{l}2f_{,ij}(n-1)g^{ij}+2\Gamma_{ij}^{a}f_{,a}g^{ij}-f_{,i}g^{ab}g_{jb,a}g^{ij}\\-f_{,j}g^{ab}g_{ib,a}g^{ij}-n\Gamma_{ij}^{d}f_{,d}g^{ij}+\frac{n}{2}f_{,d}g^{cd}g_{ab,c}g^{ab}\end{array}\right)
$$

$$
=\frac{g^{\alpha\beta}}{2(n-1)}\left(2f_{,\alpha\beta}(n-1)+f_{,d}g^{cd}\left(\begin{array}{c}g_{\alpha c,\beta}-g_{\beta\alpha,c}-g_{c\beta,\alpha}\\+\frac{n}{2}\left(2g_{\alpha\beta,c}-g_{\alpha c,\beta}-g_{\beta c,\alpha}\right)\end{array}\right)\right)=-M^{2}\cdot f \quad . \tag{251}
$$

Obviously this cannot be treated in the same – simple – way as we processed (234), but we are not going to consider this general case here. Instead, we refer to one of our follow-up papers.

9.9 Bringing in the Multiverse Theory

9.9.1 The Other Way to Potentially Solve the Spinor-Origin-Problem

While in [46] we concentrated on finding an explanation for the origin of the vector F_i or f_i (the spinor) by using inner scalar products F_i*F^i, applying lists of Ricci scalar scaling functions or simply ignoring the vector problem (as so many scientists did before), we here intend to directly apply Everett's multiverse suggestion [56] to approach this problem in a completely new way.

9.9.1.1 Towards the Classical Dirac Method by Operator Factorization

We start with a metric in 16 dimensions, principally having the form (189), which we slightly adjust as follows:

$$G_{\alpha\beta} = {}_nF[f_n]\begin{pmatrix} {}_4F_1[f_1]\cdot {}^4_1g_{\alpha\beta} & {}^40 & {}^40 & {}^40 \\ {}^40 & {}_4F_2[f_2]\cdot {}^4_2g_{\alpha\beta} & {}^40 & {}^40 \\ {}^40 & {}^40 & {}_4F_3[f_3]\cdot {}^4_3g_{\alpha\beta} & {}^40 \\ {}^40 & {}^40 & {}^40 & {}_4F_4[f_4]\cdot {}^4_4g_{\alpha\beta} \end{pmatrix}.$$
$${}_4F_i[f_i] = C_{f1i}\cdot\left(C_{fi} + f_i[t_i, x_i, y_i, z_i]\right)^2; \quad {}^40 = \begin{pmatrix} 0 & 0 & 0 & 0 \\ 0 & 0 & 0 & 0 \\ 0 & 0 & 0 & 0 \\ 0 & 0 & 0 & 0 \end{pmatrix} \tag{252}$$

For simplicity only, the sub-metrics ${}^4_ig_{\alpha\beta}$ shall have the form (9). In this case, when evaluating the Ricci scalar, we obtain sums of Laplace operators applied on the various wave functions f_i:

$$R^* = 0 = R + \sum_{m=1}^{4}\frac{F_m'}{F_m}\cdot\Delta_m f_m$$
$$\Delta_m f = \frac{1}{\sqrt{{}^4g}}\sum_{\alpha,\beta=0}^{3}\partial_\beta\left(\sqrt{{}^4g}\cdot{}^4g^{\alpha\beta}\cdot\partial_\alpha f\right). \tag{253}$$

Assuming the Ricci scalar to be of the form:

$$R = \sum_{m=1}^{4}\frac{F_m'}{F_m}\cdot\left(\pm m_m^2\right), \tag{254}$$

leads us to:

$$R^* = 0 = \sum_{m=1}^{4}\frac{F_m'}{F_m}\cdot\left(\Delta_m f_m \pm m_m^2\right). \tag{255}$$

Even in the case of improper or vanishing R these Laplace equations can easily be made Klein-Gordon equations with the right number of add-on dimensions as follows:

$$G_{\alpha\beta}={}_{n}F[f_{n}]\begin{pmatrix} {}_{5}F_{1}[f_{1}]\cdot{}_{1}^{5}g_{\alpha\beta} & {}^{5}0 & {}^{5}0 & {}^{5}0 \\ {}^{5}0 & {}_{5}F_{2}[f_{2}]\cdot{}_{2}^{5}g_{\alpha\beta} & {}^{5}0 & {}^{5}0 \\ {}^{5}0 & {}^{5}0 & {}_{5}F_{3}[f_{3}]\cdot{}_{3}^{5}g_{\alpha\beta} & {}^{5}0 \\ {}^{5}0 & {}^{5}0 & {}^{5}0 & {}_{5}F_{4}[f_{4}]\cdot{}_{4}^{5}g_{\alpha\beta} \end{pmatrix}. \tag{256}$$

$${}_{5}F_{i}[f_{i}]=C_{fli}\cdot(C_{fi}+f_{i})^{4/3};\quad {}^{5}0=\begin{pmatrix} 0&0&0&0&0\\ 0&0&0&0&0\\ 0&0&0&0&0\\ 0&0&0&0&0\\ 0&0&0&0&0 \end{pmatrix}$$

Choosing one of the 5 sub-coordinates for each Klein-Gordon equation in the right way gives us mass. For more information the reader is – for instance – referred to [4], section 6.11.2 on page 300. We see that in both cases (255) and (256) with:

$$\begin{aligned} R^{*}=0&=\sum_{m=1}^{4}\frac{F_{m}{}'}{F_{m}}\cdot\left(\Delta_{m}f_{m}\pm m_{m}^{2}\right), \\ \Delta_{m}f&=\frac{1}{\sqrt{{}^{5}g}}\sum_{\alpha,\beta=0}^{4}\partial_{\beta}\left(\sqrt{{}^{5}g}\cdot{}^{5}g^{\alpha\beta}\cdot\partial_{\alpha}f\right) \end{aligned} \tag{257}$$

the results are 4 decoupled Klein-Gordon equations with mass, which could be used to derive the Dirac equation in the classical way [27] with just the small but significant difference to the classical path, that – this time – the origin of the function vector F_i is metrically clear. The problem left would be the consideration of the different Laplace operators Δ_m, the masses m_m, and the factors $\frac{F_{m}{}'}{F_{m}}$.

Thereby the simplest path forward is the assumption that:

a) the differences for Δ_m and m_m are negligible, because they are just Everett world aspects with effectively the same coordinates and parameters, which
b) could also just hold for the wrapping functions F_i or we assume that due to (37) (or F_i as given in (256)), we can approximate constant factors $\frac{F_{m}{}'}{F_{m}}$ if $C_{fi}>f_{i}$.

Please note that with a suitable Ricci scalar R or more add-on-dimensions (c.f. [9, 10, 30]) also potentials could be added to both (255) and (257).

9.9.1.2 Towards a Factorization of the Nonlinear Differential Operator

However, things are getting a bit more interesting when sticking to the 4 sub-dimensions, but instead of choosing the F_i in accordance with (186), allowing for (239) and the eigenvalue solution to the

Laplace operators as assumed in (244). For simplicity, we chose (241) and also assume ${}_nF[f_n]=1$. In this case (222) would look as follows:

$$\begin{aligned}R^* = 0 &= R - 3\cdot\sum_{i=1}^{4}\left(\frac{F_i'}{F_i}\Delta_i f_i + \frac{f_{i,\alpha}f_{i,\beta}}{4F_i^2}\,{}_i^4g^{\alpha\beta}\left(4F_iF_i'' - 2\cdot F_i'F_i'\right)\right)\\ &= R - 3\cdot\sum_{i=1}^{4}\frac{F_i'}{F_i}\left(f_i\cdot\Delta_i f_i\right) \qquad .\\ \Delta_i f &= \frac{1}{\sqrt{{}_i^4g}}\sum_{\alpha,\beta=0}^{3}\partial_\beta\left(\sqrt{{}_i^4g}\cdot{}_i^4g^{\alpha\beta}\cdot\partial_\alpha f\right)\end{aligned} \tag{258}$$

Now, with R=0 and the evaluation technique shown above (see (243) and following), we would have an equation of the following kind:

$$0 = \sum_{i=1}^{4}\frac{F_i'}{F_i}\left(f_i\cdot\Delta_i f_i - f_{i,\alpha}f_{i,\beta}\,{}_i^4g^{\alpha\beta}\right). \tag{259}$$

Thereby we have applied the setting (242). So, now we have the vector of functions in the equation. Unfortunately, however, we have different operators, metrics ${}_i^4g^{\alpha\beta}$, and the factor $\frac{F_i'}{F_i}$ also varies with the index i. However, Taylor expanding $\frac{F_i'}{F_i}$ at f=1 and assuming an Everett-split-up of space-time into 4 metrics with geometrically equal, but multiverse distinguishable characteristics, leads us to the usual equation:

$$0 = f_i\cdot\Delta f_i - f_{i,\alpha}f_{i,\beta}g^{\alpha\beta}, \tag{260}$$

which we can use as a starting point for the derivation of the Dirac equation.

9.9.1.3 Setting the Laplace Operator Term

Going back to the metric setting (256), we obtain the following Ricci scalar:

$$\begin{aligned}R^* &= R - 4\cdot\sum_{i=1}^{4}\left(\frac{F_i'}{F_i}\Delta_i f_i + \frac{f_{i,\alpha}f_{i,\beta}}{4F_i^2}\,{}_i^5g^{\alpha\beta}\left(4F_iF_i'' - F_i'F_i'\right)\right)\\ \Delta_i f &= \frac{1}{\sqrt{{}_i^5g}}\sum_{\alpha,\beta=0}^{4}\partial_\beta\left(\sqrt{{}_i^5g}\cdot{}_i^5g^{\alpha\beta}\cdot\partial_\alpha f\right)\end{aligned} \tag{261}$$

and demand it to vanish in accordance with our scalarized Einstein-Hilbert action condition (80). Now we assume R=0 and demand either all $\Delta_i f_i = 0$ or (less restrictive)

$$R_\Delta - 4\cdot\sum_{i=1}^{4}\left(\frac{F_i'}{F_i}\Delta_i f_i\right) = 0; \quad R = R_\Delta + R_\nabla. \tag{262}$$

This leaves us with:

$$0 = R_{\nabla} - 4 \cdot \sum_{i=1}^{4} \left(\frac{f_{i,\alpha} f_{i,\beta}}{4F_i^2} {}_i^5 g^{\alpha\beta} \left(4F_i F_i'' - F_i' F_i' \right) \right) \tag{263}$$

and a variety of options to extract Dirac equations with the help of suitable settings for the still undetermined wrapping functions F_i. As before, we assume the different functions to "originate" from different Everett worlds, respectively being "somehow" always present but manifest in our world only "if having been chosen".

9.10 The Everett Spinor and the Everett Wave Function Ensemble

In all the cases above the function vector f_i was obtained from different worlds of the multiverse theory. With respect to Penrose's critique about the non-falsifiability of Everett's original approach (see [215] for more), we now see that we

a) still have the factors $\frac{F_i'}{F_i}$,

b) and the global entanglement function ${}_nF[f_n]$,

which provides a connection between the various Everett worlds and which may be detectable in some way.

Regarding Penrose's criticism, we will only get a problem when the Everett-vectors are being created via the "spinor-settings" (246) to (249). Then it is not so clear to these authors how to actually see the "universal decision-maker" (c.f. [56]) at work when selecting one certain wave function out of the ensemble.

In order to distinguish between the spinor setting ensemble (in four dimensions this ensemble consists of 4 quite fundamental solutions and these are usually the two spins up and down, and the two time-directions matter and antimatter) and the other quantum numbers (c.f. the example in the section "Example: The Everett-Schrödinger Hydrogen Atom" further below in this book), we might resort to metric sub-structures as introduced in "Bringing in the Multiverse Theory" with the example (252) for instance. Thereby it does not matter – apparently – which one, the spinor or the other quantum number ensemble, is getting the inner or the outer position. More complex nested structures are thinkable as well (like Russian dolls, for instance).

9.11 The Metrically Derived Klein-Gordon-Everett Equation

Again we could just apply the metric setting (256) already including condition (37) and obtain the following Ricci scalar when assuming ${}_nF[f_n]=1$:

$$R^* = 0 = \sum_{i=1}^{4} \frac{F_i'}{F_i} \cdot \Delta_i f_i$$
$$\Delta_i f = \frac{1}{\sqrt{{}_i^5 g}} \sum_{\alpha,\beta=0}^{4} \partial_\beta \left(\sqrt{{}_i^5 g} \cdot {}_i^5 g^{\alpha\beta} \cdot \partial_\alpha f \right), \tag{264}$$

where we may introduce the masses via one of the 5 dimensions (see [4] section 6.11.2) and solve (264) by demanding all resulting Klein-Gordon equations to vanish. Alternatively, we could leave $_nF[f_n]$ open and interpret all the various solutions to the f_i as solutions to different Everett worlds with metrics ${}_i^5 g^{\alpha\beta}$. The global volume or scaling function $_nF[f_n]$ then acts as a kind of selector, which decides which of the solutions is going to be realized in "our" metric ${}_{our}^{5} g^{\alpha\beta}$. Unfortunately, with (256) only 4 different solutions would be coded. This problem, however, can easily be remediated via the following metric extension:

$$G_{\alpha\beta} = {}_nF[f_n] \cdot g_{\alpha\beta} = {}_nF[f_n] \begin{pmatrix} {}_N F_1[f_1] \cdot {}_1^N g_{\alpha\beta} & {}^N 0 & \cdots & \cdots & \cdots & {}^N 0 \\ {}^N 0 & {}_N F_2[f_2] \cdot {}_2^N g_{\alpha\beta} & \cdots & \cdots & \cdots & \vdots \\ {}^N 0 & {}^N 0 & \ddots & \cdots & \cdots & \vdots \\ {}^N 0 & {}^N 0 & \cdots & \ddots & \cdots & \vdots \\ \cdots & \cdots & \cdots & \cdots & \ddots & \vdots \\ {}^N 0 & \cdots & \cdots & \cdots & \cdots & {}_{n/N}^N g_{\alpha\beta} \end{pmatrix}. \tag{265}$$
$${}_N F_i[f_i] = C_{f1i} \cdot \left(C_{fi} + f_i[t_i, x_i, y_i, z_i, \ldots] \right)^{\frac{4}{N-2}}; \quad {}^N 0 = \begin{pmatrix} 0 & \cdots & 0 & 0 \\ \vdots & \ddots & 0 & 0 \\ 0 & 0 & 0 & 0 \\ 0 & 0 & 0 & 0 \end{pmatrix}$$

The subsequent Ricci scalar reads:

$$R^* = 0 = R + \frac{{}_nF'}{{}_nF} \cdot \Delta_n f_n + \sum_{i=1}^{n/N} \frac{F_i'}{F_i} \cdot \Delta_i f_i$$
$$\Delta_i f = \frac{1}{\sqrt{{}_i^N g}} \sum_{\alpha,\beta=0}^{N-1} \partial_\beta \left(\sqrt{{}_i^N g} \cdot {}_i^N g^{\alpha\beta} \cdot \partial_\alpha f \right); \quad \Delta_n f = \frac{1}{\sqrt{g}} \sum_{\alpha,\beta=0}^{n-1} \partial_\beta \left(\sqrt{g} \cdot g^{\alpha\beta} \cdot \partial_\alpha f \right) \tag{266}$$

Now, once again, we can either have the usual Klein-Gordon equations for the f_i via a Ricci curvature of the following kind:

$$R = \sum_{i=1}^{n/N} R_i = \sum_{i=1}^{n/N} \frac{F_i'}{F_i} \cdot m_i^2 \cdot f_i \quad \Rightarrow \quad 0 = \frac{{}_nF'}{{}_nF} \cdot \Delta_n f_n + \sum_{i=1}^{n/N} \frac{F_i'}{F_i} \cdot \left(\Delta_i f_i + m_i^2 \cdot f_i \right), \tag{267}$$

or we obtain masses via certain coordinates of the sub-metrics as demonstrated in [4], section 6.11.2 and set R=0.

As N can be an arbitrary number, we can have functions f_i and sub-metrics ${}_i^N g^{\alpha\beta}$ for all possible solutions to the differential equation:

$$0 = \Delta_i f_i + m_i^2 \cdot f_i \xrightarrow{\Delta_i f_i = \Delta_{i+1} f_i = \Delta f_i;\ m_i^2 = m_{i+1}^2 = m^2\ \forall i} 0 = \Delta f_i + m^2 \cdot f_i. \tag{268}$$

With the global function ${}_{n}F[f_n]$ as Everett-selector (multi-world-entanglement), we will see the realization of one of the many f_i in our metric ${}^{N}_{our}g^{\alpha\beta}$. All other options may be realized "elsewhere", which is to say in those other Everett worlds. However, the global function ${}_{n}F[f_n]$, seen as a selector, might just also be interpreted as the "divine decision maker" who decides upon the finally realized f_i for the observer. Both interpretations solve the Schrödinger cat problem. With the existence of such a global function it should be possible to actually measure the entanglement with the other worlds… if they exist.

9.11.1 The General (Tensor) Linearization Problem of the Scalarized Quantum Einstein Field Equations

It should be noted that the metric Everett world approach in combination with the $F_i[f_i]$ wrapper technique allows a completely new way to linearize not only the classical Einstein field equations but also the quantum extensions for a great variety of coordinates and metrics.

The essential trick thereby is the construction of the wrapper functions F_i in such a way, that they always fulfill the following condition[16]:

$$4F_iF_i''-F_i'F_i'=0. \tag{269}$$

9.12 The Metrically Derived Schrödinger-Everett Equation

The – or at least one – Schrödinger-Everett equation can be found by applying the transition from the usual Klein-Gordon to the Schrödinger equation and by just extending this to the various time-coordinates t_i for all sub-metrics ${}^{N}_{i}g^{\alpha\beta}$.

We start from our result (222) (index i), only that, for the reason of recognition, instead of the symbol f we apply the classical Greek symbol Ψ. We have applied condition (236), respectively (37). We also, in order to be close enough to the classical Schrödinger case, assume a Minkowski-like metric time component. We write (222) as follows:

$$0=\begin{cases}\left(\Psi-C_{f0}\right)\cdot R+(1-n)\cdot\left(\Delta_{n-1}\Psi-\dfrac{\partial^2\Psi}{c^2\cdot\partial t^2}\right) & n>2\\ R+C_{f0}\cdot(1-n)\cdot\left(\Delta_{n-1}\Psi-\dfrac{\partial^2\Psi}{c^2\cdot\partial t^2}\right) & n=2\end{cases}. \tag{270}$$

We ignore the n=2-case and take it that R can be assumed to be linear to f, which is to say that the curvature of space-time is caused by and proportional to the quantum effects residing in it. This gives us:

---

[16] See also section "The Conventional Physics Realm"!

$$0=\left(\Psi-C_{f0}\right)\cdot \mathrm{R}_c\cdot\Psi+\left(1-\mathrm{n}\right)\cdot\left(\Delta_{\mathrm{n}-1}\Psi-\frac{\partial^2\Psi}{\mathrm{c}^2\cdot\partial \mathrm{t}^2}\right). \tag{271}$$

Assuming the quantum function to be small against the classical physics (an important aspect in Schrödinger's work anyway, and this also agrees with the practical observation that the space-time curvature is small as we experience our surrounding space to be mainly flat), we discard of the quadratic term and result in:

$$0=-C_{f0}\cdot \mathrm{R}_c\cdot\Psi+\left(1-\mathrm{n}\right)\cdot\left(\Delta_{\mathrm{n}-1}\Psi-\frac{\partial^2\Psi}{\mathrm{c}^2\cdot\partial \mathrm{t}^2}\right)=\left[\left(1-\mathrm{n}\right)\cdot\left(\Delta_{\mathrm{n}-1}-\frac{\partial^2}{\mathrm{c}^2\cdot\partial \mathrm{t}^2}\right)-C_{\mathrm{R}}\right]\Psi. \tag{272}$$

Then we can separate the time derivative and – for the reason of recognition in connection with Schrödinger's classical approach – we reformulate everything as follows:

$$\begin{gathered}0=-C_{f0}\cdot \mathrm{R}_c\cdot\Psi+\left(1-\mathrm{n}\right)\cdot\left(\Delta_{\mathrm{n}-1}\Psi-\frac{\partial^2\Psi}{\mathrm{c}^2\cdot\partial \mathrm{t}^2}\right)=\left[\left(1-\mathrm{n}\right)\cdot\left(\Delta_{\mathrm{n}-1}-\frac{\partial^2}{\mathrm{c}^2\cdot\partial \mathrm{t}^2}\right)-C_{\mathrm{R}}\right]\Psi \\ \xrightarrow{\left(\Delta_{\mathrm{n}-1}-\frac{\partial^2}{\mathrm{c}^2\cdot\partial \mathrm{t}^2}\right)-\frac{C_{\mathrm{R}}}{(1-\mathrm{n})}=\mathrm{C1}\cdot\frac{\partial_t^2}{\mathrm{c}^2}+\Delta_{\mathrm{n}-1}-\mathrm{M}^2} \\ \Rightarrow 0=\left[\Delta-\mathrm{M}^2\right]\Psi=\left[\left(\mathrm{C1}\cdot\frac{\partial_t^2}{\mathrm{c}^2}+\underbrace{\frac{1}{\sqrt{\mathrm{g}}}\partial_\alpha\sqrt{\mathrm{g}}\cdot \mathrm{g}^{\alpha\beta}\partial_\beta}_{\text{3D}-\Delta-\text{Operator}}\right)-\mathrm{M}^2\right]\Psi\end{gathered}. \tag{273}$$

Please note that our Greek indices α and β are running from 1 to n-1 only, and not – as usual – from 0 to n-1, because the time or 0-component has been separated. Now we introduce a function $\Psi=\Phi+\mathrm{X}$ and demand the following additional condition:

$$\partial_t\Psi=\mathrm{c}^2\cdot \mathrm{C2}\cdot(\Phi-\mathrm{X}). \tag{274}$$

Together with (273) we obtain:

$$0=-\mathrm{M}^2(\Phi+\mathrm{X})+\left(\mathrm{C1}\cdot \mathrm{C2}\cdot\partial_t(\Phi-\mathrm{X})+\frac{1}{\sqrt{\mathrm{g}}}\partial_\alpha\sqrt{\mathrm{g}}\cdot \mathrm{g}^{\alpha\beta}\partial_\beta(\Phi+\mathrm{X})\right). \tag{275}$$

The following two equations summed up would result in (275):

$$\begin{aligned}0&=-\mathrm{M}^2\Phi+\left(\mathrm{C1}\cdot \mathrm{C2}\cdot\partial_t\Phi+\frac{1}{\sqrt{\mathrm{g}}}\partial_\alpha\sqrt{\mathrm{g}}\cdot \mathrm{g}^{\alpha\beta}\partial_\beta\Phi\right)\\ 0&=-\mathrm{M}^2\mathrm{X}+\left(\frac{1}{\sqrt{\mathrm{g}}}\partial_\alpha\sqrt{\mathrm{g}}\cdot \mathrm{g}^{\alpha\beta}\partial_\beta \mathrm{X}-\mathrm{C1}\cdot \mathrm{C2}\cdot\partial_t \mathrm{X}\right)\end{aligned}. \tag{276}$$

We see that we have obtained a Schrödinger-like first derivative for the time-coordinate by the introduction of $\Psi=\Phi+\mathrm{X}$. Along the way we have found a matter and an antimatter case, which is leading us to the two Schrödinger equations (276).

According to the derivation above and our results from the section before we can directly extract the Schrödinger-Everett equation as follows:

$$R^* = 0 = R + \frac{{}_n F'}{{}_n F} \cdot \Delta_n f_n + \sum_{i=1}^{n/N} \frac{F_i'}{F_i} \cdot \left(\Delta_i f_i \pm C1_i \cdot C2_i \cdot \partial_{ti} f_i \right)$$

$$\Delta_i f = \frac{1}{\sqrt{{}_i^N g}} \sum_{\alpha,\beta=1}^{N-1} \partial_\beta \left(\sqrt{{}_i^N g} \cdot {}_i^N g^{\alpha\beta} \cdot \partial_\alpha f \right); \quad \Delta_n f = \frac{1}{\sqrt{g}} \sum_{\alpha,\beta=0}^{n-1} \partial_\beta \left(\sqrt{g} \cdot g^{\alpha\beta} \cdot \partial_\alpha f \right) \quad . \tag{277}$$

The usual Schrödinger equations for the f_i are now obtained via a Ricci curvature of the following kind:

$$R = \sum_{i=1}^{n/N} R_i = \sum_{i=1}^{n/N} \frac{F_i'}{F_i} \cdot m_i^2 \cdot f_i$$

$$\Rightarrow \quad 0 = \frac{{}_n F'}{{}_n F} \cdot \Delta_n f_n + \sum_{i=1}^{n/N} \frac{F_i'}{F_i} \cdot \left(\Delta_i f_i \pm C1_i \cdot C2_i \cdot \partial_{ti} f_i + m_i^2 \cdot f_i \right) \quad , \tag{278}$$

or we obtain masses via certain coordinates of the sub-metrics as demonstrated in [4], section 6.11.2 and set R=0.

As in the Klein-Gordon case, N can be an arbitrary number, allowing us to have functions f_i and sub-metrics ${}_i^N g^{\alpha\beta}$ for all possible solutions to the differential equation:

$$0 = \Delta_i f_i \pm C1_i \cdot C2_i \cdot \partial_{ti} f_i + m_i^2 \cdot f_i$$

$$\xrightarrow{\Delta_i f_i = \Delta_{i+1} f_i = \Delta f_i; \;\; m_i^2 = m_{i+1}^2 = m^2 \;\; C1_i \cdot C2_i \cdot \partial_{ti} f_i = C1_{i+1} \cdot C2_{i+1} \cdot \partial_{ti} f_i = C1 \cdot C2 \cdot \partial_t f_i \;\; \forall i} \quad . \tag{279}$$

$$0 = \Delta f_i \pm C1 \cdot C2 \cdot \partial_t f_i + m^2 \cdot f_i$$

The global function ${}_n F[f_n]$ would again appear as multi-world-entanglement or Everett-selector.

9.13 Example: The Everett-Schrödinger Hydrogen Atom

Again, we apply the general global or multiverse metric (265) and split it up into many sub-metrics, which should all have the ability to provide the conditions for the Schrödinger hydrogen atom.

Thereby we allow for complex exponents for our metric appearances of the radius coordinate r and by moving to N=8 dimensions and by changing the setting for the metrics ${}_k^N g^{\alpha\beta}$ according to:

$$
{}_{k}^{N}g_{\alpha\beta} = g_{\alpha\beta}^{8} = \begin{pmatrix} c^2 & 0 & 0 & 0 & \cdots & 0 \\ 0 & 1 & 0 & 0 & \cdots & 0 \\ 0 & 0 & r_k^2 & 0 & \cdots & 0 \\ 0 & 0 & 0 & r_k^2 \cdot \sin^2 \varphi_{k1} & \cdots & 0 \\ \cdots & \cdots & \cdots & \cdots & \ddots & 0 \\ 0 & 0 & 0 & 0 & 0 & g_{77} \end{pmatrix}
$$

$$
\times F_k\left[f_k\left[t_k, r_k, \vartheta_k, \varphi_{k1}, \varphi_{k2}, \varphi_{k3}, \varphi_{k4}, \varphi_{k5}\right]\right]; \quad F_k[f_k] = f_k^{\frac{2}{3}}
$$

$$
f_k\left[t_k, r_k, \vartheta_k, \varphi_{k1}, \varphi_{k2}, \varphi_{k3}, \varphi_{k4}, \varphi_{k5}\right] = f_{kt}[t] \cdot f_{kr}[r] \cdot f_{k\vartheta}[\vartheta] \cdot \prod_{i=1}^{5} f_{k\varphi_{ki}}[\varphi_{ki}];
$$

$$
g_{44} = \left(g_{\varphi_{k2}}[\varphi_{k2}]'\right)^2 \cdot r_k^{i \cdot s}; \tag{280}
$$

$$
g_{55} = \left(g_{\varphi_{k3}}[\varphi_{k3}]'\right)^2 \cdot r_k^{s};
$$

$$
g_{66} = \left(g_{\varphi_{k4}}[\varphi_{k4}]'\right)^2 \cdot r_k^{-i \cdot s};
$$

$$
g_{77} = -\left(g_{\varphi_{k5}}[\varphi_{k5}]'\right)^2 \cdot r_k^{-s};
$$

we can get the partial differential equation for the Schrödinger hydrogen problem [25, 26]. It should be pointed out that we assumed the same speed of light in vacuum for all Everett "worlds", respectively all sub-metrics ${}_{k}^{N}g^{\alpha\beta}$.

For simplicity and brevity, we omit the index k from now on.

Our next step is to apply a separation approach and to partially "fix" some of the functions as follows (for i=3, 4, 5[17]):

$$
\begin{gathered}
f_{\varphi_i}[\varphi_i] = f_{\varphi_i}\left[g_{\varphi_i}[\varphi_i]\right] \equiv f_{\varphi_i}\left[g_{\varphi_i}\right]; \\
f_{\varphi_i}\left[g_{\varphi_i}\right] = C_{-i} \cdot e^{-i \cdot A_i \cdot g_{\varphi_i}} + C_{+i} \cdot e^{+i \cdot A_i \cdot g_{\varphi_i}} = C_{ci} \cdot \cos\left[A_i \cdot g_{\varphi_i}\right] + C_{si} \cdot \sin\left[A_i \cdot g_{\varphi_i}\right] \\
\text{especially (classically):} \\
f_{\varphi 1}\left[g_{\varphi_1}\right] = C_{-1} \cdot e^{-i \cdot A_1 \cdot g_{\varphi_1}} + C_{+1} \cdot e^{+i \cdot A_1 \cdot g_{\varphi_1}} = C_{c1} \cdot \cos\left[\overbrace{A_1}^{\hat{=} m} \cdot g_{\varphi_1}\right] + C_{s1} \cdot \sin\left[\overbrace{A_1}^{\hat{=} m} \cdot g_{\varphi_1}\right]
\end{gathered}, \tag{281}
$$

$$
f_t[t] = C_{t1} \cdot \cos[c \cdot E \cdot t] + C_{t2} \cdot \sin[c \cdot E \cdot t], \tag{282}
$$

---

[17] If not used as index, "i" stands for the imaginary number $i = \sqrt{-1}$.

$$f_\vartheta[\vartheta=\varphi_0]=C_{P\vartheta}\cdot P_L^{A_1}\left[\cos[\vartheta=\varphi_0]\right]+C_{Q\vartheta}\cdot Q_L^{A_1}\left[\cos[\vartheta=\varphi_0]\right], \tag{283}$$

with the associated Legendre polynomials $P_L^{A_1}, Q_L^{A_1}$. This gives us the following differential equation for the r-dependency of f[…]:

$$R^*=0=\left(\frac{L\cdot(L+1)}{r^2}+r^{i\cdot s}\cdot A_2^2+r^s\cdot A_3^2+r^{-i\cdot s}\cdot A_4^2-r^{-s}\cdot A_5^2-\overbrace{\Delta_{3D-sphere}[r]}^{2\frac{\partial}{r\cdot\partial r}+\frac{\partial^2}{\partial r^2}}+E^2\right)\Psi$$

$$\Psi=\Psi[r,\vartheta=\varphi_0,\varphi=\varphi_1]=f_r[r]\cdot f_\vartheta[\vartheta=\varphi_0]\cdot f_\varphi[\varphi=\varphi_1] \quad . \tag{284}$$

$$\Rightarrow$$

$$0=f_\vartheta[\vartheta]\cdot f[\varphi]\cdot\left(\begin{array}{c}\frac{L\cdot(L+1)}{r^2}+r^{i\cdot s}\cdot A_2^2+r^s\cdot A_3^2+r^{-i\cdot s}\cdot A_4^2\\ -r^{-s}\cdot A_5^2-\frac{2\partial}{r\cdot\partial r}-\frac{\partial^2}{\partial r^2}+E^2\end{array}\right)f_r[r]$$

We see that for the choice of s=1 and $f_{\varphi_i}[\varphi_i]=\text{const}$ for i=2,3,4, we obtain:

$$0=f_\vartheta[\vartheta]\cdot f[\varphi]\cdot\left(\frac{L\cdot(L+1)}{r^2}-\frac{A_5^2}{r}-\frac{2\partial}{r\cdot\partial r}-\frac{\partial^2}{\partial r^2}+E^2\right)f_r[r], \tag{285}$$

which is just the classical Schrödinger hydrogen problem with the corresponding solution:

$$f_r[r]=e^{-r\cdot E}\cdot r^L\cdot\left(C_U U[-nn,2\cdot(1+L),2\cdot r\cdot\mu]+C_L\cdot L_{nn}^{1+2\cdot L}[2\cdot r\cdot\mu]\right)$$
$$nn=\frac{A_5^2}{2\cdot E}-1-L \quad . \tag{286}$$

Now we have a list of hydrogen solutions f_k with metrics ${}_k^N g^{\alpha\beta}$ and a global function ${}_nF[f_n]$, which would again appear as multi-world-entanglement or Everett-selector.

9.14 The Universal Code

The attentive reader will already have realized the following:

1. The metric equations above already ARE the fundamental recipe we were looking for in this book.
2. Any problem addressed with this recipe should be approachable and solvable because, in a universe governed by extremal principles, there can neither be a higher generality nor a more profound fundamentality. However, before anybody gets the idea that we now have THE

SOLUTION to everything… no matter what it is: The important prerequisite for the generality and fundamentality of the method is the universal reign of the extremal principle.

3. The metric equations above also present the universal code.
4. Any program written in this code is directly applicable for any REAL quantum computer or quantum gravity computer machine. In fact, it presents the code for the most general and thus, quantum gravity based Turing machine, constructable in this universe.
5. The recipe even contains the necessary ingredients for the awakening of a conscious mind and thus, a true Fundamentally Based Intelligence, FBI, instead of an Artificial Intelligence, AI. However, in order to work this out, the content presented here needs to be combined with a holistic "Mathematical Psychology" which is presented elsewhere [Q].

[Q] N. Schwarzer, "Mathematical Psychology – The World of Thoughts as a Quantum Space-Time with a Gravitational Core", Jenny Stanford Publishing, ISBN: 9789815129274

10 There Are No Particles

Now we will address the so-called wave-particle duality via a fundamentally based mathematical approach leading to the destruction of the concept of particles.

This topic provides insight into the higher dimensionality of just everything and allows for the description of the process of creation, T1, as dimensional growth with entanglement, while separation, T2, comes with disentanglement of systems of higher dimensionality into sub-systems of lower dimensionality.

10.1 Starting as Simple as Possible

We start with the assumption of a universe with just one attribute and we give this attribute the symbol x. Now we want to have a "something" or a "happening" on this very attribute and we explicitly demand this happening to be a point-like particle. In order to make it such, the "thing" has to have a position. We give this position the coordinate x_0. Now, however, we have two properties and assuming a universe governed by a minimum principle and following the procedures laid out in the theory section above (considering equations (6) to (38) totally suffices), we obtain the following equation for the Ricci scalar of our scaled metric tensor:

$$R^* = 0 = f[x, x_0]^{(2,0)} + f[x, x_0]^{(0,2)}. \tag{287}$$

The subsequent and most general solution can be found easily via:

$$\Rightarrow f[x, x_0] = f^+[x + i \cdot x_0] + f^-[x - i \cdot x_0]. \tag{288}$$

This solution, however, is no particle solution, but it is a wave. Particles could be constructed from there, but only as a superposition of waves. Hence, the result still is an ensemble of waves. Even in cases where we assume constant Ricci curvature, which we will consider further below within higher dimensional and thus, more reasonable space-times, no true and reasonable point-like particle solutions can be obtained.

We conclude that by trying to construct particle objects, we contradict the whole concept of particles, because the minimum principle leads us to waves and oscillation solutions in the moment, we intend to localize an object (and as such "context" or "environment" are of necessity infinitely dimensional thereby evading the opening assumption in both resolution and properties – see the following pages of this book).

Critics might now argue that our starting point was the assumption of a continuous attribute and that – in order to force the idea of particles – one could just choose discontinuous coordinates instead. The flaw of this "plan" lies in the fact that discontinuity does not naturally exist, but can always be constructed out of continuity via infinite limiting procedures. Thus, a metric concept can easily handle discrete coordinates (see [88, 90, 92]) and – as it was stated before – particles are only possible as compactified wave ensembles.

Another potential criticism may come from point solutions as known in potential theory as considered and discussed in [67, 182, 183]. Then indeed, one might want to revive the idea of particles, but at the cost of forcing singularities into the theory. And, as various examples for the mathematical construction of Dirac delta function shows, such singularities can again be created out of superposed wave ensembles.

10.2 Generalization of the "There Are No Particles" Hypothesis in Higher Dimensionalities

10.2.1 Entangled Positioning

After having seen that in order to have "something", one needs at least two dimensions, namely one to code an (any) attribute for the "something" and another one for coding its position on this attribute, we also realized that this perception automatically destroys the concept of particles… at least in those two dimensions.

Now we want to extend our investigation to higher dimensionalities. Therefore, in order to avoid too much cross-referencing through the whole book, we are going to repeat some essential equations.

This time we are going to start in 3 dimensions where we first saw the positioning problem in connection with the so-called quantum well in one of our previous publications [95]. When metrically deriving the one-dimensional quantum well equations directly out of the Einstein-Hilbert action [95], namely, we observed that – just as in two dimensions above – the positioning parameter of an object (happening, wave-structure, or whatever) could actually be treated as an attribute of its own. However, as attributes or properties are just dimensions in the metric picture, we had to add an additional coordinate to our system. Seeing the property coordinate y and the positioning coordinate y_0 as being entangled, we may choose our metric approach as follows:

$$g_{ij}=\overbrace{F\left[f\left[y,y_0,\varphi\right]\right]}^{\equiv F}\cdot\begin{pmatrix} g_{00} & 0 & 0 \\ 0 & g_{11} & 0 \\ 0 & 0 & g_{22} \end{pmatrix}=F\cdot\begin{pmatrix} e^{f_0[y,y_0]} & 0 & 0 \\ 0 & e^{f_0[y,y_0]} & 0 \\ 0 & 0 & g_{22}=\overbrace{k\left[y,y_0\right]}^{\equiv k} \end{pmatrix}. \tag{289}$$

$$F\left[f\left[y,y_0,\varphi\right]\right]=f\left[y,y_0,\varphi\right]^{\frac{4}{n-2}}\xrightarrow{n=3}F\left[\overbrace{f\left[y,y_0,\varphi\right]}^{\equiv f}\right]=f\left[y,y_0,\varphi\right]^{4}$$

The coordinate φ is needed for the construction of the quantum well potential. Thereby we have chosen the exponential function approach for the pairwise entanglement of y and y_0 in order to obtain a linear partial differential equation in accordance with our condition (36) and its subsequent solution (37). With the approach for the entanglement function f_0 as follows:

$$f_0[y,y_0]^{(0,2)}+f_0[y,y_0]^{(2,0)}=-8\cdot C_0$$
$$\Rightarrow f_0[y,y_0]=-2\cdot C_0\cdot(y-y_0)^2+f_{01}^{+}[y+i\cdot y_0]+f_{02}^{-}[y-i\cdot y_0], \tag{290}$$

we obtain the following Ricci scalar for our scaled metric tensor:

$$R^*=0=-8\cdot\frac{k\cdot f^{(0,0,2)}+e^{-f_0[y,y_0]}\cdot\left(f^{(0,2,0)}+f^{(2,0,0)}\right)}{f^5}-\frac{e^{-f_0[y,y_0]}\cdot\left(f_0^{(2,0)}+f_0^{(0,2)}\right)}{f^4}$$
$$\Rightarrow \tag{291}$$
$$0=\frac{k\cdot f^{(0,0,2)}+e^{-f_0[y,y_0]}\cdot\left(f^{(0,2,0)}+f^{(2,0,0)}\right)}{f}+e^{-f_0[y,y_0]}\cdot C_0$$

where we assumed a constant well-potential. Reshaping the last line of (291) into:

$$R^*=0=\frac{k\cdot f^{(0,0,2)}}{f}+e^{-f_0[y,y_0]}\cdot\left(C_0+\frac{f^{(0,2,0)}+f^{(2,0,0)}}{f}\right), \tag{292}$$

or:

$$R^*=0=e^{f_0[y,y_0]}\cdot\frac{k\cdot f^{(0,0,2)}}{f}+C_0+\frac{f^{(0,2,0)}+f^{(2,0,0)}}{f}, \tag{293}$$

shows us that we can either have an eigenvalue equation for the function f with a $e^{2\cdot C_0\cdot(y-y_0)^2+\dots}$-vanishing potential term from (293), where C_0 has to be negative, leading to:

$$R^*=0=e^{-2\cdot|C_0|\cdot(y-y_0)^2+\dots}\cdot\frac{k\cdot f^{(0,0,2)}}{f}+C_0+\frac{f^{(0,2,0)}+f^{(2,0,0)}}{f}, \tag{294}$$

or we obtain a lasting potential in space-time and have a vanishing derivative term for the function f as follows from (292):

$$R^*=0=\frac{k\cdot f^{(0,0,2)}}{f}+e^{-2\cdot C_0\cdot(y-y_0)^2+\dots}\cdot\left(C_0+\frac{f^{(0,2,0)}+f^{(2,0,0)}}{f}\right), \tag{295}$$

in which case the parameter C_0 has to be positive.

Here, however, we are not interested in solving the quantum well equation, but only in the interpretation of the positioning coordinate when treating it as "independent" degree of freedom. Thereby we used the quotation marks for the word "independent" to emphasize that the entanglement with the dimension y already leads to some restrictions with respect to y_0.

As in the subsection before, we again obtained solutions which are clearly waves and oscillations. Particles can be constructed from those as wave-superpositions, but then these particles consist of anything but, and hence, are just waves.

A completely general and simpler setting for the potential $k[y,y_0]$ can be obtained in higher numbers of dimensions as it was shown in [95]. Here we only repeat the derivation for the case n=6 (the simplest one) where we choose the metric as follows:

$$
g_{\alpha\beta}^{6}=\begin{pmatrix} e^{f_0} & 0 & 0 & 0 & 0 & 0 \\ 0 & e^{f_0} & 0 & 0 & 0 & 0 \\ 0 & 0 & g_{22} & 0 & 0 & 0 \\ 0 & 0 & 0 & g_{33} & 0 & 0 \\ 0 & 0 & 0 & 0 & g_{44} & 0 \\ 0 & 0 & 0 & 0 & 0 & g_{55} \end{pmatrix} \times F\left[f\left[y,y_0,\varphi,\varphi_1,\varphi_2,\varphi_3\right]\right];
$$
$$
F[f]=C_F\cdot\left(f+C_f\right)^{\frac{4}{n-2}}\xrightarrow{n=6}C_F\cdot\left(f+C_f\right)
$$
$$
g_{22}=\left(g_{\varphi}[\varphi]'\right)^2\cdot k\left[y,y_0\right]^{i};\quad g_{33}=\left(g_{\varphi_1}[\varphi_1]'\right)^2\cdot k\left[y,y_0\right]^{-i};
$$
$$
g_{44}=\left(g_{\varphi_2}[\varphi_2]'\right)^2\cdot k\left[y,y_0\right]^{1};\quad g_{55}=\left(g_{\varphi_3}[\varphi_3]'\right)^2\cdot k\left[y,y_0\right]^{-1};
\tag{296}
$$

but leaving all the settings and demanding f_0 to be:

$$
\begin{gathered}
f_0\left[y,y_0\right]^{(0,2)}+f_0\left[y,y_0\right]^{(2,0)}=-5\cdot C_{00} \\
\Rightarrow f_0\left[y,y_0\right]=-\frac{5}{4}\cdot C_{00}\cdot\left(y-y_0\right)^2+f_{01}^{+}\left[y+i\cdot y_0\right]+f_{02}^{-}\left[y-i\cdot y_0\right]
\end{gathered},
\tag{297}
$$

results in:

$$
\begin{gathered}
0=k\cdot f^{(0,0,0,0,0,2)}-C_{00}\cdot f+e^{-f_0}\cdot\left(f^{(0,2,0,0,0,0)}+f^{(2,0,0,0,0,0)}\right) \\
\xrightarrow{f_{\varphi 3}[\varphi_3]=C_{-i}\cdot e^{-i\cdot k_c\cdot\varphi_3}+C_{+i}\cdot e^{+i\cdot k_c\cdot\varphi_3}=C_{c2}\cdot\cos[k_c\cdot\varphi_3]+C_{s2}\cdot\sin[k_c\cdot\varphi_3]} \\
0=k\cdot k_c^2\cdot f-C_{00}\cdot f+e^{-f_0}\cdot\left(f^{(0,2,0,0,0,0)}+f^{(2,0,0,0,0,0)}\right) \\
\xrightarrow{f=f[y,y_0]} \\
0=\left(k\cdot k_c^2-C_{00}\right)\cdot f+e^{-f_0}\cdot\left(f^{(0,2)}+f^{(2,0)}\right)
\end{gathered}.
\tag{298}
$$

We see that we have now obtained a very “clean” eigenvalue equation with potential directly from our metric (296). In addition, we have a localization function f_0, for which the same reasoning holds as discussed in connection with our approach in 3 dimensions and equations (290) to (295).

By the way: when now assuming the function f to only depend on the coordinate y, we obtain the classical one-dimensional Schrödinger equation, only that we also have a positioning dependency due to f_0. Regarding typical quantum well problems with respect to such an equation, usually reading ($f_0=0$):

$$
\begin{gathered}
0=\left(k\cdot k_c^2-C_{00}\right)\cdot f+e^{-f_0}\cdot f'' \\
\xrightarrow{f_0\to\approx 0} \\
0=\left(k\cdot k_c^2-C_{00}\right)\cdot f+\left(1-f_0+\frac{f_0^2}{2}-\ldots\right)\cdot f'', \\
\Rightarrow \\
0=\left(k\cdot k_c^2-C_{00}\right)\cdot f+f''
\end{gathered}
\tag{299}
$$

we refer to the textbook literature (e.g., [24, 96, 97]). Further discussion can also be found in connection with the so-called quantum tunneling and the collapse of the wave function in [98, 99]. The interesting aspect here is that there are no particle solutions… at least not without infinite super-positioning of wave solutions or without singular solutions.

We need to examine that this disruption of the particle concept does not mean that we now must find completely new ways to obtain masses and energies. As we have already seen in [95] how the localization of an object, happening, or wave-structure brings mass and/or energy, finding new ways for the mathematical creation of masses, potentials, and energies is not of need (see sections above). In fact, the localization via the entanglement function $f_0=f_0[y,y_0]$ "localizes" either the potential term with a C_{00} being negative or the derivative terms of f when we have $C_{00}>0$ (see discussion around equations (290) to (295)). We also realize that C_{00} is just giving the characteristics of a mass or energy term in the classical Schrödinger or Klein-Gordon equation:

$$0=\left(k\cdot k_c^2-\overbrace{C_{00}}^{\text{energy/mass}}\right)\cdot f+f'', \tag{300}$$

This shows us that the functional connection (entanglement) of dimensions can produce mass or energy. In our case the entanglement is between the main property and the localization of a "happening" (**a wave and not a particle**) on this property. Without this localization our system would have no mass-energy term.

So, we not only have the interesting aspect of the killing of the particle concept when trying to localize an object, but also when giving it mass or energy. In simpler words: things can have masses, but then they are not particles in the classical sense, but wave ensembles and entanglements.

In addition, in dependence on the structure of f_0, parts of it may also appear to us as potential. So, in case $k_c=0$, we may rewrite the last line in (298) as follows:

$$0=e^{f_0}\cdot C_{00}\cdot f-\left(f^{(0,2)}+f^{(2,0)}\right), \tag{301}$$

where the first addend $e^{f_0}\cdot C_{00}\cdot f$ is the equivalent for a classical potential, while with an eigenvalue setting of:

$$f^{(0,2)}=\left(E+m\cdot c^2\right)\cdot f, \tag{302}$$

we could also have mass or energy or both. The resulting equation would then read:

$$0=\left(e^{f_0}\cdot C_{00}-E-m\cdot c^2\right)\cdot f[y]-\frac{\partial^2 f[y]}{\partial y^2}, \tag{303}$$

which is again a one-dimensional Schrödinger equation with potential, mass, and energy, completely obtained from a metric set-up and without any postulation.

10.2.2 Time-like Entangled Positioning

When repeating the evaluations above with a time-like positioning coordinate as follows:

$$g^6_{\alpha\beta} = \begin{pmatrix} e^{f_0} & 0 & 0 & 0 & 0 & 0 \\ 0 & -e^{f_0} & 0 & 0 & 0 & 0 \\ 0 & 0 & g_{22} & 0 & 0 & 0 \\ 0 & 0 & 0 & g_{33} & 0 & 0 \\ 0 & 0 & 0 & 0 & g_{44} & 0 \\ 0 & 0 & 0 & 0 & 0 & g_{55} \end{pmatrix} \times F\left[f\left[y, y_0, \varphi, \varphi_1, \varphi_2, \varphi_3\right]\right]; \quad , \tag{304}$$

$$F[f] = C_F \cdot (f + C_f)^{\frac{4}{n-2}} \xrightarrow{n=6} C_F \cdot (f + C_f)$$

$$g_{22} = \left(g_\varphi[\varphi]'\right)^2 \cdot k\left[y, y_0\right]^i; \quad g_{33} = \left(g_{\varphi_1}[\varphi_1]'\right)^2 \cdot k\left[y, y_0\right]^{-i};$$

$$g_{44} = \left(g_{\varphi_2}[\varphi_2]'\right)^2 \cdot k\left[y, y_0\right]^1; \quad g_{55} = \left(g_{\varphi_3}[\varphi_3]'\right)^2 \cdot k\left[y, y_0\right]^{-1};$$

and – as before – leaving all the settings and demanding f_0 to be:

$$f_0\left[y, y_0\right]^{(0,2)} - f_0\left[y, y_0\right]^{(2,0)} = 5 \cdot C_{00}$$
$$\Rightarrow f_0\left[y, y_0\right] = \frac{5}{2 \cdot (c_0^2 - 1)} \cdot C_{00} \cdot (y - c_0 \cdot y_0)^2 + f_{01}^+\left[y + y_0\right] + f_{02}^-\left[y - y_0\right], \tag{305}$$

this results in:

$$0 = k \cdot f^{(0,0,0,0,0,2)} - C_{00} \cdot f + e^{-f_0} \cdot \left(f^{(0,2,0,0,0,0)} + f^{(2,0,0,0,0,0)}\right)$$
$$\xrightarrow{f_{\varphi 3}[\varphi_3] = C_{-i} \cdot e^{-i \cdot k_c \cdot \varphi_3} + C_{+i} \cdot e^{+i \cdot k_c \cdot \varphi_3} = C_{c2} \cdot \cos[k_c \cdot \varphi_3] + C_{s2} \cdot \sin[k_c \cdot \varphi_3]}$$
$$0 = k \cdot k_c^2 \cdot f - C_{00} \cdot f + e^{-f_0} \cdot \left(f^{(0,2,0,0,0,0)} + f^{(2,0,0,0,0,0)}\right) \quad . \tag{306}$$
$$\xrightarrow{f = f[y, y_0]}$$
$$0 = \left(k \cdot k_c^2 - C_{00}\right) \cdot f + e^{-f_0} \cdot \left(f^{(0,2)} + f^{(2,0)}\right)$$

10.3 Particle-Wave-Duality or the Death of Particles

We realize that for both cases (297) and (305) we obtain clearly particle-like compactness via the terms:

$$e^{-?\cdot(y - c_0 \cdot y_0)^2}; e^{-?\cdot(y - y_0)^2}, \tag{307}$$

but we do not truly obtain particles, because we also have wave-like behavior via the characteristic function $f_{01}^+[\ldots]; f_{02}^-[\ldots]$. Even the functions (307) are only the result of the assumption of constant residuals from other dimensions and/or constant Ricci curvature and the subsequent "particles" are in fact just Gaussian bell-shaped distributions. Thus, the apparent particles are still just waves with only a particle-like appearance due to a compactification of superposed waves leading to Gaussian shaped space-time distributions. Like the so-called Dirac delta function, which is only a mathematical trick to obtain a certain point-like distribution and which can be constructed by the means of suitable limiting procedures or by superposed waves, particles are only superposed waves.

10.4 Laplace, Klein-Gordon or Dirac

In classical Quantum Theory the particle concept has been applied in connection with the Klein-Gordon, the Schrödinger, and the Dirac equation. Here we want to show that we can obtain and handle these equations without the need of the particle idea. Along the way, we will find some interesting cognition about the nature of energy and masses.

In order to avoid criss-cross-referencing through the whole book, we here need some repetition in order to work out the pure wave-character of nature in connection with the fundamental quantum equations, directly extracted from the Einstein-Hilbert action and thus, the Hamilton extremal principle. This will lead to a bit of redundancy also with certain essential text pieces from above.

By leaving the entanglement function between the property y and the position coordinate y_0 arbitrary (here for our example in n=6 from the previous subsection):

$$g_{\alpha\beta}^{6}=\begin{pmatrix} F_0[f_0]\cdot g_{000} & F_0[f_0]\cdot g_{010} & 0 & 0 & 0 & 0 \\ F_0[f_0]\cdot g_{010} & F_0[f_0]\cdot g_{011} & 0 & 0 & 0 & 0 \\ 0 & 0 & g_{22} & 0 & 0 & 0 \\ 0 & 0 & 0 & g_{33} & 0 & 0 \\ 0 & 0 & 0 & 0 & g_{44} & 0 \\ 0 & 0 & 0 & 0 & 0 & g_{55} \end{pmatrix}\times F\left[f\left[y,y_0,\varphi,\varphi_1,\varphi_2,\varphi_3\right]\right];$$

$$F[f]=C_F\cdot\left(f+C_f\right)^{\frac{4}{n-2}}\xrightarrow{n=6}C_F\cdot\left(f+C_f\right);\quad g_{0ij}=\begin{pmatrix} g_{000} & g_{010} \\ g_{010} & g_{011} \end{pmatrix}$$

$$g_{22}=\left(g_{\varphi}[\varphi]'\right)^2\cdot k\left[y,y_0\right]^{i};\quad g_{33}=\left(g_{\varphi_1}[\varphi_1]'\right)^2\cdot k\left[y,y_0\right]^{-i};$$

$$g_{44}=\left(g_{\varphi_2}[\varphi_2]'\right)^2\cdot k\left[y,y_0\right]^{1};\quad g_{55}=\left(g_{\varphi_3}[\varphi_3]'\right)^2\cdot k\left[y,y_0\right]^{-1};\qquad ,(308)$$

we can also construct suitable conditions for the derivation of the Dirac equation. Instead of fixing the function F_0 via condition (36) as we did in (296) and (304), thereby avoiding the nonlinear terms, we obtain in the general case:

$$R_0^* \cdot F_0 = R_0 - \left(\begin{array}{c} \frac{1}{2F_0} \left(\begin{array}{l} 2F_{0,ij}(n-1) g_0{}^{ij} + 2\Gamma_{0ij}^{a} F_{0,a} g_0{}^{ij} \\ -F_{0,i} g_0{}^{ab} g_{0jb,a} g_0{}^{ij} - F_{,j} g_0{}^{ab} g_{0ib,a} g_0{}^{ij} \\ -n\Gamma_{0ij}^{d} F_{0,d} g_0{}^{ij} + \frac{n}{2} F_{,d} g_0{}^{cd} g_{0ab,c} g_0{}^{ab} \end{array} \right) \\ + \frac{F_{0,i} \cdot F_{0,j}}{4F_0^2} g_0{}^{ij} \left((n-6)(n-1) \right) \end{array} \right)$$

$$\xrightarrow{n=2}$$

$$= R_0 - \left(\frac{F_0'}{2F_0} \left(\begin{array}{l} 2f_{0,ij} g_0{}^{ij} + 2\Gamma_{0ij}^{a} f_{0,a} g_0{}^{ij} \\ -f_{0,i} g_0{}^{ab} g_{0jb,a} g_0{}^{ij} - f_{0,j} g_0{}^{ab} g_{0ib,a} g_0{}^{ij} \\ -2\Gamma_{0ij}^{d} f_{0,d} g_0{}^{ij} + f_{0,d} g_0{}^{cd} g_{0ab,c} g_0{}^{ab} \end{array} \right) + \frac{f_{0,i} \cdot f_{0,j}}{F_0^2} g_0{}^{ij} \left(F_0 F_0'' - (F_0')^2 \right) \right)$$

$$= R_0 - \left(\begin{array}{c} \frac{F_0'}{2F_0} \left(2f_{0,ij} g_0{}^{ij} - f_{0,d} g_0{}^{ab} g_0{}^{cd} \left(g_{0cb,a} + g_{0cb,a} - g_{0ab,c} \right) \right) \\ + \frac{f_{0,i} \cdot f_{0,j}}{F_0^2} g_0{}^{ij} \left(F_0 F_0'' - (F_0')^2 \right) \end{array} \right) , \quad (309)$$

$$= R_0 - \left(\begin{array}{c} \frac{F_0'}{2F_0} \left(2f_{0,ij} g_0{}^{ij} - f_{0,d} g_0{}^{ab} g_0{}^{cd} \left(g_{0cb,a} + g_{0cb,a} \overbrace{+ g_{0ca,b} - g_{0ca,b}}^{=0} - g_{0ab,c} \right) \right) \\ + \frac{f_{0,i} \cdot f_{0,j}}{F_0^2} g_0{}^{ij} \left(F_0 F_0'' - (F_0')^2 \right) \end{array} \right)$$

$$= R_0 - \left(\begin{array}{c} \frac{F_0'}{2F_0} \left(2f_{0,ij} g_0{}^{ij} - f_{0,d} g_0{}^{ab} g_0{}^{cd} \left(g_{0cb,a} + g_{0ca,b} - g_{0ab,c} + g_{0cb,a} - g_{0ca,b} \right) \right) \\ + \frac{f_{0,i} \cdot f_{0,j}}{F_0^2} g_0{}^{ij} \left(F_0 F_0'' - (F_0')^2 \right) \end{array} \right)$$

$$= R_0 - \left(\begin{array}{c} \frac{F_0'}{F_0} \left(f_{0,ab} g_0{}^{ab} - f_{0,d} g_0{}^{ab} \Gamma_{0ab}^{d} + f_{0,d} g_0{}^{ab} g_0{}^{cd} \left(g_{0cb,a} - g_{0ca,b} \right) \right) \\ + \frac{f_{0,i} \cdot f_{0,j}}{F_0^2} g_0{}^{ij} \left(F_0 F_0'' - (F_0')^2 \right) \end{array} \right)$$

$$= R_0 - \left(\frac{F_0'}{F_0} \left(\Delta_0 f_0 - f_{0,d} g_0{}^{ab} g_0{}^{cd} \left(g_{0cb,a} - g_{0ca,b} \right) \right) + \frac{f_{0,i} \cdot f_{0,j}}{F_0^2} g_0{}^{ij} \left(F_0 F_0'' - (F_0')^2 \right) \right) . \quad (310)$$

From the last line in (309) we immediately see that condition (36) and metrics of the type (9) would leave us with:

$$R_0^* \cdot F_0 = R_0 - \frac{F_0'}{F_0} \Delta_0 f_0 = R_0 - C_{f0} \cdot \Delta_0 f_0, \tag{311}$$

which results in the Laplace equation in completely Ricci-free sub-spaces:

$$\Delta_0 f_0 = 0. \tag{312}$$

Now we add in the Dirac starting point, who took for granted that [27, 28]:

a) the space-time can be assumed Ricci flat and thus, R=0.
b) the Laplace operator term in (309) (last line) has an eigenvalue μ^2.

Also setting F[f]=f gives us:

$$R_0^* \cdot F_0 = R_0 - \frac{1}{f_0}\left(\Delta_0 f_0 - \frac{f_{0,i} \cdot f_{0,j}}{f_0} g_0^{ij}\right) \xrightarrow{R_0=R_0^*=0} 0 = f_0 \Delta_0 f_0 - f_{0,i} \cdot f_{0,j} g_0^{ij}. \tag{313}$$

At this point we assume the function f_0 to be some kind of list or vector f_{0i}, but not in a metric sense (for more see [46]), and we write (313) as follows:

$$0 = \mu^2 \cdot f_{0i}^{\ 2} - f_{0i,\alpha} f_{0i,\beta} g_0^{\alpha\beta}. \tag{314}$$

Repeating our Dirac-derivation from the corresponding previous sections of this book and applying it on the resulting scalar quantum gravity equations (313) and (314) makes it clear that the reason for the Dirac approach in the quantum gravity or metric picture is not motivated by a problematic probability discussion (as the classical Dirac approach [27] was), but by an attempt to linearize equation (314). After all, nature seems to consist of superposable solutions and this requires linear differential equations to dominate the scene.

Now we apply the known relation between the Dirac matrices (note that we here only give the Dirac matrices in the Cartesian, respectively the Minkowski case for a two-dimensional space and – for completeness only – a four-dimensional space-time, respectively):

$$\gamma^0 = \begin{pmatrix} 0 & 1 \\ -1 & 0 \end{pmatrix}; \quad \gamma^1 = \begin{pmatrix} 0 & i \\ i & 0 \end{pmatrix}; \quad I = \begin{pmatrix} 1 & \\ & 1 \end{pmatrix}, \tag{315}$$

$$\begin{gathered} \gamma^0 = \begin{pmatrix} 1 & & & \\ & 1 & & \\ & & -1 & \\ & & & -1 \end{pmatrix}; \quad \gamma^1 = \begin{pmatrix} & & & 1 \\ & & 1 & \\ & -1 & & \\ -1 & & & \end{pmatrix} \\ \gamma^2 = \begin{pmatrix} & & & -i \\ & & i & \\ & i & & \\ -i & & & \end{pmatrix}; \quad \gamma^3 = \begin{pmatrix} & & 1 & \\ & & & -1 \\ -1 & & & \\ & 1 & & \end{pmatrix}; \quad I = \begin{pmatrix} 1 & & & \\ & 1 & & \\ & & 1 & \\ & & & 1 \end{pmatrix} \end{gathered}, \tag{316}$$

(note that all empty slots are standing for zeros) and the metric tensor, reading:

$$g_0{}^{\alpha\beta}\cdot I=\frac{\gamma^\alpha\gamma^\beta+\gamma^\beta\gamma^\alpha}{2},\tag{317}$$

and put (314) into the following form:

$$\begin{aligned}
&0=\left(-f_{0i,\alpha}f_{0i,\beta}g_0{}^{\alpha\beta}+\mu^2\cdot f_{0i}{}^2\right)\cdot I\\
\Rightarrow\quad = \quad &\Rightarrow -f_{0i,\alpha}f_{0i,\beta}\left(\frac{\gamma^\alpha\gamma^\beta+\gamma^\beta\gamma^\alpha}{2}\right)+\frac{I+I}{2}\cdot\mu^2\cdot f_{0i}{}^2=0\\
&\Rightarrow -f_{0i,\alpha}f_{0i,\beta}\gamma^\alpha\gamma^\beta+I\cdot\mu^2\cdot f_{0i}{}^2\\
=&\left(-f_{0i,\alpha}\gamma^\alpha f_{0i,\beta}\gamma^\beta+I\cdot\mu\cdot f_{0i}\overbrace{\left(f_{0i,\beta}\gamma^\beta-f_{0i,\alpha}\gamma^\alpha\right)}^{f_{0i,\beta}\gamma^\beta-f_{0i,\alpha}\gamma^\alpha=f_{0i,\gamma}\gamma^\gamma-f_{0i,\gamma}\gamma^\gamma=0}+I\cdot\mu^2\cdot f_{0i}{}^2\right).\\
&=\left(f_{0i,\alpha}\gamma^\alpha+I\cdot\mu\cdot f_{0i}\right)\left(f_{0i,\beta}\gamma^\beta-I\cdot\mu\cdot f_{0i}\right)=0\\
&\left(f_{0i,\beta}\gamma^\beta-I\cdot\mu\cdot f_{0i}\right)=0\\
&\Rightarrow -f_{0i,\alpha}f_{0i,\beta}\gamma^\beta\gamma^\alpha+I\cdot\mu^2\cdot f_{0i}{}^2=0\\
&\Rightarrow\left(f_{0i,\alpha}\gamma^\alpha+I\cdot\mu\cdot f_{0i}\right)\left(f_{0i,\beta}\gamma^\beta-I\cdot\mu\cdot f_{0i}\right)=0
\end{aligned}\tag{318}$$

This gives us the Dirac equation almost in its classical form [24, 27, 28]:

$$0=f_{0i,\beta}\gamma^\beta-I\cdot\mu\cdot f_{0i}.\tag{319}$$

We should emphasize that the Dirac matrices (316) are only valid for Cartesian coordinates, while in curvilinear coordinates quite some adjustments are needed [100].

As stated in our previous Dirac-like derivation in this book, in order to make this consistent with the decomposition of the Klein-Gordon equation, we have to treat the f_{0i} as scalars, which allows us to rewrite (318):

$$\begin{aligned}
&\left\{f_{0i,\alpha},f_{0i,\beta}\right\}=\left\{\nabla_\alpha f_{0i},\nabla_\beta f_{0i}\right\}=\left\{f_{0i;\alpha},f_{0i;\beta}\right\}\Rightarrow 0=\left(-f_{0i;\alpha}f_{0i;\beta}g_0{}^{\alpha\beta}+\mu^2\cdot f_{0i}{}^2\right)\cdot I\\
&\Rightarrow\quad = \quad\Rightarrow -f_{0i;\alpha}f_{0i;\beta}\left(\frac{\gamma^\alpha\gamma^\beta+\gamma^\beta\gamma^\alpha}{2}\right)+\frac{I^2+I^2}{2}\cdot\mu^2\cdot f_{0i}{}^2=0\\
&\Rightarrow -f_{0i;\alpha}f_{0i,\beta}\gamma^\alpha\gamma^\beta+I^2\cdot\mu^2\cdot f_{0i}{}^2=0\\
&\Rightarrow\left(f_{0i;\alpha}\gamma^\alpha+I\cdot\mu\cdot f_{0i}\right)\left(f_{0i;\beta}\gamma^\beta-I\cdot\mu\cdot f_{0i}\right)=0\\
&\Rightarrow -f_{0i;\alpha}f_{0i;\beta}\gamma^\beta\gamma^\alpha+I^2\cdot\mu^2\cdot f_{0i}{}^2=0\\
&\Rightarrow\left(f_{0i;\alpha}\gamma^\alpha+i\cdot I\cdot m\cdot f_{0i}\right)\left(f_{0i;\beta}\gamma^\beta-I\cdot\mu\cdot f_{0i}\right)=0
\end{aligned}\quad.\tag{320}$$

The advantage of the Dirac approach clearly is that here the structure of the equation with the quaternions already assures the satisfaction of the eigenvalue equation for the Laplace operator term, which is to say the Klein-Gordon condition that we used above in (314):

$$\Delta f_{0i}=\mu^2\cdot f_{0i}.\tag{321}$$

This can be "proven" as follows:

At first, we write (321) in the following way:

$$\Delta f_{0i} = g_0^{\alpha\beta}\nabla_\alpha\nabla_\beta f_{0i} = \mu^2 \cdot f_{0i}. \tag{322}$$

Now we apply (317) again and obtain:

$$\begin{aligned}
&g_0^{\alpha\beta}\nabla_\alpha\nabla_\beta f_{0i} - \mu^2 \cdot f_i = 0 \\
&0 = \left(g_0^{\alpha\beta}\nabla_\alpha\nabla_\beta f_{0i} - \mu^2 \cdot f_{0i}\right)\cdot I \\
\Rightarrow \; = \; &\Rightarrow \left(\frac{\gamma^\alpha\gamma^\beta + \gamma^\beta\gamma^\alpha}{2}\right)\nabla_\alpha\nabla_\beta f_{0i} - \frac{I^2 + I^2}{2}\cdot\mu^2 \cdot f_{0i} = 0 \\
&\Rightarrow \gamma^\alpha\gamma^\beta\nabla_\alpha\nabla_\beta f_{0i} - I^2 \cdot \mu^2 \cdot f_{0i} = 0 \\
&\Rightarrow \left(\gamma^\alpha\nabla_\alpha + I\cdot\mu\right)\left(\gamma^\beta\nabla_\beta - I\cdot\mu\right) f_{0i} = 0 \\
&\Rightarrow \gamma^\beta\gamma^\alpha\nabla_\alpha\nabla_\beta f_{0i} - I^2 \cdot \mu^2 \cdot f_{0i} = 0 \\
&\Rightarrow \left(\gamma^\alpha\nabla_\alpha + I\cdot\mu\right)\left(\gamma^\beta\nabla_\beta - I\cdot\mu\right) f_{0i} = 0
\end{aligned} \quad . \tag{323}$$

Since it is completely sufficient that the functions f_i fulfill the inner operator in the fifth and last line of equation (323), we can demand:

$$\left(\gamma^\beta\nabla_\beta - I\cdot\mu\right) f_{0i} = 0 \;\; \Rightarrow \;\; \gamma^\beta\nabla_\beta f_{0i} - I\cdot\mu\cdot f_{0i} = 0 \tag{324}$$

and thus, again, have "almost" reproduced our Dirac equation (319).

We see, that with the setting for F_0 and the eigenvalue condition for the Laplace operator (321), we automatically did not only obtain the Dirac equation via (314), but also "created" mass and / energy via the eigenvalue parameter μ as connector between the Laplace operator and the squared Nabla-like operator.

But how could we also obtain the Klein-Gordon equation as this, too, should be obtainable from the general approach without the need to evolve towards the Dirac equation?

Well, the problem is easily solved via a simple adjustment of the functional wrapper $F_0[f_0]$ in (309). We know from [4], chapter 11, that with F given as:

$$4FF'' + F'F'(n-6) = 0 \;\; \Rightarrow \;\; F = C_F \cdot e^{f_i \cdot C_f}; \;\; n = 2, \tag{325}$$

and by setting this F into (309), we result in:

$$R_0^* \cdot F_0 = R_0 - \frac{F_0'}{F_0}\Delta_0 f_0 = R_0 - C_{f0}\cdot\Delta_0 f_0 \xrightarrow{R_0^*=0} 0 = R_0 - C_{f0}\cdot\Delta_0 f_0 . \tag{326}$$

Now we simply assume R_0 not to be zero as in the Dirac approach, but:

$$R_0 = C_{f0}\cdot\mu^2\cdot f_0 \;\; \Rightarrow \;\; 0 = \mu^2 \cdot f_0 - \Delta_0 f_0 , \tag{327}$$

and can directly finish the derivation via (323) without any incompatibility problem at all, because due to (325) the nonlinear term in (309) has disappeared. Instead of (327), we might also set:

$$R_0 = C_{f0} \cdot \mu^2 \quad \Rightarrow \quad 0 = \mu^2 - \Delta_0 f_0, \tag{328}$$

giving us the exponential behavior from the "entangled positioning" section and thus, the wave-particle dualism again. The latter is – in essence – nothing else but the coffin nail for the particle concept.

The approach also opens up the possibility to introduce a potential function V via a suitable setting for R_0 as follows:

$$R_0 = C_{f0} \cdot \left(\mu^2 - V\right) \cdot f_0 \quad \Rightarrow \quad 0 = \left(\mu^2 - V\right) \cdot f_0 - \Delta_0 f_0, \tag{329}$$

$$R_0 = C_{f0} \cdot \left(\mu^2 - V \cdot f_0\right) \quad \Rightarrow \quad 0 = \left(\mu^2 - V \cdot f_0\right) - \Delta_0 f_0, \tag{330}$$

where, of course, we can also evolve the Dirac equation by just following the classical Dirac path of operational factorization (see [28]).

10.4.1 More General Options

Even though the setting of F[f] via (325) provides us with a nice compact derivation, we should – just for completeness – not ignore other possibilities, which, as it was shown in our previous paper [43] and has been represented in various ways in chapters 9 and 10 of [4], can be obtained via the degrees of freedom offering themselves in form of the wrapper function F[f].

For instance: Knowing that, instead of condition (325), we could also have any other condition in (309) like e.g., (with an arbitrary function $H'[f] = \frac{\partial H[f]}{\partial f} = H'$ – for convenience we here omit the index i for the list of functions f_i):

$$\frac{4FF'' + (F')^2 (n-6)}{4F} = F' \cdot H' \quad \Rightarrow \quad F = C_{f1} \cdot e^{C_{f0} \cdot \int_1^f e^{H[\phi]} d\phi} \qquad n = 2, \tag{331}$$

leading us to:

$$\begin{aligned} R_0^* \cdot F_0 = R_0 - \frac{F_0'}{F_0} \Delta f_0 - \frac{f_{0,\alpha} f_{0,\beta}}{F_0} g_0^{\alpha\beta} H' F_0' \\ \xrightarrow{R_0^* = 0} \qquad\qquad . \\ R_0 = \frac{F_0'}{F_0}\left(f_0 \cdot \Delta f_0 + f_{0,\alpha} f_{0,\beta} g_0^{\alpha\beta} H'\right) = C_{f0} \cdot \left(f_0 \cdot \Delta f_0 + f_{0,\alpha} f_{0,\beta} g_0^{\alpha\beta} H'\right) \end{aligned} \tag{332}$$

We will not discuss this generalization in this book.

10.5 From Scalar Mass and Energy to a Mass-Energy-Vector or –Matrix?

With the particle idea being killed, we may now revisit the concept of masses and investigate their true character within a jittering universe of waves and oscillations. Especially the innovation T2, as a mass transport mechanism, may have eluded proper explanation with respect to the conventional physics realm simply due to an imperfect or incomplete perception of the quantity mass. Classically, either postulated or described via the so-called Higgs mechanism [58], mass is always treated as a scalar.

But what if this perception is wrong?

And there is another reason why we should dive deeper into the matter:

As elaborated at the beginning of this book, namely, the consideration of the evolution of the key technologies T1 and T2 in a holistic way within their market environment, reveals an essential feature of these inventions the moment they will have reached a certain scale. By considering the building of structures and their individual growth as a process of dimensional increase and the reverse process as dimensional decrease, any market situation automatically favors technologies which have the ability to imbibe as many degrees of freedom as possible for steering and variation of the process. Even though we are not truly talking about living technologies (at least when considering the innovations isolated), we can still see the action of Darwin's selection law, because the processes are quasi alive, simply because they are organic processes within a socioeconomic organism, passively pushed to existence by the individuals living there. Hence, Darwin's rule of the survival of the fittest [70] applied to a socioeconomic system requires the essential innovations to have a competitive edge. Only this will guarantee the necessary flexibility and adaptivity for optimum performance in a rapidly changing (evolving) socioeconomic environment. Thereby, the technology itself can (and probably will) contribute to the evolutionary pressure on an almost variable scale, accessible by the owner. As already said above, this is just self-interaction at its best. It is similar to the Higgs mechanism [58], but unlike the latter not of scalar, but of tensorial character and when understood and controlled it can be of fundamental use to the owner and exploiter of the invention… hence, with this innovation, the owner actually has much more than just a key technology, but also a key to its own positioning within the socioeconomic system and his existence in there. By adding in the understanding about critical processes and phase transitions in socioeconomic spaces [8, 202], which are losing the dominating scales on a variety of attributes in such transitional states and knowing that certain technologies can influence scale lengths on a selection of attributes, one might (or even has to) discuss the social aspects of these inventions, too. The deposition and the polar-selective technology, which is to say T1 and T2, respectively, clearly have the potential for such a holistic creativity. But to set this potential free, we have to manage a better understanding of mass transport and mass formation processes in general and we start with the question: What are masses and are they truly scalars?

Starting with the last line in (309) and adding (321), we try a separation of the factors via base vectors as follows:

$$
\begin{aligned}
0 &= -f_{0i,\alpha} f_{0i,\beta} g_0^{\alpha\beta} + \mu^2 \cdot f_{0i}^{\ 2} \\
0 &= -f_{0i,\alpha} f_{0i,\beta} \mathbf{e}_0^{\ \alpha} \cdot \mathbf{e}_0^{\ \beta} + \mu^2 \cdot f_{0i}^{\ 2} \\
-f_{0i,\alpha} f_{0i,\beta} \mathbf{e}_0^{\ \alpha} \cdot \mathbf{e}_0^{\ \beta} + \mu^2 \cdot f_{0i}^{\ 2} &+ \overbrace{\left(f_{0i,\beta} \mathbf{e}^{\beta} - f_{0i,\alpha} \mathbf{e}^{\alpha} \right)}^{=0} \cdot \mu \cdot f_i \\
&= \left(f_{0i,\alpha} \mathbf{e}_0^{\ \alpha} + \mu \cdot f_{0i} \right) \left(f_{0i,\beta} \mathbf{e}_0^{\ \beta} - \mu \cdot f_{0i} \right)
\end{aligned} \quad . \tag{333}
$$

We see that there is a problem with the different ranks of the differential and the non-differential term in the two factors in the last line. The simplest way to overcome this problem would be to make μ a vector $\boldsymbol{\mu}$ and rewrite the evaluation above as follows:

$$
\begin{aligned}
0 &= -f_{0i,\alpha} f_{0i,\beta} g_0^{\alpha\beta} + \mu^2 \cdot f_{0i}^{\ 2} \\
0 &= -f_{0i,\alpha} f_{0i,\beta} \mathbf{e}_0^{\ \alpha} \cdot \mathbf{e}_0^{\ \beta} + \overbrace{\boldsymbol{\mu} \cdot \boldsymbol{\mu}}^{=\mu^2} \cdot f_{0i}^{\ 2} \\
-f_{0i,\alpha} f_{0i,\beta} \mathbf{e}_0^{\ \alpha} \cdot \mathbf{e}_0^{\ \beta} + \boldsymbol{\mu} \cdot \boldsymbol{\mu} \cdot f_{0i}^{\ 2} &+ \overbrace{\left(f_{0i,\beta} \mathbf{e}_0^{\ \beta} - f_{0i,\alpha} \mathbf{e}_0^{\ \alpha} \right)}^{=0} \cdot \boldsymbol{\mu} \cdot f_{0i} \\
&= \left(f_{0i,\alpha} \mathbf{e}_0^{\ \alpha} + \boldsymbol{\mu} \cdot f_{0i} \right) \left(f_{0i,\beta} \mathbf{e}_0^{\ \beta} - \boldsymbol{\mu} \cdot f_{0i} \right)
\end{aligned} \quad . \tag{334}
$$

Now we have obtained a Dirac equation with mass-energy "base-vectors" instead of the classical scalars.

10.5.1 Generalization of the Mass-Energy Base-Vectors or Matrices

Considering (334), we realize that the list or vector character of f is not of need anymore. Instead of the list or vector of the mass-energy parameter μ has taken on the necessary rank to make the equation above intrinsically consistent and allowing us to rewrite it as follows:

$$
\begin{aligned}
0 &= -f_{0,\alpha} f_{0,\beta} g_0^{\alpha\beta} + \mu^2 \cdot f_0^{\ 2} \\
0 &= -f_{0,\alpha} f_{0,\beta} \mathbf{e}_0^{\ \alpha} \cdot \mathbf{e}_0^{\ \beta} + \overbrace{\boldsymbol{\mu} \cdot \boldsymbol{\mu}}^{=\mu^2} \cdot f_0^{\ 2} \\
-f_{0,\alpha} f_{0,\beta} \mathbf{e}_0^{\ \alpha} \cdot \mathbf{e}_0^{\ \beta} + \boldsymbol{\mu} \cdot \boldsymbol{\mu} \cdot f_0^{\ 2} &+ \overbrace{\left(f_{0,\beta} \mathbf{e}_0^{\ \beta} - f_{0,\alpha} \mathbf{e}_0^{\ \alpha} \right)}^{=0} \cdot \boldsymbol{\mu} \cdot f_0 \\
&= \left(f_{0,\alpha} \mathbf{e}_0^{\ \alpha} + \boldsymbol{\mu} \cdot f_0 \right) \left(f_{0,\beta} \mathbf{e}_0^{\ \beta} - \boldsymbol{\mu} \cdot f_0 \right)
\end{aligned} \quad . \tag{335}
$$

This would be just another way to solve the problem of:

- keeping the functions F and f, respectively, F_0 and f_0 (for the 2-D cases) scalars in order to make the scaling of the metric approach to work and leave the tensor character unchanged (c.f. (7)) and
- still obtain a rank consistent first order differential equation of Dirac-like character.

Other options are discussed in [46]. However, it should be explicitly pointed out that this does not make the Dirac spinor obsolete.

It just makes it what it should have been right from its introduction in 1928, namely, a list or set of solutions.

A similar operation could be performed with the Dirac matrices by applying (317) again, which would give us:

$$
\begin{aligned}
&0=\left(-f_{0,\alpha}f_{0,\beta}g_{0}^{\alpha\beta}+\mu^{2}\cdot f_{0}^{2}\right)\cdot I\\
&\Rightarrow\ =\ \Rightarrow -f_{0,\alpha}f_{0,\beta}\left(\frac{\gamma^{\alpha}\gamma^{\beta}+\gamma^{\beta}\gamma^{\alpha}}{2}\right)+\frac{I+I}{2}\cdot\mu^{2}\cdot f_{0}^{2}=0\\
&\Rightarrow -f_{0,\alpha}f_{0,\beta}\gamma^{\alpha}\gamma^{\beta}+\overbrace{\mathbf{M}\cdot\mathbf{M}}^{I\cdot\mu^{2}}\cdot f_{0}^{2}\\
&=\left(-f_{0,\alpha}\gamma^{\alpha}f_{0,\beta}\gamma^{\beta}+\mathbf{M}\cdot f_{0}\cdot\overbrace{\left(f_{0,\beta}\gamma^{\beta}-f_{0,\alpha}\gamma^{\alpha}\right)}^{f_{0i,\beta}\gamma^{\beta}-f_{0i,\alpha}\gamma^{\alpha}=f_{0i,\gamma}\gamma^{\gamma}-f_{0i,\gamma}\gamma^{\gamma}=0}+I\cdot\mu^{2}\cdot f_{0}^{2}\right).\\
&=\left(f_{0,\alpha}\gamma^{\alpha}+\mathbf{M}\cdot f_{0}\right)\left(f_{0i,\beta}\gamma^{\beta}-\mathbf{M}\cdot f_{0}\right)=0\\
&\left(f_{0,\beta}\gamma^{\beta}-\mathbf{M}\cdot f_{0}\right)=0\\
&\Rightarrow -f_{0,\alpha}f_{0,\beta}\gamma^{\beta}\gamma^{\alpha}+I\cdot\mu^{2}\cdot f_{0}^{2}=0\\
&\Rightarrow\left(f_{0,\alpha}\gamma^{\alpha}+\mathbf{M}\cdot f_{0}\right)\left(f_{0,\beta}\gamma^{\beta}-\mathbf{M}\cdot f_{0}\right)=0
\end{aligned}
\tag{336}
$$

Again, we have no need for the list of functions f_i anymore but instead have to "live" with a mass-energy matrix **M** instead of a scalar μ.

10.6 More 2-D Entanglement and Subsequent Duality and Particle-Killing

The attentive reader may have already realized that we can generalize our two-dimensional entanglement in the following way also for higher dimensionalities:

$$g^n_{\alpha\beta}=\begin{pmatrix} F_0[f_0]\cdot g_{000} & F_0[f_0]\cdot g_{010} & \cdots & 0 & 0 & \cdots \\ F_0[f_0]\cdot g_{010} & F_0[f_0]\cdot g_{011} & 0 & 0 & 0 & \cdots \\ \vdots & \cdots & \ddots & 0 & 0 & \cdots \\ 0 & \cdots & 0 & F_i[f_i]\cdot g_{i00} & F_i[f_i]\cdot g_{i10} & \cdots \\ 0 & \cdots & 0 & F_i[f_i]\cdot g_{i10} & F_i[f_i]\cdot g_{i11} & \cdots \\ \vdots & \cdots & \vdots & \vdots & \vdots & \ddots \end{pmatrix}$$

$$\times F\left[f\left[y,y_0,\ldots,\varphi_i,\varphi_{i0},\ldots\right]\right]; \tag{337}$$

$$F[f]=C_F\cdot(f+C_f)^{\frac{4}{n-2}};\quad g_{0ij}=\begin{pmatrix} g_{000} & g_{010} \\ g_{010} & g_{011}\end{pmatrix};\quad g_{iij}=\begin{pmatrix} g_{i00} & g_{i10} \\ g_{i10} & g_{i11}\end{pmatrix}$$

With the settings from section “Laplace, Klein-Gordon or Dirac” for the wrapping function F_i, we could again create the right conditions for Laplace, Klein-Gordon or Dirac equations and the subsequent solutions in two dimensions. These solutions, which are all carrying the wave-particle dualism, do then make up the metric core for the super scaling with the function F[f] for the total metric $g^n_{\alpha\beta}$. With our current choice of F[f] in (337) and the assumption of a generally Ricci-free space-time, the final equation to solve would be:

$$\xrightarrow{g^n_{\alpha\beta}=g_{\alpha\beta}}$$

$$R^*=\frac{1}{F}\left(R-\frac{1}{2F}\begin{pmatrix} 2F_{,ij}(n-1)g^{ij}+2\Gamma^a_{ij}F_{,a}g^{ij} \\ -F_{,i}g^{ab}g_{jb,a}g^{ij}-F_{,j}g^{ab}g_{ib,a}g^{ij} \\ -n\Gamma^d_{ij}F_{,d}g^{ij}+\frac{n}{2}F_{,d}\gamma^{cd}g_{ab,c}g^{ab}\end{pmatrix}-\frac{F_{,i}\cdot F_{,j}}{4F^2}g^{ij}((n-6)(n-1))\right)$$

$$\xrightarrow{R^*=0}$$

$$0=R-\left(\begin{array}{c}\frac{F'}{2F}\begin{pmatrix} 2f_{,ij}(n-1)g^{ij}+2\Gamma^a_{ij}f_{,a}g^{ij} \\ -f_{,i}g^{ab}g_{jb,a}g^{ij}-f_{,j}g^{ab}g_{ib,a}g^{ij} \\ -n\Gamma^d_{ij}f_{,d}g^{ij}+\frac{n}{2}f_{,d}g^{cd}g_{ab,c}g^{ab}\end{pmatrix} \\ +(n-1)\frac{f_{,i}\cdot f_{,j}}{4F^2}g^{ij}\left(4FF''+(F')^2(n-6)\right)\end{array}\right)$$

$$\xrightarrow{F[f]=C_F\cdot(f+C_f)^{\frac{4}{n-2}}}$$

$$0=R-\frac{2}{n-2}\cdot\frac{1}{(f+C_f)}\begin{pmatrix} 2f_{,ij}(n-1)g^{ij}+2\Gamma^a_{ij}f_{,a}g^{ij} \\ -f_{,i}g^{ab}g_{jb,a}g^{ij}-f_{,j}g^{ab}g_{ib,a}g^{ij} \\ -n\Gamma^d_{ij}f_{,d}g^{ij}+\frac{n}{2}f_{,d}g^{cd}g_{ab,c}g^{ab}\end{pmatrix}. \tag{338}$$

In our following publications we will investigate higher dimensional sub-solutions and their intrinsic wave-particle dualism.

10.7 Bearing on "Quantum Gravity Marketing" and "Quantum Psychology"

We start with the individual psychology and ask the following four questions:

Is it possible that our feelings – which sometimes, when dealing with a bigger problem, where thoughts seem split up and form almost independent "entities" – have something to do with our findings here?

Is it possible that thoughts – when being in the process of creation do "feel" more like waves, while true ideas "haptically" manifest themselves as particles with mass and energy – have some bearing on the matter we discussed in this paper?

Is the conscious mind just a huge collection of quantum gravity processes with wave-particle dualities and has it evolved into the ability to create the right conditions to maintain the quantum gravity environment for allowing these states to exist in a fairly stable manner (long enough to have the impression of possessing consciousness)?

And why do we think that the quantum approach is the right one for all mind things?

Well, if we were wrong about the quantum-mind-connection, the Dunning-Kruger dependency would never have come out of the generalized Planck radiation law (see [14]) regarding the emittance of confidence, and the penetration of alien ideas into a well-protected mind could not be explained by quantum tunneling [98], and the collapse of an ensemble of thoughts to an idea or decision would probably not feel like a sudden flash of the mind [99].

Now we can easily see how all this – of course – is also related to marketing and thus, of much broader interest and with many more fields of applications than the lesser holistic reader might think. But as already said at the beginning of this book: possessing a technology of metamorphological power forces the owner of this very technology automatically to always also consider (model, simulate, foresee) the influence of his "gamechanger" on the socioeconomic environment as a whole. Anything else might not just lead to undesired effects (undesired by the innovation owner with respect to his social standing), but also to societal faulting, which potentially destroys the market basis for the key technology. The yeast fungi in a sugar solution, which slowly poison their own ecosystem, are an excellent example.

However, in the interest of brevity, we will keep the "Quantum Psychology of Markets" out of this book. The connection undoubtedly merits a separate consideration, contemplation, discussion, and – YES – publication (e.g., [56, 82]). This, however, shall not hinder the interested and skilled reader to already start working on the new and important field of "Quantum Marketing and Influencing".

10.8 Losing the Particle Idea – A Few Conclusions

What classically appears as a wave function in Quantum Theory is just the volume or scale fluctuation of the metric in a Quantum Gravity Theory. Thereby the classical wave function determines the metric volume perturbation and vanishes in the far distance of the quantum object under consideration. As the metric respectively, its volume usually is much bigger than the fluctuations of the same and as the quantum effects tend to wear off rather quickly, Quantum Theory only becomes important at smaller scales (usually, which is to say under ordinary energy, temperature, and gravity conditions).

We have shown that the so-called wave-particle duality shows itself directly in the simplest quantum gravity solutions as sub-structures of the metric tensor. Along the way, the idea of the particle reveals itself as an – in principle – unnecessary side-kick of compactified wave ensembles.

There are only waves and thus, consequently, everything of higher complexity can be decomposed into basic oscillations and ground waves and the latter can be used to compose things of higher complexity and structure. In essence, this just means that the general recipe for composition and decomposition – of everything (here of course with respect to the analysis of our key technologies T1 and T2) – requires nothing more than a generalized Fourier formation and a generalized Fourier analysis, respectively.

11 Why We Should Now Kill Schrödinger's Cat

De facto, we will not kill the cat, but only the corresponding thought experiment. In other words, we are going to deliver the poor cat from its unjust fate of an existence in an infinite limbo state.

This topic will be considered here because the protection of our two innovations T1 and T2 depends to some extend on the fact that, similar to the Schrödinger cat thought experiment, any "opening of the box" destroys the experiment and the superposed information within[18]. The Schrödinger cat problem is therefore of great importance for the protection of the innovations.

11.1 How Quantum Gravity Solves the Problem of the Wave Function Collapse

11.1.1 The Collapse of the Wave Function

Now we have (at least) three fundamentally derived scalar quantum gravity equations, namely (35), (107), and (117). All these equations can be linearized via conditions (36), (108), and (119), respectively. After this linearization process we obtain the standard quantum equations either directly or for most standard metrics (e.g., (9) and (111) for the cases (36) and (108), respectively).

This brings us to the following conjecture:

The collapse of the wave functions process is just a transition from linearity to nonlinearity, thereby forcing the wave function to leave its state of superposition of eigenfunctions and to take on the state of only one "observable" solution. It is the fragility of linearity, being only possible via a delicate setting for the metric "wrapper" F[f] of the wave function via conditions (36), (108), and (119), which, when the system is nudged (e.g., by the measurement) and pushed out of its comfortable stupor in a metrically restricted state allowing for the linear equations (38), (110), and (118) into the much higher dimensional reality with the nonlinear equations (35), (107), and (117). The "nudging" destroys the frail setting, which allows for the occurrence of linearity in the first place and forces the system onto a nonlinear partial differential apparatus. This causes the collapse and the destruction of the coherence of the eigenfunction ensemble. Actually, one might just put it like this: The nudging wakes the system up and forces it to make a decision.

And what about Schrödinger's cat?

Well, in order to achieve the coherence of a system as big as a cat, including Schrödinger's box and the complicated killing mechanism, one had to cool down everything so close to the absolute zero that the cat would be dead anyway. Thus, the question does not arise at all.

That being said, however, it must be pointed out that in "artificial space-times" quantum conditions can be created by the inhabitants. This has been worked out as a key aspect of consciousness.

[18] Please note: Information cannot get lost in this universe. Thus, correctly one should better formulate that the information being critical about the T1-T2-experiments is rendered inaccessible when the box is been opened.

Consequently, any owner of a transformative technology could try to construct artificial quantum or effective quasi quantum conditions in the environment of his technology that any "opening of the box" immediately renders the information within, which is to say the secrets about the innovation are rendered quite fundamentally inaccessible.

11.1.1.1 T2 as a Collapse of Wave Function Application

Processes which would switch between the states of coherent linear superposition and nonlinear individuality allow for astounding effects similar to the calculation wonders of quantum computers. Applied in cycles with tact and clocked suitably in the two states, will then obviously work as quantum thermodynamic machines, allowing for ordering and self-organization of self-assembling processes, which defy, if not to say mock the second law of thermodynamics. T2 is just such a process.

11.2 A Bit of Repetition and a Few Consequences Regarding the Deliverance of Schrödinger's Cat

Regarding our transformative key technologies, be it with respect to their protection in a competitive and potentially hostile socioeconomic environment or with respect to their physical description, we should summarize the following from what was elaborated and discussed above:

What classically appears as wave function in Quantum Theory is just the volume or scale fluctuation of the metric in a Quantum Gravity Theory. Thereby the classical wave function determines the metric volume perturbation and vanishes in the far distance of the quantum object under consideration. As the metric, respectively, its volume usually is much bigger than the fluctuations of the same and as the quantum effects tend to wear off rather quickly, Quantum Theory only becomes important at smaller scales (usually, which is to say under ordinary energy, temperature, and gravity conditions).

What classically appears as potential is just the entanglement of dimensions.

We also conclude that for any space or space-time of properties, attributes, degrees of freedom, or – YES – dimensions, apparently no matter what it is, the following holds:

a) It can produce coherent states due to possible settings for the volume part of the metric, allowing for linear partial differential equations and thus, additivity of eigen solutions for the wave functions. These eigen solutions, respectively, the ensembles of eigen solutions are just perturbations existing on these very properties, attributes, degrees of freedom or – YES – dimensions in accordance with our linearized quantum gravity field equations.
b) When the system is "nudged" (perturbed), the frail setting, which allowed for the linearity of the quantum gravity field equations, is destroyed and the additivity is lost. The system "has to make a decision" about which eigen state to exist in and the wave function collapses, because – after the nudging – it has to satisfy a nonlinear field equation.

The important finding, when deriving the wave function collapse via our metric approach, is the generality of the effect. It is by no means restricted to small particles and quantum fields. But by adjusting the parameters (Planck constant and mass) to the experimental findings, we see that indeed only "things" with small masses and a limited number of degrees of freedom have a chance of showing significant coherence in order to achieve the delicate states allowing for quantum gravity linearity. Schrödinger's cat can therefore never be realized, because coherence of a system as big as a cat requires cooling the "thing" down to absolute zero, thereby killing the cat for sure and rendering the question about the combined dead-alive-state of the Schrödinger cat completely meaningless.

It should be pointed out that, in addition to the linearity-nonlinearity-transition, also our metric Everett approach with the "universal selector" or "decision maker" would allow for the explanation of the wave function collapse. We refer to the corresponding sections above in this book.

12 From the Three Towers to the Three Generations

In this section we will see that when applying the Bianchi identity on a scalarized Quantum Gravity Physics realm we obtain three solutions. With this recognition we can solve the problem of the three generations of elementary particles. However, as the Bianchi identity is just a fundamental mathematical rule and nothing physical, which is to say nothing restricted to particle physics or physics at all, the 3-generation-outcome holds for any system and not just elementary particles. Here we investigate this problem in order to obtain the means for the discussion of the possibility of explaining the wondersome behavior of the selection agent in T2 via a permanent change between "generations" or "generational states". Thereby it could be assumed that the selectivity is just a switch between one generation in the bonded state with the target material and another generation in the unbonded state. Just like with the examples of charged leptons, the unbonded state would be correlated to the electron with low mass, while the bonded state correlates with the muon or the tauon with significantly higher masses.

Along the way, we will also – fundamentally – derive the metric origin of matter and energy. This is of need for our transformative technologies T1 and T2 as they deal with matter. And by insisting on the most fundamental possible starting point for our attempt to describe the innovations mathematically later on, we – naturally – have to know how to metrically derive matter and energy. As a consequence of this derivation, we will have to adjust, if not to destroy the covariance principle… at least in its dogmatic classical interpretation [203].

12.1 Using Coordinate Transformation to Obtain Other Degrees of Freedom

When considering the transformation rules of tensors, the classical Einstein-Hilbert action kernel:

$$\delta W = 0 = \delta \int_V d^n x \sqrt{-g} \cdot R \tag{339}$$

would only be affected with respect to the determinant term, because we have:

$$\begin{aligned}
\delta W = 0 &= \delta \int_V d^n x \sqrt{-g} \cdot R = \delta \int_V d^n x \sqrt{-g} \cdot R_{\alpha\beta} g^{\alpha\beta} \\
&= \delta \int_V d^n x \sqrt{-\det\left[\frac{\partial x^a}{\partial x^\alpha}\frac{\partial x^b}{\partial x^\beta} g_{ab}\right]} \cdot \frac{\partial x^a}{\partial x^\alpha}\frac{\partial x^b}{\partial x^\beta} R_{ab} g^{cd} \frac{\partial x^\alpha}{\partial x^c}\frac{\partial x^\beta}{\partial x^d} \cdot \\
&= \delta \int_V d^n x \sqrt{-\det\left[\frac{\partial x^a}{\partial x^\alpha}\frac{\partial x^b}{\partial x^\beta} g_{ab}\right]} \cdot R_{ab} g^{ab}
\end{aligned} \tag{340}$$

Describing the transformation matrix $\frac{\partial x^a}{\partial x^\alpha}$ via a function term $\frac{\partial x^a}{\partial x^\alpha}=\frac{\partial G^a}{\partial x^\alpha}=G^a_{\ ,\alpha}$, we can write:

$$\delta W=0=\delta\int_V d^n x\sqrt{-g}\cdot R=\delta\int_V d^n x\sqrt{-\det\left[G^a_{\ ,\alpha}G^b_{\ ,\beta}g_{ab}\right]}\cdot R\,. \tag{341}$$

We may reformulate (340) by just using the Jacobi-determinate as follows:

$$\delta W=0=\delta\int_V d^n x\sqrt{-g}\cdot R=\delta\int_V d^n x\left|\det\left[G^a_{\ ,\alpha}\right]\right|\cdot R_{ab}g^{ab}\,. \tag{342}$$

With the determinant being given via the Levi-Civita symbol as follows:

$$\det\left[G^a_{\ ,\alpha}\right]=\frac{1}{n!}\cdot\varepsilon_{i_1\ldots i_n}\varepsilon^{j_1\ldots j_n}G^{i_1}_{\ ,j_1}\cdot\ldots\cdot G^{i_n}_{\ ,j_n}, \tag{343}$$

the functional derivative or variation with respect to the function vector G^a would lead to:

$$\delta W=\delta_{G^a}\int_V d^n x\left|\det\left[G^a_{\ ,\alpha}\right]\right|\cdot R=-\left(\left(\frac{d\left|\det\left[G^a_{\ ,\alpha}\right]\right|}{d\left(G^a_{\ ,\alpha}\right)}\right)\cdot R\right)_{,\alpha}. \tag{344}$$

To give an example, in two dimensions this would lead to the equations:

$$\begin{aligned}
\delta W&=\delta_{G^a}\int_V d^n x\left|\det\left[G^a_{\ ,\alpha}\right]\right|\cdot R=-\left(\left(\frac{d\left|\det\left[G^a_{\ ,\alpha}\right]\right|}{d\left(G^a_{\ ,\alpha}\right)}\right)\cdot R\right)_{,\alpha}\\
&\xrightarrow{n=2;\ G^1_{\ ,2}=G^2_{\ ,1}}\\
&=-\left(\begin{pmatrix}\frac{d\left|\det\left[G^a_{\ ,\alpha}\right]\right|}{d\left(G^1_{\ ,1}\right)} & \frac{d\left|\det\left[G^a_{\ ,\alpha}\right]\right|}{d\left(G^1_{\ ,2}\right)}\\ \frac{d\left|\det\left[G^a_{\ ,\alpha}\right]\right|}{d\left(G^1_{\ ,2}\right)} & \frac{d\left|\det\left[G^a_{\ ,\alpha}\right]\right|}{d\left(G^2_{\ ,2}\right)}\end{pmatrix}_a^\alpha\cdot R\right)_{,\alpha}\\
&=-\left(\begin{pmatrix}\frac{d\left|G^1_{\ ,1}G^2_{\ ,2}-\left(G^1_{\ ,2}\right)^2\right|}{d\left(G^1_{\ ,1}\right)} & \frac{d\left|G^1_{\ ,1}G^2_{\ ,2}-\left(G^1_{\ ,2}\right)^2\right|}{d\left(G^1_{\ ,2}\right)}\\ \frac{d\left|G^1_{\ ,1}G^2_{\ ,2}-\left(G^1_{\ ,2}\right)^2\right|}{d\left(G^1_{\ ,2}\right)} & \frac{d\left|G^1_{\ ,1}G^2_{\ ,2}-\left(G^1_{\ ,2}\right)^2\right|}{d\left(G^2_{\ ,2}\right)}\end{pmatrix}_a^\alpha\cdot R\right)_{,\alpha}\\
&=-\left(\begin{pmatrix}G^2_{\ ,2} & -2\cdot G^1_{\ ,2}\\ -2\cdot G^1_{\ ,2} & G^1_{\ ,1}\end{pmatrix}_a^\alpha\cdot R\right)_{,\alpha}
\end{aligned} \tag{345}$$

If we prefer to rewrite everything with the determinant of the metric tensor, the variation equation would read:

$$\delta W = 0 = \delta_{G^a|G^b} \int_V d^n x \sqrt{-\det\left[G^a{}_{,\alpha} G^b{}_{,\beta} g_{ab}\right]} \cdot R$$
$$= -\left(\left(\frac{d\sqrt{-\det\left[G^a{}_{,\alpha} G^b{}_{,\beta} g_{ab}\right]}}{d\left(G^a{}_{,\alpha}\right)} + \frac{d\sqrt{-\det\left[G^a{}_{,\alpha} G^b{}_{,\beta} g_{ab}\right]}}{d\left(G^b{}_{,\beta}\right)}\right) \cdot R\right)_{,\alpha|\beta}$$
$$= -\left(\left(\sqrt{\det\left[G^b{}_{,\beta}\right]} \frac{d\sqrt{\det\left[G^a{}_{,\alpha}\right]}}{d\left(G^a{}_{,\alpha}\right)} + \sqrt{\det\left[G^a{}_{,\alpha}\right]} \frac{d\sqrt{\det\left[G^b{}_{,\beta}\right]}}{d\left(G^b{}_{,\beta}\right)}\right) \cdot \sqrt{-g} \cdot R\right)_{,\alpha|\beta} . \tag{346}$$
$$\Rightarrow 0 = \left(\begin{array}{c} \left(\left(\frac{d\sqrt{\det\left[G^a{}_{,\alpha}\right]}}{d\left(G^a{}_{,\alpha}\right)}\right)_a^\alpha \cdot \sqrt{-g \det\left[G^b{}_{,\beta}\right]} \cdot R\right)_{,\alpha} \\ + \left(\left(\frac{d\sqrt{\det\left[G^b{}_{,\beta}\right]}}{d\left(G^b{}_{,\beta}\right)}\right)_b^\beta \cdot \sqrt{-g\sqrt{\det\left[G^a{}_{,\alpha}\right]}} \cdot R\right)_{,\beta} \end{array}\right)$$

Again, using the example in 2 dimensions with n=2, we now obtain:

$$\delta W = 0 = \delta_{G^a|G^b} \int_V d^n x \sqrt{-\det\left[G^a{}_{,\alpha} G^b{}_{,\beta} g_{ab}\right]} \cdot R$$
$$= \left(\begin{array}{c} \left(\frac{1}{2\sqrt{\det\left[G^a{}_{,\alpha}\right]}} \begin{pmatrix} G^2{}_{,2} & -2 \cdot G^1{}_{,2} \\ -2 \cdot G^1{}_{,2} & G^1{}_{,1} \end{pmatrix}_a^\alpha \cdot \sqrt{-g \det\left[G^b{}_{,\beta}\right]} \cdot R\right)_{,\alpha} \\ + \left(\frac{1}{2\sqrt{\det\left[G^b{}_{,\beta}\right]}} \begin{pmatrix} G^2{}_{,2} & -2 \cdot G^1{}_{,2} \\ -2 \cdot G^1{}_{,2} & G^1{}_{,1} \end{pmatrix}_b^\beta \cdot \sqrt{-g\sqrt{\det\left[G^a{}_{,\alpha}\right]}} \cdot R\right)_{,\beta} \end{array}\right)$$
$$\Rightarrow 0 = \frac{1}{2} \left(\begin{array}{c} \left(\begin{pmatrix} G^2{}_{,2} & -2 \cdot G^1{}_{,2} \\ -2 \cdot G^1{}_{,2} & G^1{}_{,1} \end{pmatrix}_a^\alpha \cdot \sqrt{-g} \cdot R\right)_{,\alpha} \\ + \left(\begin{pmatrix} G^2{}_{,2} & -2 \cdot G^1{}_{,2} \\ -2 \cdot G^1{}_{,2} & G^1{}_{,1} \end{pmatrix}_b^\beta \cdot \sqrt{-g} \cdot R\right)_{,\beta} \end{array}\right)$$
$$= \left(\begin{pmatrix} G^2{}_{,2} & -2 \cdot G^1{}_{,2} \\ -2 \cdot G^1{}_{,2} & G^1{}_{,1} \end{pmatrix}_a^\alpha \cdot \sqrt{-g} \cdot R\right)_{,\alpha} . \tag{347}$$

12.2 What Is Matter?

We realize that in n dimensions, we obtain n equations for the function G^a. These equations are of third order, because of the fact that the Ricci scalar already contains up to second order derivatives and we are reminded of the following well-known condition for the Einstein field equations:

$$0 = G^{\alpha\beta}{}_{;\beta} = \left(R^{\alpha\beta} - \frac{R}{2} g^{\alpha\beta} \right)_{;\beta} = G_{\alpha}{}^{\beta}{}_{;\beta} = \left(R_{\alpha}^{\beta} - \frac{R}{2} \delta_{\alpha}^{\beta} \right)_{;\beta}. \tag{348}$$

Please note that only in (348) we – exceptionally – used the symbol $G^{\alpha\beta}$ for the Einstein tensor and not for the scaled metric tensor.

This condition can directly be obtained from the Bianchi identity for the Riemann curvature tensor, reading:

$$0 = R_{\alpha\beta\gamma\delta;\varepsilon} + R_{\alpha\beta\varepsilon\gamma;\delta} + R_{\alpha\beta\delta\varepsilon;\gamma}, \tag{349}$$

via contraction with the metric tensor as follows:

$$0 = \left(R_{\alpha\beta\gamma\delta;\varepsilon} + R_{\alpha\beta\varepsilon\gamma;\delta} + R_{\alpha\beta\delta\varepsilon;\gamma} \right) g^{\alpha\delta} g^{\beta\gamma} = \xrightarrow{g^{\alpha\delta}{}_{;\lambda}=0} = 2 \cdot R^{\alpha}{}_{\varepsilon;\alpha} - R_{;\varepsilon}. \tag{350}$$

In other words, our variation result (344) and (346) are just different forms of the condition (generalized conservation law) (348).

Applying this onto our scaled metric variation result (our quantum Einstein field equations) from (29), after bringing them into mixed form, we would have to demand, respectively obtain the additional condition:

$$0=\left(\begin{array}{c}
R^{\beta}_{\alpha;\beta}-\frac{\delta^{\beta}_{\alpha}}{2}R_{;\beta}+\frac{\delta^{\beta}_{\alpha}}{2}\left(\begin{array}{c}\frac{1}{2F}\left(\begin{array}{c}2F_{,ij}(n-1)g^{ij}+2\Gamma^{a}_{ij}F_{,a}g^{ij}\\ -F_{,i}g^{ab}g_{jb,a}g^{ij}-F_{,j}g^{ab}g_{ib,a}g^{ij}\\ -n\Gamma^{d}_{ij}F_{,d}g^{ij}+\frac{n}{2}F_{,d}g^{cd}g_{ab,c}g^{ab}\end{array}\right)\\ +\frac{F_{,i}\cdot F_{,j}}{4F^{2}}g^{ij}\left((n-6)(n-1)\right)\end{array}\right)_{;\beta}\\
+\frac{F_{;\beta}}{2F^{2}}\left(g^{\beta\gamma}\left(\begin{array}{c}F_{,\alpha\gamma}(n-2)+F_{,ab}g_{\alpha\gamma}g^{ab}\\ +F_{,a}g^{ab}\left(g_{\gamma b,\alpha}-g_{\gamma\alpha,b}\right)-F_{,\alpha}g^{ab}g_{\gamma b,a}-F_{,\gamma}g^{ab}g_{\alpha b,a}\\ +F_{,d}g^{cd}\frac{1}{2}n\left(\frac{2}{n}g_{\alpha c,\gamma}-g_{\alpha c,\gamma}-g_{\gamma c,\alpha}+g_{\alpha\gamma,c}+\frac{1}{n}g_{\alpha\gamma}g_{ab,c}g^{ab}\right)\\ -\frac{1}{2F}\left(F_{,\alpha}\cdot F_{,\gamma}(3n-6)+g_{\alpha\gamma}F_{,c}F_{,d}g^{cd}(4-n)\right)\end{array}\right)\right)\\
-\frac{1}{2F}\left(g^{\beta\gamma}\left(\begin{array}{c}F_{,\alpha\gamma}(n-2)+F_{,ab}g_{\alpha\gamma}g^{ab}\\ +F_{,a}g^{ab}\left(g_{\gamma b,\alpha}-g_{\gamma\alpha,b}\right)-F_{,\alpha}g^{ab}g_{\gamma b,a}-F_{,\gamma}g^{ab}g_{\alpha b,a}\\ +F_{,d}g^{cd}\frac{1}{2}n\left(\frac{2}{n}g_{\alpha c,\gamma}-g_{\alpha c,\gamma}-g_{\gamma c,\alpha}+g_{\alpha\gamma,c}+\frac{1}{n}g_{\alpha\gamma}g_{ab,c}g^{ab}\right)\\ -\frac{1}{2F}\left(F_{,\alpha}\cdot F_{,\gamma}(3n-6)+g_{\alpha\gamma}F_{,c}F_{,d}g^{cd}(4-n)\right)\end{array}\right)\right)_{;\beta}\\
+\kappa T^{\beta}_{\alpha;\beta}+\Lambda\cdot\overbrace{\left(F\cdot g_{\alpha b}\cdot\frac{g^{b\beta}}{F}\right)_{;\beta}}^{\delta^{\beta}_{\alpha;\beta}=0}
\end{array}\right). \quad (351)$$

Important note: Thereby, the covariant derivative has been evaluated with respect to the metric tensor $G_{\alpha\beta}$ and not $g_{\alpha\beta}$. Covariant derivation on the basis of the unscaled metric, gives:

$$
0=\left(\begin{array}{c}
\overbrace{R^{\beta}_{\alpha;\beta}-\frac{\delta^{\beta}_{\alpha}}{2}R_{;\beta}}^{=0,\ \text{Bianchi}}+\frac{\delta^{\beta}_{\alpha}}{2}\left(\frac{1}{2F}\begin{pmatrix}2F_{,ij}(n-1)g^{ij}+2\Gamma^{a}_{ij}F_{,a}g^{ij}\\ -F_{,i}g^{ab}g_{jb,a}g^{ij}-F_{,j}g^{ah}g_{ib,a}g^{ij}\\ -n\Gamma^{d}_{ij}F_{,d}g^{ij}+\frac{n}{2}F_{,d}g^{cd}g_{ab,c}g^{ab}\end{pmatrix}+\frac{F_{,i}\cdot F_{,j}}{4F^{2}}g^{ij}\left((n-6)(n-1)\right)\right)_{,\beta}\\
+\frac{F_{,\beta}}{2F^{2}}\left(g^{\beta\gamma}\begin{pmatrix}F_{,\alpha\gamma}(n-2)+F_{,ab}g_{\alpha\gamma}g^{ab}\\ +F_{,a}g^{ab}\left(g_{\gamma b,\alpha}-g_{\gamma\alpha,b}\right)-F_{,\alpha}g^{ab}g_{\gamma b,a}-F_{,\gamma}g^{ab}g_{\alpha b,a}\\ +F_{,d}g^{cd}\frac{1}{2}n\left(\frac{2}{n}g_{\alpha c,\gamma}-g_{\alpha c,\gamma}-g_{\gamma c,\alpha}+g_{\alpha\gamma,c}+\frac{1}{n}g_{\alpha\gamma}g_{ab,c}g^{ab}\right)\\ -\frac{1}{2F}\left(F_{,\alpha}\cdot F_{,\gamma}(3n-6)+g_{\alpha\gamma}F_{,c}F_{,d}g^{cd}(4-n)\right)\end{pmatrix}\right)\\
-\frac{1}{2F}\left(g^{\beta\gamma}\begin{pmatrix}F_{,\alpha\gamma}(n-2)+F_{,ab}g_{\alpha\gamma}g^{ab}\\ +F_{,a}g^{ab}\left(g_{\gamma b,\alpha}-g_{\gamma\alpha,b}\right)-F_{,\alpha}g^{ab}g_{\gamma b,a}-F_{,\gamma}g^{ab}g_{\alpha b,a}\\ +F_{,d}g^{cd}\frac{1}{2}n\left(\frac{2}{n}g_{\alpha c,\gamma}-g_{\alpha c,\gamma}-g_{\gamma c,\alpha}+g_{\alpha\gamma,c}+\frac{1}{n}g_{\alpha\gamma}g_{ab,c}g^{ab}\right)\\ -\frac{1}{2F}\left(F_{,\alpha}\cdot F_{,\gamma}(3n-6)+g_{\alpha\gamma}F_{,c}F_{,d}g^{cd}(4-n)\right)\end{pmatrix}\right)_{;\beta}\\
+\kappa T^{\beta}_{\alpha;\beta}+\Lambda\cdot\overbrace{\left(F\cdot g_{\alpha b}\cdot\frac{g^{b\beta}}{F}\right)_{;\beta}}^{\delta^{\beta}_{\alpha;\beta}=0}
\end{array}\right)
\tag{352}
$$

and makes the Ricci terms disappear.

While the classical covariant derivative with respect to $g_{\alpha\beta}$ acts as follows on a mixed tensor:

$$
\begin{array}{c}
R^{\beta}_{\alpha;\gamma}=R^{\beta}_{\alpha,\gamma}+\Gamma^{\beta}_{\gamma a}R^{a}_{\alpha}-\Gamma^{a}_{\gamma\alpha}R^{\beta}_{a}\quad\Rightarrow\quad R^{\beta}_{\alpha;\beta}=R^{\beta}_{\alpha,\beta}+\Gamma^{\beta}_{\beta a}R^{a}_{\alpha}-\Gamma^{a}_{\beta\alpha}R^{\beta}_{a}\\
\Gamma^{\gamma}_{\alpha\beta}=\frac{g^{\gamma a}}{2}\left(g_{\alpha a,\beta}-g_{\alpha c,\gamma}-g_{\gamma c,\alpha}\right)
\end{array},
\tag{353}
$$

the covariant derivative with respect to $G_{\alpha\beta}$ evaluates as follows when being applied on a mixed tensor:

$$
\begin{array}{c}
R^{\beta}_{\alpha;\gamma}=R^{\beta}_{\alpha,\gamma}+\Gamma^{*\beta}_{\gamma a}R^{a}_{\alpha}-\Gamma^{*a}_{\gamma\alpha}R^{\beta}_{a}\quad\Rightarrow\quad R^{\beta}_{\alpha;\beta}=R^{\beta}_{\alpha,\beta}+\Gamma^{*\beta}_{\beta a}R^{a}_{\alpha}-\Gamma^{*a}_{\beta\alpha}R^{\beta}_{a}\\
\Gamma^{*\gamma}_{\alpha\beta}=\frac{G^{\gamma a}}{2}\left(G_{\alpha a,\beta}-G_{\alpha c,\gamma}-G_{\gamma c,\alpha}\right)
\end{array}.
\tag{354}
$$

The corresponding new Christoffel symbols are given via:

$$
\begin{aligned}
\Gamma^{*\gamma}_{\alpha\beta} &= \frac{g^{\gamma\sigma}}{2\cdot F[f]}\left(\left[F[f]\cdot g_{\sigma\alpha}\right]_{,\beta}+\left[F[f]\cdot g_{\sigma\beta}\right]_{,\alpha}-\left[F[f]\cdot g_{\alpha\beta}\right]_{,\sigma}\right)\\
&= \frac{g^{\gamma\sigma}}{2\cdot F}\left(\left[F\cdot g_{\sigma\alpha}\right]_{,\beta}+\left[F\cdot g_{\sigma\beta}\right]_{,\alpha}-\left[F\cdot g_{\alpha\beta}\right]_{,\sigma}\right)\\
&= \frac{g^{\gamma\sigma}}{2}\left(g_{\sigma\alpha,\beta}+g_{\sigma\beta,\alpha}-g_{\alpha\beta,\sigma}\right)+\frac{g^{\gamma\sigma}F'}{2\cdot F}\left(f_{,\beta}\cdot g_{\sigma\alpha}+f_{,\alpha}\cdot g_{\sigma\beta}-f_{,\sigma}\cdot g_{\alpha\beta}\right)\cdot\\
&= \Gamma^{\gamma}_{\alpha\beta}+\frac{F'}{2\cdot F}\left(f_{,\beta}\cdot\delta^{\gamma}_{\alpha}+f_{,\alpha}\cdot\delta^{\gamma}_{\beta}-f_{,\sigma}\cdot g^{\gamma\sigma}g_{\alpha\beta}\right)\\
&\equiv \Gamma^{\gamma}_{\alpha\beta}+\Gamma^{**\gamma}_{\alpha\beta}
\end{aligned}
\tag{355}
$$

This makes (354) to:

$$
\begin{aligned}
& R^{\beta}_{\alpha;\gamma}=R^{\beta}_{\alpha,\gamma}+\left(\Gamma^{\beta}_{\gamma a}+\Gamma^{**\beta}_{\gamma a}\right)R^{a}_{\alpha}-\left(\Gamma^{a}_{\gamma\alpha}+\Gamma^{**\beta}_{\gamma a}\right)R^{\beta}_{a}\\
\Rightarrow\quad & R^{\beta}_{\alpha;\beta}=R^{\beta}_{\alpha,\beta}+\left(\Gamma^{\beta}_{\beta a}+\Gamma^{**\beta}_{\beta a}\right)R^{a}_{\alpha}-\left(\Gamma^{a}_{\beta\alpha}+\Gamma^{**a}_{\beta\alpha}\right)R^{\beta}_{a}
\end{aligned}
\tag{356}
$$

Assuming vanishing cosmological constant and matter Lagrange density, which in the classical approaches were just postulated anyway, immediately leads to:

$$
0=\left(\begin{array}{c}
R^{\beta}_{\alpha;\beta}-\frac{\delta^{\beta}_{\alpha}}{2}R_{;\beta}+\frac{\delta^{\beta}_{\alpha}}{2}\left(\begin{array}{c}\frac{1}{2F}\left(\begin{array}{c}2F_{,ij}(n-1)g^{ij}+2\Gamma^{a}_{ij}F_{,a}g^{ij}\\ -F_{,i}g^{ab}g_{jb,a}g^{ij}-F_{,j}g^{ab}g_{ib,a}g^{ij}\\ -n\Gamma^{d}_{ij}F_{,d}g^{ij}+\frac{n}{2}F_{,d}g^{cd}g_{ab,c}g^{ab}\end{array}\right)\\ +\frac{F_{,i}\cdot F_{,j}}{4F^{2}}g^{ij}\left((n-6)(n-1)\right)\end{array}\right)_{;\beta}\\
+\frac{F_{;\beta}}{2F^{2}}\left(g^{\beta\gamma}\left(\begin{array}{c}F_{,\alpha\gamma}(n-2)+F_{,ab}g_{\alpha\gamma}g^{ab}\\ +F_{,a}g^{ab}\left(g_{\gamma b,\alpha}-g_{\gamma\alpha,b}\right)-F_{,\alpha}g^{ab}g_{\gamma b,a}-F_{,\gamma}g^{ab}g_{\alpha b,a}\\ +F_{,d}g^{cd}\frac{1}{2}n\left(\frac{2}{n}g_{\alpha c,\gamma}-g_{\alpha c,\gamma}-g_{\gamma c,\alpha}+g_{\alpha\gamma,c}+\frac{1}{n}g_{\alpha\gamma}g_{ab,c}g^{ab}\right)\\ -\frac{1}{2F}\left(F_{,\alpha}\cdot F_{,\gamma}(3n-6)+g_{\alpha\gamma}F_{,c}F_{,d}g^{cd}(4-n)\right)\end{array}\right)\right)\\
-\frac{1}{2F}\left(g^{\beta\gamma}\left(\begin{array}{c}F_{,\alpha\gamma}(n-2)+F_{,ab}g_{\alpha\gamma}g^{ab}\\ +F_{,a}g^{ab}\left(g_{\gamma b,\alpha}-g_{\gamma\alpha,b}\right)-F_{,\alpha}g^{ab}g_{\gamma b,a}-F_{,\gamma}g^{ab}g_{\alpha b,a}\\ +F_{,d}g^{cd}\frac{1}{2}n\left(\frac{2}{n}g_{\alpha c,\gamma}-g_{\alpha c,\gamma}-g_{\gamma c,\alpha}+g_{\alpha\gamma,c}+\frac{1}{n}g_{\alpha\gamma}g_{ab,c}g^{ab}\right)\\ -\frac{1}{2F}\left(F_{,\alpha}\cdot F_{,\gamma}(3n-6)+g_{\alpha\gamma}F_{,c}F_{,d}g^{cd}(4-n)\right)\end{array}\right)\right)_{;\beta}
\end{array}\right).
\tag{357}
$$

Now taking into account the contracted Bianchi identity from (350) (remember, this was with respect to the unscaled metric $g_{\alpha\beta}$), we obtain:

$$R^{\beta}_{\alpha;\beta}-\frac{\delta^{\beta}_{\alpha}}{2}R_{;\beta}=R^{\beta}_{\alpha,\beta}+\left(\Gamma^{\beta}_{\beta a}+\Gamma^{**\beta}_{\beta a}\right)R^{a}_{\alpha}-\left(\Gamma^{a}_{\beta\alpha}+\Gamma^{**a}_{\beta\alpha}\right)R^{\beta}_{a}-\frac{\delta^{\beta}_{\alpha}}{2}R_{;\beta}$$

$$=\overbrace{R^{\beta}_{\alpha,\beta}+\Gamma^{\beta}_{\beta a}R^{a}_{\alpha}-\Gamma^{a}_{\beta\alpha}R^{\beta}_{a}-\frac{\delta^{\beta}_{\alpha}}{2}R_{;\beta}}^{=0}+\Gamma^{**\beta}_{\beta a}R^{a}_{\alpha}-\Gamma^{**a}_{\beta\alpha}R^{\beta}_{a}\quad\Rightarrow$$

$$0=\left(\begin{array}{l}\Gamma^{**\beta}_{\beta a}R^{a}_{\alpha}-\Gamma^{**a}_{\beta\alpha}R^{\beta}_{a}+\frac{\delta^{\beta}_{\alpha}}{2}\left(\frac{1}{2F}\left(\begin{array}{l}2F_{,ij}(n-1)g^{ij}+2\Gamma^{a}_{ij}F_{,a}g^{ij}\\-F_{,i}g^{ab}g_{jb,a}g^{ij}-F_{,j}g^{ab}g_{ib,a}g^{ij}\\-n\Gamma^{d}_{ij}F_{,d}g^{ij}+\frac{n}{2}F_{,d}g^{cd}g_{ab,c}g^{ab}\end{array}\right)+\frac{F_{,i}\cdot F_{,j}}{4F^{2}}g^{ij}\left((n-6)(n-1)\right)\right)_{;\beta}\\
+\frac{F_{;\beta}}{2F^{2}}\left(g^{\beta\gamma}\left(\begin{array}{l}F_{,\alpha\gamma}(n-2)+F_{,ab}g_{\alpha\gamma}g^{ab}\\+F_{,a}g^{ab}\left(g_{\gamma b,\alpha}-g_{\gamma\alpha,b}\right)-F_{,\alpha}g^{ab}g_{\gamma b,a}-F_{,\gamma}g^{ab}g_{\alpha b,a}\\+F_{,d}g^{cd}\frac{1}{2}n\left(\frac{2}{n}g_{\alpha c,\gamma}-g_{\alpha c,\gamma}-g_{\gamma c,\alpha}+g_{\alpha\gamma,c}+\frac{1}{n}g_{\alpha\gamma}g_{ab,c}g^{ab}\right)\\-\frac{1}{2F}\left(F_{,\alpha}\cdot F_{,\gamma}(3n-6)+g_{\alpha\gamma}F_{,c}F_{,d}g^{cd}(4-n)\right)\end{array}\right)\right)\\
-\frac{1}{2F}\left(g^{\beta\gamma}\left(\begin{array}{l}F_{,\alpha\gamma}(n-2)+F_{,ab}g_{\alpha\gamma}g^{ab}\\+F_{,a}g^{ab}\left(g_{\gamma b,\alpha}-g_{\gamma\alpha,b}\right)-F_{,\alpha}g^{ab}g_{\gamma b,a}-F_{,\gamma}g^{ab}g_{\alpha b,a}\\+F_{,d}g^{cd}\frac{1}{2}n\left(\frac{2}{n}g_{\alpha c,\gamma}-g_{\alpha c,\gamma}-g_{\gamma c,\alpha}+g_{\alpha\gamma,c}+\frac{1}{n}g_{\alpha\gamma}g_{ab,c}g^{ab}\right)\\-\frac{1}{2F}\left(F_{,\alpha}\cdot F_{,\gamma}(3n-6)+g_{\alpha\gamma}F_{,c}F_{,d}g^{cd}(4-n)\right)\end{array}\right)\right)_{;\beta}\end{array}\right).\qquad(358)$$

Thus, we are left with scaling function terms and their derivatives plus a funny residual from the mixed Einstein tensor, which also results from the scaling function. By considering all the scaling function terms as matter components (effective matter for an observer like ourselves) with:

$$\kappa T_{\alpha}^{\beta}=\left(\begin{array}{c}\frac{\delta_{\alpha}^{\beta}}{2}\left(\begin{array}{c}\frac{1}{2F}\left(\begin{array}{c}2F_{,ij}(n-1)g^{ij}+2\Gamma_{ij}^{a}F_{,a}g^{ij}\\ -F_{,i}g^{ab}g_{jb,a}g^{ij}-F_{,j}g^{ab}g_{ib,a}g^{ij}\\ -n\Gamma_{ij}^{d}F_{,d}g^{ij}+\frac{n}{2}F_{,d}g^{cd}g_{ab,c}g^{ab}\end{array}\right)\\ +\frac{F_{,i}\cdot F_{,j}}{4F^{2}}g^{ij}((n-6)(n-1))\end{array}\right)\\ -\frac{1}{2F}\left(g^{\beta\gamma}\left(\begin{array}{c}F_{,\alpha\gamma}(n-2)+F_{,ab}g_{\alpha\gamma}g^{ab}\\ +F_{,a}g^{ab}\left(g_{\gamma b,\alpha}-g_{\gamma\alpha,b}\right)-F_{,\alpha}g^{ab}g_{\gamma b,a}-F_{,\gamma}g^{ab}g_{\alpha b,a}\\ +F_{,d}g^{cd}\frac{1}{2}n\left(\frac{2}{n}g_{\alpha c,\gamma}-g_{\alpha c,\gamma}-g_{\gamma c,\alpha}+g_{\alpha\gamma,c}+\frac{1}{n}g_{\alpha\gamma}g_{ab,c}g^{ab}\right)\\ -\frac{1}{2F}\left(F_{,\alpha}\cdot F_{,\gamma}(3n-6)+g_{\alpha\gamma}F_{,c}F_{,d}g^{cd}(4-n)\right)\end{array}\right)\right)\end{array}\right) \tag{359}$$

we see that, while (358) would just be a conservation law, reading:

$$\Gamma_{\beta a}^{**\beta}R_{\alpha}^{a}-\Gamma_{\beta\alpha}^{**a}R_{a}^{\beta}+T_{\alpha;\beta}^{\beta}=0\,, \tag{360}$$

this would by no means mirror the classical conservation law (e.g., see [101], p. 269), which was:

$$T_{\alpha;\beta}^{\beta}=0\,. \tag{361}$$

Thereby the classical equation refers to a covariant derivative with respect to the unscaled metric tensor $g_{\alpha\beta}$ and not the scaled one $G_{\alpha\beta}$.

However, it must be pointed out again that, as the covariant derivative here has to be performed with respect to the scaled metric, we have to evaluate the following:

$$\begin{array}{c}T_{\alpha;\gamma}^{\beta}=T_{\alpha,\gamma}^{\beta}+\left(\Gamma_{\gamma a}^{\beta}+\Gamma_{\gamma a}^{**\beta}\right)T_{\alpha}^{a}-\left(\Gamma_{\gamma\alpha}^{a}+\Gamma_{\gamma a}^{**\beta}\right)T_{a}^{\beta}\\ \Rightarrow\quad T_{\alpha;\beta}^{\beta}=T_{\alpha,\beta}^{\beta}+\left(\Gamma_{\beta a}^{\beta}+\Gamma_{\beta a}^{**\beta}\right)T_{\alpha}^{a}-\left(\Gamma_{\beta\alpha}^{a}+\Gamma_{\beta\alpha}^{**a}\right)T_{a}^{\beta}\end{array}. \tag{362}$$

We conclude that when assuming scaling factors as matter sources, the classical condition (361) (e.g., see [101], p. 269) does not suffice. With (358), respectively (360) providing the correct form, we might like to demand, when insisting on a classical form like (361), the following separate conditions for our energy momentum tensor (for the reason of simplicity, we temporarily set κ=1):

$$0=R^{\beta}_{\alpha;\beta}-\frac{\delta^{\beta}_{\alpha}}{2}R_{;\beta}+T^{\beta}_{\alpha;\beta}=\left(\begin{array}{c}R^{\beta}_{\alpha,\beta}+\left(\Gamma^{\beta}_{\beta a}+\Gamma^{**\beta}_{\beta a}\right)R^{a}_{\alpha}-\left(\Gamma^{a}_{\beta\alpha}+\Gamma^{**a}_{\beta\alpha}\right)R^{\beta}_{a}-\frac{\delta^{\beta}_{\alpha}}{2}R_{;\beta}\\+T^{\beta}_{\alpha,\beta}+\left(\Gamma^{\beta}_{\beta a}+\Gamma^{**\beta}_{\beta a}\right)T^{a}_{\alpha}-\left(\Gamma^{a}_{\beta\alpha}+\Gamma^{**a}_{\beta\alpha}\right)T^{\beta}_{a}\end{array}\right)$$

$$=\left(\begin{array}{c}\overbrace{R^{\beta}_{\alpha,\beta}+\Gamma^{\beta}_{\beta a}R^{a}_{\alpha}-\Gamma^{a}_{\beta\alpha}R^{\beta}_{a}-\frac{\delta^{\beta}_{\alpha}}{2}R_{;\beta}}^{=0,\text{ Bianchi}}+\Gamma^{**\beta}_{\beta a}R^{a}_{\alpha}-\Gamma^{**a}_{\beta\alpha}R^{\beta}_{a}\\+T^{\beta}_{\alpha,\beta}+\left(\Gamma^{\beta}_{\beta a}+\Gamma^{**\beta}_{\beta a}\right)T^{a}_{\alpha}-\left(\Gamma^{a}_{\beta\alpha}+\Gamma^{**a}_{\beta\alpha}\right)T^{\beta}_{a}\end{array}\right)$$

$$\Rightarrow 0=\Gamma^{**\beta}_{\beta a}R^{a}_{\alpha}-\Gamma^{**a}_{\beta\alpha}R^{\beta}_{a}+T^{\beta}_{\alpha,\beta}+\left(\Gamma^{\beta}_{\beta a}+\Gamma^{**\beta}_{\beta a}\right)T^{a}_{\alpha}-\left(\Gamma^{a}_{\beta\alpha}+\Gamma^{**a}_{\beta\alpha}\right)T^{\beta}_{a}$$

$$\xrightarrow{\text{we want}}$$

$$=\left(\overbrace{\Gamma^{**\beta}_{\beta a}R^{a}_{\alpha}-\Gamma^{**a}_{\beta\alpha}R^{\beta}_{a}+\Gamma^{**\beta}_{\beta a}T^{a}_{\alpha}-\Gamma^{**a}_{\beta\alpha}T^{\beta}_{a}}^{=0,\text{ scaling}}+\overbrace{T^{\beta}_{\alpha,\beta}+\Gamma^{\beta}_{\beta a}T^{a}_{\alpha}-\Gamma^{a}_{\beta\alpha}T^{\beta}_{a}}^{=0,\text{ classical}}\right)$$

$$\Rightarrow\left\{\begin{array}{c}0=\Gamma^{**\beta}_{\beta a}R^{a}_{\alpha}-\Gamma^{**a}_{\beta\alpha}R^{\beta}_{a}+\Gamma^{**\beta}_{\beta a}T^{a}_{\alpha}-\Gamma^{**a}_{\beta\alpha}T^{\beta}_{a}\\0=T^{\beta}_{\alpha,\beta}+\Gamma^{\beta}_{\beta a}T^{a}_{\alpha}-\Gamma^{a}_{\beta\alpha}T^{\beta}_{a}\end{array}\right. . \quad (363)$$

We could easily have such a condition, however, when, instead of applying the covariant derivative with respect to the scaled metric, we simply derivate with respect to the unscaled metric. This would be much more reasonable anyway, because, when demanding the postulated matter via the artificially added Lagrange density to vanish, the quantized (scaled) Einstein field equations would be given via (29) and, by using the abbreviation (359), in mixed form reads:

$$0=R^{*\beta}_{\alpha}-\frac{\delta^{\beta}_{\alpha}}{2}R^{*}=R^{\beta}_{\alpha}-\frac{\delta^{\beta}_{\alpha}}{2}R+T^{\beta}_{\alpha}. \quad (364)$$

Thus, applying the covariant derivative with respect to the scaled metric would be the derivative of a zero, which does not make much sense. Things are pretty much different when we apply the covariant derivative with respect to the unscaled metric, because this leads us to:

$$0=R^{*\beta}_{\alpha\,;\beta}-\frac{\delta^{\beta}_{\alpha}}{2}R^{*}_{\;;\beta}=\overbrace{R^{\beta}_{\alpha;\beta}-\frac{\delta^{\beta}_{\alpha}}{2}R_{;\beta}}^{=0,\text{ Bianchi}}+T^{\beta}_{\alpha;\beta}\quad\Rightarrow\quad T^{\beta}_{\alpha;\beta}=0. \quad (365)$$

Now the scaling has not only provided us with matter, but also gave us the classical conservation law.

We are going to discuss this aspect a bit more in our next subsection "The Evolution of Matter" within the section "New Matter? …". But before we come to that we will investigate the evolution potential of the matter term, its behavior in the case of $n\rightarrow\infty$, and will also try to solve the so-called 3-generation problem of elementary particles.

12.3 The Evolution of Matter

Further below in this book, especially in connection with the diffusion equation and the importance of the understanding of Sigmoid curves and their presence also in the processes of our innovations T1 and T2, we will permanently stumble over so-called evolution equations. Thereby, we also have to take into account the influence of self-organization which, too, leads us automatically to evolution and its many derivatives and appearances with respect to all forms of matter. This we will consider in the section here.

From condition (365) it becomes clear that the Bianchi identity just hands us a general evolution equation for the matter being contained in any arbitrary system. Simply by extracting the time derivative from (358), which has to be adjusted in accordance with the covariant derivative being performed based on the unscaled instead of the scaled metric:

$$
\begin{aligned}
R^{\beta}_{\alpha;\beta}-\frac{\delta^{\beta}_{\alpha}}{2}R_{;\beta} &= R^{\beta}_{\alpha,\beta}+\Gamma^{\beta}_{\beta a}R^{a}_{\alpha}-\Gamma^{a}_{\beta\alpha}R^{\beta}_{a}-\frac{\delta^{\beta}_{\alpha}}{2}R_{;\beta}\\
&= \overbrace{R^{\beta}_{\alpha,\beta}+\Gamma^{\beta}_{\beta a}R^{a}_{\alpha}-\Gamma^{a}_{\beta\alpha}R^{\beta}_{a}-\frac{\delta^{\beta}_{\alpha}}{2}R_{;\beta}}^{=0} \quad\Rightarrow
\end{aligned}
$$

$$
0=\left(\begin{array}{l}
\frac{\delta^{\beta}_{\alpha}}{2}\left(\begin{array}{c}
\frac{1}{2F}\left(\begin{array}{c}
2F_{,ij}(n-1)g^{ij}+2\Gamma^{a}_{ij}F_{,a}g^{ij}\\
-F_{,i}g^{ab}g_{jb,a}g^{ij}-F_{,j}g^{ab}g_{ib,a}g^{ij}\\
-n\Gamma^{d}_{ij}F_{,d}g^{ij}+\frac{n}{2}F_{,d}g^{cd}g_{ab,c}g^{ab}
\end{array}\right)\\
+\frac{F_{,i}\cdot F_{,j}}{4F^{2}}g^{ij}\left((n-6)(n-1)\right)
\end{array}\right)_{;\beta}\\
+\frac{F_{,\beta}}{2F^{2}}\left(g^{\beta\gamma}\left(\begin{array}{c}
F_{,\alpha\gamma}(n-2)+F_{,ab}g_{\alpha\gamma}g^{ab}\\
+F_{,a}g^{ab}\left(g_{\gamma b,\alpha}-g_{\gamma\alpha,b}\right)-F_{,\alpha}g^{ab}g_{\gamma b,a}-F_{,\gamma}g^{ab}g_{\alpha b,a}\\
+F_{,d}g^{cd}\frac{1}{2}n\left(\frac{2}{n}g_{\alpha c,\gamma}-g_{\alpha c,\gamma}-g_{\gamma c,\alpha}+g_{\alpha\gamma,c}+\frac{1}{n}g_{\alpha\gamma}g_{ab,c}g^{ab}\right)\\
-\frac{1}{2F}\left(F_{,\alpha}\cdot F_{,\gamma}(3n-6)+g_{\alpha\gamma}F_{,c}F_{,d}g^{cd}(4-n)\right)
\end{array}\right)\right)\\
-\frac{1}{2F}\left(g^{\beta\gamma}\left(\begin{array}{c}
F_{,\alpha\gamma}(n-2)+F_{,ab}g_{\alpha\gamma}g^{ab}\\
+F_{,a}g^{ab}\left(g_{\gamma b,\alpha}-g_{\gamma\alpha,b}\right)-F_{,\alpha}g^{ab}g_{\gamma b,a}-F_{,\gamma}g^{ab}g_{\alpha b,a}\\
+F_{,d}g^{cd}\frac{1}{2}n\left(\frac{2}{n}g_{\alpha c,\gamma}-g_{\alpha c,\gamma}-g_{\gamma c,\alpha}+g_{\alpha\gamma,c}+\frac{1}{n}g_{\alpha\gamma}g_{ab,c}g^{ab}\right)\\
-\frac{1}{2F}\left(F_{,\alpha}\cdot F_{,\gamma}(3n-6)+g_{\alpha\gamma}F_{,c}F_{,d}g^{cd}(4-n)\right)
\end{array}\right)\right)_{;\beta}
\end{array}\right), \qquad (366)
$$

respectively, (365) we obtain:

$$
\begin{gathered}
T^{\beta}_{\alpha;\beta} = 0 = T^{\beta}_{\alpha,\beta} + \Gamma^{\beta}_{\beta\delta} T^{\delta}_{\alpha} - \Gamma^{\delta}_{\beta\alpha} T^{\beta}_{\delta} \\
\Rightarrow T^{0}_{\alpha,0} = -\Gamma^{0}_{0\delta} T^{\delta}_{\alpha} + \Gamma^{\delta}_{0\alpha} T^{0}_{\delta} - T^{b}_{\alpha,b} - \Gamma^{b}_{b\delta} T^{\delta}_{\alpha} + \Gamma^{\delta}_{b\alpha} T^{b}_{\delta}, \\
\alpha,\beta,\delta = 0,1,\ldots,n-1; \quad b = 1,\ldots,n-1
\end{gathered}
\tag{367}
$$

which seems to be a first order evolution equation. However, when inserting the corresponding term for the matter tensor as given in (359) we see, that, depending on the complexity of the system under investigation, we in fact have a third order evolution equation. Third order equations, however, in general have three solutions and here we are already starting to stumble over the connection with the so-called 3-generation problem of elementary particles.

The situation somewhat simplifies in cases where at least the metric is static (independent on t) and only the volume or wave functions show time dependency.

In general, however, we are not truly satisfied with the situation, because (366) or (365) do not seem to allow us any reasonably superposable solutions. In order to investigate this fact more closely, we bring the matter equation in a different form (regarding the evaluation we refer to the appendix C in [7]):

$$
\kappa T^{\beta}_{\alpha} = \left(\begin{array}{c}
\frac{\delta^{\beta}_{\alpha}}{2} \left(\begin{array}{c}
\frac{F'}{2F} \left(\begin{array}{c}
2\Gamma^{a}_{ij} f_{,a} g^{ij} - f_{,i} g^{ab} g_{jb,a} g^{ij} - f_{,j} g^{ab} g_{ib,a} g^{ij} \\
-n\Gamma^{d}_{ij} f_{,d} g^{ij} + \frac{n}{2} f_{,d} g^{cd} g_{ab,c} g^{ab}
\end{array} \right) \\
+F' f_{,ab} \frac{(n-1)}{F} g^{ab}
\end{array} \right) \\
-\frac{1}{2F} \left(\begin{array}{c}
g^{\beta\gamma} \left(\begin{array}{c}
F' \left(\begin{array}{c}
f_{,a} g^{ab} \left(g_{\gamma b,\alpha} - g_{\gamma\alpha,b} \right) - f_{,\alpha} g^{ab} g_{\gamma b,a} - f_{,\gamma} g^{ab} g_{\alpha b,a} \\
+f_{,d} g^{cd} \frac{1}{2} n \left(\begin{array}{c}
\frac{2}{n} g_{\alpha c,\gamma} - g_{\alpha c,\gamma} - g_{\gamma c,\alpha} \\
+g_{\alpha\gamma,c} + \frac{1}{n} g_{\alpha\gamma} g_{ab,c} g^{ab}
\end{array} \right)
\end{array} \right) \\
+F' f_{,\alpha\gamma} (n-2) + F' f_{,ab} g_{\alpha\gamma} g^{ab}
\end{array} \right) \\
+\frac{g^{\beta\gamma} f_{,\alpha} \cdot f_{,\gamma}}{2F} \left((6-3n)(F')^{2} + 2FF''(n-2) \right) \\
+\frac{f_{,a} f_{,b} \delta^{\beta}_{\alpha} g^{ab}}{4F} \left(4FF''(2-n) + (F')^{2} \left(2(n-4) - (n-1)(n-6) \right) \right)
\end{array} \right)
\end{array} \right),
\tag{368}
$$

where we have the internal structure of F[f] taken into account (but still being left general) and find the linear and the nonlinear (red) derivative terms of the wave function f well separated.

12.3.1 The "Infinity Options Principle"

Almost as a by-product we can now derive conditions for the construction of universes without quantum uncertainty. This might be very useful when being interested in absolute certainty… for a change. Unfortunately, so the immediate critical argument, we cannot construct universes. Fine, point well taken,… but we can create systems and try to isolate them as much as possible from the rest, thereby obtaining a "universe of our liking"… or at least close to it. Now, being interested in achieving absolute certainty about certain things in this – our – universe, we want to derive the conditions where we can get rid of all quantum induced uncertainty. We will find that for all scalar quantum equations such conditions exist and could be activated, but this – most unfortunately – is only possible in universes of infinite dimensionality. Nevertheless, with respect to our key technologies and their positioning, attractiveness, and robustness inside a socioeconomic space-time (hence, society), we now know that intelligence ABOUT EVERYTHING is the key for gaining as much certainty as possible for any activity and measure regarding our market performance.

Here now the necessary derivation:

We see, that when demanding both nonlinear terms showing the same F-dependency factors via:

$$(6-3n)(F')^2+2FF''(n-2)=4FF''(2-n)+(F')^2\left(2(n-4)-(n-1)(n-6)\right), \tag{369}$$

which gives us:

$$F[f]=C_{f1}\cdot\left(f\cdot\left(14+(n-8)n\right)+2\cdot C_{f0}\cdot(3-2n)\right)^{\frac{4n-6}{14+(n-8)n}}, \tag{370}$$

we would only obtain a linear differential field equation from (368) for n=3/2,2,19/7, and – most interestingly – n=∞. This reminds us of our "infinity options principle" as introduced and investigated in [4]. Having seen and explained at the beginning of our book, namely, that in principle all degrees of freedom of a system are to be considered as dimensions, we have to conclude that even (on first sight) simple models or systems are of relatively high dimension. Thus, the investigation of the matter equation of an arbitrary system in the case of $n\to\infty$ is very important. Take alone any system within our very own universe. As such a system never is truly closed and separated from the rest of the universe, we – in principle – always can consider any system as being of huge dimensionality. The fact that some dimensions of such systems are more dominant than others does not contradict our "infinity options principle". For the setting (369) the resulting matter equation (368) would read:

$$\kappa T_{\alpha}^{\beta}=\frac{2}{f}\begin{pmatrix}\delta_{\alpha}^{\beta}\left(\frac{1}{2}\left(\frac{1}{2}f_{,d}g^{cd}g_{ab,c}g^{ab}-\Gamma_{ij}^{d}f_{,d}g^{ij}\right)+f_{,ab}g^{ab}\right)\\-g^{\beta\gamma}\left(f_{,d}g^{cd}\frac{1}{2}\left(g_{\alpha\gamma,c}-g_{\alpha c,\gamma}-g_{\gamma c,\alpha}\right)+f_{,\alpha\gamma}\right)\end{pmatrix}. \tag{371}$$

Thereby we have assumed that neither the Ricci scalar nor the Ricci tensor nor the functions f depend on the dimension n, at least not in an effective dependency of an exponent n^p with $p\geq1$. We also have to demand that none of the contraction sums in (368) results in anything growing faster than n.

On the other hand, when demanding

$$\begin{aligned}&0=(6-3n)(F')^2+2FF''(n-2)\\&\Rightarrow F[f]=C_{f1}\cdot\left(f\cdot(8-3n)-2\cdot C_{f0}\right)^{\frac{2}{8-3n}},\end{aligned} \tag{372}$$

the limit of n→∞ results in:

$$\kappa T_{\alpha}^{\beta}=\frac{1}{3f}\left(\begin{array}{c}\delta_{\alpha}^{\beta}\left(\frac{1}{2}\left(\Gamma_{ij}^{d}f_{,d}g^{ij}-\frac{1}{2}f_{,d}g^{cd}g_{ab,c}g^{ab}\right)-f_{,ab}g^{ab}\right)\\+g^{\beta\gamma}\left(f_{,d}g^{cd}\frac{1}{2}\left(g_{\alpha\gamma,c}-g_{\alpha c,\gamma}-g_{\gamma c,\alpha}\right)+f_{,\alpha\gamma}\right)-7\frac{f_{,a}f_{,b}\delta_{\alpha}^{\beta}g^{ab}}{6f}\end{array}\right). \tag{373}$$

We see that we would end up with a scalar nonlinear derivative term for the function f. Such an equation, as it was shown in [7] in the section "First Order Evolution Equations from the Einstein-Hilbert Action – Option B revisited: The Dirac-Evolution Equation for Curved Space-Times", could be linearized via an extended and slightly generalized Dirac approach.

Finally, we have the condition:

$$\begin{aligned}&0=4FF''(2-n)+(F')^2\left(2(n-4)-(n-1)(n-6)\right)\\&\Rightarrow F[f]=C_{f1}\cdot\left(f\cdot(n-3)-4\cdot C_{f0}\right)^{\frac{4}{n-3}}\end{aligned}, \tag{374}$$

which results in:

$$\kappa\cdot\lim_{n\to\infty}T_{\alpha}^{\beta}=\frac{4}{f}\left(\begin{array}{c}\frac{\delta_{\alpha}^{\beta}}{2}\left(\frac{1}{2}\left(-\Gamma_{ij}^{d}f_{,d}g^{ij}+\frac{1}{2}f_{,d}g^{cd}g_{ab,c}g^{ab}\right)+f_{,ab}g^{ab}\right)\\-\frac{1}{2}\left(g^{\beta\gamma}\left(\left(f_{,d}g^{cd}\frac{1}{2}\left(-g_{\alpha c,\gamma}-g_{\gamma c,\alpha}+g_{\alpha\gamma,c}\right)\right)+f_{,\alpha\gamma}\right)\right)\end{array}\right). \tag{375}$$

Simplification gives:

$$\kappa T_{\alpha}^{\beta}=\frac{2}{f}\left(\begin{array}{c}\delta_{\alpha}^{\beta}\left(\frac{1}{2}\left(\frac{1}{2}f_{,d}g^{cd}g_{ab,c}g^{ab}-\Gamma_{ij}^{d}f_{,d}g^{ij}\right)+f_{,ab}g^{ab}\right)\\-g^{\beta\gamma}\left(f_{,d}g^{cd}\frac{1}{2}\left(g_{\alpha\gamma,c}-g_{\alpha c,\gamma}-g_{\gamma c,\alpha}\right)+f_{,\alpha\gamma}\right)\end{array}\right), \tag{376}$$

which is just the result from (371) again.

We immediately realize that for the conditions (374) and (369) the "infinity options principle" provides us with an eigenvalue field equation of the type:

$$0=\begin{pmatrix}\delta_{\alpha}^{\beta}\left(\frac{1}{2}\left(\frac{1}{2}f_{,d}g^{cd}g_{ab,c}g^{ab}-\Gamma_{ij}^{d}f_{,d}g^{ij}\right)+f_{,ab}g^{ab}\right)\\ -g^{\beta\gamma}\left(f_{,d}g^{cd}\frac{1}{2}\left(g_{\alpha\gamma,c}-g_{\alpha c,\gamma}-g_{\gamma c,\alpha}\right)+f_{,\alpha\gamma}\right)\end{pmatrix}-f\cdot\frac{\kappa}{2}T_{\alpha}^{\beta},\qquad(377)$$

where the matter tensor can be extracted from the contracted Bianchi identity (360) leading us to:

$$\left(\frac{\kappa}{2}T_{\alpha}^{\beta}\right)_{;\beta}=\frac{\kappa}{2}T_{\alpha;\beta}^{\beta}=\begin{pmatrix}\frac{\delta_{\alpha}^{\beta}}{f}\left(\frac{1}{2}\left(\frac{1}{2}f_{,d}g^{cd}g_{ab,c}g^{ab}-\Gamma_{ij}^{d}f_{,d}g^{ij}\right)+f_{,ab}g^{ab}\right)\\ -\frac{g^{\beta\gamma}}{f}\left(f_{,d}g^{cd}\frac{1}{2}\left(g_{\alpha\gamma,c}-g_{\alpha c,\gamma}-g_{\gamma c,\alpha}\right)+f_{,\alpha\gamma}\right)\end{pmatrix}_{;\beta}=0$$

$$\Rightarrow 0=\begin{pmatrix}\left(\frac{1}{2\cdot f}\left(\frac{1}{2}f_{,d}g^{cd}g_{ab,c}g^{ab}-\Gamma_{ij}^{d}f_{,d}g^{ij}\right)+f_{,ab}g^{ab}\right)_{;\alpha}\\ -\left(\frac{g^{\beta\gamma}}{f}\left(f_{,d}g^{cd}\frac{1}{2}\left(g_{\alpha\gamma,c}-g_{\alpha c,\gamma}-g_{\gamma c,\alpha}\right)+f_{,\alpha\gamma}\right)\right)_{;\beta}\end{pmatrix}$$

$$=\begin{pmatrix}\left(\frac{1}{2\cdot f}\left(\frac{1}{2}f_{,d}g^{cd}g_{ab,c}g^{ab}-\Gamma_{ij}^{d}f_{,d}g^{ij}\right)+f_{,ab}g^{ab}\right)_{,\alpha}\\ -\left(\frac{g^{\beta\gamma}}{f}\left(f_{,d}g^{cd}\frac{1}{2}\left(g_{\alpha\gamma,c}-g_{\alpha c,\gamma}-g_{\gamma c,\alpha}\right)+f_{,\alpha\gamma}\right)\right)_{,\beta}\\ -\Gamma_{\beta\delta}^{\beta}\left(\frac{g^{\delta\gamma}}{f}\left(f_{,d}g^{cd}\frac{1}{2}\left(g_{\alpha\gamma,c}-g_{\alpha c,\gamma}-g_{\gamma c,\alpha}\right)+f_{,\alpha\gamma}\right)\right)\end{pmatrix}.\qquad(378)$$

Applying the decomposition from (367) leads us to the following complex matter evolution field equation:

$$
0=\begin{pmatrix}
\begin{pmatrix}
\frac{\delta^{\beta}_{\alpha}}{f}\left(\frac{1}{2}\left(\frac{1}{2}f_{,d}g^{cd}g_{ab,c}g^{ab}-\Gamma^{d}_{ij}f_{,d}g^{ij}\right)+f_{,ab}g^{ab}\right)\\
-\frac{g^{\beta\gamma}}{f}\left(f_{,d}g^{cd}\frac{1}{2}\left(g_{\alpha\gamma,c}-g_{\alpha c,\gamma}-g_{\gamma c,\alpha}\right)+f_{,\alpha\gamma}\right)
\end{pmatrix}_{,0}\\
+\Gamma^{0}_{0\delta}\begin{pmatrix}
\frac{\delta^{\delta}_{\alpha}}{f}\left(\frac{1}{2}\left(\frac{1}{2}f_{,d}g^{cd}g_{ab,c}g^{ab}-\Gamma^{d}_{ij}f_{,d}g^{ij}\right)+f_{,ab}g^{ab}\right)\\
-\frac{g^{\delta\gamma}}{f}\left(f_{,d}g^{cd}\frac{1}{2}\left(g_{\alpha\gamma,c}-g_{\alpha c,\gamma}-g_{\gamma c,\alpha}\right)+f_{,\alpha\gamma}\right)
\end{pmatrix}\\
-\Gamma^{\delta}_{0\alpha}\begin{pmatrix}
\frac{\delta^{0}_{\delta}}{f}\left(\frac{1}{2}\left(\frac{1}{2}f_{,d}g^{cd}g_{ab,c}g^{ab}-\Gamma^{d}_{ij}f_{,d}g^{ij}\right)+f_{,ab}g^{ab}\right)\\
-\frac{g^{0\gamma}}{f}\left(f_{,d}g^{cd}\frac{1}{2}\left(g_{\delta\gamma,c}-g_{\delta c,\gamma}-g_{\gamma c,\delta}\right)+f_{,\delta\gamma}\right)
\end{pmatrix}\\
+\begin{pmatrix}
\frac{\delta^{b}_{\alpha}}{f}\left(\frac{1}{2}\left(\frac{1}{2}f_{,d}g^{cd}g_{ab,c}g^{ab}-\Gamma^{d}_{ij}f_{,d}g^{ij}\right)+f_{,ab}g^{ab}\right)\\
-\frac{g^{b\gamma}}{f}\left(f_{,d}g^{cd}\frac{1}{2}\left(g_{\alpha\gamma,c}-g_{\alpha c,\gamma}-g_{\gamma c,\alpha}\right)+f_{,\alpha\gamma}\right)
\end{pmatrix}_{,b}\\
+\Gamma^{b}_{b\delta}\begin{pmatrix}
\frac{\delta^{\delta}_{\alpha}}{f}\left(\frac{1}{2}\left(\frac{1}{2}f_{,d}g^{cd}g_{ab,c}g^{ab}-\Gamma^{d}_{ij}f_{,d}g^{ij}\right)+f_{,ab}g^{ab}\right)\\
-\frac{g^{\delta\gamma}}{f}\left(f_{,d}g^{cd}\frac{1}{2}\left(g_{\alpha\gamma,c}-g_{\alpha c,\gamma}-g_{\gamma c,\alpha}\right)+f_{,\alpha\gamma}\right)
\end{pmatrix}\\
-\Gamma^{\delta}_{b\alpha}\begin{pmatrix}
\frac{\delta^{b}_{\delta}}{f}\left(\frac{1}{2}\left(\frac{1}{2}f_{,d}g^{cd}g_{ab,c}g^{ab}-\Gamma^{d}_{ij}f_{,d}g^{ij}\right)+f_{,ab}g^{ab}\right)\\
-\frac{g^{b\gamma}}{f}\left(f_{,d}g^{cd}\frac{1}{2}\left(g_{\delta\gamma,c}-g_{\delta c,\gamma}-g_{\gamma c,\delta}\right)+f_{,\delta\gamma}\right)
\end{pmatrix}
\end{pmatrix}. \tag{379}
$$

$$\alpha,\beta,\delta=0,1,\ldots,n-1;\quad b=1,\ldots,n-1$$

In the case of condition (372), n→∞ gives us the following perfect starting point for our Dirac factorization:

$$
0=3\cdot f^{2}\cdot\kappa T^{\beta}_{\alpha}-f\cdot\begin{pmatrix}
\delta^{\beta}_{\alpha}\left(\frac{1}{2}\left(\Gamma^{d}_{ij}f_{,d}g^{ij}-\frac{1}{2}f_{,d}g^{cd}g_{ab,c}g^{ab}\right)-f_{,ab}g^{ab}\right)\\
+g^{\beta\gamma}\left(f_{,d}g^{cd}\frac{1}{2}\left(g_{\alpha\gamma,c}-g_{\alpha c,\gamma}-g_{\gamma c,\alpha}\right)+f_{,\alpha\gamma}\right)
\end{pmatrix}+\frac{7}{6}\delta^{\beta}_{\alpha}f_{,a}f_{,b}g^{ab}, \tag{380}
$$

only that this time, in contrast to the classical Dirac starting point (which is the scalar Klein-Gordon equation) we do not have a scalar, but a tensor field equation.

On the other hand, direct application of the contracted Bianchi identity (360) on (373) yields:

$$\kappa T^{\beta}_{\alpha;\beta}=0=\left(\frac{1}{f}\left(\begin{array}{l}\delta^{\beta}_{\alpha}\left(\frac{1}{2}\left(\Gamma^{d}_{ij}f_{,d}g^{ij}-\frac{1}{2}f_{,d}g^{cd}g_{ab,c}g^{ab}\right)-f_{,ab}g^{ab}\right)\\ +g^{\beta\gamma}\left(f_{,d}g^{cd}\frac{1}{2}\left(g_{\alpha\gamma,c}-g_{\alpha c,\gamma}-g_{\gamma c,\alpha}\right)+f_{,\alpha\gamma}\right)\end{array}\right)-\frac{7}{6}\frac{f_{,a}f_{,b}\delta^{\beta}_{\alpha}g^{ab}}{f^{2}}\right)_{;\beta}$$

$$=\left(\begin{array}{l}\left(\frac{1}{2\cdot f}\left(\Gamma^{d}_{ij}f_{,d}g^{ij}-\frac{1}{2}f_{,d}g^{cd}g_{ab,c}g^{ab}\right)-f_{,ab}g^{ab}\right)_{;\alpha}\\ +\left(\frac{g^{\beta\gamma}}{f}\left(f_{,d}g^{cd}\frac{1}{2}\left(g_{\alpha\gamma,c}-g_{\alpha c,\gamma}-g_{\gamma c,\alpha}\right)+f_{,\alpha\gamma}\right)\right)_{;\beta}\end{array}\right)-\frac{7}{6}\left(g^{ab}\frac{f_{,a}f_{,b}}{f^{2}}\right)_{;\alpha}. \tag{381}$$

Thereby we have used the fact that the covariant derivate of the metric tensor gives zero. Also taking into account that for a scalar the covariant derivative equals the ordinary derivative allows us some further evolution of the last equation:

$$0=\left(\begin{array}{l}\left(\frac{1}{2\cdot f}\left(\Gamma^{d}_{ij}f_{,d}g^{ij}-\frac{1}{2}f_{,d}g^{cd}g_{ab,c}g^{ab}\right)-f_{,ab}g^{ab}\right)_{,\alpha}\\ +\left(\frac{g^{\beta\gamma}}{f}\left(f_{,d}g^{cd}\frac{1}{2}\left(g_{\alpha\gamma,c}-g_{\alpha c,\gamma}-g_{\gamma c,\alpha}\right)+f_{,\alpha\gamma}\right)\right)_{,\beta}\\ +\Gamma^{\beta}_{\beta\delta}\left(\frac{g^{\delta\gamma}}{f}\left(f_{,d}g^{cd}\frac{1}{2}\left(g_{\alpha\gamma,c}-g_{\alpha c,\gamma}-g_{\gamma c,\alpha}\right)+f_{,\alpha\gamma}\right)\right)\end{array}\right)-\frac{7}{6}\left(g^{ab}\frac{f_{,a}f_{,b}}{f^{2}}\right)_{,\alpha}. \tag{382}$$

12.3.2 Towards a Solution to the 3-Generation Particle Problem

Assuming a “particle at rest” wave function of Dirac style [27], we expect solutions with time-dependencies of the following kind (c… speed of light in vacuum, m… rest mess, $\hbar$… reduced Planck constant):

$$f=f\left[\pm i\cdot\mu\cdot t\right]=f\left[\pm i\cdot\frac{m\cdot c^{2}}{\hbar}\cdot t\right]. \tag{383}$$

Expansion of equation (366) leads to a third order polynomial equation in μ, which can be written in the following form:

$$0=C_{1}\cdot\mu-C_{2}\cdot\mu^{2}+\mu^{3}. \tag{384}$$

Unfortunately, this equation does not contain a constant term with respect to μ and thus, for the time being we here introduce such a term and just postpone the explanation where it could come from to a little later in this section:

$$0 = \boxed{-C_0} + C_1 \cdot \mu - C_2 \cdot \mu^2 + \mu^3 . \tag{385}$$

Assuming that the wave functions are of such character (e.g., exponential, like within the original Dirac particle at rest solution [27]), we should expect the coefficients C_i to be constants or at least quasi constants within our time-scale.

The general solution to a three-order polynomial could be given via the following product form:

$$\begin{gathered}(\mu-\mu_1)\cdot(\mu-\mu_2)\cdot(\mu-\mu_3)= \\ \mu^3-\mu^2\cdot(\mu_1+\mu_2+\mu_3)+\mu\cdot(\mu_1\mu_2+\mu_1\mu_3+\mu_2\mu_3)-\mu_1\mu_2\mu_3\end{gathered} . \tag{386}$$

Comparing the last line of (386) with (384) gives us:

$$\begin{aligned}C_2 &= \mu_1+\mu_2+\mu_3 \\ C_1 &= \mu_1\mu_2+\mu_1\mu_3+\mu_2\mu_3 \\ C_0 &= \mu_1\mu_2\mu_3\end{aligned} . \tag{387}$$

Thus, in principle, the 3-generation problem would be solved, but we find two flaws, respectively, missing aspects:

A) Equation (366) does not have a constant term.

B) A proper particle at rest solution, including the metric, still needs to be found.

We note that problem A is not really a problem, because even though (366) does not sport a constant term, not on first and shallow sight anyway, so could – perhaps – a more general form give us what we need (remember that (366) is only a simplification of (352)). We would only need to apply a different form for the “cosmological constant”, neither truly being a constant, nor being connected with a derivative of the function F after variation and subsequent differentiation. Problem B, however, still requires some work, because not only the functions F and f need to be found, but, as already said above, also the corresponding metric tensor $g_{\alpha\beta}$.

Nevertheless, we can state that we have found a way for the solution of the 3-generation particle problem and are now going to apply the technique of reverse engineering to find out more.

12.3.2.1 Reverse Engineering

We use (352) as our most general starting point and by assuming the matter, which was artificially added by Hilbert and Einstein via the Lagrange matter density L_M or the energy momentum tensor T_{α}^{β}, to vanish, plus also taking into account that the contracted Bianchi identity gives $R_{\alpha;\beta}^{\beta}-\frac{\delta_{\alpha}^{\beta}}{2}R_{;\beta}=0$ (the covariant derivation has to be performed with respect to the unscaled metric $g_{\alpha\beta}$) and we obtain:

$$0=\left(\begin{array}{l}
\delta_{\alpha}^{\beta}\cdot\left(\Lambda\cdot\Phi_{;\beta}+\frac{1}{2}\left(\begin{array}{c}
\frac{1}{2F}\left(\begin{array}{c}
2F_{,ij}(n-1)g^{ij}+2\Gamma_{ij}^{a}F_{,a}g^{ij}\\
-F_{,i}g^{ab}g_{jb,a}g^{ij}-F_{,j}g^{ab}g_{ib,a}g^{ij}\\
-n\Gamma_{ij}^{d}F_{,d}g^{ij}+\frac{n}{2}F_{,d}g^{cd}g_{ab,c}g^{ab}
\end{array}\right)\\
+\frac{F_{,i}\cdot F_{,j}}{4F^{2}}g^{ij}\left((n-6)(n-1)\right)
\end{array}\right)\right)_{;\beta}\\
+\frac{F_{;\beta}}{2F^{2}}g^{\beta\gamma}\left(\begin{array}{c}
F_{,\alpha\gamma}(n-2)+F_{,ab}g_{\alpha\gamma}g^{ab}\\
+F_{,a}g^{ab}\left(g_{\gamma b,\alpha}-g_{\gamma\alpha,b}\right)-F_{,\alpha}g^{ab}g_{\gamma b,a}-F_{,\gamma}g^{ab}g_{\alpha b,a}\\
+F_{,d}g^{cd}\frac{1}{2}n\left(\frac{2}{n}g_{\alpha c,\gamma}-g_{\alpha c,\gamma}-g_{\gamma c,\alpha}+g_{\alpha\gamma,c}+\frac{1}{n}g_{\alpha\gamma}g_{ab,c}g^{ab}\right)\\
-\frac{1}{2F}\left(F_{,\alpha}\cdot F_{,\gamma}(3n-6)+g_{\alpha\gamma}F_{,c}F_{,d}g^{cd}(4-n)\right)
\end{array}\right)\\
-\frac{1}{2F}\left(g^{\beta\gamma}\left(\begin{array}{c}
F_{,\alpha\gamma}(n-2)+F_{,ab}g_{\alpha\gamma}g^{ab}\\
+F_{,a}g^{ab}\left(g_{\gamma b,\alpha}-g_{\gamma\alpha,b}\right)-F_{,\alpha}g^{ab}g_{\gamma b,a}-F_{,\gamma}g^{ab}g_{\alpha b,a}\\
+F_{,d}g^{cd}\frac{1}{2}n\left(\frac{2}{n}g_{\alpha c,\gamma}-g_{\alpha c,\gamma}-g_{\gamma c,\alpha}+g_{\alpha\gamma,c}+\frac{1}{n}g_{\alpha\gamma}g_{ab,c}g^{ab}\right)\\
-\frac{1}{2F}\left(F_{,\alpha}\cdot F_{,\gamma}(3n-6)+g_{\alpha\gamma}F_{,c}F_{,d}g^{cd}(4-n)\right)
\end{array}\right)\right)_{;\beta}
\end{array}\right). \tag{388}$$

Thereby we introduced the general function Φ as to be somehow related to a "cosmological constant" Λ. As this function is assumed to be a scalar function, we can simplify and reshape (388) as follows:

$$
0=\left(\begin{array}{c}
\Lambda\cdot\Phi_{,\alpha}+\frac{\delta_{\alpha}^{\beta}}{2}\left(\begin{array}{c}\frac{1}{2F}\left(\begin{array}{c}2F_{,ij}(n-1)g^{ij}+2\Gamma_{ij}^{a}F_{,a}g^{ij}\\-F_{,i}g^{ab}g_{jb,a}g^{ij}-F_{,j}g^{ab}g_{ib,a}g^{ij}\\-n\Gamma_{ij}^{d}F_{,d}g^{ij}+\frac{n}{2}F_{,d}g^{cd}g_{ab,c}g^{ab}\end{array}\right)\\+\frac{F_{,i}\cdot F_{,j}}{4F^{2}}g^{ij}\left((n-6)(n-1)\right)\end{array}\right)_{;\beta}\\
+\frac{F_{;\beta}}{2F^{2}}\left(g^{\beta\gamma}\left(\begin{array}{c}F_{,\alpha\gamma}(n-2)+F_{,ab}g_{\alpha\gamma}g^{ab}\\+F_{,a}g^{ab}\left(g_{\gamma b,\alpha}-g_{\gamma\alpha,b}\right)-F_{,\alpha}g^{ab}g_{\gamma b,a}-F_{,\gamma}g^{ab}g_{\alpha b,a}\\+F_{,d}g^{cd}\frac{1}{2}n\left(\frac{2}{n}g_{\alpha c,\gamma}-g_{\alpha c,\gamma}-g_{\gamma c,\alpha}+g_{\alpha\gamma,c}+\frac{1}{n}g_{\alpha\gamma}g_{ab,c}g^{ab}\right)\\-\frac{1}{2F}\left(F_{,\alpha}\cdot F_{,\gamma}(3n-6)+g_{\alpha\gamma}F_{,c}F_{,d}g^{cd}(4-n)\right)\end{array}\right)\right)\\
-\frac{1}{2F}\left(g^{\beta\gamma}\left(\begin{array}{c}F_{,\alpha\gamma}(n-2)+F_{,ab}g_{\alpha\gamma}g^{ab}\\+F_{,a}g^{ab}\left(g_{\gamma b,\alpha}-g_{\gamma\alpha,b}\right)-F_{,\alpha}g^{ab}g_{\gamma b,a}-F_{,\gamma}g^{ab}g_{\alpha b,a}\\+F_{,d}g^{cd}\frac{1}{2}n\left(\frac{2}{n}g_{\alpha c,\gamma}-g_{\alpha c,\gamma}-g_{\gamma c,\alpha}+g_{\alpha\gamma,c}+\frac{1}{n}g_{\alpha\gamma}g_{ab,c}g^{ab}\right)\\-\frac{1}{2F}\left(F_{,\alpha}\cdot F_{,\gamma}(3n-6)+g_{\alpha\gamma}F_{,c}F_{,d}g^{cd}(4-n)\right)\end{array}\right)\right)_{;\beta}
\end{array}\right). \tag{389}
$$

In order to simplify matters for the moment, we assumed a space-time with a huge number of dimensions and applied the infinite option principle with the setting (374) for the function F[f]. Thus, we obtain the following:

$$
\begin{aligned}
0 &= \begin{pmatrix}
\Lambda\cdot\Phi_{,\alpha} - \dfrac{2f_{;\beta}}{f^2}\begin{pmatrix}
\delta^{\beta}_{\alpha}\left(\dfrac{1}{2}\left(\dfrac{1}{2}f_{,d}g^{cd}g_{ab,c}g^{ab}-\Gamma^{d}_{ij}f_{,d}g^{ij}\right)+f_{,ab}g^{ab}\right)\\
-g^{\beta\gamma}\left(f_{,d}g^{cd}\dfrac{1}{2}\left(g_{\alpha\gamma,c}-g_{\alpha c,\gamma}-g_{\gamma c,\alpha}\right)+f_{,\alpha\gamma}\right)
\end{pmatrix}\\
+\dfrac{2}{f}\begin{pmatrix}
\delta^{\beta}_{\alpha}\left(\dfrac{1}{2}\left(\dfrac{1}{2}f_{,d}g^{cd}g_{ab,c}g^{ab}-\Gamma^{d}_{ij}f_{,d}g^{ij}\right)+f_{,ab}g^{ab}\right)\\
-g^{\beta\gamma}\left(f_{,d}g^{cd}\dfrac{1}{2}\left(g_{\alpha\gamma,c}-g_{\alpha c,\gamma}-g_{\gamma c,\alpha}\right)+f_{,\alpha\gamma}\right)
\end{pmatrix}_{;\beta}
\end{pmatrix}\\
&= \begin{pmatrix}
\Lambda\cdot\Phi_{,\alpha} - \dfrac{2}{f^2}\begin{pmatrix}
f_{,\alpha}\left(\dfrac{1}{2}\left(\dfrac{1}{2}f_{,d}g^{cd}g_{ab,c}g^{ab}-\Gamma^{d}_{ij}f_{,d}g^{ij}\right)+f_{,ab}g^{ab}\right)\\
-f_{,\beta}g^{\beta\gamma}\left(f_{,d}g^{cd}\dfrac{1}{2}\left(g_{\alpha\gamma,c}-g_{\alpha c,\gamma}-g_{\gamma c,\alpha}\right)+f_{,\alpha\gamma}\right)
\end{pmatrix}\\
+\dfrac{2}{f}\begin{pmatrix}
\delta^{\beta}_{\alpha}\left(\dfrac{1}{2}\left(\dfrac{1}{2}f_{,d}g^{cd}g_{ab,c}g^{ab}-\Gamma^{d}_{ij}f_{,d}g^{ij}\right)+f_{,ab}g^{ab}\right)_{;\beta}\\
-\overbrace{g^{\beta\gamma}{}_{;\beta}}^{=0}\left(f_{,d}g^{cd}\dfrac{1}{2}\left(g_{\alpha\gamma,c}-g_{\alpha c,\gamma}-g_{\gamma c,\alpha}\right)+f_{,\alpha\gamma}\right)\\
-g^{\beta\gamma}\left(f_{,d}g^{cd}\dfrac{1}{2}\left(g_{\alpha\gamma,c}-g_{\alpha c,\gamma}-g_{\gamma c,\alpha}\right)+f_{,\alpha\gamma}\right)_{;\beta}
\end{pmatrix}
\end{pmatrix}\\
&= \begin{pmatrix}
\Lambda\cdot\Phi_{,\alpha} - \dfrac{2}{f^2}\begin{pmatrix}
f_{,\alpha}\left(\dfrac{1}{2}\left(\dfrac{1}{2}f_{,d}g^{cd}g_{ab,c}g^{ab}-\Gamma^{d}_{ij}f_{,d}g^{ij}\right)+f_{,ab}g^{ab}\right)\\
-f_{,\beta}g^{\beta\gamma}\left(f_{,d}g^{cd}\dfrac{1}{2}\left(g_{\alpha\gamma,c}-g_{\alpha c,\gamma}-g_{\gamma c,\alpha}\right)+f_{,\alpha\gamma}\right)
\end{pmatrix}\\
+\dfrac{2}{f}\begin{pmatrix}
\left(\dfrac{1}{2}\left(\dfrac{1}{2}f_{,d}g^{cd}g_{ab,c}g^{ab}-\Gamma^{d}_{ij}f_{,d}g^{ij}\right)+f_{,ab}g^{ab}\right)_{,\alpha}\\
-g^{\beta\gamma}\left(f_{,d}g^{cd}\dfrac{1}{2}\left(g_{\alpha\gamma,c}-g_{\alpha c,\gamma}-g_{\gamma c,\alpha}\right)+f_{,\alpha\gamma}\right)_{;\beta}
\end{pmatrix}
\end{pmatrix}.
\end{aligned}
\tag{390}
$$

We immediately realize that only with a cosmological term of the type:

$$
\Lambda\cdot\Phi_{,\alpha}=\frac{\Lambda_A}{f}\cdot a_{\alpha}+\frac{\Lambda_B}{f^2}\cdot b_{\alpha}
\tag{391}
$$

we would result in a complete polynomial of third order with respect to our particle at rest-mass-setting corresponding to our approach from above (c.f. (383)). Thereby, as we can extract from our "particle at rest" approach, the components a_0 and b_0 should not depend on f or any derivatives of f. The resulting third order equation would read:

$$0=\left(\begin{array}{c}\frac{\Lambda_A}{f}\cdot a_\alpha+\frac{\Lambda_B}{f^2}\cdot b_\alpha-\frac{2}{f^2}\left(\begin{array}{c}f_{,\alpha}\left(\frac{1}{2}\left(\frac{1}{2}f_{,d}g^{cd}g_{ab,c}g^{ab}-\Gamma^d_{ij}f_{,d}g^{ij}\right)+f_{,ab}g^{ab}\right)\\-f_{,\beta}g^{\beta\gamma}\left(f_{,d}g^{cd}\frac{1}{2}\left(g_{\alpha\gamma,c}-g_{\alpha c,\gamma}-g_{\gamma c,\alpha}\right)+f_{,\alpha\gamma}\right)\end{array}\right)\\+\frac{2}{f}\left(\begin{array}{c}\left(\frac{1}{2}\left(\frac{1}{2}f_{,d}g^{cd}g_{ab,c}g^{ab}-\Gamma^d_{ij}f_{,d}g^{ij}\right)+f_{,ab}g^{ab}\right)_{,\alpha}\\-g^{\beta\gamma}\left(f_{,d}g^{cd}\frac{1}{2}\left(g_{\alpha\gamma,c}-g_{\alpha c,\gamma}-g_{\gamma c,\alpha}\right)+f_{,\alpha\gamma}\right)_{;\beta}\end{array}\right)\end{array}\right)\tag{392}$$

and generalizes as follows in the case of arbitrary n and F[f]:

$$0=\left(\begin{array}{c}\frac{\Lambda_A}{F}\cdot a_\alpha+\frac{\Lambda_B}{F^2}\cdot b_\alpha+\frac{\delta^\beta_\alpha}{2}\left(\begin{array}{c}\frac{1}{2F}\left(\begin{array}{c}2F_{,ij}(n-1)g^{ij}+2\Gamma^a_{ij}F_{,a}g^{ij}\\-F_{,i}g^{ab}g_{jb,a}g^{ij}-F_{,j}g^{ab}g_{ib,a}g^{ij}\\-n\Gamma^d_{ij}F_{,d}g^{ij}+\frac{n}{2}F_{,d}g^{cd}g_{ab,c}g^{ab}\end{array}\right)\\+\frac{F_{,i}\cdot F_{,j}}{4F^2}g^{ij}\left((n-6)(n-1)\right)\end{array}\right)_{;\beta}\\+\frac{F_{;\beta}}{2F^2}\left(g^{\beta\gamma}\left(\begin{array}{c}F_{,\alpha\gamma}(n-2)+F_{,ab}g_{\alpha\gamma}g^{ab}\\+F_{,a}g^{ab}\left(g_{\gamma b,\alpha}-g_{\gamma\alpha,b}\right)-F_{,\alpha}g^{ab}g_{\gamma b,a}-F_{,\gamma}g^{ab}g_{\alpha b,a}\\+F_{,d}g^{cd}\frac{1}{2}n\left(\frac{2}{n}g_{\alpha c,\gamma}-g_{\alpha c,\gamma}-g_{\gamma c,\alpha}+g_{\alpha\gamma,c}+\frac{1}{n}g_{\alpha\gamma}g_{ab,c}g^{ab}\right)\\-\frac{1}{2F}\left(F_{,\alpha}\cdot F_{,\gamma}(3n-6)+g_{\alpha\gamma}F_{,c}F_{,d}g^{cd}(4-n)\right)\end{array}\right)\right)\\-\frac{1}{2F}\left(g^{\beta\gamma}\left(\begin{array}{c}F_{,\alpha\gamma}(n-2)+F_{,ab}g_{\alpha\gamma}g^{ab}\\+F_{,a}g^{ab}\left(g_{\gamma b,\alpha}-g_{\gamma\alpha,b}\right)-F_{,\alpha}g^{ab}g_{\gamma b,a}-F_{,\gamma}g^{ab}g_{\alpha b,a}\\+F_{,d}g^{cd}\frac{1}{2}n\left(\frac{2}{n}g_{\alpha c,\gamma}-g_{\alpha c,\gamma}-g_{\gamma c,\alpha}+g_{\alpha\gamma,c}+\frac{1}{n}g_{\alpha\gamma}g_{ab,c}g^{ab}\right)\\-\frac{1}{2F}\left(F_{,\alpha}\cdot F_{,\gamma}(3n-6)+g_{\alpha\gamma}F_{,c}F_{,d}g^{cd}(4-n)\right)\end{array}\right)\right)_{;\beta}\end{array}\right).\tag{393}$$

12.3.2.2 Further Reverse Engineering Towards the Kernel of the Einstein-Hilbert Action

In case that (393) indeed provides the key for the solution of the 3-generation problem of elementary particles, we would not need to dig any further. But, so the academic question which immediately pops up, why has the cosmological constant term, which – if (393) was correct – cannot be a constant anymore but must be related somehow to the wave function F[f], not revealed itself as such a function earlier?

So, only out of principle interest, when further deriving the corresponding kernel of the Hilbert integral (Einstein-Hilbert action) we could just start with the assumption that a transformation of the kind:

$$\delta W=0=\delta_{G_{\alpha\beta}}\int_V d^n x\sqrt{-g\cdot F^n}\cdot(-2\cdot\Theta\cdot\Lambda)\simeq\delta_{\Gamma_{\alpha\beta}}\int_V d^n x\sqrt{-g\cdot\Theta^2\cdot F^n}\cdot(-2\cdot\Lambda)$$

$$\xrightarrow{\Gamma_{\alpha\beta}=G_{\alpha\beta}\cdot\Theta^{\frac{2}{n}}=g_{\alpha\beta}\cdot F\cdot\Theta^{\frac{2}{n}}}$$

$$=\int_V d^n x\sqrt{-g\cdot\Theta^2\cdot F^n}\cdot\Lambda\cdot\Gamma_{\alpha\beta}\delta\Gamma^{\alpha\beta}=\int_V d^n x\sqrt{-g\cdot\Theta^2\cdot F^n}\cdot\Lambda\cdot G_{\alpha\beta}\cdot\Theta^{\frac{2}{n}}\delta\Gamma^{\alpha\beta} \tag{394}$$

$$=\int_V d^n x\sqrt{-g\cdot\Theta^2\cdot F^n}\cdot\Lambda\cdot g_{\alpha\beta}\cdot F\cdot\Theta^{\frac{2}{n}}\delta\Gamma^{\alpha\beta}$$

would only be of small error when being combined with the rest of the kernel of the Einstein-Hilbert action in the form:

$$\delta W=0=\delta\int_V d^n x\left(\sqrt{-g}\cdot\Phi_R[R]\right)=\delta\int_V d^n x\left(\sqrt{-g}\cdot(R-2\Lambda+L_M)\right) \tag{395}$$

(note L_M=0). Together with (391) this gives us:

$$\Lambda\cdot\left(F\cdot\Theta^{\frac{2}{n}}\right)_{,\alpha}=\Lambda\cdot\Phi_{,\alpha}=\frac{\Lambda_A}{f}\cdot a_\alpha+\frac{\Lambda_B}{f^2}\cdot b_\alpha. \tag{396}$$

The error being made can be estimated from the following calculation:

$$
\begin{aligned}
&\delta_{\Gamma_{\alpha\beta}}\int_V d^n x\sqrt{-g\cdot\Theta^2\cdot F^n}\cdot(-2\cdot\Lambda)=\\
&=\int_V d^n x\sqrt{-g\cdot\Theta^2\cdot F^n}\cdot\Lambda\cdot\Gamma_{\alpha\beta}\delta\Gamma^{\alpha\beta}=\int_V d^n x\sqrt{-g\cdot\Theta^2\cdot F^n}\cdot\Lambda\cdot g_{\alpha\beta}\cdot F\cdot\Theta^{\frac{2}{n}}\delta\Gamma^{\alpha\beta}\\
&=\int_V d^n x\sqrt{-g\cdot\Theta^2\cdot F^n}\cdot\Lambda\cdot g_{\alpha\beta}\cdot F\cdot\Theta^{\frac{2}{n}}\left(g^{\alpha\beta}\delta\left(F\cdot\Theta^{\frac{2}{n}}\right)^{-1}+\frac{\delta g^{\alpha\beta}}{F\cdot\Theta^{\frac{2}{n}}}\right)\\
&=\int_V d^n x\sqrt{-g\cdot\Theta^2\cdot F^n}\cdot\Lambda\cdot g_{\alpha\beta}\cdot F\cdot\Theta^{\frac{2}{n}}\left(G^{\alpha\beta}\delta\Theta^{-\frac{2}{n}}+\frac{\delta G^{\alpha\beta}}{\Theta^{\frac{2}{n}}}\right)\\
&=\int_V d^n x\sqrt{-g\cdot\Theta^2\cdot F^n}\cdot\Lambda\cdot G_{\alpha\beta}\cdot\Theta^{\frac{2}{n}}\left(G^{\alpha\beta}\delta\Theta^{-\frac{2}{n}}+\frac{\delta G^{\alpha\beta}}{\Theta^{\frac{2}{n}}}\right)\\
&=\int_V d^n x\sqrt{-g\cdot\Theta^2\cdot F^n}\cdot\Lambda\cdot G_{\alpha\beta}\cdot\left(\Theta^{\frac{2}{n}}\cdot G^{\alpha\beta}\delta\Theta^{-\frac{2}{n}}+\delta G^{\alpha\beta}\right)
\end{aligned}
\quad , \qquad (397)
$$

which leads us to:

$$
\begin{aligned}
&=\int_V d^n x\sqrt{-g\cdot\Theta^2\cdot F^n}\cdot\Lambda\cdot G_{\alpha\beta}\cdot\left(\Theta^{\frac{2}{n}}\cdot G^{\alpha\beta}\delta\Theta^{-\frac{2}{n}}+\delta G^{\alpha\beta}\right)\\
&\Rightarrow\delta W=0=\delta_{G_{\alpha\beta}}\int_V d^n x\sqrt{-g\cdot F^n}\cdot(-2\cdot\Theta\cdot\Lambda)=\int_V d^n x\sqrt{-g\cdot F^n}\cdot\Theta\cdot\Lambda\delta G^{\alpha\beta}+?\\
&=\delta_{\Gamma_{\alpha\beta}}\int_V d^n x\sqrt{-g\cdot\Theta^2\cdot F^n}\cdot(-2\cdot\Lambda)-\varepsilon\cdot\int_V d^n x\sqrt{-g\cdot\Theta^2\cdot F^n}\cdot\Lambda\cdot g_{\alpha\beta}\cdot F\left(\Theta^{\frac{2}{n}}\cdot G^{\alpha\beta}\delta\Theta^{-\frac{2}{n}}\right)
\end{aligned}
. \qquad (398)
$$

$$
\begin{aligned}
&\Rightarrow\delta W=0=\delta_{G_{\alpha\beta}}\int_V d^n x\sqrt{-g\cdot F^n}\cdot(-2\cdot\Theta\cdot\Lambda)\simeq\int_V d^n x\sqrt{-g\cdot F^n}\cdot\Lambda\cdot\Theta\cdot G_{\alpha\beta}\delta G^{\alpha\beta}\\
&\Rightarrow\delta W=0=\delta_{g_{\alpha\beta}}\int_V d^n x\sqrt{-g\cdot F^n}\cdot(-2\cdot\Theta\cdot\Lambda)\simeq\int_V d^n x\sqrt{-g\cdot F^n}\cdot\Lambda\cdot\Theta\cdot g_{\alpha\beta}\delta g^{\alpha\beta}
\end{aligned}
$$

Thus, the function in connection with the cosmological constant would also be treated like a constant during the variation with respect to the metric tensor. This is in perfect agreement with the results from [6] (c.f. Eq. (45) above). On the other hand, one could also write:

$$
\begin{aligned}
\delta W=0=\delta_{G_{\alpha\beta}}\int_V d^n x\sqrt{-g\cdot F^n}\cdot(-2\cdot\Lambda)&=\int_V d^n x\sqrt{-g\cdot F^n}\cdot\Lambda\cdot G_{\alpha\beta}\delta G^{\alpha\beta}\\
&=\int_V d^n x\sqrt{-g\cdot F^n}\cdot\Lambda\cdot F\cdot g_{\alpha\beta}\delta\left(\frac{g^{\alpha\beta}}{F}\right)\\
&=\int_V d^n x\sqrt{-g\cdot F^n}\cdot\Lambda\cdot F\cdot g_{\alpha\beta}\left(g^{\alpha\beta}\delta F^{-1}+\frac{\delta g^{\alpha\beta}}{F}\right)\\
&=\int_V d^n x\sqrt{-g\cdot F^n}\cdot\Lambda\cdot g_{\alpha\beta}\left(F\cdot g^{\alpha\beta}\delta F^{-1}+\delta g^{\alpha\beta}\right)\\
&=\int_V d^n x\sqrt{-g\cdot F^n}\cdot\Lambda\cdot\left(n\cdot F\cdot\delta F^{-1}+g_{\alpha\beta}\delta g^{\alpha\beta}\right)
\end{aligned}
\quad . \qquad (399)
$$

This would force us to a discussion about the individual sizes of the variation terms $\delta F^{-1}, g_{\alpha\beta}\delta g^{\alpha\beta}$, something we would like to avoid as it leads to a parameter setting that could only be obtained experimentally. It also brings in the question about which variation is the correct choice: the one with respect to $g_{\alpha\beta}$, $G_{\alpha\beta}$ or even $\Gamma_{\alpha\beta}$?

Differently and – if one so will – more thoroughly, we may want to perform the variation in (399) as follows:

$$\begin{aligned}\delta W = 0 = \delta_{G_{\alpha\beta}} \int_V d^n x \sqrt{-g \cdot F^n} \cdot (-2\cdot\Lambda) = \delta_{F[f]\cdot g_{\alpha\beta}} \int_V d^n x \sqrt{-g \cdot F[f]^n} \cdot (-2\cdot\Lambda) \\ = \delta_{F[f]\cdot g_{\alpha\beta}} \int_V d^n x \sqrt{-G} \cdot (-2\cdot\Lambda) = \\ P(F)\cdot\delta_{g_{\alpha\beta}} \int_V d^n x \sqrt{-G} \cdot (-2\cdot\Lambda) + Q(g_{\alpha\beta})\cdot\delta_{F[f]} \int_V d^n x \sqrt{-G} \cdot (-2\cdot\Lambda)\end{aligned}, \quad (400)$$

or:

$$\begin{aligned}\delta W = 0 = \delta_{G_{\alpha\beta}} \int_V d^n x \sqrt{-g \cdot F^n} \cdot (-2\cdot\Lambda) \\ = \delta_{g_{\alpha\beta}} \int_V d^n x \sqrt{-G} \cdot p(F) \cdot (-2\cdot\Lambda) + \delta_{F[f]} \int_V d^n x \sqrt{-G} \cdot q(g_{\alpha\beta}) \cdot (-2\cdot\Lambda)\end{aligned}. \quad (401)$$

In the simplest case we might be able to set p=q=1 and obtain:

$$\begin{aligned}\delta W = 0 = \delta_{G_{\alpha\beta}} \int_V d^n x \sqrt{-g \cdot F^n} \cdot (-2\cdot\Lambda) \\ = \delta_{F[f]\cdot g_{\alpha\beta}} \int_V d^n x \sqrt{-g \cdot F[f]^n} \cdot (-2\cdot\Lambda) \\ = \delta_{g_{\alpha\beta}} \int_V d^n x \cdot F[f] \cdot \sqrt{-g \cdot F^n} \cdot (-2\cdot\Lambda) + \delta_{F[f]} \int_V d^n x \sqrt{-g \cdot F^n} \cdot (-2\cdot\Lambda) \\ = \int_V d^n x \sqrt{-g \cdot F^n} \cdot \Lambda \cdot g_{\alpha\beta}\delta g^{\alpha\beta} - 2\cdot\delta_F \int_V d^n x \cdot \sqrt{-g \cdot F^n} \cdot \Lambda \\ = \int_V d^n x \sqrt{-g \cdot F^n} \cdot \Lambda \cdot g_{\alpha\beta}\delta g^{\alpha\beta} - 2\cdot\delta_F \int_V d^n x \cdot \sqrt{-g} \cdot \overbrace{F^{\frac{n}{2}}}^{\equiv\Omega[F]} \cdot \Lambda \\ = \int_V d^n x \sqrt{-g \cdot F^n} \cdot \Lambda \cdot g_{\alpha\beta}\delta g^{\alpha\beta} - n\cdot \int_V d^n x \cdot \sqrt{-g} \cdot F^{\frac{n}{2}-1} \cdot \Lambda\,\delta F\end{aligned}. \quad (402)$$

On the other side we know that:

$$\begin{aligned}\delta W = 0 = \delta_{G_{\alpha\beta}} \int_V d^n x \sqrt{-g \cdot F^n} \cdot (-2\cdot\Lambda) = \int_V d^n x \sqrt{-g \cdot F^n} \cdot \Lambda \cdot G^{\alpha\beta}\delta G_{\alpha\beta} \\ \xrightarrow{G^{\alpha\beta}\delta G_{\alpha\beta} + G_{\alpha\beta}\delta G^{\alpha\beta} = 0} \\ \delta W = 0 = \delta_{G^{\alpha\beta}} \int_V d^n x \sqrt{-g \cdot F^n} \cdot (-2\cdot\Lambda) = -\int_V d^n x \sqrt{-g \cdot F^n} \cdot \Lambda \cdot G^{\alpha\beta}\delta G_{\alpha\beta}\end{aligned} \quad (403)$$

and thus:

$$\begin{aligned}
\delta W = 0 = \delta_{G_{\alpha\beta}} \int_V d^n x \sqrt{-g \cdot F^n} \cdot (-2 \cdot \Lambda) &= \int_V d^n x \sqrt{-g \cdot F^n} \cdot \Lambda \cdot G^{\alpha\beta} \delta G_{\alpha\beta} \\
&= \int_V d^n x \sqrt{-g \cdot F^n} \cdot \Lambda \cdot F^{-1} \cdot g^{\alpha\beta} \delta\left(F \cdot g_{\alpha\beta}\right) \\
&= \int_V d^n x \sqrt{-g \cdot F^n} \cdot \Lambda \cdot g^{\alpha\beta} \delta g_{\alpha\beta} + \int_V d^n x \sqrt{-g \cdot F^n} \cdot \Lambda \cdot F^{-1} \cdot g^{\alpha\beta} g_{\alpha\beta} \delta F \\
&= \int_V d^n x \sqrt{-g \cdot F^n} \cdot \Lambda \cdot g^{\alpha\beta} \delta g_{\alpha\beta} + \int_V d^n x \sqrt{-g \cdot F^n} \cdot \Lambda \cdot F^{-1} \cdot n \cdot \delta F
\end{aligned}$$

$$\xrightarrow{G^{\alpha\beta}\delta G_{\alpha\beta} + G_{\alpha\beta}\delta G^{\alpha\beta} = 0}$$

$$\begin{aligned}
\delta W = 0 = \delta_{G^{\alpha\beta}} \int_V d^n x \sqrt{-g \cdot F^n} \cdot (-2 \cdot \Lambda) &= -\int_V d^n x \sqrt{-g \cdot F^n} \cdot \Lambda \cdot G_{\alpha\beta} \delta G^{\alpha\beta} \\
&= -\int_V d^n x \sqrt{-g \cdot F^n} \cdot \Lambda \cdot F \cdot g_{\alpha\beta} \delta\left(F^{-1} \cdot g^{\alpha\beta}\right) \\
&= -\int_V d^n x \sqrt{-g \cdot F^n} \cdot \Lambda \cdot g_{\alpha\beta} \delta g^{\alpha\beta} - \int_V d^n x \sqrt{-g \cdot F^n} \cdot \Lambda \cdot F \cdot g_{\alpha\beta} g^{\alpha\beta} \delta\left(F^{-1}\right) \\
&= -\int_V d^n x \sqrt{-g \cdot F^n} \cdot \Lambda \cdot g_{\alpha\beta} \delta g^{\alpha\beta} - \int_V d^n x \sqrt{-g \cdot F^n} \cdot \Lambda \cdot F \cdot n \cdot \delta\left(F^{-1}\right) \\
&= -\int_V d^n x \sqrt{-g \cdot F^n} \cdot \Lambda \cdot g_{\alpha\beta} \delta g^{\alpha\beta} + \int_V d^n x \sqrt{-g \cdot F^n} \cdot \Lambda \cdot F^{-1} \cdot n \cdot \delta F
\end{aligned} \quad . \tag{404}$$

We see that the results from (404) do not agree with (402).

The moment another function Θ shows its presence in connection with the cosmological constant, things would become even more complicated, because we have to write:

$$\begin{aligned}
\delta W = 0 &= \delta_{G_{\alpha\beta}} \int_V d^n x \sqrt{-g \cdot F^n} \cdot (-2 \cdot \Theta \cdot \Lambda) \\
&= \delta_{F[f] \cdot g_{\alpha\beta}} \int_V d^n x \sqrt{-g \cdot F[f]^n} \cdot \left(-2 \cdot \Theta[F[f]] \cdot \Lambda\right) \\
&= \delta_{g_{\alpha\beta}} \int_V d^n x \cdot F[f] \cdot \sqrt{-g \cdot F^n} \cdot (-2 \cdot \Theta \cdot \Lambda) + \delta_{F[f]} \int_V d^n x \sqrt{-g \cdot F^n} \cdot (-2 \cdot \Theta \cdot \Lambda) \cdot \\
&= \int_V d^n x \sqrt{-g \cdot F^n} \cdot \Theta \cdot \Lambda \cdot F[f] \cdot g_{\alpha\beta} \delta g^{\alpha\beta} - 2 \cdot \delta_F \int_V d^n x \cdot \sqrt{-g \cdot F^n} \cdot \Theta \cdot \Lambda \\
&= \int_V d^n x \sqrt{-g \cdot F^n} \cdot \Theta \cdot \Lambda \cdot F[f] \cdot g_{\alpha\beta} \delta g^{\alpha\beta} - 2 \cdot \delta_F \int_V d^n x \cdot \sqrt{-g} \cdot \overbrace{\Theta \cdot F^{\frac{n}{2}}}^{\equiv \Omega[F]} \cdot \Lambda
\end{aligned} \tag{405}$$

Now we treat the second term via the technique of the functional derivative and write:

$$\delta_F\int_V d^nx\cdot\sqrt{-g}\cdot\overbrace{\Theta\cdot F^{\frac{n}{2}}}^{\equiv\Omega[F]}\cdot\Lambda=\int_V d^nx\cdot\lim_{\in\to 0}\frac{\Lambda}{\in}\left(\begin{array}{c}\int_V d^nx\cdot\left(\sqrt{-g}\cdot\Omega\left[F[\mathbf{x}]+\in\cdot\delta[\mathbf{x}-\mathbf{y}]\right]\right)\\ -\int_V d^nx\cdot\left(\sqrt{-g}\cdot\Omega[F[\mathbf{x}]]\right)\end{array}\right)_{\mathbf{y}\to\mathbf{x}}$$

$$=\int_V d^nx\cdot\sqrt{-g}\cdot\Lambda\cdot\frac{\partial\Omega}{\partial F}=\int_V d^nx\cdot\sqrt{-g}\cdot\Lambda\cdot\Omega' \quad . \tag{406}$$

This means, when the function Θ is a function of F[f], then variation with respect to the metric tensor $G_{\alpha\beta}$ leads to very complicated split-ups of the kind (400) or (401) and thus, we ask:

Is there a simpler way to find the correct approach for the kernel of the Einstein-Hilbert action that solves the 3-generation problem of elementary particles?

12.3.2.3 Just a Few Trials

Remembering the Ricci scalar of a scaled metric of the kind:

$$G_{\alpha\beta}=g_{\alpha\beta}\cdot F[f], \tag{407}$$

would be (extracted from the integrand:

$$\delta W=0=\delta\int_V d^nx\left(\sqrt{-g\cdot F^n}\times\left(\left(\begin{array}{c}\left(R-\frac{1}{2F}\left(\begin{array}{c}2F_{,\alpha\beta}(n-1)g^{\alpha\beta}+2\Gamma^a_{\alpha\beta}F_{,a}g^{\alpha\beta}\\ -F_{,\alpha}g^{ab}g_{\beta b,a}g^{\alpha\beta}-F_{,\beta}g^{ab}g_{\alpha b,a}g^{\alpha\beta}\\ -n\Gamma^d_{\alpha\beta}F_{,d}g^{\alpha\beta}+\frac{n}{2}F_{,d}g^{cd}g_{ab,c}g^{ab}\end{array}\right)\\ -\frac{F_{,\alpha}\cdot F_{,\beta}}{4F^2}g^{\alpha\beta}\left((n-6)(n-1)\right)\end{array}\right)\frac{1}{F}\right)\\ -2\Lambda+L_M\right)\right), \tag{408}$$

regarding the derivation see [4], chapter 8 or appendix D in here):

$$R^*=\left(\underbrace{R-\frac{1}{2F}\left(\begin{array}{c}2F_{,\alpha\beta}(n-1)g^{\alpha\beta}+2\Gamma^a_{\alpha\beta}F_{,a}g^{\alpha\beta}\\ -F_{,\alpha}g^{ab}g_{\beta b,a}g^{\alpha\beta}-F_{,\beta}g^{ab}g_{\alpha b,a}g^{\alpha\beta}\\ -n\Gamma^d_{\alpha\beta}F_{,d}g^{\alpha\beta}+\frac{n}{2}F_{,d}g^{cd}g_{ab,c}g^{ab}\end{array}\right)-\frac{F_{,\alpha}\cdot F_{,\beta}}{4F^2}g^{\alpha\beta}\left((n-6)(n-1)\right)}_{\equiv -D_{Ricci-G}F}\right)\frac{1}{F}, \tag{409}$$

we see that in the case of:

$$D_{Ricci-G}F=0, \tag{410}$$

(409) simplifies to:

$$R^*=\frac{R}{F} \tag{411}$$

and consequently, also gives:

$$R^{*}_{\alpha\beta}G^{\alpha\beta}=\boxed{R^{*}=\frac{R}{F}}=\frac{R_{\alpha\beta}g^{\alpha\beta}}{F}=R_{\alpha\beta}G^{\alpha\beta}\quad\Rightarrow\quad R^{*}_{\alpha\beta}=R_{\alpha\beta}. \tag{412}$$

From there it directly follows that a scaled metric of type (407) with a scaling factor F[f] fulfilling condition (410) results in the following vacuum Einstein field equations:

$$R^{*}_{\alpha\beta}-\frac{1}{2}R^{*}G_{\alpha\beta}=R_{\alpha\beta}-\frac{1}{2}\frac{R}{F}\cdot F\cdot g_{\alpha\beta}=R_{\alpha\beta}-\frac{1}{2}R\cdot g_{\alpha\beta}=0, \tag{413}$$

which are just the classical field equations without a metric scaling factor, if it weren't for the fact that they also require the fulfillment of condition (410). Thereby we have performed the variation under the integral in (408) with respect to the scaled metric tensor $G_{\alpha\beta}$.

Assuming a cosmological constant without condition (410) renders the field equations to:

$$R^{*}_{\alpha\beta}-\frac{1}{2}R^{*}G_{\alpha\beta}+\Lambda G_{\alpha\beta}=R^{*}_{\alpha\beta}-\frac{1}{2}R^{*}F\cdot g_{\alpha\beta}+\Lambda\cdot F\cdot g_{\alpha\beta}=0. \tag{414}$$

Now we introduce another metric:

$$\Gamma_{\alpha\beta}=\Psi[\psi]\cdot G_{\alpha\beta} \tag{415}$$

and by demanding:

$$D_{Ricci-\Gamma}\Psi[\psi]=0, \tag{416}$$

and performing the variation under the corresponding Einstein-Hilbert action which now reads:

$$\delta W=0=\delta\int_V d^n x\left(\sqrt{-\Gamma}\times\left(\overbrace{\left(\frac{R^{*}}{\Psi}-\overbrace{\left(\frac{1}{2\Psi^2}\begin{pmatrix}2\Psi_{,\alpha\beta}(n-1)G^{\alpha\beta}+2\Gamma^{a}_{\alpha\beta}\Psi_{,a}G^{\alpha\beta}\\-\Psi_{,\alpha}G^{ab}G_{\beta b,a}G^{\alpha\beta}-\Psi_{,\beta}G^{ab}G_{\alpha b,a}G^{\alpha\beta}\\-n\Gamma^{d}_{\alpha\beta}\Psi_{,d}G^{\alpha\beta}+\frac{n}{2}\Psi_{,d}G^{cd}G_{ab,c}G^{ab}\end{pmatrix}+\frac{\Psi_{,\alpha}\cdot\Psi_{,\beta}}{4\Psi^3}G^{\alpha\beta}((n-6)(n-1))\right)}^{\equiv\frac{1}{\Psi}D_{Ricci-\Gamma}\Psi=0}\right)}^{R^{**}}-2\Lambda\right)\right), \tag{417}$$

$$\text{with: }\quad\Gamma=G\cdot\Psi^{n}=g\cdot F^{n}\cdot\Psi^{n}$$

with respect to the new metric tensor $\Gamma_{\alpha\beta}$, results in the following field equations:

$$\begin{aligned}R^{**}_{\alpha\beta}-\frac{1}{2}R^{**}\Gamma_{\alpha\beta}+\Lambda\Gamma_{\alpha\beta}&=R^{*}_{\alpha\beta}-\frac{1}{2}\frac{R^{*}}{\Psi}\Psi G_{\alpha\beta}+\Lambda\Gamma_{\alpha\beta}\\&=R^{*}_{\alpha\beta}-\frac{1}{2}R^{*}F\cdot g_{\alpha\beta}+\Lambda\cdot F\cdot\Psi\cdot g_{\alpha\beta}=0\end{aligned}. \tag{418}$$

Ignoring the fact that this equation already gives zero, bringing it into mixed form, inserting the corresponding terms for the scaled Ricci scalar and tensor, and subjecting all to a contracted covariant derivative with respect to the double scaled metric tensor $\Gamma_{\alpha\beta}$ with the Christoffel symbol to be obtained from:

$$
\begin{aligned}
\Gamma^{*\gamma}_{\alpha\beta} &= \frac{g^{\gamma\sigma}}{2\cdot F[f]\Psi[\psi]}\begin{pmatrix} \left[F[f]\Psi[\psi]\cdot g_{\sigma\alpha}\right]_{,\beta} + \left[F[f]\Psi[\psi]\cdot g_{\sigma\beta}\right]_{,\alpha} \\ -\left[F[f]\Psi[\psi]\cdot g_{\alpha\beta}\right]_{,\sigma} \end{pmatrix} \\
&= \frac{g^{\gamma\sigma}}{2\cdot F\cdot\Psi}\left(\left[F\cdot\Psi\cdot g_{\sigma\alpha}\right]_{,\beta} + \left[F\cdot\Psi\cdot g_{\sigma\beta}\right]_{,\alpha} - \left[F\cdot\Psi\cdot g_{\alpha\beta}\right]_{,\sigma}\right) \\
&= \begin{pmatrix} \frac{g^{\gamma\sigma}}{2}\left(g_{\sigma\alpha,\beta} + g_{\sigma\beta,\alpha} - g_{\alpha\beta,\sigma}\right) \\ +\frac{g^{\gamma\sigma}F'}{2\cdot F}\left(f_{,\beta}\cdot g_{\sigma\alpha} + f_{,\alpha}\cdot g_{\sigma\beta} - f_{,\sigma}\cdot g_{\alpha\beta}\right) + \frac{g^{\gamma\sigma}\Psi'}{2\cdot\Psi}\left(\psi_{,\beta}\cdot g_{\sigma\alpha} + \psi_{,\alpha}\cdot g_{\sigma\beta} - \psi_{,\sigma}\cdot g_{\alpha\beta}\right) \end{pmatrix} \\
&= \Gamma^{\gamma}_{\alpha\beta} + \frac{F'}{2\cdot F}\left(f_{,\beta}\cdot\delta^{\gamma}_{\alpha} + f_{,\alpha}\cdot\delta^{\gamma}_{\beta} - f_{,\sigma}\cdot g^{\gamma\sigma}g_{\alpha\beta}\right) + \frac{\Psi'}{2\cdot\Psi}\left(\psi_{,\beta}\cdot\delta^{\gamma}_{\alpha} + \psi_{,\alpha}\cdot\delta^{\gamma}_{\beta} - \psi_{,\sigma}\cdot g^{\gamma\sigma}g_{\alpha\beta}\right) \\
&\equiv \Gamma^{\gamma}_{\alpha\beta} + \Gamma^{**\gamma}_{\alpha\beta} + \Gamma^{***\gamma}_{\alpha\beta} \\
&\text{with: } F' = \frac{\partial F[f]}{\partial f};\quad \Psi' = \frac{\partial\Psi[\psi]}{\partial\psi}
\end{aligned}
\quad , (419)
$$

leads to:

$$
0 = \begin{pmatrix}
\Gamma^{**\beta}_{\beta a}R^{a}_{\alpha} + \Gamma^{***\beta}_{\beta a}R^{a}_{\alpha} - \Gamma^{**a}_{\beta\alpha}R^{\beta}_{a} - \Gamma^{***a}_{\beta\alpha}R^{\beta}_{a} \\
+\frac{\delta^{\beta}_{\alpha}}{2}\begin{pmatrix} \frac{1}{2F}\begin{pmatrix} 2F_{,ij}(n-1)g^{ij} + 2\Gamma^{a}_{ij}F_{,a}g^{ij} \\ -F_{,i}g^{ab}g_{jb,a}g^{ij} - F_{,j}g^{ab}g_{ib,a}g^{ij} \\ -n\Gamma^{d}_{ij}F_{,d}g^{ij} + \frac{n}{2}F_{,d}g^{cd}g_{ab,c}g^{ab} \end{pmatrix} \\ +\frac{F_{,i}\cdot F_{,j}}{4F^2}g^{ij}\left((n-6)(n-1)\right) \end{pmatrix}_{;\beta} \\
+\frac{F_{;\beta}}{2F^2}\left(g^{\beta\gamma}\begin{pmatrix} F_{,\alpha\gamma}(n-2) + F_{,ab}g_{\alpha\gamma}g^{ab} \\ +F_{,a}g^{ab}\left(g_{\gamma b,\alpha} - g_{\gamma\alpha,b}\right) - F_{,\alpha}g^{ab}g_{\gamma b,a} - F_{,\gamma}g^{ab}g_{\alpha b,a} \\ +F_{,d}g^{cd}\frac{1}{2}n\left(\frac{2}{n}g_{\alpha c,\gamma} - g_{\alpha c,\gamma} - g_{\gamma c,\alpha} + g_{\alpha\gamma,c} + \frac{1}{n}g_{\alpha\gamma}g_{ab,c}g^{ab}\right) \\ -\frac{1}{2F}\left(F_{,\alpha}\cdot F_{,\gamma}(3n-6) + g_{\alpha\gamma}F_{,c}F_{,d}g^{cd}(4-n)\right) \end{pmatrix}\right) \\
-\frac{1}{2F}\left(g^{\beta\gamma}\begin{pmatrix} F_{,\alpha\gamma}(n-2) + F_{,ab}g_{\alpha\gamma}g^{ab} \\ +F_{,a}g^{ab}\left(g_{\gamma b,\alpha} - g_{\gamma\alpha,b}\right) - F_{,\alpha}g^{ab}g_{\gamma b,a} - F_{,\gamma}g^{ab}g_{\alpha b,a} \\ +F_{,d}g^{cd}\frac{1}{2}n\left(\frac{2}{n}g_{\alpha c,\gamma} - g_{\alpha c,\gamma} - g_{\gamma c,\alpha} + g_{\alpha\gamma,c} + \frac{1}{n}g_{\alpha\gamma}g_{ab,c}g^{ab}\right) \\ -\frac{1}{2F}\left(F_{,\alpha}\cdot F_{,\gamma}(3n-6) + g_{\alpha\gamma}F_{,c}F_{,d}g^{cd}(4-n)\right) \end{pmatrix}\right)_{;\beta}
\end{pmatrix}. \quad (420)
$$

Apparently, at least so our hope, here now the terms $\Gamma_{\beta a}^{**\beta}R_{\alpha}^{a}+\Gamma_{\beta a}^{***\beta}R_{\alpha}^{a}-\Gamma_{\beta\alpha}^{**a}R_{a}^{\beta}-\Gamma_{\beta\alpha}^{***a}R_{a}^{\beta}$ in the last line make the difference and provide us with the correct third order polynomial equation, including the necessary constant term for the particle at rest approach when attacking the 3-generation particle problem. Thereby, even though being perfectly aware of the fact that we just produced a triviality, because what we wrote is just:

$$R_{\alpha\ ;\beta}^{**\beta}-\frac{\delta_{\alpha}^{\beta}}{2}R^{**}{}_{;\beta}=0, \tag{421}$$

which is the contracted Bianchi identity, our hope is nourished by the complexity of the inner form of the scaled Einstein tensor, which perhaps allows for some intrinsic solutions or zeros. We find:

$$0=\left(\begin{array}{c}
\Gamma_{\beta a}^{**\beta}R_{\alpha}^{a}+\Gamma_{\beta a}^{***\beta}R_{\alpha}^{a}-\Gamma_{\beta\alpha}^{**a}R_{a}^{\beta}-\Gamma_{\beta\alpha}^{***a}R_{a}^{\beta}\\
+\frac{1}{2}\left(\begin{array}{c}
\frac{1}{2F}\left(\begin{array}{c}
2F_{,ij}\left(n-1\right)g^{ij}+2\Gamma_{ij}^{a}F_{,a}g^{ij}\\
-F_{,i}g^{ab}g_{jb,a}g^{ij}-F_{,j}g^{ab}g_{ib,a}g^{ij}\\
-n\Gamma_{ij}^{d}F_{,d}g^{ij}+\frac{n}{2}F_{,d}g^{cd}g_{ab,c}g^{ab}
\end{array}\right)\\
+\frac{F_{,i}\cdot F_{,j}}{4F^{2}}g^{ij}\left(\left(n-6\right)\left(n-1\right)\right)
\end{array}\right)_{;\alpha}\\
+\frac{F_{;\beta}}{2F^{2}}\left(g^{\beta\gamma}\left(\begin{array}{c}
F_{,\alpha\gamma}\left(n-2\right)+F_{,ab}g_{\alpha\gamma}g^{ab}\\
+F_{,a}g^{ab}\left(g_{\gamma b,\alpha}-g_{\gamma\alpha,b}\right)-F_{,\alpha}g^{ab}g_{\gamma b,a}-F_{,\gamma}g^{ab}g_{\alpha b,a}\\
+F_{,d}g^{cd}\frac{1}{2}n\left(\frac{2}{n}g_{\alpha c,\gamma}-g_{\alpha c,\gamma}-g_{\gamma c,\alpha}+g_{\alpha\gamma,c}+\frac{1}{n}g_{\alpha\gamma}g_{ab,c}g^{ab}\right)\\
-\frac{1}{2F}\left(F_{,\alpha}\cdot F_{,\gamma}\left(3n-6\right)+g_{\alpha\gamma}F_{,c}F_{,d}g^{cd}\left(4-n\right)\right)
\end{array}\right)\right)\\
-\frac{1}{2F}\left(g^{\beta\gamma}\left(\begin{array}{c}
F_{,\alpha\gamma}\left(n-2\right)+F_{,ab}g_{\alpha\gamma}g^{ab}\\
+F_{,a}g^{ab}\left(g_{\gamma b,\alpha}-g_{\gamma\alpha,b}\right)-F_{,\alpha}g^{ab}g_{\gamma b,a}-F_{,\gamma}g^{ab}g_{\alpha b,a}\\
+F_{,d}g^{cd}\frac{1}{2}n\left(\frac{2}{n}g_{\alpha c,\gamma}-g_{\alpha c,\gamma}-g_{\gamma c,\alpha}+g_{\alpha\gamma,c}+\frac{1}{n}g_{\alpha\gamma}g_{ab,c}g^{ab}\right)\\
-\frac{1}{2F}\left(F_{,\alpha}\cdot F_{,\gamma}\left(3n-6\right)+g_{\alpha\gamma}F_{,c}F_{,d}g^{cd}\left(4-n\right)\right)
\end{array}\right)\right)_{;\beta}
\end{array}\right). \tag{422}$$

When applying the infinite option principle with the setting (374) for the function F[f], things are getting a bit more interesting, because this last equation simplifies dramatically to:

$$\text{with: } 0=4FF''(2-n)+(F')^2(2(n-4)-(n-1)(n-6))$$

$$\Rightarrow F[f]=C_{f1}\cdot(f\cdot(n-3)-4\cdot C_{f0})^{\frac{4}{n-3}}$$

$$\Rightarrow$$

$$0=\left(\begin{array}{c}\lim\limits_{n\to\infty}\left(\Gamma^{**\beta}_{\beta a}R^a_\alpha+\Gamma^{***\beta}_{\beta a}R^a_\alpha-\Gamma^{**a}_{\beta\alpha}R^\beta_a-\Gamma^{***a}_{\beta\alpha}R^\beta_a+\Lambda\cdot\left(F_{,\alpha}\cdot\Psi+F\cdot\Psi_{,\alpha}\right)\right)\\ -\frac{2}{f^2}\left(\begin{array}{c}f_{,\alpha}\left(\frac{1}{2}\left(\frac{1}{2}f_{,d}g^{cd}g_{ab,c}g^{ab}-\Gamma^d_{ij}f_{,d}g^{ij}\right)+f_{,ab}g^{ab}\right)\\ -f_{,\beta}g^{\beta\gamma}\left(f_{,d}g^{cd}\frac{1}{2}\left(g_{\alpha\gamma,c}-g_{\alpha c,\gamma}-g_{\gamma c,\alpha}\right)+f_{,\alpha\gamma}\right)\end{array}\right)\\ +\frac{2}{f}\left(\begin{array}{c}\left(\frac{1}{2}\left(\frac{1}{2}f_{,d}g^{cd}g_{ab,c}g^{ab}-\Gamma^d_{ij}f_{,d}g^{ij}\right)+f_{,ab}g^{ab}\right)_{,\alpha}\\ -\left(g^{\beta\gamma}\left(f_{,d}g^{cd}\frac{1}{2}\left(g_{\alpha\gamma,c}-g_{\alpha c,\gamma}-g_{\gamma c,\alpha}\right)+f_{,\alpha\gamma}\right)\right)_{;\beta}\end{array}\right)\end{array}\right), \tag{423}$$

via:

$$=\left(\begin{array}{c}\lim\limits_{n\to\infty}\left(\begin{array}{c}\frac{F'}{2\cdot F}\left(\begin{array}{c}\left(f_{,\beta}\cdot\delta^\beta_a+f_{,a}\cdot\delta^\beta_\beta-f_{,\sigma}\cdot g^{\beta\sigma}g_{a\beta}\right)R^a_\alpha\\ -\left(f_{,\beta}\cdot\delta^a_\alpha+f_{,\alpha}\cdot\delta^a_\beta-f_{,\sigma}\cdot g^{a\sigma}g_{\alpha\beta}\right)R^\beta_a\end{array}\right)+\Gamma^{***\beta}_{\beta a}R^a_\alpha-\Gamma^{***a}_{\beta\alpha}R^\beta_a\\ +\Lambda\cdot\left(C_{f1}\cdot\frac{4}{n-3}\cdot(f+C_{f0})^{\frac{4}{n-3}-1}\cdot f_{,\alpha}\cdot\Psi+C_{f1}\cdot(f+C_{f0})^{\frac{4}{n-3}}\cdot\Psi_{,\alpha}\right)\end{array}\right)\\ -\frac{2}{f^2}\left(\begin{array}{c}f_{,\alpha}\left(\frac{1}{2}\left(\frac{1}{2}f_{,d}g^{cd}g_{ab,c}g^{ab}-\Gamma^d_{ij}f_{,d}g^{ij}\right)+f_{,ab}g^{ab}\right)\\ -f_{,\beta}g^{\beta\gamma}\left(f_{,d}g^{cd}\frac{1}{2}\left(g_{\alpha\gamma,c}-g_{\alpha c,\gamma}-g_{\gamma c,\alpha}\right)+f_{,\alpha\gamma}\right)\end{array}\right)\\ +\frac{2}{f}\left(\begin{array}{c}\left(\frac{1}{2}\left(\frac{1}{2}f_{,d}g^{cd}g_{ab,c}g^{ab}-\Gamma^d_{ij}f_{,d}g^{ij}\right)+f_{,ab}g^{ab}\right)_{,\alpha}\\ -\left(g^{\beta\gamma}\left(f_{,d}g^{cd}\frac{1}{2}\left(g_{\alpha\gamma,c}-g_{\alpha c,\gamma}-g_{\gamma c,\alpha}\right)+f_{,\alpha\gamma}\right)\right)_{;\beta}\end{array}\right)\end{array}\right), \tag{424}$$

and finally:

$$\Rightarrow 0=\left(\begin{array}{c}\frac{2}{f}f_{,a}R^{a}_{\alpha}+\lim_{n\to\infty}\left(\Gamma^{***\beta}_{\beta a}R^{a}_{\alpha}-\Gamma^{***a}_{\beta\alpha}R^{\beta}_{a}+\Lambda\cdot C_{f1}\cdot\Psi_{,\alpha}\right)\\ -\frac{2}{f^{2}}\left(\begin{array}{c}f_{,\alpha}\left(\frac{1}{2}\left(\frac{1}{2}f_{,d}g^{cd}g_{ab,c}g^{ab}-\Gamma^{d}_{ij}f_{,d}g^{ij}\right)+f_{,ab}g^{ab}\right)\\ -f_{,\beta}g^{\beta\gamma}\left(f_{,d}g^{cd}\frac{1}{2}\left(g_{\alpha\gamma,c}-g_{\alpha c,\gamma}-g_{\gamma c,\alpha}\right)+f_{,\alpha\gamma}\right)\end{array}\right)\\ +\frac{2}{f}\left(\begin{array}{c}\left(\frac{1}{2}\left(\frac{1}{2}f_{,d}g^{cd}g_{ab,c}g^{ab}-\Gamma^{d}_{ij}f_{,d}g^{ij}\right)+f_{,ab}g^{ab}\right)_{,\alpha}\\ -\left(g^{\beta\gamma}\left(f_{,d}g^{cd}\frac{1}{2}\left(g_{\alpha\gamma,c}-g_{\alpha c,\gamma}-g_{\gamma c,\alpha}\right)+f_{,\alpha\gamma}\right)\right)_{;\beta}\end{array}\right)\end{array}\right), \tag{425}$$

$$\Rightarrow 0=\left(\begin{array}{c}\frac{2}{f}f_{,a}R^{a}_{\alpha}+\lim_{n\to\infty}\Lambda\cdot C_{f1}\cdot\Psi_{,\alpha}\\ +\lim_{n\to\infty}\frac{\Psi'}{2\cdot\Psi}\left(\begin{array}{c}\left(\psi_{,\beta}\cdot\delta^{\beta}_{a}+\psi_{,a}\cdot\delta^{\beta}_{\beta}-\psi_{,\sigma}\cdot g^{\beta\sigma}g_{a\beta}\right)R^{a}_{\alpha}\\ -\left(\psi_{,\beta}\cdot\delta^{a}_{\alpha}+\psi_{,\alpha}\cdot\delta^{a}_{\beta}-\psi_{,\sigma}\cdot g^{a\sigma}g_{\alpha\beta}\right)R^{\beta}_{a}\end{array}\right)\\ -\frac{2}{f^{2}}\left(\begin{array}{c}f_{,\alpha}\left(\frac{1}{2}\left(\frac{1}{2}f_{,d}g^{cd}g_{ab,c}g^{ab}-\Gamma^{d}_{ij}f_{,d}g^{ij}\right)+f_{,ab}g^{ab}\right)\\ -f_{,\beta}g^{\beta\gamma}\left(f_{,d}g^{cd}\frac{1}{2}\left(g_{\alpha\gamma,c}-g_{\alpha c,\gamma}-g_{\gamma c,\alpha}\right)+f_{,\alpha\gamma}\right)\end{array}\right)\\ +\frac{2}{f}\left(\begin{array}{c}\left(\frac{1}{2}\left(\frac{1}{2}f_{,d}g^{cd}g_{ab,c}g^{ab}-\Gamma^{d}_{ij}f_{,d}g^{ij}\right)+f_{,ab}g^{ab}\right)_{,\alpha}\\ -\left(g^{\beta\gamma}\left(f_{,d}g^{cd}\frac{1}{2}\left(g_{\alpha\gamma,c}-g_{\alpha c,\gamma}-g_{\gamma c,\alpha}\right)+f_{,\alpha\gamma}\right)\right)_{;\beta}\end{array}\right)\end{array}\right), \tag{426}$$

$$\Rightarrow 0=\begin{pmatrix} \frac{2}{f}f_{,a}R^{a}_{\alpha}+\lim\limits_{n\to\infty}\Lambda\cdot C_{f1}\cdot\Psi_{,\alpha} \\ +\lim\limits_{n\to\infty}\frac{\Psi'}{2\cdot\Psi}\begin{pmatrix} \left(\psi_{,\beta}\cdot\delta^{\beta}_{a}+\psi_{,a}\cdot n-\psi_{,\sigma}\cdot\delta^{\sigma}_{a}\right)R^{a}_{\alpha} \\ -\left(\psi_{,\beta}\cdot\delta^{a}_{\alpha}+\psi_{,\alpha}\cdot\delta^{a}_{\beta}-\psi_{,\sigma}\cdot g^{a\sigma}g_{\alpha\beta}\right)R^{\beta}_{a} \end{pmatrix} \\ -\frac{2}{f^{2}}\begin{pmatrix} f_{,\alpha}\left(\frac{1}{2}\left(\frac{1}{2}f_{,d}g^{cd}g_{ab,c}g^{ab}-\Gamma^{d}_{ij}f_{,d}g^{ij}\right)+f_{,ab}g^{ab}\right) \\ -f_{,\beta}g^{\beta\gamma}\left(f_{,d}g^{cd}\frac{1}{2}\left(g_{\alpha\gamma,c}-g_{\alpha c,\gamma}-g_{\gamma c,\alpha}\right)+f_{,\alpha\gamma}\right) \end{pmatrix} \\ +\frac{2}{f}\begin{pmatrix} \left(\frac{1}{2}\left(\frac{1}{2}f_{,d}g^{cd}g_{ab,c}g^{ab}-\Gamma^{d}_{ij}f_{,d}g^{ij}\right)+f_{,ab}g^{ab}\right)_{,\alpha} \\ -\left(g^{\beta\gamma}\left(f_{,d}g^{cd}\frac{1}{2}\left(g_{\alpha\gamma,c}-g_{\alpha c,\gamma}-g_{\gamma c,\alpha}\right)+f_{,\alpha\gamma}\right)\right)_{;\beta} \end{pmatrix} \end{pmatrix}. \tag{427}$$

As the last equation still sports a term with n (see second line), we fix the function $\Psi[\psi]$ via the condition:

$$\text{with}:\quad 0=4\Psi''\Psi+(\Psi')^{2}(n-6)\Rightarrow\Psi[\psi]=C_{\psi1}\cdot\left(\psi\cdot(n-2)-4\cdot C_{\psi0}\right)^{\frac{4}{n-2}}, \tag{428}$$

which linearizes the $D_{\text{Ricci-}\Gamma}$ operator in (417), because this now reads:

$$D_{Ricci-\Gamma}\Psi=\left(\begin{array}{c}\frac{1}{2\Psi}\left(\begin{array}{c}2\Psi_{,\alpha\beta}(n-1)G^{\alpha\beta}+2\Gamma^{a}_{\alpha\beta}\Psi_{,a}G^{\alpha\beta}\\ -\Psi_{,\alpha}G^{ab}G_{\beta b,a}G^{\alpha\beta}-\Psi_{,\beta}G^{ab}G_{\alpha b,a}G^{\alpha\beta}\\ -n\Gamma^{d}_{\alpha\beta}\Psi_{,d}G^{\alpha\beta}+\frac{n}{2}\Psi_{,d}G^{cd}G_{ab,c}G^{ab}\end{array}\right)\\ -\frac{\Psi_{,\alpha}\cdot\Psi_{,\beta}}{4\Psi^{2}}G^{\alpha\beta}((n-6)(n-1))\end{array}\right)\xrightarrow{\Psi=\Psi[\psi]}$$

$$=\frac{1}{2\Psi}\left(\begin{array}{c}2\left(\Psi'\psi_{,\alpha\beta}+\Psi''\Psi'\psi_{,\alpha}\psi_{,\beta}\right)(n-1)G^{\alpha\beta}\\ +2\Gamma^{a}_{\alpha\beta}\Psi_{,a}G^{\alpha\beta}\\ -\Psi_{,\alpha}G^{ab}G_{\beta b,a}G^{\alpha\beta}-\Psi_{,\beta}G^{ab}G_{\alpha b,a}G^{\alpha\beta}\\ -n\Gamma^{d}_{\alpha\beta}\Psi_{,d}G^{\alpha\beta}+\frac{n}{2}\Psi_{,d}G^{cd}G_{ab,c}G^{ab}\end{array}\right)-\frac{\Psi_{,\alpha}\cdot\Psi_{,\beta}}{4\Psi^{2}}G^{\alpha\beta}((n-6)(n-1))$$

$$=\left(\begin{array}{c}\frac{1}{2\Psi}\left(\begin{array}{c}2\Psi'\psi_{,\alpha\beta}(n-1)G^{\alpha\beta}\\ +2\Gamma^{a}_{\alpha\beta}\Psi_{,a}G^{\alpha\beta}\\ -\Psi_{,\alpha}G^{ab}G_{\beta b,a}G^{\alpha\beta}-\Psi_{,\beta}G^{ab}G_{\alpha b,a}G^{\alpha\beta}\\ -n\Gamma^{d}_{\alpha\beta}\Psi_{,d}G^{\alpha\beta}+\frac{n}{2}\Psi_{,d}G^{cd}G_{ab,c}G^{ab}\end{array}\right)\\ +(n-1)\frac{\psi_{,\alpha}\cdot\psi_{,\beta}}{4\Psi^{2}}G^{\alpha\beta}\underbrace{\left(4\Psi''\Psi-(\Psi')^{2}(n-6)\right)}_{=0}\end{array}\right),\quad(429)$$

and thus, results in:

$$\Rightarrow 0=\left(\begin{array}{c}\frac{2}{f}f_{,a}R^{a}_{\alpha}+\frac{2}{\psi}\psi_{,a}R^{a}_{\alpha}\\ -\frac{2}{f^{2}}\left(\begin{array}{c}f_{,\alpha}\left(\frac{1}{2}\left(\frac{1}{2}f_{,d}g^{cd}g_{ab,c}g^{ab}-\Gamma^{d}_{ij}f_{,d}g^{ij}\right)+f_{,ab}g^{ab}\right)\\ -f_{,\beta}g^{\beta\gamma}\left(f_{,d}g^{cd}\frac{1}{2}\left(g_{\alpha\gamma,c}-g_{\alpha c,\gamma}-g_{\gamma c,\alpha}\right)+f_{,\alpha\gamma}\right)\end{array}\right)\\ +\frac{2}{f}\left(\begin{array}{c}\left(\frac{1}{2}\left(\frac{1}{2}f_{,d}g^{cd}g_{ab,c}g^{ab}-\Gamma^{d}_{ij}f_{,d}g^{ij}\right)+f_{,ab}g^{ab}\right)_{,\alpha}\\ -\left(g^{\beta\gamma}\left(f_{,d}g^{cd}\frac{1}{2}\left(g_{\alpha\gamma,c}-g_{\alpha c,\gamma}-g_{\gamma c,\alpha}\right)+f_{,\alpha\gamma}\right)\right)_{;\beta}\end{array}\right)\end{array}\right)\quad(430)$$

for equation (427).

Using (371) as abbreviation in the form:

$$T^{\beta}_{\infty\alpha}=\frac{2}{\kappa f}\left(\begin{array}{c}\delta^{\beta}_{\alpha}\left(\frac{1}{2}\left(\frac{1}{2}f_{,d}g^{cd}g_{ab,c}g^{ab}-\Gamma^{d}_{ij}f_{,d}g^{ij}\right)+f_{,ab}g^{ab}\right)\\ -g^{\beta\gamma}\left(f_{,d}g^{cd}\frac{1}{2}\left(g_{\alpha\gamma,c}-g_{\alpha c,\gamma}-g_{\gamma c,\alpha}\right)+f_{,\alpha\gamma}\right)\end{array}\right) \tag{431}$$

and applying (362) with the full expansion (419) plus the settings for F and Ψ from (423) and (428) respectively, we can rewrite (430) as follows:

$$\begin{aligned}0&=\left(\begin{array}{c}\frac{2}{f}f_{,a}R^{a}_{\alpha}+\frac{2}{\psi}\psi_{,a}R^{a}_{\alpha}\\ +\kappa\lim\limits_{n\to\infty}\left(T^{\beta}_{\alpha,\beta}+\left(\Gamma^{\beta}_{\beta a}+\Gamma^{**\beta}_{\beta a}+\Gamma^{***\beta}_{\beta a}\right)T^{a}_{\alpha}-\left(\Gamma^{a}_{\beta\alpha}+\Gamma^{**a}_{\beta\alpha}+\Gamma^{***a}_{\beta\alpha}\right)T^{\beta}_{a}\right)\end{array}\right)\\ &=\left(\begin{array}{c}\frac{2}{f}f_{,a}R^{a}_{\alpha}+\frac{2}{\psi}\psi_{,a}R^{a}_{\alpha}\\ +\kappa\left(T^{\beta}_{\alpha,\beta}+\left(\Gamma^{\beta}_{\beta a}+\frac{2}{f}f_{,a}+\frac{2}{\psi}\psi_{,a}\right)T^{a}_{\alpha}-\Gamma^{a}_{\beta\alpha}T^{\beta}_{a}\right)\end{array}\right)\end{aligned} \tag{432}$$

We see that, even though many potentially constant terms vanished (caused by the infinity option setting for F and F and Ψ from (423) and (428), respectively), we still have a constant term and can therefore derive a third order polynomial equation for the masses of particles at rest.

Together with our simple assumption for the function f as a particle at rest from (383), namely, the last equation would now give us the solution to the masses for the particles. As it is a polynomial equation of third order, we can expect to obtain three solutions.

If it weren't for the fact that we have based our calculation on a triviality (c.f. (89)), this would answer the question about the occurrence of the three particle generations. Nevertheless, we now want to discuss a few examples. Thereby, not having a complete solution of the quantum Einstein field equations also sporting the right particle at rest dependency (383), we simply try a few approximations. Assuming a non-vanishing Ricci tensor, but derivatives of the unscaled metric tensor to be small of higher order (Cartesian dominated), we could further evaluate (432):

$$g_{ab,c}\approx 0\Rightarrow T^{\beta}_{\infty\alpha}=\frac{2}{\kappa f}\left(\delta^{\beta}_{\alpha}f_{,ab}g^{ab}-g^{\beta\gamma}f_{,\alpha\gamma}\right)$$

$$\Rightarrow$$

$$0=\left(\begin{array}{c}\frac{2}{f}f_{,a}R^{a}_{\alpha}+\frac{2}{\psi}\psi_{,a}R^{a}_{\alpha}\\ +\kappa\left(\left(\frac{2}{\kappa f}\left(\delta^{\beta}_{\alpha}f_{,ab}g^{ab}-g^{\beta\gamma}f_{,\alpha\gamma}\right)\right)_{,\beta}+\left(\frac{2}{f}f_{,a}+\frac{2}{\psi}\psi_{,a}\right)\left(\frac{2}{\kappa f}\left(\delta^{a}_{\alpha}f_{,db}g^{db}-g^{a\gamma}f_{,\alpha\gamma}\right)\right)\right)\end{array}\right). \tag{433}$$

$$=\left(\begin{array}{c}\frac{2}{f}f_{,a}R^{a}_{\alpha}+\frac{2}{\psi}\psi_{,a}R^{a}_{\alpha}\\ +\left(\frac{2}{f}\left(\delta^{\beta}_{\alpha}f_{,ab}g^{ab}-g^{\beta\gamma}f_{,\alpha\gamma}\right)\right)_{,\beta}+\left(\frac{2}{f}f_{,a}+\frac{2}{\psi}\psi_{,a}\right)\left(\frac{2}{f}\left(\delta^{a}_{\alpha}f_{,db}g^{db}-g^{a\gamma}f_{,\alpha\gamma}\right)\right)\end{array}\right)$$

In the case of a perfectly flat space (vanishing Ricci tensor) this simplifies to:

$$f=f[t]$$

$$\Rightarrow$$

$$0=\left(\frac{2}{f}\left(\delta^{\beta}_{\alpha}f_{,tt}g^{00}-g^{00}f_{,tt}\right)\right)_{,\beta}+\left(\frac{2}{f}f_{,t}+\frac{2}{\psi}\psi_{,t}\right)\left(\frac{2}{f}\left(\delta^{0}_{0}f_{,tt}g^{00}-g^{00}f_{,tt}\right)\right)$$

$$=\left(\begin{array}{c}\left(-\frac{f_{,t}}{f^{2}}\left(\delta^{0}_{0}f_{,tt}g^{00}-g^{00}f_{,tt}\right)\right)+\frac{2}{f}\left(\delta^{0}_{0}f_{,ttt}g^{00}-g^{00}f_{,ttt}\right)\\ +\left(\frac{2}{f}f_{,t}+\frac{2}{\psi}\psi_{,t}\right)\left(\frac{2}{f}\left(\delta^{0}_{0}f_{,tt}g^{00}-g^{00}f_{,tt}\right)\right)\end{array}\right)$$

$$=\left(\begin{array}{c}\left(-\frac{f_{,t}}{f^{2}}\left(f_{,tt}g^{00}-g^{00}f_{,tt}\right)\right)+\frac{2}{f}\left(f_{,ttt}g^{00}-g^{00}f_{,ttt}\right)\\ +\left(\frac{2}{f}f_{,t}+\frac{2}{\psi}\psi_{,t}\right)\left(\frac{2}{f}\left(f_{,tt}g^{00}-g^{00}f_{,tt}\right)\right)\end{array}\right). \tag{434}$$

Unfortunately, as the terms with $\left(\delta^{0}_{0}g^{00}-g^{00}\right)$ are all giving zero, equation (434) is meaningless for us.

Sticking to the infinity option principle but leaving the metric components arbitrary, changes the situation for (432) as follows:

$$0=\begin{pmatrix} \frac{2}{f}f_{,t}R^0_\alpha+\frac{2}{\psi}\psi_{,a}R^a_\alpha \\ -\frac{2}{f^2}\begin{pmatrix} \mathbf{e}_\alpha f_{,0}\left(\frac{1}{2}\left(\frac{1}{2}f_{,0}g^{c0}g_{ab,c}g^{ab}-\Gamma^0_{ij}f_{,0}g^{ij}\right)+f_{,00}g^{00}\right) \\ -f_{,0}\left(g^{0\gamma}f_{,0}g^{c0}\frac{1}{2}\left(g_{\alpha\gamma,c}-g_{\alpha c,\gamma}-g_{\gamma c,\alpha}\right)+g^{00}f_{,00}\mathbf{e}_\alpha\right)\end{pmatrix} \\ +\frac{2}{f}\begin{pmatrix} \left(\frac{1}{2}\left(\frac{1}{2}f_{,0}g^{c0}g_{ab,c}g^{ab}-\Gamma^0_{ij}f_{,0}g^{ij}\right)+f_{,00}g^{00}\right)_{,\alpha} \\ -\left(g^{\beta\gamma}f_{,d}g^{cd}\frac{1}{2}\left(g_{\alpha\gamma,c}-g_{\alpha c,\gamma}-g_{\gamma c,\alpha}\right)+g^{\beta 0}f_{,00}\mathbf{e}_\alpha\right)_{;\beta}\end{pmatrix}\end{pmatrix}$$

$$=\begin{pmatrix} \frac{2}{f}f_{,t}R^0_\alpha+\frac{2}{\psi}\psi_{,a}R^a_\alpha \\ -\frac{2}{f^2}\begin{pmatrix} \mathbf{e}_\alpha f_{,0}\left(\frac{1}{2}\left(\frac{1}{2}f_{,0}g^{c0}g_{ab,c}g^{ab}-\Gamma^0_{ij}f_{,0}g^{ij}\right)+f_{,00}g^{00}\right) \\ -f_{,0}\left(g^{0\gamma}f_{,0}g^{c0}\frac{1}{2}\left(g_{\alpha\gamma,c}-g_{\alpha c,\gamma}-g_{\gamma c,\alpha}\right)+g^{00}f_{,00}\mathbf{e}_\alpha\right)\end{pmatrix} \\ +\frac{2}{f}\begin{pmatrix} \frac{\mathbf{e}_\alpha}{2}f_{,00}\left(\frac{1}{2}g^{c0}g_{ab,c}g^{ab}-\Gamma^0_{ij}g^{ij}\right)+f_{,000}\mathbf{e}_\alpha g^{00} \\ +\frac{1}{2}f_{,0}\left(\frac{1}{2}g^{c0}g_{ab,c}g^{ab}-\Gamma^0_{ij}g^{ij}\right)_{,\alpha}+f_{,00}g^{00}{}_{,\alpha} \\ -\left(g^{\beta\gamma}f_{,d}g^{cd}\frac{1}{2}\left(g_{\alpha\gamma,c}-g_{\alpha c,\gamma}-g_{\gamma c,\alpha}\right)+g^{\beta 0}f_{,00}\mathbf{e}_\alpha\right)_{,\beta} \\ -\left(\Gamma^\beta_{\beta a}+\Gamma^{**\beta}_{\beta a}+\Gamma^{***\beta}_{\beta a}\right)\left(g^{a\gamma}f_{,d}g^{cd}\frac{1}{2}\left(g_{\alpha\gamma,c}-g_{\alpha c,\gamma}-g_{\gamma c,\alpha}\right)+g^{a0}f_{,00}\mathbf{e}_\alpha\right) \\ +\left(\Gamma^a_{\beta\alpha}+\Gamma^{**a}_{\beta\alpha}+\Gamma^{***a}_{\beta\alpha}\right)\left(g^{\beta\gamma}f_{,d}g^{cd}\frac{1}{2}\left(g_{a\gamma,c}-g_{ac,\gamma}-g_{\gamma c,a}\right)+g^{\beta 0}f_{,00}\mathbf{e}_a\right)\end{pmatrix}\end{pmatrix}. \quad (435)$$

Inserting (383) in the Dirac-like form:

$$f=f[M\cdot t]=f[\pm i\cdot\mu\cdot t]=f\left[\pm i\cdot\frac{m\cdot c^2}{\hbar}\cdot t\right]=C_f\cdot e^{M\cdot t} \quad (436)$$

results in:

$$
0=\begin{pmatrix}
2MR_{\alpha}^{0}+\frac{2}{\psi}\psi_{,a}R_{\alpha}^{a}\\
-2M^{2}\begin{pmatrix}
\mathbf{e}_{\alpha}\left(\frac{1}{2}\left(\frac{1}{2}g^{c0}g_{ab,c}g^{ab}-\Gamma_{ij}^{0}g^{ij}\right)+Mg^{00}\right)\\
-\left(g^{0\gamma}g^{c0}\frac{1}{2}\left(g_{\alpha\gamma,c}-g_{\alpha c,\gamma}-g_{\gamma c,\alpha}\right)+g^{00}M\mathbf{e}_{\alpha}\right)
\end{pmatrix}\\
+2\begin{pmatrix}
\begin{pmatrix}
\frac{1}{2}\begin{pmatrix}
\frac{M}{2}\begin{pmatrix}
M\mathbf{e}_{\alpha}g^{c0}g_{ab,c}g^{ab}+g^{c0}{}_{,\alpha}g_{ab,c}g^{ab}\\
+g^{c0}g_{ab,c,\alpha}g^{ab}+g^{c0}g_{ab,c}g^{ab}{}_{,\alpha}
\end{pmatrix}\\
-M\left(\Gamma_{ij}^{0}M\mathbf{e}_{\alpha}g^{ij}-\Gamma_{ij,\alpha}^{0}g^{ij}-\Gamma_{ij}^{0}g^{ij}{}_{,\alpha}\right)
\end{pmatrix}\\
+M^{3}\mathbf{e}_{\alpha}g^{00}+M^{2}g^{00}{}_{,\alpha}
\end{pmatrix}\\
-M\left(g^{c0}\frac{1}{2}\left(g_{\alpha\gamma,c}-g_{\alpha c,\gamma}-g_{\gamma c,\alpha}\right)g^{\beta\gamma}{}_{,\beta}+g^{\beta0}{}_{,\beta}M\mathbf{e}_{\alpha}\right)\\
-M^{2}\left(g^{c0}\frac{1}{2}\left(g_{\alpha\gamma,c}-g_{\alpha c,\gamma}-g_{\gamma c,\alpha}\right)g^{0\gamma}+g^{00}M\mathbf{e}_{\alpha}\right)\\
-Mg^{\beta\gamma}\left(g^{c0}\frac{1}{2}\left(g_{\alpha\gamma,c}-g_{\alpha c,\gamma}-g_{\gamma c,\alpha}\right)\right)_{,\beta}\\
-M\left(\Gamma_{\beta a}^{\beta}+\Gamma_{\beta a}^{**\beta}+\Gamma_{\beta a}^{***\beta}\right)\left(g^{a\gamma}g^{c0}\frac{1}{2}\left(g_{\alpha\gamma,c}-g_{\alpha c,\gamma}-g_{\gamma c,\alpha}\right)+g^{a0}M\mathbf{e}_{\alpha}\right)\\
+M\left(\Gamma_{\beta\alpha}^{a}+\Gamma_{\beta\alpha}^{**a}+\Gamma_{\beta\alpha}^{***a}\right)\left(g^{\beta\gamma}g^{c0}\frac{1}{2}\left(g_{a\gamma,c}-g_{ac,\gamma}-g_{\gamma c,a}\right)+g^{\beta0}M\mathbf{e}_{a}\right)
\end{pmatrix}
\end{pmatrix}. \qquad (437)
$$

Please note that in (437) and following we only use the vector $\mathbf{e}_{\alpha}$ to point out that the pure t-dependency of our wave function f demands an outcome only for the derivative with respect to t and thus, we should in principle write $\mathbf{e}_{\alpha}=\mathbf{e}_{0(\alpha)}$, meaning that only the 0-component does not vanish.

Simplification of (437) gives the following polynomial equation for M:

$$
0=\left(\begin{array}{l}
\frac{2}{\psi}\psi_{,a}R_{\alpha}^{a}-2M^{3}g^{00}\mathbf{e}_{\alpha}\\
+M\left(\begin{array}{l}
2R_{\alpha}^{0}+\left(\Gamma_{\beta\alpha}^{a}+\Gamma_{\beta\alpha}^{**a}+\Gamma_{\beta\alpha}^{***a}\right)g^{\beta\gamma}g^{c0}\left(g_{a\gamma,c}-g_{ac,\gamma}-g_{\gamma c,a}\right)\\
-\left(\Gamma_{\beta a}^{\beta}+\Gamma_{\beta a}^{**\beta}+\Gamma_{\beta a}^{***\beta}\right)g^{a\gamma}g^{c0}\left(g_{\alpha\gamma,c}-g_{\alpha c,\gamma}-g_{\gamma c,\alpha}\right)\\
-2g^{\beta\gamma}\left(g^{c0}\frac{1}{2}\left(g_{\alpha\gamma,c}-g_{\alpha c,\gamma}-g_{\gamma c,\alpha}\right)\right)_{,\beta}\\
-g^{\beta\gamma}{}_{,\beta}g^{c0}\left(g_{\alpha\gamma,c}-g_{\alpha c,\gamma}-g_{\gamma c,\alpha}\right)+\Gamma_{ij,\alpha}^{0}g^{ij}+\Gamma_{ij}^{0}g^{ij}{}_{,\alpha}\\
+\frac{1}{2}\left(g^{c0}{}_{,\alpha}g_{ab,c}g^{ab}+g^{c0}g_{ab,c,\alpha}g^{ab}+g^{c0}g_{ab,c}g^{ab}{}_{,\alpha}\right)
\end{array}\right)\\
+M^{2}\left(\begin{array}{l}
2\left(\Gamma_{\beta\alpha}^{0}+\Gamma_{\beta\alpha}^{**0}+\Gamma_{\beta\alpha}^{***0}\right)g^{\beta0}-2\left(\Gamma_{\beta a}^{\beta}+\Gamma_{\beta a}^{**\beta}+\Gamma_{\beta a}^{***\beta}\right)g^{a0}\mathbf{e}_{\alpha}-2g^{\beta0}{}_{,\beta}\mathbf{e}_{\alpha}\\
-\mathbf{e}_{\alpha}\left(\frac{1}{2}g^{c0}g_{ab,c}g^{ab}-\Gamma_{ij}^{0}g^{ij}\right)+g^{0\gamma}\left(g^{c0}\left(g_{\alpha\gamma,c}-g_{\alpha c,\gamma}-g_{\gamma c,\alpha}\right)\right)\\
-g^{0\gamma}g^{c0}\left(g_{\alpha\gamma,c}-g_{\alpha c,\gamma}-g_{\gamma c,\alpha}\right)+2g^{00}{}_{,\alpha}-\Gamma_{ij}^{0}\mathbf{e}_{\alpha}g^{ij}+\frac{1}{2}\mathbf{e}_{\alpha}g^{c0}g_{ab,c}g^{ab}
\end{array}\right)
\end{array}\right). \quad (438)
$$

We know that the general solution to a three-order polynomial could be given via the following product form:

$$
\begin{array}{c}
\left(M-M_{1}\right)\cdot\left(M-M_{2}\right)\cdot\left(M-M_{3}\right)=\\
M^{3}-M^{2}\cdot\left(M_{1}+M_{2}+M_{3}\right)+M\cdot\left(M_{1}M_{2}+M_{1}M_{3}+M_{2}M_{3}\right)-M_{1}M_{2}M_{3}
\end{array}. \quad (439)
$$

Comparing the last line of (439) with (438) gives us:

$$M_1M_2M_3\mathbf{e}_\alpha = \frac{1}{\psi}\frac{\psi_{,a}R^a_\alpha}{g^{00}}$$

$$\left(M_1M_2 + M_1M_3 + M_2M_3\right)\mathbf{e}_\alpha$$
$$= -\frac{1}{2g^{00}}\begin{pmatrix} 2R^0_\alpha + \left(\Gamma^a_{\beta\alpha} + \Gamma^{**a}_{\beta\alpha} + \Gamma^{***a}_{\beta\alpha}\right)g^{\beta\gamma}g^{c0}\left(g_{a\gamma,c} - g_{ac,\gamma} - g_{\gamma c,a}\right) \\ -\left(\Gamma^\beta_{\beta a} + \Gamma^{**\beta}_{\beta a} + \Gamma^{***\beta}_{\beta a}\right)g^{a\gamma}g^{c0}\left(g_{\alpha\gamma,c} - g_{\alpha c,\gamma} - g_{\gamma c,\alpha}\right) \\ -2g^{\beta\gamma}\left(g^{c0}\frac{1}{2}\left(g_{\alpha\gamma,c} - g_{\alpha c,\gamma} - g_{\gamma c,\alpha}\right)\right)_{,\beta} \\ -g^{\beta\gamma}{}_{,\beta}g^{c0}\left(g_{\alpha\gamma,c} - g_{\alpha c,\gamma} - g_{\gamma c,\alpha}\right) + \Gamma^0_{ij,\alpha}g^{ij} + \Gamma^0_{ij}g^{ij}{}_{,\alpha} \\ +\frac{1}{2}\left(g^{c0}{}_{,\alpha}g_{ab,c}g^{ab} + g^{c0}g_{ab,c,\alpha}g^{ab} + g^{c0}g_{ab,c}g^{ab}{}_{,\alpha}\right) \end{pmatrix}$$

$$\left(M_1 + M_2 + M_3\right)\mathbf{e}_\alpha$$
$$= \frac{1}{2g^{00}}\begin{pmatrix} 2\left(\Gamma^0_{\beta\alpha} + \Gamma^{**0}_{\beta\alpha} + \Gamma^{***0}_{\beta\alpha}\right)g^{\beta 0} - 2\left(\Gamma^\beta_{\beta a} + \Gamma^{**\beta}_{\beta a} + \Gamma^{***\beta}_{\beta a}\right)g^{a0}\mathbf{e}_\alpha - 2g^{\beta 0}{}_{,\beta}\mathbf{e}_\alpha \\ -\mathbf{e}_\alpha\left(\frac{1}{2}g^{c0}g_{ab,c}g^{ab} - \Gamma^0_{ij}g^{ij}\right) + g^{0\gamma}\left(g^{c0}\left(g_{\alpha\gamma,c} - g_{\alpha c,\gamma} - g_{\gamma c,\alpha}\right)\right) \\ -g^{0\gamma}g^{c0}\left(g_{\alpha\gamma,c} - g_{\alpha c,\gamma} - g_{\gamma c,\alpha}\right) + 2g^{00}{}_{,\alpha} - \Gamma^0_{ij}\mathbf{e}_\alpha g^{ij} + \frac{1}{2}\mathbf{e}_\alpha g^{c0}g_{ab,c}g^{ab} \end{pmatrix}, \tag{440}$$

which are the equations for the determination of the three masses of the 3 particle generations. Our only and rather unfortunate problem is that we do not know the metric which would give us the correct spatial and timely distribution of our elementary particles.

As we are still in the approximation regime of the infinite option principle, we can simplify (437) with the help of (432):

$$\lim_{n\to\infty}\left(\Gamma^{a}_{\beta\alpha}+\Gamma^{**a}_{\beta\alpha}+\Gamma^{***a}_{\beta\alpha}\right)=\Gamma^{a}_{\beta\alpha}$$

$$\lim_{n\to\infty}\left(\Gamma^{\beta}_{\beta a}+\Gamma^{**\beta}_{\beta a}+\Gamma^{***\beta}_{\beta a}\right)=\Gamma^{\beta}_{\beta a}+\frac{2}{f}f_{,a}+\frac{2}{\psi}\psi_{,a}$$

$$\Rightarrow$$

$$0=\left(\begin{array}{c}
2MR^{0}_{\alpha}+\frac{2}{\psi}\psi_{,a}R^{a}_{\alpha} \\
-2M^{2}\left(\begin{array}{c}
\mathbf{e}_{\alpha}\left(\frac{1}{2}\left(\frac{1}{2}g^{c0}g_{ab,c}g^{ab}-\Gamma^{0}_{ij}g^{ij}\right)+Mg^{00}\right) \\
-\left(g^{0\gamma}g^{c0}\frac{1}{2}\left(g_{\alpha\gamma,c}-g_{\alpha c,\gamma}-g_{\gamma c,\alpha}\right)+g^{00}M\mathbf{e}_{\alpha}\right)
\end{array}\right) \\
+2\left(\begin{array}{c}
\frac{1}{2}\left(\begin{array}{c}
\frac{M}{2}\left(\begin{array}{c}
M\mathbf{e}_{\alpha}g^{c0}g_{ab,c}g^{ab}+g^{c0}{}_{,\alpha}g_{ab,c}g^{ab} \\
+g^{c0}g_{ab,c,\alpha}g^{ab}+g^{c0}g_{ab,c}g^{ab}{}_{,\alpha}
\end{array}\right) \\
-M\left(\Gamma^{0}_{ij}M\mathbf{e}_{\alpha}g^{ij}-\Gamma^{0}_{ij,\alpha}g^{ij}-\Gamma^{0}_{ij}g^{ij}{}_{,\alpha}\right) \\
+M^{3}\mathbf{e}_{\alpha}g^{00}+M^{2}g^{00}{}_{,\alpha}
\end{array}\right) \\
-M\left(g^{c0}\frac{1}{2}\left(g_{\alpha\gamma,c}-g_{\alpha c,\gamma}-g_{\gamma c,\alpha}\right)g^{\beta\gamma}{}_{,\beta}+g^{\beta0}{}_{,\beta}M\mathbf{e}_{\alpha}\right) \\
-M^{2}\left(g^{c0}\frac{1}{2}\left(g_{\alpha\gamma,c}-g_{\alpha c,\gamma}-g_{\gamma c,\alpha}\right)g^{0\gamma}+g^{00}M\mathbf{e}_{\alpha}\right) \\
-Mg^{\beta\gamma}\left(g^{c0}\frac{1}{2}\left(g_{\alpha\gamma,c}-g_{\alpha c,\gamma}-g_{\gamma c,\alpha}\right)\right)_{,\beta} \\
-M\left(\Gamma^{\beta}_{\beta a}+2M\mathbf{e}_{a}+\frac{2}{\psi}\psi_{,a}\right)\left(g^{a\gamma}g^{c0}\frac{1}{2}\left(g_{\alpha\gamma,c}-g_{\alpha c,\gamma}-g_{\gamma c,\alpha}\right)+g^{a0}M\mathbf{e}_{\alpha}\right) \\
+M\Gamma^{a}_{\beta\alpha}\left(g^{\beta\gamma}g^{c0}\frac{1}{2}\left(g_{a\gamma,c}-g_{ac,\gamma}-g_{\gamma c,a}\right)+g^{\beta0}M\mathbf{e}_{a}\right)
\end{array}\right)
\end{array}\right). \quad (441)$$

This changes the equation above and (440) as follows:

$$
0=\left(\begin{array}{c}
\frac{2}{\psi}\psi_{,a}R^{a}_{\alpha}-6M^{3}g^{00}\mathbf{e}_{\alpha}\\
+M\left(\begin{array}{l}
2R^{0}_{\alpha}+\Gamma^{a}_{\beta\alpha}g^{\beta\gamma}g^{c0}\left(g_{a\gamma,c}-g_{ac,\gamma}-g_{\gamma c,a}\right)\\
-\left(\Gamma^{\beta}_{\beta a}+\frac{2}{\psi}\psi_{,a}\right)g^{a\gamma}g^{c0}\left(g_{\alpha\gamma,c}-g_{\alpha c,\gamma}-g_{\gamma c,\alpha}\right)\\
-2g^{\beta\gamma}\left(g^{c0}\frac{1}{2}\left(g_{\alpha\gamma,c}-g_{\alpha c,\gamma}-g_{\gamma c,\alpha}\right)\right)_{,\beta}\\
-g^{\beta\gamma}{}_{,\beta}g^{c0}\left(g_{\alpha\gamma,c}-g_{\alpha c,\gamma}-g_{\gamma c,\alpha}\right)+\Gamma^{0}_{ij,\alpha}g^{ij}+\Gamma^{0}_{ij}g^{ij}{}_{,\alpha}\\
+\frac{1}{2}\left(g^{c0}{}_{,\alpha}g_{ab,c}g^{ab}+g^{c0}g_{ab,c,\alpha}g^{ab}+g^{c0}g_{ab,c}g^{ab}{}_{,\alpha}\right)
\end{array}\right)\\
+M^{2}\left(\begin{array}{c}
2\Gamma^{0}_{\beta\alpha}g^{\beta0}-2\left(\Gamma^{\beta}_{\beta a}+\frac{2}{\psi}\psi_{,a}\right)g^{a0}\mathbf{e}_{\alpha}-2g^{\beta0}{}_{,\beta}\mathbf{e}_{\alpha}\\
-2g^{0\gamma}g^{c0}\left(g_{\alpha\gamma,c}-g_{\alpha c,\gamma}-g_{\gamma c,\alpha}\right)\\
-\mathbf{e}_{\alpha}\left(\frac{1}{2}g^{c0}g_{ab,c}g^{ab}-\Gamma^{0}_{ij}g^{ij}\right)+g^{0\gamma}\left(g^{c0}\left(g_{\alpha\gamma,c}-g_{\alpha c,\gamma}-g_{\gamma c,\alpha}\right)\right)\\
-g^{0\gamma}g^{c0}\left(g_{\alpha\gamma,c}-g_{\alpha c,\gamma}-g_{\gamma c,\alpha}\right)+2g^{00}{}_{,\alpha}-\Gamma^{0}_{ij}\mathbf{e}_{\alpha}g^{ij}+\frac{1}{2}\mathbf{e}_{\alpha}g^{c0}g_{ab,c}g^{ab}
\end{array}\right)
\end{array}\right), \tag{442}
$$

$$M_1M_2M_3\mathbf{e}_\alpha = \frac{1}{\psi}\frac{\psi_{,a}R^a_\alpha}{3g^{00}}$$

$$\left(M_1M_2 + M_1M_3 + M_2M_3\right)\mathbf{e}_\alpha$$

$$= -\frac{1}{6g^{00}}\begin{pmatrix} 2R^0_\alpha + \Gamma^a_{\beta\alpha}g^{\beta\gamma}g^{c0}\left(g_{a\gamma,c} - g_{ac,\gamma} - g_{\gamma c,a}\right) \\ -\left(\Gamma^\beta_{\beta a} + \frac{2}{\psi}\psi_{,a}\right)g^{a\gamma}g^{c0}\left(g_{\alpha\gamma,c} - g_{\alpha c,\gamma} - g_{\gamma c,\alpha}\right) \\ -2g^{\beta\gamma}\left(g^{c0}\frac{1}{2}\left(g_{\alpha\gamma,c} - g_{\alpha c,\gamma} - g_{\gamma c,\alpha}\right)\right)_{,\beta} \\ -g^{\beta\gamma}{}_{,\beta}g^{c0}\left(g_{\alpha\gamma,c} - g_{\alpha c,\gamma} - g_{\gamma c,\alpha}\right) + \Gamma^0_{ij,\alpha}g^{ij} + \Gamma^0_{ij}g^{ij}{}_{,\alpha} \\ +\frac{1}{2}\left(g^{c0}{}_{,\alpha}g_{ab,c}g^{ab} + g^{c0}g_{ab,c,\alpha}g^{ab} + g^{c0}g_{ab,c}g^{ab}{}_{,\alpha}\right) \end{pmatrix}$$

$$\left(M_1 + M_2 + M_3\right)\mathbf{e}_\alpha$$

$$= \frac{1}{6g^{00}}\begin{pmatrix} 2\Gamma^0_{\beta\alpha}g^{\beta0} - 2\left(\Gamma^\beta_{\beta a} + \frac{2}{\psi}\psi_{,a}\right)g^{a0}\mathbf{e}_\alpha - 2g^{\beta0}{}_{,\beta}\mathbf{e}_\alpha - 2g^{0\gamma}g^{c0}\left(g_{\alpha\gamma,c} - g_{\alpha c,\gamma} - g_{\gamma c,\alpha}\right) \\ -\mathbf{e}_\alpha\left(\frac{1}{2}g^{c0}g_{ab,c}g^{ab} - \Gamma^0_{ij}g^{ij}\right) + g^{0\gamma}\left(g^{c0}\left(g_{\alpha\gamma,c} - g_{\alpha c,\gamma} - g_{\gamma c,\alpha}\right)\right) \\ -g^{0\gamma}g^{c0}\left(g_{\alpha\gamma,c} - g_{\alpha c,\gamma} - g_{\gamma c,\alpha}\right) + 2g^{00}{}_{,\alpha} - \Gamma^0_{ij}\mathbf{e}_\alpha g^{ij} + \frac{1}{2}\mathbf{e}_\alpha g^{c0}g_{ab,c}g^{ab} \end{pmatrix}. \quad (443)$$

Taking into account the meaning of the vector $\mathbf{e}_\alpha = \mathbf{e}_{0(\alpha)}$ makes (443) to:

$$M_1M_2M_3 = \frac{1}{\psi}\frac{\psi_{,a}R_0^a}{3g^{00}}$$

$$\left(M_1M_2 + M_1M_3 + M_2M_3\right)$$

$$= -\frac{1}{6g^{00}}\begin{pmatrix} 2R_\alpha^0 + \Gamma_{\beta 0}^a g^{\beta\gamma} g^{c0}\left(g_{a\gamma,c} - g_{ac,\gamma} - g_{\gamma c,a}\right) \\ -\left(\Gamma_{\beta a}^\beta + \frac{2}{\psi}\psi_{,a}\right) g^{a\gamma} g^{c0}\left(g_{0\gamma,c} - g_{0c,\gamma} - g_{0c,\alpha}\right) \\ -2g^{\beta\gamma}\left(g^{c0}\frac{1}{2}\left(g_{0\gamma,c} - g_{0c,\gamma} - g_{\gamma c,0}\right)\right)_{,\beta} \\ -g^{\beta\gamma}{}_{,\beta} g^{c0}\left(g_{0\gamma,c} - g_{0c,\gamma} - g_{\gamma c,0}\right) + \Gamma_{ij,0}^0 g^{ij} + \Gamma_{ij}^0 g^{ij}{}_{,0} \\ +\frac{1}{2}\left(g^{c0}{}_{,0} g_{ab,c} g^{ab} + g^{c0} g_{ab,c,0} g^{ab} + g^{c0} g_{ab,c} g^{ab}{}_{,0}\right) \end{pmatrix}$$

$$\left(M_1 + M_2 + M_3\right)$$

$$= \frac{1}{6g^{00}}\begin{pmatrix} 2\Gamma_{\beta 0}^0 g^{\beta 0} - 2\left(\Gamma_{\beta a}^\beta + \frac{2}{\psi}\psi_{,a}\right) g^{a0} - 2g^{\beta 0}{}_{,\beta} - 2g^{0\gamma} g^{c0}\left(g_{0\gamma,c} - g_{0c,\gamma} - g_{\gamma c,0}\right) \\ -\left(\frac{1}{2} g^{c0} g_{ab,c} g^{ab} - \Gamma_{ij}^0 g^{ij}\right) + g^{0\gamma}\left(g^{c0}\left(g_{0\gamma,c} - g_{0c,\gamma} - g_{\gamma c,0}\right)\right) \\ -g^{0\gamma} g^{c0}\left(g_{0\gamma,c} - g_{0c,\gamma} - g_{\gamma c,0}\right) + 2g^{00}{}_{,\alpha} - \Gamma_{ij}^0 g^{ij} + \frac{1}{2} g^{c0} g_{ab,c} g^{ab} \end{pmatrix}$$

and for $\alpha > 0$:

$$0 = \begin{pmatrix} 2R_\alpha^0 + \Gamma_{\beta\alpha}^a g^{\beta\gamma} g^{c0}\left(g_{a\gamma,c} - g_{ac,\gamma} - g_{\gamma c,a}\right) - \left(\Gamma_{\beta a}^\beta + \frac{2}{\psi}\psi_{,a}\right) g^{a\gamma} g^{c0}\left(g_{\alpha\gamma,c} - g_{\alpha c,\gamma} - g_{\gamma c,\alpha}\right) \\ -2g^{\beta\gamma}\left(g^{c0}\frac{1}{2}\left(g_{\alpha\gamma,c} - g_{\alpha c,\gamma} - g_{\gamma c,\alpha}\right)\right)_{,\beta} - g^{\beta\gamma}{}_{,\beta} g^{c0}\left(g_{\alpha\gamma,c} - g_{\alpha c,\gamma} - g_{\gamma c,\alpha}\right) + \Gamma_{ij,\alpha}^0 g^{ij} \\ +\frac{1}{2}\left(g^{c0}{}_{,\alpha} g_{ab,c} g^{ab} + g^{c0} g_{ab,c,\alpha} g^{ab} + g^{c0} g_{ab,c} g^{ab}{}_{,\alpha}\right) + \Gamma_{ij}^0 g^{ij}{}_{,\alpha} \end{pmatrix}$$

$$0 = \begin{pmatrix} 2\Gamma_{\beta\alpha}^0 g^{\beta 0} - 2g^{0\gamma} g^{c0}\left(g_{\alpha\gamma,c} - g_{\alpha c,\gamma} - g_{\gamma c,\alpha}\right) + g^{0\gamma}\left(g^{c0}\left(g_{\alpha\gamma,c} - g_{\alpha c,\gamma} - g_{\gamma c,\alpha}\right)\right) \\ -g^{0\gamma} g^{c0}\left(g_{\alpha\gamma,c} - g_{\alpha c,\gamma} - g_{\gamma c,\alpha}\right) + 2g^{00}{}_{,\alpha} \end{pmatrix} \quad (444)$$

and thus, degenerates them to 3 scalar equations plus 2 vector equations running from α=1 to α=∞.

But what happens in the cases of finite n?

Leaving F and Ψ open, we obtain the following equation from (422)

$$
0=\left(\begin{array}{l}
\overbrace{\frac{F'}{2\cdot F}f_{,a}nR^{a}_{\alpha}}^{\Gamma^{**\beta}_{\beta a}}+\overbrace{\frac{\Psi'}{2\cdot\Psi}\psi_{,a}nR^{a}_{\alpha}}^{\Gamma^{***\beta}_{\beta a}}-\overbrace{\frac{F'}{2\cdot F}\left(f_{,\beta}\cdot\delta^{a}_{\alpha}+f_{,\alpha}\cdot\delta^{a}_{\beta}-f_{,\sigma}\cdot g^{a\sigma}g_{\alpha\beta}\right)}^{\Gamma^{**a}_{\beta\alpha}}R^{\beta}_{a} \\
-\overbrace{\frac{\Psi'}{2\cdot\Psi}\left(\psi_{,\beta}\cdot\delta^{a}_{\alpha}+\psi_{,\alpha}\cdot\delta^{a}_{\beta}-\psi_{,\sigma}\cdot g^{a\sigma}g_{\alpha\beta}\right)}^{\Gamma^{***a}_{\beta\alpha}}R^{\beta}_{a} \\
+\left(\begin{array}{c}
\frac{F'}{4F}\left(\begin{array}{c}
2f_{,ij}(n-1)g^{ij}+2\Gamma^{a}_{ij}f_{,a}g^{ij} \\
-f_{,i}g^{ab}g_{jb,a}g^{ij}-f_{,j}g^{ab}g_{ib,a}g^{ij} \\
-n\Gamma^{d}_{ij}f_{,d}g^{ij}+\frac{n}{2}f_{,d}g^{cd}g_{ab,c}g^{ab}
\end{array}\right) \\
+(n-1)\frac{f_{,i}\cdot f_{,j}}{8F^{2}}g^{ij}\left((n-6)(F')^{2}+4FF''\right)
\end{array}\right)_{,\alpha} \\
+\frac{(F')^{2}f_{,\beta}}{2F^{2}}g^{\beta\gamma}\left(\begin{array}{c}
f_{,\alpha\gamma}(n-2)+f_{,ab}g_{\alpha\gamma}g^{ab} \\
+f_{,a}g^{ab}\left(g_{\gamma b,\alpha}-g_{\gamma\alpha,b}\right)-f_{,\alpha}g^{ab}g_{\gamma b,a}-f_{,\gamma}g^{ab}g_{\alpha b,a} \\
+f_{,d}g^{cd}\frac{1}{2}n\left(\frac{2}{n}g_{\alpha c,\gamma}-g_{\alpha c,\gamma}-g_{\gamma c,\alpha}+g_{\alpha\gamma,c}+\frac{1}{n}g_{\alpha\gamma}g_{ab,c}g^{ab}\right) \\
+f_{,\alpha}f_{,\gamma}\left(\frac{F''}{F'}(n-2)-\frac{F'}{2F}(3n-6)\right) \\
+g_{\alpha\gamma}f_{,c}f_{,d}g^{cd}\left(\frac{F''}{F'}-\frac{F'}{2F}(4-n)\right)
\end{array}\right) \\
-\frac{F'}{2F}\left(g^{\beta\gamma}\left(\begin{array}{c}
f_{,\alpha\gamma}(n-2)+f_{,ab}g_{\alpha\gamma}g^{ab} \\
+f_{,a}g^{ab}\left(g_{\gamma b,\alpha}-g_{\gamma\alpha,b}\right)-f_{,\alpha}g^{ab}g_{\gamma b,a}-f_{,\gamma}g^{ab}g_{\alpha b,a} \\
+f_{,d}g^{cd}\frac{1}{2}n\left(\frac{2}{n}g_{\alpha c,\gamma}-g_{\alpha c,\gamma}-g_{\gamma c,\alpha}+g_{\alpha\gamma,c}+\frac{1}{n}g_{\alpha\gamma}g_{ab,c}g^{ab}\right) \\
+f_{,\alpha}f_{,\gamma}\left(\frac{F''}{F'}(n-2)-\frac{F'}{2F}(3n-6)\right) \\
+g_{\alpha\gamma}f_{,c}f_{,d}g^{cd}\left(\frac{F''}{F'}-\frac{F'}{2F}(4-n)\right)
\end{array}\right)\right)_{;\beta}
\end{array}\right). \tag{445}
$$

Pulling out the scalar terms, which would give just ordinary derivatives, by using the results from [7] appendix C, allows to compactify the product of first derivatives in f:

$$
0=\left(\begin{array}{c}
\overbrace{\frac{F'}{2\cdot F}f_{,a}n}^{\Gamma^{**\beta}_{\beta a}}R^{a}_{\alpha}+\overbrace{\frac{\Psi'}{2\cdot\Psi}\psi_{,a}n}^{\Gamma^{***\beta}_{\beta a}}R^{a}_{\alpha}-\overbrace{\frac{F'}{2\cdot F}\left(f_{,\beta}\cdot\delta^{a}_{\alpha}+f_{,\alpha}\cdot\delta^{a}_{\beta}-f_{,\sigma}\cdot g^{a\sigma}g_{\alpha\beta}\right)}^{\Gamma^{**a}_{\beta\alpha}}R^{\beta}_{a}\\
-\overbrace{\frac{\Psi'}{2\cdot\Psi}\left(\psi_{,\beta}\cdot\delta^{a}_{\alpha}+\psi_{,\alpha}\cdot\delta^{a}_{\beta}-\psi_{,\sigma}\cdot g^{a\sigma}g_{\alpha\beta}\right)}^{\Gamma^{***a}_{\beta\alpha}}R^{\beta}_{a}\\
+\frac{1}{2}\left(\begin{array}{c}
\frac{F'}{2F}\left(\begin{array}{c}
2\Gamma^{a}_{ij}f_{,a}g^{ij}-f_{,i}g^{ab}g_{jb,a}g^{ij}-f_{,j}g^{ab}g_{ib,a}g^{ij}-f_{,d}g^{cd}g_{ab,c}g^{ab}\\
-n\Gamma^{d}_{ij}f_{,d}g^{ij}+\frac{n}{2}f_{,d}g^{cd}g_{ab,c}g^{ab}+2f_{,ab}\left(n-2\right)g^{ab}
\end{array}\right)\\
-\frac{f_{,a}f_{,b}g^{ab}}{4F^{2}}\left(4FF''(2-n)+(F')^{2}\left(2(n-4)-(n-1)(n-6)\right)\right)
\end{array}\right)_{,\alpha}\\
-\left(\frac{g^{\beta\gamma}}{2F}\left(\begin{array}{c}
F'\left(\begin{array}{c}
f_{,a}g^{ab}\left(g_{\gamma b,\alpha}-g_{\gamma\alpha,b}\right)-f_{,\alpha}g^{ab}g_{\gamma b,a}-f_{,\gamma}g^{ab}g_{\alpha b,a}\\
+f_{,d}g^{cd}\frac{1}{2}n\left(\frac{2}{n}g_{\alpha c,\gamma}-g_{\alpha c,\gamma}-g_{\gamma c,\alpha}+g_{\alpha\gamma,c}\right)
\end{array}\right)\\
+F'f_{,\alpha\gamma}\left(n-2\right)\\
+\frac{f_{,\alpha}\cdot f_{,\gamma}}{2F}\left((6-3n)(F')^{2}+2FF''(n-2)\right)
\end{array}\right)\right)_{;\beta}
\end{array}\right). \tag{446}
$$

As we can only get rid of one of the squared first order derivatives, we chose condition (372) and set the constant $C_{f0}=0$. Then the last line in (446) disappears and we obtain:

$$\frac{F'}{F}=\frac{2}{f\cdot(8-3n)}$$

$$\frac{4FF''(2-n)+(F')^2(2(n-4)-(n-1)(n-6))}{4F^2}=-\frac{(n-2)(7n-19)}{\left(f\cdot(3n-8)\right)^2}$$

$$\Rightarrow$$

$$0=\left(\begin{array}{c}\frac{f_{,a}nR_{\alpha}^{a}}{f\cdot(8-3n)}+\frac{\Psi'}{2\cdot\Psi}\psi_{,a}nR_{\alpha}^{a}-\frac{1}{f\cdot(8-3n)}\left(f_{,\beta}\cdot\delta_{\alpha}^{a}+f_{,\alpha}\cdot\delta_{\beta}^{a}-f_{,\sigma}\cdot g^{a\sigma}g_{\alpha\beta}\right)R_{a}^{\beta}\\ -\frac{\Psi'}{2\cdot\Psi}\left(\psi_{,\beta}\cdot\delta_{\alpha}^{a}+\psi_{,\alpha}\cdot\delta_{\beta}^{a}-\psi_{,\sigma}\cdot g^{a\sigma}g_{\alpha\beta}\right)R_{a}^{\beta}\\ +\frac{1}{2}\left(\begin{array}{c}\frac{1}{f\cdot(8-3n)}\left(\begin{array}{c}2\Gamma_{ij}^{a}f_{,a}g^{ij}-f_{,i}g^{ab}g_{jb,a}g^{ij}-f_{,j}g^{ab}g_{ib,a}g^{ij}-f_{,d}g^{cd}g_{ab,c}g^{ab}\\ -n\Gamma_{ij}^{d}f_{,d}g^{ij}+\frac{n}{2}f_{,d}g^{cd}g_{ab,c}g^{ab}+2f_{,ab}(n-2)g^{ab}\end{array}\right)\\ +f_{,a}f_{,b}g^{ab}\frac{(n-2)(7n-19)}{\left(f\cdot(3n-8)\right)^2}\end{array}\right)_{,\alpha}\\ \underbrace{-\left(\frac{g^{\beta\gamma}}{f\cdot(8-3n)}\left(\begin{array}{c}f_{,a}g^{ab}\left(g_{\gamma b,\alpha}-g_{\gamma\alpha,b}\right)-f_{,\alpha}g^{ab}g_{\gamma b,a}-f_{,\gamma}g^{ab}g_{\alpha b,a}\\ +f_{,d}g^{cd}\frac{1}{2}n\left(\frac{2}{n}g_{\alpha c,\gamma}-g_{\alpha c,\gamma}-g_{\gamma c,\alpha}+g_{\alpha\gamma,c}\right)+f_{,\alpha\gamma}(n-2)\end{array}\right)\right)_{;\beta}}_{=A_{\alpha;\beta}^{\beta}}\end{array}\right). \quad (447)$$

Choosing condition (374) and setting the constant $C_{f0}=0$ on the other hand gives us:

$$\frac{F'}{F}=\frac{4}{f\cdot(n-3)}$$

$$\frac{(6-3n)(F')^2+2FF''(n-2)}{4F^2}=\frac{38-14n}{\left(f\left(n-3\right)\right)^2}$$

$$\Rightarrow$$

$$0=\begin{pmatrix}\frac{2f_{,a}nR^{a}_{\alpha}}{f\cdot(n-3)}+\frac{\Psi'}{2\cdot\Psi}\psi_{,a}nR^{a}_{\alpha}-\frac{2}{f\cdot(n-3)}\left(f_{,\beta}\cdot\delta^{a}_{\alpha}+f_{,\alpha}\cdot\delta^{a}_{\beta}-f_{,\sigma}\cdot g^{a\sigma}g_{\alpha\beta}\right)R^{\beta}_{a}\\ -\frac{\Psi'}{2\cdot\Psi}\left(\psi_{,\beta}\cdot\delta^{a}_{\alpha}+\psi_{,\alpha}\cdot\delta^{a}_{\beta}-\psi_{,\sigma}\cdot g^{a\sigma}g_{\alpha\beta}\right)R^{\beta}_{a}\\ +\left(\frac{1}{f\cdot(n-3)}\begin{pmatrix}2\Gamma^{a}_{ij}f_{,a}g^{ij}-f_{,i}g^{ab}g_{jb,a}g^{ij}-f_{,j}g^{ab}g_{ib,a}g^{ij}-f_{,d}g^{cd}g_{ab,c}g^{ab}\\ -n\Gamma^{d}_{ij}f_{,d}g^{ij}+\frac{n}{2}f_{,d}g^{cd}g_{ab,c}g^{ab}+2f_{,ab}(n-2)g^{ab}\end{pmatrix}\right)_{,\alpha}\\ \underbrace{-\left(g^{\beta\gamma}\left(\frac{2}{f\cdot(n-3)}\begin{pmatrix}f_{,a}g^{ab}\left(g_{\gamma b,\alpha}-g_{\gamma\alpha,b}\right)-f_{,\alpha}g^{ab}g_{\gamma b,a}-f_{,\gamma}g^{ab}g_{\alpha b,a}\\ +f_{,d}g^{cd}\frac{1}{2}n\left(\frac{2}{n}g_{\alpha c,\gamma}-g_{\alpha c,\gamma}-g_{\gamma c,\alpha}+g_{\alpha\gamma,c}\right)\\ f_{,\alpha\gamma}(n-2)\end{pmatrix}\\ +f_{,\alpha}\cdot f_{,\gamma}\frac{38-14n}{\left(f\left(n-3\right)\right)^2}\right)\right)_{;\beta}}_{=B^{\beta}_{\alpha;\beta}}\end{pmatrix}. \tag{448}$$

Reminder: Please note that the covariant derivative in (447) and (448) is with respect to the double scaled metric tensor $\Gamma_{\alpha\beta}$ and therefore, has to be performed as follows:

$$A^{\beta}_{\alpha;\beta}=\begin{pmatrix}A^{\beta}_{\alpha,\beta}+\Gamma^{\beta}_{a\beta}A^{a}_{\alpha}-\Gamma^{a}_{\alpha\beta}A^{\beta}_{a}+\frac{f_{,a}nA^{a}_{\alpha}}{f\cdot(8-3n)}+\frac{\Psi'}{2\cdot\Psi}\psi_{,a}nA^{a}_{\alpha}\\ -\frac{1}{f\cdot(8-3n)}\left(f_{,\beta}\cdot\delta^{a}_{\alpha}+f_{,\alpha}\cdot\delta^{a}_{\beta}-f_{,\sigma}\cdot g^{a\sigma}g_{\alpha\beta}\right)A^{\beta}_{a}\\ -\frac{\Psi'}{2\cdot\Psi}\left(\psi_{,\beta}\cdot\delta^{a}_{\alpha}+\psi_{,\alpha}\cdot\delta^{a}_{\beta}-\psi_{,\sigma}\cdot g^{a\sigma}g_{\alpha\beta}\right)A^{\beta}_{a}\end{pmatrix}$$

$$B^{\beta}_{\alpha;\beta}=\begin{pmatrix}B^{\beta}_{\alpha,\beta}+\Gamma^{\beta}_{a\beta}B^{a}_{\alpha}-\Gamma^{a}_{\alpha\beta}B^{\beta}_{a}+\frac{f_{,a}nB^{a}_{\alpha}}{f\cdot(8-3n)}+\frac{\Psi'}{2\cdot\Psi}\psi_{,a}nB^{a}_{\alpha}\\ -\frac{1}{f\cdot(8-3n)}\left(f_{,\beta}\cdot\delta^{a}_{\alpha}+f_{,\alpha}\cdot\delta^{a}_{\beta}-f_{,\sigma}\cdot g^{a\sigma}g_{\alpha\beta}\right)B^{\beta}_{a}\\ -\frac{\Psi'}{2\cdot\Psi}\left(\psi_{,\beta}\cdot\delta^{a}_{\alpha}+\psi_{,\alpha}\cdot\delta^{a}_{\beta}-\psi_{,\sigma}\cdot g^{a\sigma}g_{\alpha\beta}\right)B^{\beta}_{a}\end{pmatrix}. \tag{449}$$

For our particle at rest approach from (383) in the Dirac-like form (436) both equations, (447) and (448), give third order polynomials in M and thus "solve" the 3-generation problem in arbitrary dimensional space-times, except for n=3 in the setting (374), because this is leading to singularities in equation (448). We wrote the word "solve" in quotation marks, because we are still using the improper starting point (89) and thus, extracted (447) and (448) only as some kind of intrinsic solution without a suitable mathematical reasoning behind. In [7], appendix D it will be demonstrated that this does not lead to consistent results.

12.3.2.4 Simpler Than Expected

In order to sort these inconsistencies out, we use our definition for the energy momentum tensor from (359), set the cosmological constant and the classical, which is to say postulated matter L_M to zero, and then have the following quantum Einstein field equations:

$$0=\left(\begin{array}{c}\boxed{R_{\alpha\beta}-\frac{g_{\alpha\beta}}{2}R}+\frac{g_{\alpha\beta}}{2}\left(\begin{array}{c}\frac{1}{2F}\left(\begin{array}{c}2F_{,ij}(n-1)g^{ij}+2\Gamma_{ij}^{a}F_{,a}g^{ij}\\ -F_{,i}g^{ab}g_{jb,a}g^{ij}-F_{,j}g^{ab}g_{ib,a}g^{ij}\\ -n\Gamma_{ij}^{d}F_{,d}g^{ij}+\frac{n}{2}F_{,d}g^{cd}g_{ab,c}g^{ab}\end{array}\right)\\ +\frac{F_{,i}\cdot F_{,j}}{4F^{2}}g^{ij}\left((n-6)(n-1)\right)\end{array}\right)\\ -\frac{1}{2F}\left(\begin{array}{c}F_{,\alpha\beta}(n-2)+F_{,ab}g_{\alpha\beta}g^{ab}\\ +F_{,a}g^{ab}\left(g_{\beta b,\alpha}-g_{\beta\alpha,b}\right)-F_{,\alpha}g^{ab}g_{\beta b,a}-F_{,\beta}g^{ab}g_{\alpha b,a}\\ +F_{,d}g^{cd}\frac{1}{2}n\left(\frac{2}{n}g_{\alpha c,\beta}-g_{\alpha c,\beta}-g_{\beta c,\alpha}+g_{\alpha\beta,c}+\frac{1}{n}g_{\alpha\beta}g_{ab,c}g^{ab}\right)\\ -\frac{1}{2F}\left(F_{,\alpha}\cdot F_{,\beta}(3n-6)+g_{\alpha\beta}F_{,c}F_{,d}g^{cd}(4-n)\right)\end{array}\right)\\ +\kappa T_{\alpha\beta}+\Lambda\cdot F\cdot g_{\alpha\beta}\end{array}\right) \tag{450}$$

in their mixed form:

$$0=\left(\begin{array}{c}R_{\alpha}^{\beta}-\frac{\delta_{\alpha}^{\beta}}{2}R+\frac{\delta_{\alpha}^{\beta}}{2}\left(\begin{array}{c}\frac{1}{2F}\left(\begin{array}{c}2F_{,ij}(n-1)g^{ij}+2\Gamma_{ij}^{a}F_{,a}g^{ij}\\-F_{,i}g^{ah}g_{jb,a}g^{ij}-F_{,j}g^{ab}g_{ib,a}g^{ij}\\-n\Gamma_{ij}^{d}F_{,d}g^{ij}+\frac{n}{2}F_{,d}g^{cd}g_{ab,c}g^{ab}\end{array}\right)\\+\frac{F_{,i}\cdot F_{,j}}{4F^{2}}g^{ij}((n-6)(n-1))\end{array}\right)\\-\frac{1}{2F}\left(g^{\beta\gamma}\left(\begin{array}{c}F_{,\alpha\gamma}(n-2)+F_{,ab}g_{\alpha\gamma}g^{ab}\\+F_{,a}g^{ab}\left(g_{\gamma b,\alpha}-g_{\gamma\alpha,b}\right)-F_{,\alpha}g^{ab}g_{\gamma b,a}-F_{,\gamma}g^{ab}g_{\alpha b,a}\\+F_{,d}g^{cd}\frac{1}{2}n\left(\frac{2}{n}g_{\alpha c,\gamma}-g_{\alpha c,\gamma}-g_{\gamma c,\alpha}+g_{\alpha\gamma,c}+\frac{1}{n}g_{\alpha\gamma}g_{ab,c}g^{ab}\right)\\-\frac{1}{2F}\left(F_{,\alpha}\cdot F_{,\gamma}(3n-6)+g_{\alpha\gamma}F_{,c}F_{,d}g^{cd}(4-n)\right)\end{array}\right)\right)\end{array}\right). \tag{451}$$

$$=R_{\alpha}^{\beta}-\frac{\delta_{\alpha}^{\beta}}{2}R+\kappa T_{\alpha}^{\beta}=R_{\alpha}^{\beta}-\frac{\delta_{\alpha}^{\beta}}{2}R+\frac{1}{2F}\backslash\!\!\!T_{\alpha}^{\beta}\left[g_{\alpha\beta};F\right]$$

For completeness, we here also give the result without the affine connections (regarding the evaluation see appendix D), with more simplifications, and purely with the derivatives of the metric tensor:

$$0=\left(\begin{array}{c}R_{\alpha}^{\beta}-\frac{\delta_{\alpha}^{\beta}}{2}R+\frac{\delta_{\alpha}^{\beta}}{2}\left(\begin{array}{c}\frac{1}{2F}\left(2g^{\alpha\beta}F_{,\alpha\beta}(n-1)+F_{,d}g^{cd}g^{\alpha\beta}\left((n-1)g_{\alpha\beta,c}-ng_{\alpha c,\beta}\right)\right)\\+\frac{F_{,i}\cdot F_{,j}}{4F^{2}}g^{ij}((n-6)(n-1))\end{array}\right)\\-\frac{1}{2F}\left(g^{\beta\gamma}\left(\begin{array}{c}F_{,\alpha\gamma}(n-2)+F_{,ab}g_{\alpha\gamma}g^{ab}\\+F_{,a}g^{ab}\left(g_{\gamma b,\alpha}-g_{\gamma\alpha,b}\right)-F_{,\alpha}g^{ab}g_{\gamma b,a}-F_{,\gamma}g^{ab}g_{\alpha b,a}\\+F_{,d}g^{cd}\frac{1}{2}n\left(\frac{2}{n}g_{\alpha c,\gamma}-g_{\alpha c,\gamma}-g_{\gamma c,\alpha}+g_{\alpha\gamma,c}+\frac{1}{n}g_{\alpha\gamma}g_{ab,c}g^{ab}\right)\\-\frac{1}{2F}\left(F_{,\alpha}\cdot F_{,\gamma}(3n-6)+g_{\alpha\gamma}F_{,c}F_{,d}g^{cd}(4-n)\right)\end{array}\right)\right)\end{array}\right). \tag{452}$$

$$=R_{\alpha}^{\beta}-\frac{\delta_{\alpha}^{\beta}}{2}R+\kappa T_{\alpha}^{\beta}=R_{\alpha}^{\beta}-\frac{\delta_{\alpha}^{\beta}}{2}R+\frac{1}{2F}\backslash\!\!\!T_{\alpha}^{\beta}\left[g_{\alpha\beta};F\right]$$

Thereby we have used the fact that our energy momentum tensor, which just results out of the scaling of the metric tensor $g_{\alpha\beta}$ could just be interpreted as an operator acting on the function scaling F by the means of certain differentiations based on the metric tensor $g_{\alpha\beta}$. We used the backslash notation in order to emphasize the operator character. Applying a second factor as introduced in (415), leading us to a new metric tensor:

$$\Gamma_{\alpha\beta} = \Psi[\psi] \cdot G_{\alpha\beta} = \Psi[\psi] \cdot F[f] \cdot g_{\alpha\beta} \tag{453}$$

automatically yields the following quantum Einstein field equations, which we also give in mixed form:

$$0 = R_{\alpha}^{\beta} - \frac{\delta_{\alpha}^{\beta}}{2} R + \frac{\not{\partial}_{\alpha}^{\beta}\left[g_{\alpha\beta}; F \cdot \Psi\right]}{2F \cdot \Psi} = R_{\alpha}^{\beta} - \frac{\delta_{\alpha}^{\beta}}{2} R + \frac{\not{\partial}_{\alpha}^{\beta}\left[g_{\alpha\beta}; F\right]}{2F} + \frac{\not{\partial}_{\alpha}^{\beta}\left[G_{\alpha\beta}; \Psi\right]}{2\Psi}. \tag{454}$$

When forming the covariant derivative from this last expression which is based on the unscaled metric tensor $g_{\alpha\beta}$, we obtain the following equation:

$$\begin{aligned}
0 &= \left(R_{\alpha}^{\beta} - \frac{\delta_{\alpha}^{\beta}}{2} R + \frac{\not{\partial}_{\alpha}^{\beta}\left[g_{\alpha\beta}; F \cdot \Psi\right]}{2F \cdot \Psi} \right)_{;\beta} = \left(R_{\alpha}^{\beta} - \frac{\delta_{\alpha}^{\beta}}{2} R + \frac{\not{\partial}_{\alpha}^{\beta}\left[g_{\alpha\beta}; F\right]}{2F} + \frac{\not{\partial}_{\alpha}^{\beta}\left[G_{\alpha\beta}; \Psi\right]}{2\Psi} \right)_{;\beta} \\
&= \overbrace{\left(R_{\alpha}^{\beta} - \frac{\delta_{\alpha}^{\beta}}{2} R \right)_{;\beta}}^{=0,\ \text{Bianchi}} + \left(\frac{\not{\partial}_{\alpha}^{\beta}\left[g_{\alpha\beta}; F\right]}{2F} + \frac{\not{\partial}_{\alpha}^{\beta}\left[G_{\alpha\beta}; \Psi\right]}{2\Psi} \right)_{;\beta} \\
\Rightarrow &\left(\frac{\not{\partial}_{\alpha}^{\beta}\left[g_{\alpha\beta}; F\right]}{2F} + \frac{\not{\partial}_{\alpha}^{\beta}\left[G_{\alpha\beta}; \Psi\right]}{2\Psi} \right)_{;\beta} = \left(\frac{\not{\partial}_{\alpha}^{\beta}\left[g_{\alpha\beta}; F\right]}{2F} + \frac{\not{\partial}_{\alpha}^{\beta}\left[F \cdot g_{\alpha\beta}; \Psi\right]}{2\Psi} \right)_{;\beta} = 0
\end{aligned}, \tag{455}$$

which is just the conservation law for our scaling factors.

Thereby it is clear that in situations where we would already have the vacuum Einstein field equations to be fulfilled, which is to say, where we find that:

$$R_{\alpha}^{\beta} - \frac{\delta_{\alpha}^{\beta}}{2} R = 0, \tag{456}$$

the whole derivative procedure in (455) would not make any sense, because due to (456), we would automatically also have:

$$\frac{\not{\partial}_{\alpha}^{\beta}\left[g_{\alpha\beta}; F \cdot \Psi\right]}{2F \cdot \Psi} = 0, \tag{457}$$

which renders the condition (455) trivial and meaningless.

Thus, (455) does only deliver us additional information about the matter terms $\frac{\not{\partial}_{\alpha}^{\beta}\left[g_{\alpha\beta}; F \cdot \Psi\right]}{2F \cdot \Psi}$ when $R_{\alpha}^{\beta} - \frac{\delta_{\alpha}^{\beta}}{2} R$ does not give zero, hence the vacuum equation (456) is not fulfilled. In other words: In order to make (455) meaningful, there has to be matter. This may sound like a triviality and in fact it is, but this triviality is rendered quite important when seeing its mathematical origin, where it culminates to the simple conclusion that only curved space-times of the type $R_{\alpha}^{\beta} - \frac{\delta_{\alpha}^{\beta}}{2} R \neq 0$ can hold any detectable matter giving three particle solutions. This still would be a triviality to say, but only when ignoring the fact that a space-time with a simple scaled metric of type (407) and the resulting field equations:

$$0 = R_{\alpha}^{*\beta} - \frac{\delta_{\alpha}^{\beta}}{2} R^{*}, \tag{458}$$

which are apparently also just vacuum field equations, do contain matter, namely (also c.f. (454)):

$$0 = R_{\alpha}^{*\beta} - \frac{\delta_{\alpha}^{\beta}}{2} R^{*} = R_{\alpha}^{\beta} - \frac{\delta_{\alpha}^{\beta}}{2} R + \frac{\overbrace{\not{R}_{\alpha}^{\beta}\left[g_{\alpha\beta};F\right]}^{\text{matter}}}{2F}. \tag{459}$$

It looks like that it somehow depends on the "perspective" of the observer whether he "wants to see" matter or just space-time. This aspect will be investigated and discussed further below (see subsection "New Matter?..." and [7]).

Going back to equation (455), we are now demanding the first operator term in (455) to give a constant vector:

$$\left(\frac{\not{R}_{\alpha}^{\beta}\left[g_{\alpha\beta};F\right]}{2F}\right)_{;\beta} = C_{\alpha}, \tag{460}$$

which would just result in an eigenvalue equation for F of the type:

$$\begin{aligned}\frac{1}{2}\cdot\left(\frac{\not{R}_{\alpha}^{\beta}\left[g_{\alpha\beta};F\right]_{;\beta}}{F} - \frac{F_{;\beta}}{F^{2}}\not{R}_{\alpha}^{\beta}\left[g_{\alpha\beta};F\right]\right) = \frac{1}{2F}\cdot\left(\not{R}_{\alpha}^{\beta}\left[g_{\alpha\beta};F\right]_{;\beta} - \frac{F_{,\beta}}{F}\not{R}_{\alpha}^{\beta}\left[g_{\alpha\beta};F\right]\right) = C_{\alpha} \\ \Rightarrow\left(\not{R}_{\alpha}^{\beta}\left[g_{\alpha\beta};F\right]_{;\beta} - \frac{F_{,\beta}}{F}\not{R}_{\alpha}^{\beta}\left[g_{\alpha\beta};F\right]\right) = 2FC_{\alpha}\end{aligned} \tag{461}$$

and the following equation for the function Ψ:

$$\begin{aligned}C_{\alpha} + \left(\frac{\not{R}_{\alpha}^{\beta}\left[F\cdot g_{\alpha\beta};\Psi\right]}{2\Psi}\right)_{;\beta} = C_{\alpha} + \frac{1}{2\Psi}\left(\not{R}_{\alpha}^{\beta}\left[F\cdot g_{\alpha\beta};\Psi\right]_{;\beta} - \Psi_{,\beta}\frac{\not{R}_{\alpha}^{\beta}\left[F\cdot g_{\alpha\beta};\Psi\right]}{\Psi}\right) = 0 \\ \Rightarrow \\ 2\Psi C_{\alpha} = \left(\Psi_{,\beta}\frac{\not{R}_{\alpha}^{\beta}\left[F\cdot g_{\alpha\beta};\Psi\right]}{\Psi} - \not{R}_{\alpha}^{\beta}\left[F\cdot g_{\alpha\beta};\Psi\right]_{;\beta}\right)\end{aligned}, \tag{462}$$

also being an eigenvalue equation, we might be able to extract suitable solutions.

The vector in (460) could also be introduced as a vector of functions:

$$\left(\frac{\not{R}_{\alpha}^{\beta}\left[g_{\alpha\beta};F\right]}{2F}\right)_{;\beta} = V_{\alpha} \tag{463}$$

and then generalizes the eigen equations above as follows:

$$\left(\not{R}_{\alpha}^{\beta}\left[g_{\alpha\beta};F\right]_{;\beta} - \frac{F_{,\beta}}{F}\not{R}_{\alpha}^{\beta}\left[g_{\alpha\beta};F\right]\right) = 2FV_{\alpha}, \tag{464}$$

$$\left(\Psi_{,\beta}\frac{\not{R}_{\alpha}^{\beta}\left[F\cdot g_{\alpha\beta};\Psi\right]}{\Psi} - \not{R}_{\alpha}^{\beta}\left[F\cdot g_{\alpha\beta};\Psi\right]_{;\beta}\right) = 2\Psi V_{\alpha}. \tag{465}$$

This way, one might get the impression that the 3-generation problem could be solved, but apart from the fact that the corresponding evaluations are still very cumbersome even in the simplest space-times (see [7], appendix E), we also have the "little snag" of a linear dependency between the two functions F and Ψ. In the chapter "Uncertainties One Cannot Get Rid Of" it was shown how this problem can be solved. Here now, we will discuss the apparent dependence of the appearance of matter on the perspective of the observer and the critical dimensional size of systems.

12.3.3 New Matter? About a Theory of Perspectivity

Ignoring the cosmological constant and using our definition for the energy momentum tensor from (359) and (451), we have the following quantum Einstein field equations (450) in their mixed form:

$$\left(\begin{array}{c} R_{\alpha}^{\beta}-\frac{\delta_{\alpha}^{\beta}}{2}R+\frac{\delta_{\alpha}^{\beta}}{2}\left(\begin{array}{c}\frac{1}{2F}\left(\begin{array}{c}2F_{,ij}\left(n-1\right)g^{ij}+2\Gamma_{ij}^{a}F_{,a}g^{ij}\\ -F_{,i}g^{ab}g_{jb,a}g^{ij}-F_{,j}g^{ab}g_{ib,a}g^{ij}\\ -n\Gamma_{ij}^{d}F_{,d}g^{ij}+\frac{n}{2}F_{,d}g^{cd}g_{ab,c}g^{ab}\end{array}\right)\\ +\frac{F_{,i}\cdot F_{,j}}{4F^{2}}g^{ij}\left(\left(n-6\right)\left(n-1\right)\right)\end{array}\right)\\ -\frac{1}{2F}\left(g^{\beta\gamma}\left(\begin{array}{c}F_{,\alpha\gamma}\left(n-2\right)+F_{,ab}g_{\alpha\gamma}g^{ab}\\ +F_{,a}g^{ab}\left(g_{\gamma b,\alpha}-g_{\gamma\alpha,b}\right)-F_{,\alpha}g^{ab}g_{\gamma b,a}-F_{,\gamma}g^{ab}g_{\alpha b,a}\\ +F_{,d}g^{cd}\frac{1}{2}n\left(\frac{2}{n}g_{\alpha c,\gamma}-g_{\alpha c,\gamma}-g_{\gamma c,\alpha}+g_{\alpha\gamma,c}+\frac{1}{n}g_{\alpha\gamma}g_{ab,c}g^{ab}\right)\\ -\frac{1}{2F}\left(F_{,\alpha}\cdot F_{,\gamma}\left(3n-6\right)+g_{\alpha\gamma}F_{,c}F_{,d}g^{cd}\left(4-n\right)\right)\end{array}\right)\right)\end{array}\right)$$

$$=R_{\alpha}^{*\beta}-\frac{\delta_{\alpha}^{\beta}}{2}R^{*}=R_{\alpha}^{\beta}-\frac{\delta_{\alpha}^{\beta}}{2}R+\frac{\overbrace{\aleph_{\alpha}^{\beta}\left[g_{\alpha\beta};F\right]}^{\text{matter}}}{2F}=R_{\alpha}^{\beta}-\frac{\delta_{\alpha}^{\beta}}{2}R+\kappa T_{\alpha}^{\beta}\quad. \tag{466}$$

Applying the contracted covariant derivative with respect to the scaled metric tensor, thereby taking into account (356) and (362) leads to:

$$\left(R_{\alpha}^{\beta}-\frac{\delta_{\alpha}^{\beta}}{2}R+\kappa T_{\alpha}^{\beta}\right)_{;\beta}=\left(\begin{array}{c}R_{\alpha,\beta}^{\beta}+\left(\Gamma_{\beta a}^{\beta}+\Gamma_{\beta a}^{**\beta}\right)R_{\alpha}^{a}-\left(\Gamma_{\beta\alpha}^{a}+\Gamma_{\beta\alpha}^{**a}\right)R_{a}^{\beta}-\frac{1}{2}R_{,\alpha}\\ +\kappa\left(T_{\alpha,\beta}^{\beta}+\left(\Gamma_{\beta a}^{\beta}+\Gamma_{\beta a}^{**\beta}\right)T_{\alpha}^{a}-\left(\Gamma_{\beta\alpha}^{a}+\Gamma_{\beta\alpha}^{**a}\right)T_{a}^{\beta}\right)\end{array}\right)$$

$$=\left(\begin{array}{c}\overbrace{R_{\alpha,\beta}^{\beta}+\Gamma_{\beta a}^{\beta}R_{\alpha}^{a}-\Gamma_{\beta\alpha}^{a}R_{a}^{\beta}-\frac{1}{2}R_{,\alpha}}^{=0\,(\text{Bianchi})}+\Gamma_{\beta a}^{**\beta}R_{\alpha}^{a}-\Gamma_{\beta\alpha}^{**a}R_{a}^{\beta}\\ +\kappa\left(T_{\alpha,\beta}^{\beta}+\left(\Gamma_{\beta a}^{\beta}+\Gamma_{\beta a}^{**\beta}\right)T_{\alpha}^{a}-\left(\Gamma_{\beta\alpha}^{a}+\Gamma_{\beta\alpha}^{**a}\right)T_{a}^{\beta}\right)\end{array}\right)\quad. \tag{467}$$

$$=\Gamma_{\beta a}^{**\beta}R_{\alpha}^{a}-\Gamma_{\beta\alpha}^{**a}R_{a}^{\beta}+\kappa\left(T_{\alpha,\beta}^{\beta}+\left(\Gamma_{\beta a}^{\beta}+\Gamma_{\beta a}^{**\beta}\right)T_{\alpha}^{a}-\left(\Gamma_{\beta\alpha}^{a}+\Gamma_{\beta\alpha}^{**a}\right)T_{a}^{\beta}\right)$$

We realize that our temporary definition for the energy momentum tensor from (359) is not consistent with the condition (360). We already discussed this in the partitions following (359) and realized that it strongly depends on the character of the covariant derivative, meaning whether it is based on the scaled or the unscaled metric tensor, whether we "see" matter in the field equations or not. The simplest assumption one could make here would be to say that any dimension in addition to the time t and 3 spatial dimensions x, y and z is contributing or rather can contribute to some form of matter in dependence on the entanglement with the main 4 dimensions t, x, y, z. In [7], appendix F we demonstrated how we can obtain mass in this way (see also section "Transition to the "Classical" Physics…" in this book).

Thus, when moving back to the question "What is matter?", we have to conclude that, when only considering the simple scale extension of the Einstein-Hilbert action and its result in (466), we see no matter the moment "we would be living on the scaled metric". Then our covariant derivative on the Einstein field equations reads:

$$\left(R_{\alpha}^{*\beta}-\frac{\delta_{\alpha}^{\beta}}{2}R^{*}\right)_{;G\beta}=0 \tag{468}$$

and shows no matter. When, however, "living" on the unscaled metric and consequently performing the covariant derivative also based on this unscaled metric, we would see the following:

$$\left(R_{\alpha}^{*\beta}-\frac{\delta_{\alpha}^{\beta}}{2}R^{*}\right)_{;g\beta}=\overbrace{\left(R_{\alpha}^{*\beta}-\frac{\delta_{\alpha}^{\beta}}{2}R^{*}\right)_{;g\beta}}^{=0,\ \text{Bianchi}}+\left(\frac{\overbrace{\aleph_{\alpha}^{\beta}\left[g_{\alpha\beta};F\right]}^{\text{matter}}}{2F}\right)_{;g\beta}=0 \tag{469}$$

and thus, have matter plus the conservation law:

$$\left(\frac{\overbrace{\aleph_{\alpha}^{\beta}\left[g_{\alpha\beta};F\right]}^{\text{matter}}}{2F}\right)_{;g\beta}=0\cdot \tag{470}$$

Thereby, when considering much more complex scaling options, as introduced in chapter 11 of [4] or equation (184) in here, for instance, we can have a great variety of matter forms just by our metric scaling procedure. Another such option for formal matter creation was introduced in the Chapter "Uncertainties One Cannot Get Rid Of" above in this book.

For the moment we can conclude that the realization of matter depends on the perspective of the observer. This opens up new possibilities about the protection of innovations from greedy eyes. The necessary recipe requires metric wrappers F[f] for the system to be protected, which do not appear as scaling options at all. The greedy-eyed observer will then tend to analyze (a process which is realized via the mathematical process of covariant derivation) the system with respect to the scaled metric and will therefore see nothing but empty space. The system there is in plain sight and does not appear to be protected or hidden in any way. In order to make it even more complicated for the greedy-eyed observers, one can also apply the wrapping technique to the subsystems as mathematically shown in connection with the Everett universe approaches in the sections above (mainly see the section "The Theory"). With this extension even honeypot and ant-trap strategies can be constructed and mathematically created, handled, and optimized.

12.4 The Third Tower and the Three Generations of the Mind

Because of its importance and relevance to the whole, we here repeat our corresponding considerations from [56]. Thereby, we will keep the generality. The connection to our innovations T1 and T2 was already given above.

It was shown in [13] and partially also already in the subsections above that each system, the moment we can describe it via a Riemann space-time theory embedded in a Hamilton minimum principle of Einstein-Hilbert style [1], has 3 principle states resulting from a fundamental differential equation of third order for this very system (each system!).

As this finding has no restriction whatsoever with respect to the character of the system in question it must also hold for systems with intelligence. Hence, one automatically must ask the question whether there are generational states also for the mind… and what might consciousness has to do with it. As the latter part of the question probably is as old as mankind (figure 35), we will here go to some length to try to answer it. At least – so the authors are convinced – the moment the first human beings started to use fire to warm themselves and spend the evenings huddled together around the fireplace, sooner or later the questions must have been asked:

"What is it that makes us feel our existence?"

"What is this thing that makes me realize that I am and that mirrors this funny and sometimes fuzzy thing I call 'self' or 'myself'?!"

In other words:

The question of "what actually is consciousness?" is most likely at least one million years old.

Fig. 35: What is consciousness? [A] (see [102] for more)

Even though we may not be able to answer the question about "what actually is consciousness" in a satisfying manner, we could still try to investigate some of the properties a conscious system should have. Thereby we will simply use mathematical identities in order to find the most general laws for such "intelligent entities". In this chapter we will show that one of such principles lays in the fact that the so-called 3-generation problem of elementary particles [103] also has a bearing on the fundamental states intelligent systems can exist in. Assuming that only a quantum gravity or Theory-of-Everything-approach can describe a conscious entity, we already have our fundamental starting point. However, as the 3-generation problem of elementary particles [103] is been considered one of the most important unsolved problems in physics, we cannot just claim to have solved it (or rather think to have found a path to solve this problem) and then move on to interpret the impact of our finding on other fields, but we better present everything necessary to justify our claim and allow its falsification. And this is exactly what we will do in this chapter even though it might lead to some redundancy within the book as a whole.

12.4.1 Repetition: A Fundamental Starting Point

Again, our starting point shall be the Hamilton principle in form of the Einstein-Hilbert action [1]:

$$\delta W = 0 = \delta \int_V d^n x \left(\sqrt{-g} \cdot \Phi_R [R] \right) = \delta \int_V d^n x \left(\sqrt{-g} \cdot (R - 2\Lambda + L_M) \right). \tag{471}$$

Here R gives the Ricci scalar and g denotes the determinant of the metric tensor $g_{\alpha\beta}$. The term L_M stands for the matter Lagrange density, which was postulated by both Hilbert and Einstein, and Λ gives the cosmological constant. Assuming that we can extract a functional scaling or wrapping factor F[f], with f being a function of all coordinates, from the metric tensor in the following way:

$$G_{\alpha\beta} = g_{\alpha\beta} \cdot F[f], \tag{472}$$

it was shown in [2, 4] that we do not only obtain the Einstein field equations [1, 3], but also the most important classical quantum equations. Here, we only briefly repeat the essentials.

Setting (472) into (471) we obtain the following Einstein-Hilbert action [4]:

$$\delta W = 0 = \delta \int_V d^n x \left(\sqrt{-g \cdot F^n} \times \left(\left(\begin{array}{c} R - \frac{1}{2F} \left(\begin{array}{c} 2F_{,\alpha\beta} (n-1) g^{\alpha\beta} + 2\Gamma^a_{\alpha\beta} F_{,a} g^{\alpha\beta} \\ -F_{,\alpha} g^{ab} g_{\beta b,a} g^{\alpha\beta} - F_{,\beta} g^{ab} g_{\alpha b,a} g^{\alpha\beta} \\ -n\Gamma^d_{\alpha\beta} F_{,d} g^{\alpha\beta} + \frac{n}{2} F_{,d} g^{cd} g_{ab,c} g^{ab} \end{array} \right) \\ - \frac{F_{,\alpha} \cdot F_{,\beta}}{4F^2} g^{\alpha\beta} \left((n-6)(n-1) \right) \end{array} \right) \frac{1}{F} \\ -2\Lambda + L_M \right) \right), \tag{473}$$

or simplified and without the affine connection terms:

$$\delta W=0=\delta\int\limits_V d^nx\left(\sqrt{-g\cdot F^n}\times\left(\begin{pmatrix}R-\frac{1}{2F}\begin{pmatrix}2g^{\alpha\beta}F_{,\alpha\beta}(n-1)\\+F_{,d}g^{cd}g^{\alpha\beta}\left((n-1)g_{\alpha\beta,c}-ng_{\alpha c,\beta}\right)\end{pmatrix}\\-\frac{F_{,\alpha}\cdot F_{,\beta}}{4F^2}g^{\alpha\beta}\left((n-6)(n-1)\right)\end{pmatrix}\frac{1}{F}\\-2\Lambda+L_M\right)\right). \tag{474}$$

Thereby, in order to keep things simple and still general enough, one can often restrict the metric selection to diagonal metrics with the following properties:

$$g_{ij}=\begin{pmatrix}g_{00}&\cdots&0\\\vdots&\ddots&\vdots\\0&\cdots&g_{n-1n-1}\end{pmatrix};\quad g_{jj,j}=0, \tag{475}$$

which gives an extended Hilbert integral:

$$\delta W=0=\delta\int\limits_V d^nx\left(\sqrt{-g\cdot F^n}\times\left(\begin{pmatrix}R+\frac{F'}{F}(1-n)\Delta f\\+\frac{f_{,\alpha}f_{,\beta}g^{\alpha\beta}(1-n)}{4F^2}\left(4FF''+(F')^2(n-6)\right)\end{pmatrix}\frac{1}{F}\\-2\Lambda+L_M\right)\right). \tag{476}$$

In this section, however, we will not use this simplification. After performing the variation in (473), we obtain the following quantum gravity field equations:

$$0=\begin{pmatrix}\boxed{R_{\alpha\beta}-\frac{g_{\alpha\beta}}{2}R}+\frac{g_{\alpha\beta}}{2}\begin{pmatrix}\frac{1}{2F}\begin{pmatrix}2F_{,ij}(n-1)g^{ij}+2\Gamma_{ij}^{a}F_{,a}g^{ij}\\-F_{,i}g^{ab}g_{jb,a}g^{ij}-F_{,j}g^{ab}g_{ib,a}g^{ij}\\-n\Gamma_{ij}^{d}F_{,d}g^{ij}+\frac{n}{2}F_{,d}g^{cd}g_{ab,c}g^{ab}\end{pmatrix}\\+\frac{F_{,i}\cdot F_{,j}}{4F^2}g^{ij}\left((n-6)(n-1)\right)\end{pmatrix}\\-\frac{1}{2F}\begin{pmatrix}F_{,\alpha\beta}(n-2)+F_{,ab}g_{\alpha\beta}g^{ab}\\+F_{,a}g^{ab}\left(g_{\beta b,\alpha}-g_{\beta\alpha,b}\right)-F_{,\alpha}g^{ab}g_{\beta b,a}-F_{,\beta}g^{ab}g_{\alpha b,a}\\+F_{,d}g^{cd}\frac{1}{2}n\left(\frac{2}{n}g_{\alpha c,\beta}-g_{\alpha c,\beta}-g_{\beta c,\alpha}+g_{\alpha\beta,c}+\frac{1}{n}g_{\alpha\beta}g_{ab,c}g^{ab}\right)\\-\frac{1}{2F}\left(F_{,\alpha}\cdot F_{,\beta}(3n-6)+g_{\alpha\beta}F_{,c}F_{,d}g^{cd}(4-n)\right)\end{pmatrix}\\+\kappa T_{\alpha\beta}+\Lambda\cdot F\cdot g_{\alpha\beta}\end{pmatrix}, \tag{477}$$

respectively:

$$
0=\left(\begin{array}{c}
\boxed{R_{\alpha\beta}-\frac{g_{\alpha\beta}}{2}R}+\frac{g_{\alpha\beta}}{2}\left(\begin{array}{c}\frac{1}{2F}\left(2g^{\alpha\beta}F_{,\alpha\beta}(n-1)+F_{,d}g^{cd}g^{\alpha\beta}\left((n-1)g_{\alpha\beta,c}-ng_{\alpha c,\beta}\right)\right)\\+\frac{F_{,i}\cdot F_{,j}}{4F^{2}}g^{ij}\left((n-6)(n-1)\right)\end{array}\right)\\
-\frac{1}{2F}\left(\begin{array}{c}F_{,\alpha\beta}(n-2)+F_{,ab}g_{\alpha\beta}g^{ab}\\+F_{,a}g^{ab}\left(g_{\beta b,\alpha}-g_{\beta\alpha,b}\right)-F_{,\alpha}g^{ab}g_{\beta b,a}-F_{,\beta}g^{ab}g_{\alpha b,a}\\+F_{,d}g^{cd}\frac{1}{2}n\left(\frac{2}{n}g_{\alpha c,\beta}-g_{\alpha c,\beta}-g_{\beta c,\alpha}+g_{\alpha\beta,c}+\frac{1}{n}g_{\alpha\beta}g_{ab,c}g^{ab}\right)\\-\frac{1}{2F}\left(F_{,\alpha}\cdot F_{,\beta}(3n-6)+g_{\alpha\beta}F_{,c}F_{,d}g^{cd}(4-n)\right)\end{array}\right)\\
+\kappa T_{\alpha\beta}+\Lambda\cdot F\cdot g_{\alpha\beta}
\end{array}\right). \quad (478)
$$

We immediately recognize the classical Einstein field equations (red box) inside our new ones. It was shown in [2, 4] that the scaling terms in (477) are giving the Klein-Gordon, Dirac, and Schrödinger equation (regarding the latter see also [5]). Under the variational integral, the integrand (477) has to be multiplied with the variated metric tensor $\delta G^{\alpha\beta}=\delta\left(\frac{g^{\alpha\beta}}{F}\right)$ and therefore could be split up into two equations:

$$
0=\left(\begin{array}{c}
\boxed{R_{\alpha\beta}-\frac{g_{\alpha\beta}}{2}R}+\frac{g_{\alpha\beta}}{2}\left(\begin{array}{c}
\frac{1}{2F}\left(\begin{array}{c}
2F_{,ij}(n-1)g^{ij}+2\Gamma_{ij}^{a}F_{,a}g^{ij}\\
-F_{,i}g^{ab}g_{jb,a}g^{ij}-F_{,j}g^{ab}g_{ib,a}g^{ij}\\
-n\Gamma_{ij}^{d}F_{,d}g^{ij}+\frac{n}{2}F_{,d}g^{cd}g_{ab,c}g^{ab}
\end{array}\right)\\
+\frac{F_{,i}\cdot F_{,j}}{4F^{2}}g^{ij}\left((n-6)(n-1)\right)
\end{array}\right)\\
-\frac{1}{2F}\left(\begin{array}{c}
F_{,\alpha\beta}(n-2)+F_{,ab}g_{\alpha\beta}g^{ab}\\
+F_{,a}g^{ab}\left(g_{\beta b,\alpha}-g_{\beta\alpha,b}\right)-F_{,\alpha}g^{ab}g_{\beta b,a}-F_{,\beta}g^{ab}g_{\alpha b,a}\\
+F_{,d}g^{cd}\frac{1}{2}n\left(\frac{2}{n}g_{\alpha c,\beta}-g_{\alpha c,\beta}-g_{\beta c,\alpha}+g_{\alpha\beta,c}+\frac{1}{n}g_{\alpha\beta}g_{ab,c}g^{ab}\right)\\
-\frac{1}{2F}\left(F_{,\alpha}\cdot F_{,\beta}(3n-6)+g_{\alpha\beta}F_{,c}F_{,d}g^{cd}(4-n)\right)
\end{array}\right)\\
+\kappa T_{\alpha\beta}+\Lambda\cdot F\cdot g_{\alpha\beta}
\end{array}\right)\delta G^{\alpha\beta}
$$

$$
=\left(\begin{array}{c}
\boxed{R_{\alpha\beta}-\frac{g_{\alpha\beta}}{2}R}+\frac{g_{\alpha\beta}}{2}\left(\begin{array}{c}
\frac{1}{2F}\left(\begin{array}{c}
2F_{,ij}(n-1)g^{ij}+2\Gamma_{ij}^{a}F_{,a}g^{ij}\\
-F_{,i}g^{ab}g_{jb,a}g^{ij}-F_{,j}g^{ab}g_{ib,a}g^{ij}\\
-n\Gamma_{ij}^{d}F_{,d}g^{ij}+\frac{n}{2}F_{,d}g^{cd}g_{ab,c}g^{ab}
\end{array}\right)\\
+\frac{F_{,i}\cdot F_{,j}}{4F^{2}}g^{ij}\left((n-6)(n-1)\right)
\end{array}\right)\\
-\frac{1}{2F}\left(\begin{array}{c}
F_{,\alpha\beta}(n-2)+F_{,ab}g_{\alpha\beta}g^{ab}\\
+F_{,a}g^{ab}\left(g_{\beta b,\alpha}-g_{\beta\alpha,b}\right)-F_{,\alpha}g^{ab}g_{\beta b,a}-F_{,\beta}g^{ab}g_{\alpha b,a}\\
+F_{,d}g^{cd}\frac{1}{2}n\left(\frac{2}{n}g_{\alpha c,\beta}-g_{\alpha c,\beta}-g_{\beta c,\alpha}+g_{\alpha\beta,c}+\frac{1}{n}g_{\alpha\beta}g_{ab,c}g^{ab}\right)\\
-\frac{1}{2F}\left(F_{,\alpha}\cdot F_{,\beta}(3n-6)+g_{\alpha\beta}F_{,c}F_{,d}g^{cd}(4-n)\right)
\end{array}\right)\\
+\kappa T_{\alpha\beta}+\Lambda\cdot F\cdot g_{\alpha\beta}
\end{array}\right)\left(\begin{array}{c}
g^{\alpha\beta}\delta\left(\frac{1}{F}\right)\\
+\frac{\delta g^{\alpha\beta}}{F}
\end{array}\right),\quad(479)
$$

$$0=\left(\left(1-\frac{n}{2}\right)\left(R-\left(\frac{1}{2F}\begin{pmatrix}2F_{,ij}(n-1)g^{ij}+2\Gamma_{ij}^{a}F_{,a}g^{ij}\\-F_{,i}g^{ab}g_{jb,a}g^{ij}-F_{,j}g^{ab}g_{ib,a}g^{ij}\\-n\Gamma_{ij}^{d}F_{,d}g^{ij}+\frac{n}{2}F_{,d}g^{cd}g_{ab,c}g^{ab}\end{pmatrix}+\frac{F_{,i}\cdot F_{,j}}{4F^{2}}g^{ij}\left((n-6)(n-1)\right)\right)\right)+\Lambda\cdot F\cdot n\right)\delta\left(\frac{1}{F}\right)$$

$$0=\begin{pmatrix}R_{\alpha\beta}-\frac{g_{\alpha\beta}}{2}R+\frac{g_{\alpha\beta}}{2}\left(\frac{1}{2F}\begin{pmatrix}2F_{,ij}(n-1)g^{ij}+2\Gamma_{ij}^{a}F_{,a}g^{ij}\\-F_{,i}g^{ab}g_{jb,a}g^{ij}-F_{,j}g^{ab}g_{ib,a}g^{ij}\\-n\Gamma_{ij}^{d}F_{,d}g^{ij}+\frac{n}{2}F_{,d}g^{cd}g_{ab,c}g^{ab}\end{pmatrix}+\frac{F_{,i}\cdot F_{,j}}{4F^{2}}g^{ij}\left((n-6)(n-1)\right)\right)\\-\frac{1}{2F}\begin{pmatrix}F_{,\alpha\beta}(n-2)+F_{,ab}g_{\alpha\beta}g^{ab}\\+F_{,a}g^{ab}\left(g_{\beta b,\alpha}-g_{\beta\alpha,b}\right)-F_{,\alpha}g^{ab}g_{\beta b,a}-F_{,\beta}g^{ab}g_{\alpha b,a}\\+F_{,d}g^{cd}\frac{1}{2}n\left(\frac{2}{n}g_{\alpha c,\beta}-g_{\alpha c,\beta}-g_{\beta c,\alpha}+g_{\alpha\beta,c}+\frac{1}{n}g_{\alpha\beta}g_{ab,c}g^{ab}\right)\\-\frac{1}{2F}\left(F_{,\alpha}\cdot F_{,\beta}(3n-6)+g_{\alpha\beta}F_{,c}F_{,d}g^{cd}(4-n)\right)\end{pmatrix}\\+\kappa T_{\alpha\beta}+\Lambda\cdot F\cdot g_{\alpha\beta}\end{pmatrix}\frac{\delta g^{\alpha\beta}}{F}. \tag{480}$$

While we consider the second equation of the pair in (480) generalized or quantum Einstein field equations, we use the first one to derive the classical quantum equations (Klein-Gordon, Dirac, and Schrödinger) for a generally curved space-time. Demanding the "gravity" to be weak as follows:

$$\delta G^{\alpha\beta}=G^{\alpha\beta}\cdot\delta_{0}+\overbrace{G^{ab}\delta_{ab}^{\alpha\beta}}^{\text{Gravity}}\xrightarrow{\forall\,\delta_{ab}^{\alpha\beta}\ll\delta_{0}}=\frac{g^{\alpha\beta}}{F}\cdot\delta_{0}, \tag{481}$$

we are left with the scalar and purely metrically obtained quantum equation, which, when excluding the trivial case n=2, reads:

$$0=\left(R-\left(\frac{1}{2F}\begin{pmatrix}2F_{,ij}(n-1)g^{ij}+2\Gamma_{ij}^{a}F_{,a}g^{ij}\\-F_{,i}g^{ab}g_{jb,a}g^{ij}-F_{,j}g^{ab}g_{ib,a}g^{ij}\\-n\Gamma_{ij}^{d}F_{,d}g^{ij}+\frac{n}{2}F_{,d}g^{cd}g_{ab,c}g^{ab}\end{pmatrix}+\frac{F_{,i}\cdot F_{,j}}{4F^{2}}g^{ij}\left((n-6)(n-1)\right)\right)\right)+\Lambda\cdot F\cdot n. \tag{482}$$

From there, which is to say, based on the latter equation, all classically important equations can be obtained.

Going back to the variational integral (471), but keeping the scaled metric gives us:

$$\delta_G W = 0 = \delta_G \int_V d^n x \left(\sqrt{-G} \cdot \Phi \cdot R^* \right)$$

$$\delta W = 0 = \left[\begin{array}{c} \int_V d^n x \left(\sqrt{-G} \cdot \left(\begin{array}{c} \Phi \cdot R^*_{\mu\nu} - \frac{1}{2} \Phi \cdot R^* \cdot G_{\mu\nu} \\ + \Lambda G_{\mu\nu} - \left(\nabla_\mu \nabla_\nu - G_{\mu\nu} \Delta_G \right) \Phi \end{array} \right) \right) \delta G^{\mu\nu} \\ - \int_{\text{Surface}} d^{n-1} y \left(\sqrt{|h|} \cdot \varepsilon \cdot \Phi \cdot N^\lambda h^{\mu\nu} \partial_\lambda \left(\delta G_{\mu\nu} \right) \right) \end{array} \right]. \tag{483}$$

Thereby we have assumed:

$$\delta W = 0 = \delta \int_V d^n x \left(\sqrt{-G} \cdot \left(\Phi \cdot R^* - \overbrace{2\Lambda + L_M}^{=0} \right) \right) \tag{484}$$

and used the results from [6] within the derivation.

The usual, and rather tedious, path forward would be the discussion of the last equation and its investigation with respect to all its constituents. In previous publications (e.g., [9, 10, 11 – 13, 15 – 23, …, especially 56]), however, we have shown how to avoid such cumbersome equations (especially the surface term). It was derived in [19] that the resulting variational integral would look as follows:

$$\delta W = 0 = \int_V d^n x \left(\sqrt{-G} \cdot \left(\begin{array}{c} \Phi \cdot R^*_{\mu\nu} - \frac{1}{2} \Phi \cdot R^* G_{\mu\nu} + \Lambda G_{\mu\nu} - \left(\nabla_\mu \nabla_\nu - G_{\mu\nu} \Delta_G \right) \Phi \\ + \left\{ \begin{array}{ll} 0 & \ldots\text{"vacuum"} \\ \kappa T_{\mu\nu} & \ldots\text{postulated matter} \end{array} \right\} \Phi \end{array} \right) \right) \delta G^{\mu\nu}. \tag{485}$$

Expansion and avoiding the matter terms yields:

$$
0=\int_{V_G} d^n x\sqrt{-G}\left(\left(\begin{array}{c}
R_{\alpha\beta}-\dfrac{g_{\alpha\beta}}{2}\left(\begin{array}{c}
R-\dfrac{1}{2F}\left(\begin{array}{c}
2F_{,ij}(n-1)g^{ij}+2\Gamma^{a}_{ij}F_{,a}g^{ij}\\
-F_{,i}g^{ab}g_{jb,a}g^{ij}-F_{,j}g^{ab}g_{ib,a}g^{ij}\\
-n\Gamma^{d}_{ij}F_{,d}g^{ij}+\dfrac{n}{2}F_{,d}g^{cd}g_{ab,c}g^{ab}
\end{array}\right)\\
-\dfrac{F_{,i}\cdot F_{,j}}{4F^2}g^{ij}\left((n-7)(n-2)\right)\\
+\dfrac{1}{F}\left(F_{,ab}g^{ab}+F_{,d}g^{cd}\dfrac{1}{2}g_{ab,c}g^{ab}\right)
\end{array}\right)\\
-\dfrac{1}{2F}\left(\begin{array}{c}
F_{,\alpha\beta}(n-2)+F_{,a}g^{ab}\left(g_{\beta b,\alpha}-g_{\beta\alpha,b}\right)\\
-F_{,\alpha}g^{ab}g_{\beta b,a}-F_{,\beta}g^{ab}g_{\alpha b,a}\\
+F_{,d}g^{cd}\dfrac{1}{2}n\left(\dfrac{2}{n}g_{\alpha c,\beta}-g_{\alpha c,\beta}-g_{\beta c,\alpha}+g_{\alpha\beta,c}\right)\\
-\dfrac{1}{2F}\left(F_{,\alpha}\cdot F_{,\beta}(3n-6)\right)
\end{array}\right)\\
-\dfrac{\left(\partial_\alpha\Phi_{,\beta}-\Gamma^{*\gamma}_{\alpha\beta}\Phi_{,\gamma}-\dfrac{F\cdot g_{\alpha\beta}}{\sqrt{-gF^n}}\partial_\mu\left(\sqrt{-gF^n}\cdot\dfrac{g^{\mu\nu}}{F}\Phi_{,\nu}\right)\right)}{\Phi}
\end{array}\right)\delta G^{\alpha\beta}\right), \qquad (486)
$$

respectively, without the affine connections:

$$
0=\int_{V_G} d^n x\sqrt{-G}\left(\left(\begin{array}{c}
R_{\alpha\beta}-\dfrac{g_{\alpha\beta}}{2}\left(\begin{array}{c}
R-\dfrac{1}{2F}\left(\begin{array}{c}
2g^{\alpha\beta}F_{,\alpha\beta}(n-1)\\
+F_{,d}g^{cd}g^{\alpha\beta}\left((n-1)g_{\alpha\beta,c}-ng_{\alpha c,\beta}\right)
\end{array}\right)\\
-\dfrac{F_{,i}\cdot F_{,j}}{4F^2}g^{ij}\left((n-7)(n-2)\right)\\
+\dfrac{1}{F}\left(F_{,ab}g^{ab}+F_{,d}g^{cd}\dfrac{1}{2}g_{ab,c}g^{ab}\right)
\end{array}\right)\\
-\dfrac{1}{2F}\left(\begin{array}{c}
F_{,\alpha\beta}(n-2)+F_{,a}g^{ab}\left(g_{\beta b,\alpha}-g_{\beta\alpha,b}\right)\\
-F_{,\alpha}g^{ab}g_{\beta b,a}-F_{,\beta}g^{ab}g_{\alpha b,a}\\
+F_{,d}g^{cd}\dfrac{1}{2}n\left(\dfrac{2}{n}g_{\alpha c,\beta}-g_{\alpha c,\beta}-g_{\beta c,\alpha}+g_{\alpha\beta,c}\right)\\
-\dfrac{1}{2F}\left(F_{,\alpha}\cdot F_{,\beta}(3n-6)\right)
\end{array}\right)\\
-\dfrac{\left(\partial_\alpha\Phi_{,\beta}-\Gamma^{*\gamma}_{\alpha\beta}\Phi_{,\gamma}-\dfrac{F\cdot g_{\alpha\beta}}{\sqrt{-gF^n}}\partial_\mu\left(\sqrt{-gF^n}\cdot\dfrac{g^{\mu\nu}}{F}\Phi_{,\nu}\right)\right)}{\Phi}
\end{array}\right)\delta G^{\alpha\beta}\right). \qquad (487)
$$

It provides a great variety of intrinsic solutions (see [19]) of which we will in here only consider one option in order to deal with the so-called 3-generation problem (and, as explained at the beginning of this section, not just with respect to the three generations of elementary particles but to every system including the system of the mind).

12.4.2 Why Every System Has the Potential to Come in Three Generations or Why Every Psychosis Has Three States of Excitement

Taking (485), dividing by the function Φ, expanding the scaled Ricci tensor and scalar in accordance with the quantum Einstein field equations (477) in their mixed form from:

$$0=\left(\begin{array}{c} R_{\alpha\beta}-\frac{g_{\alpha\beta}}{2}R+\frac{g_{\alpha\beta}}{2}\left(\begin{array}{c}\frac{1}{2F}\left(\begin{array}{c}2F_{,ij}(n-1)g^{ij}+2\Gamma_{ij}^{a}F_{,a}g^{ij}\\ -F_{,i}g^{ab}g_{jb,a}g^{ij}-F_{,j}g^{ab}g_{ib,a}g^{ij}\\ -n\Gamma_{ij}^{d}F_{,d}g^{ij}+\frac{n}{2}F_{,d}g^{cd}g_{ab,c}g^{ab}\end{array}\right)\\ +\frac{F_{,i}\cdot F_{,j}}{4F^{2}}g^{ij}\left((n-6)(n-1)\right)\end{array}\right)\\ -\frac{1}{2F}\left(\begin{array}{c}F_{,\alpha\beta}(n-2)+F_{,ab}g_{\alpha\beta}g^{ab}\\ +F_{,a}g^{ab}\left(g_{\beta b,\alpha}-g_{\beta\alpha,b}\right)-F_{,\alpha}g^{ab}g_{\beta b,a}-F_{,\beta}g^{ab}g_{\alpha b,a}\\ +F_{,d}g^{cd}\frac{1}{2}n\left(\frac{2}{n}g_{\alpha c,\beta}-g_{\alpha c,\beta}-g_{\beta c,\alpha}+g_{\alpha\beta,c}+\frac{1}{n}g_{\alpha\beta}g_{ab,c}g^{ab}\right)\\ -\frac{1}{2F}\left(F_{,\alpha}\cdot F_{,\beta}(3n-6)+g_{\alpha\beta}F_{,c}F_{,d}g^{cd}(4-n)\right)\end{array}\right)\\ +\kappa T_{\alpha\beta}+\Lambda\cdot F\cdot g_{\alpha\beta}\end{array}\right), \tag{488}$$

we obtain:

$$R_{\alpha}^{\beta}-\frac{\delta_{\alpha}^{\beta}}{2}R+\frac{1}{2F}\not{\partial}_{\alpha}^{\beta}\left[g_{\alpha\beta};F\right]-\frac{\left(G^{\beta\mu}\nabla_{\mu}\nabla_{\alpha}-\delta_{\alpha}^{\beta}\Delta_{G}\right)\Phi}{\Phi}=0 \tag{489}$$

with:

$$\frac{1}{2F}\aleph_{\alpha}^{\beta}\left[g_{\alpha\beta};F\right]=\frac{1}{2F}\left(\begin{array}{c}\frac{\delta_{\alpha}^{\beta}}{2}\left(\begin{array}{c}\left(\begin{array}{c}2F_{,ij}(n-1)g^{ij}+2\Gamma_{ij}^{a}F_{,a}g^{ij}\\ -F_{,i}g^{ab}g_{jb,a}g^{ij}-F_{,j}g^{ab}g_{ib,a}g^{ij}\\ -n\Gamma_{ij}^{d}F_{,d}g^{ij}+\frac{n}{2}F_{,d}g^{cd}g_{ab,c}g^{ab}\end{array}\right)\\ +\frac{F_{,i}\cdot F_{,j}}{2F}g^{ij}\left((n-6)(n-1)\right)\end{array}\right)\\ -g^{\beta\gamma}\left(\begin{array}{c}F_{,\alpha\gamma}(n-2)+F_{,ab}g_{\alpha\gamma}g^{ab}\\ +F_{,a}g^{ab}\left(g_{\gamma b,\alpha}-g_{\gamma\alpha,b}\right)-F_{,\alpha}g^{ab}g_{\gamma b,a}-F_{,\gamma}g^{ab}g_{\alpha b,a}\\ +F_{,d}g^{cd}\frac{1}{2}n\left(\begin{array}{c}\frac{2}{n}g_{\alpha c,\gamma}-g_{\alpha c,\gamma}-g_{\gamma c,\alpha}\\ +g_{\alpha\gamma,c}+\frac{1}{n}g_{\alpha\gamma}g_{ab,c}g^{ab}\end{array}\right)\\ -\frac{1}{2F}\left(F_{,\alpha}\cdot F_{,\gamma}(3n-6)+g_{\alpha\gamma}F_{,c}F_{,d}g^{cd}(4-n)\right)\end{array}\right)\end{array}\right), \quad (490)$$

or alternatively:

$$\frac{1}{2F}\aleph_{\alpha}^{\beta}\left[g_{\alpha\beta};F\right]=\frac{1}{2F}\left(\begin{array}{c}\frac{\delta_{\alpha}^{\beta}}{2}\left(\begin{array}{c}\left(2g^{\alpha\beta}F_{,\alpha\beta}(n-1)+F_{,d}g^{cd}g^{\alpha\beta}\left((n-1)g_{\alpha\beta,c}-ng_{\alpha c,\beta}\right)\right)\\ +\frac{F_{,i}\cdot F_{,j}}{2F}g^{ij}\left((n-6)(n-1)\right)\end{array}\right)\\ -g^{\beta\gamma}\left(\begin{array}{c}F_{,\alpha\gamma}(n-2)+F_{,ab}g_{\alpha\gamma}g^{ab}\\ +F_{,a}g^{ab}\left(g_{\gamma b,\alpha}-g_{\gamma\alpha,b}\right)-F_{,\alpha}g^{ab}g_{\gamma b,a}-F_{,\gamma}g^{ab}g_{\alpha b,a}\\ +F_{,d}g^{cd}\frac{1}{2}n\left(\begin{array}{c}\frac{2}{n}g_{\alpha c,\gamma}-g_{\alpha c,\gamma}-g_{\gamma c,\alpha}\\ +g_{\alpha\gamma,c}+\frac{1}{n}g_{\alpha\gamma}g_{ab,c}g^{ab}\end{array}\right)\\ -\frac{1}{2F}\left(F_{,\alpha}\cdot F_{,\gamma}(3n-6)+g_{\alpha\gamma}F_{,c}F_{,d}g^{cd}(4-n)\right)\end{array}\right)\end{array}\right) \quad (491)$$

and:

$$\begin{array}{c}\nabla_{\mu}V_{\alpha}=\partial_{\mu}V_{\alpha}-\Gamma_{\mu\alpha}^{*\gamma}V_{\gamma};\quad \Gamma_{\mu\alpha}^{*\gamma}=\Gamma_{\mu\alpha}^{\gamma}+\frac{F'}{2\cdot F}\left(f_{,\mu}\cdot\delta_{\alpha}^{\gamma}+f_{,\alpha}\cdot\delta_{\mu}^{\gamma}-f_{,\sigma}\cdot g^{\gamma\sigma}g_{\alpha\mu}\right)\\ \Delta_{G}\Phi=\frac{1}{\sqrt{-G}}\partial_{\mu}\left(\sqrt{-G}\cdot G^{\mu\nu}\Phi_{,\nu}\right)=\frac{1}{\sqrt{-gF^{n}}}\partial_{\mu}\left(\sqrt{-gF^{n}}\cdot\frac{g^{\mu\nu}}{F}\Phi_{,\nu}\right);\quad \Phi_{,\nu}=\partial_{\nu}\Phi\end{array}. \quad (492)$$

Please note that the Laplace operator for the scaled metric could also be given as follows:

$$
\begin{aligned}
\Delta_G \Phi &= G^{\mu\nu}\left(\Phi_{,\mu\nu} - \Gamma^{*\gamma}_{\mu\nu}\Phi_{,\gamma}\right) \\
&= \frac{g^{\mu\nu}}{F}\left(\Phi_{,\mu\nu} - \left(\Gamma^{\gamma}_{\mu\nu} + \frac{F'}{2\cdot F}\left(f_{,\mu}\cdot\delta^{\gamma}_{\nu} + f_{,\nu}\cdot\delta^{\gamma}_{\mu} - f_{,\sigma}\cdot g^{\gamma\sigma}g_{\nu\mu}\right)\right)\Phi_{,\gamma}\right) \\
&= \frac{1}{F}\left(g^{\mu\nu}\Phi_{,\mu\nu} - \left(g^{\mu\nu}\Gamma^{\gamma}_{\mu\nu} + \frac{F'}{2\cdot F}\left(f_{,\mu}\cdot g^{\mu\gamma} + f_{,\nu}\cdot g^{\gamma\nu} - n\cdot f_{,\sigma}\cdot g^{\gamma\sigma}\right)\right)\Phi_{,\gamma}\right) \\
&= \frac{1}{F}\left(g^{\mu\nu}\Phi_{,\mu\nu} - \left(g^{\mu\nu}\Gamma^{\gamma}_{\mu\nu} + \frac{F'}{2\cdot F}\left(2\cdot f_{,\mu}\cdot g^{\mu\gamma} - n\cdot f_{,\sigma}\cdot g^{\gamma\sigma}\right)\right)\Phi_{,\gamma}\right)
\end{aligned}
\quad (493)
$$

Inserting the above into (488) results in:

$$
\begin{aligned}
0 &= \left(\begin{array}{c}
R^{\beta}_{\alpha} - \frac{\delta^{\beta}_{\alpha}}{2}R + \frac{1}{2F}\not{\delta}^{\beta}_{\alpha}\left[g_{\alpha\beta};F\right] \\
-\frac{\left(\begin{array}{c}
g^{\beta\mu}\left(\Phi_{,\mu\alpha} - \Gamma^{\gamma}_{\mu\alpha}\Phi_{,\gamma} + \frac{F'}{2\cdot F}\left(f_{,\mu}\cdot\delta^{\gamma}_{\alpha} + f_{,\alpha}\cdot\delta^{\gamma}_{\mu} - f_{,\sigma}\cdot g^{\gamma\sigma}g_{\alpha\mu}\right)\Phi_{,\gamma}\right) \\
-\delta^{\beta}_{\alpha}\left(g^{\mu\nu}\Phi_{,\mu\nu} - \left(g^{\mu\nu}\Gamma^{\gamma}_{\mu\nu} + \frac{F'}{2\cdot F}\left(2\cdot f_{,\mu}\cdot g^{\mu\gamma} - n\cdot f_{,\sigma}\cdot g^{\gamma\sigma}\right)\right)\Phi_{,\gamma}\right)
\end{array}\right)}{F\cdot\Phi}
\end{array}\right) \\
&= \left(\begin{array}{c}
R^{\beta}_{\alpha} - \frac{\delta^{\beta}_{\alpha}}{2}R + \frac{1}{2F}\not{\delta}^{\beta}_{\alpha}\left[g_{\alpha\beta};F\right] \\
-\frac{\left(\begin{array}{c}
g^{\beta\mu}\Phi_{,\mu\alpha} - g^{\beta\mu}\Gamma^{\gamma}_{\mu\alpha}\Phi_{,\gamma} + \frac{F'}{2\cdot F}\left(g^{\beta\mu}\left(f_{,\mu}\cdot\Phi_{,\alpha} + f_{,\alpha}\cdot\Phi_{,\mu}\right) - \delta^{\beta}_{\alpha}f_{,\sigma}\cdot g^{\gamma\sigma}\Phi_{,\gamma}\right) \\
-\delta^{\beta}_{\alpha}\left(g^{\mu\nu}\Phi_{,\mu\nu} - \left(g^{\mu\nu}\Gamma^{\gamma}_{\mu\nu} + \frac{F'}{2\cdot F}\left(2\cdot f_{,\mu}\cdot g^{\mu\gamma} - n\cdot f_{,\sigma}\cdot g^{\gamma\sigma}\right)\right)\Phi_{,\gamma}\right)
\end{array}\right)}{F\cdot\Phi}
\end{array}\right)
\end{aligned}
\quad (494)
$$

Now we have two reasonable opportunities to apply the covariant derivative and to use the Bianchi identity onto (494). At first, we form the covariant derivative with respect to the scaled metric, which gives:

$$
\begin{aligned}
0 &= \nabla_\beta \left(\begin{array}{c} R^\beta_\alpha - \frac{\delta^\beta_\alpha}{2} R + \frac{1}{2F} \aleph^\beta_\alpha \left[g_{\alpha\beta}; F \right] \\ - \frac{\left(\begin{array}{c} g^{\beta\mu}\Phi_{,\mu\alpha} - g^{\beta\mu}\Gamma^\gamma_{\mu\alpha}\Phi_{,\gamma} + \frac{F'}{2\cdot F}\left(g^{\beta\mu}\left(f_{,\mu}\cdot\Phi_{,\alpha} + f_{,\alpha}\cdot\Phi_{,\mu}\right) - \delta^\beta_\alpha f_{,\sigma}\cdot g^{\gamma\sigma}\Phi_{,\gamma}\right) \\ -\delta^\beta_\alpha \left(g^{\mu\nu}\Phi_{,\mu\nu} - \left(g^{\mu\nu}\Gamma^\gamma_{\mu\nu} + \frac{F'}{2\cdot F}\left(2\cdot f_{,\mu}\cdot g^{\mu\gamma} - n\cdot f_{,\sigma}\cdot g^{\gamma\sigma}\right)\right)\Phi_{,\gamma}\right) \end{array} \right)}{F\cdot\Phi} \end{array} \right) \\
&= \left(\begin{array}{c} \overbrace{\nabla_\beta \left(R^\beta_\alpha - \frac{\delta^\beta_\alpha}{2} R + \frac{1}{2F} \aleph^\beta_\alpha \left[g_{\alpha\beta}; F \right] \right)}^{=0;\ \text{Bianchi}} \\ -\nabla_\beta \frac{\left(\begin{array}{c} g^{\beta\mu}\Phi_{,\mu\alpha} - g^{\beta\mu}\Gamma^\gamma_{\mu\alpha}\Phi_{,\gamma} + \frac{F'}{2\cdot F}\left(g^{\beta\mu}\left(f_{,\mu}\cdot\Phi_{,\alpha} + f_{,\alpha}\cdot\Phi_{,\mu}\right) - \delta^\beta_\alpha f_{,\sigma}\cdot g^{\gamma\sigma}\Phi_{,\gamma}\right) \\ -\delta^\beta_\alpha \left(g^{\mu\nu}\Phi_{,\mu\nu} - \left(g^{\mu\nu}\Gamma^\gamma_{\mu\nu} + \frac{F'}{2\cdot F}\left(2\cdot f_{,\mu}\cdot g^{\mu\gamma} - n\cdot f_{,\sigma}\cdot g^{\gamma\sigma}\right)\right)\Phi_{,\gamma}\right) \end{array} \right)}{F\cdot\Phi} \end{array} \right) \\
\Rightarrow 0 &= \nabla_\beta \frac{\left(\begin{array}{c} g^{\beta\mu}\Phi_{,\mu\alpha} - g^{\beta\mu}\Gamma^\gamma_{\mu\alpha}\Phi_{,\gamma} + \frac{F'}{2\cdot F}\left(g^{\beta\mu}\left(f_{,\mu}\cdot\Phi_{,\alpha} + f_{,\alpha}\cdot\Phi_{,\mu}\right) - \delta^\beta_\alpha f_{,\sigma}\cdot g^{\gamma\sigma}\Phi_{,\gamma}\right) \\ -\delta^\beta_\alpha \left(g^{\mu\nu}\Phi_{,\mu\nu} - \left(g^{\mu\nu}\Gamma^\gamma_{\mu\nu} + \frac{F'}{2\cdot F}\left(2\cdot f_{,\mu}\cdot g^{\mu\gamma} - n\cdot f_{,\sigma}\cdot g^{\gamma\sigma}\right)\right)\Phi_{,\gamma}\right) \end{array} \right)}{F\cdot\Phi} .
\end{aligned}
\tag{495}
$$

And second, we can also covariant derivate (494) based on the unscaled metric, which results in:

$$
0 = \left(\begin{array}{c} R^\beta_\alpha - \frac{\delta^\beta_\alpha}{2} R + \frac{1}{2F} \aleph^\beta_\alpha \left[g_{\alpha\beta}; F \right] \\ - \frac{\left(\begin{array}{c} g^{\beta\mu}\left(\Phi_{,\mu\alpha} - \Gamma^\gamma_{\mu\alpha}\Phi_{,\gamma} + \frac{F'}{2\cdot F}\left(f_{,\mu}\cdot\delta^\gamma_\alpha + f_{,\alpha}\cdot\delta^\gamma_\mu - f_{,\sigma}\cdot g^{\gamma\sigma} g_{\alpha\mu}\right)\Phi_{,\gamma}\right) \\ -\delta^\beta_\alpha \left(g^{\mu\nu}\Phi_{,\mu\nu} - \left(g^{\mu\nu}\Gamma^\gamma_{\mu\nu} + \frac{F'}{2\cdot F}\left(2\cdot f_{,\mu}\cdot g^{\mu\gamma} - n\cdot f_{,\sigma}\cdot g^{\gamma\sigma}\right)\right)\Phi_{,\gamma}\right) \end{array} \right)}{F\cdot\Phi} \end{array} \right)_{;\beta} .
\tag{496}
$$

Thereby the term $\left(R^\beta_\alpha - \frac{\delta^\beta_\alpha}{2} R \right)_{;\beta}$ vanishes, which can directly be obtained from the Bianchi identity for the Riemann curvature tensor, reading:

$$
0 = R_{\alpha\beta\gamma\delta;\varepsilon} + R_{\alpha\beta\varepsilon\gamma;\delta} + R_{\alpha\beta\delta\varepsilon;\gamma} ,
\tag{497}
$$

via contraction with the metric tensor as follows:

$$
0 = \left(R_{\alpha\beta\gamma\delta;\varepsilon} + R_{\alpha\beta\varepsilon\gamma;\delta} + R_{\alpha\beta\delta\varepsilon;\gamma} \right) g^{\alpha\delta} g^{\beta\gamma} = \xrightarrow{g^{\alpha\delta}{}_{;\lambda}=0} = 2\cdot R^\alpha{}_{\varepsilon;\alpha} - R_{;\varepsilon} .
\tag{498}
$$

This yields:

$$0=\left(\begin{array}{c}\overbrace{\left(R_{\alpha}^{\beta}-\frac{\delta_{\alpha}^{\beta}}{2}R\right)_{;\beta}}^{=0;\,\text{Bianchi}}+\left(\frac{1}{2F}\aleph_{\alpha}^{\beta}\left[g_{\alpha\beta};F\right]\right)_{;\beta}\\ -\left(\frac{\left(\begin{array}{l}g^{\beta\mu}\Phi_{,\mu\alpha}-g^{\beta\mu}\Gamma^{\gamma}_{\mu\alpha}\Phi_{,\gamma}+\frac{F'}{2\cdot F}\left(g^{\beta\mu}\left(f_{,\mu}\cdot\Phi_{,\alpha}+f_{,\alpha}\cdot\Phi_{,\mu}\right)-\delta^{\beta}_{\alpha}f_{,\sigma}\cdot g^{\gamma\sigma}\Phi_{,\gamma}\right)\\ -\delta^{\beta}_{\alpha}\left(g^{\mu\nu}\Phi_{,\mu\nu}-\left(g^{\mu\nu}\Gamma^{\gamma}_{\mu\nu}+\frac{F'}{2\cdot F}\left(2\cdot f_{,\mu}\cdot g^{\mu\gamma}-n\cdot f_{,\sigma}\cdot g^{\gamma\sigma}\right)\right)\Phi_{,\gamma}\right)\end{array}\right)}{F\cdot\Phi}\right)_{;\beta}\end{array}\right)$$

$$=\left(\begin{array}{c}\left(\frac{1}{2F}\aleph_{\alpha}^{\beta}\left[g_{\alpha\beta};F\right]\right)_{;\beta}\\ -\left(\frac{\left(\begin{array}{l}g^{\beta\mu}\Phi_{,\mu\alpha}-g^{\beta\mu}\Gamma^{\gamma}_{\mu\alpha}\Phi_{,\gamma}+\frac{F'}{2\cdot F}\left(g^{\beta\mu}\left(f_{,\mu}\cdot\Phi_{,\alpha}+f_{,\alpha}\cdot\Phi_{,\mu}\right)-\delta^{\beta}_{\alpha}f_{,\sigma}\cdot g^{\gamma\sigma}\Phi_{,\gamma}\right)\\ -\delta^{\beta}_{\alpha}\left(g^{\mu\nu}\Phi_{,\mu\nu}-\left(g^{\mu\nu}\Gamma^{\gamma}_{\mu\nu}+\frac{F'}{2\cdot F}\left(2\cdot f_{,\mu}\cdot g^{\mu\gamma}-n\cdot f_{,\sigma}\cdot g^{\gamma\sigma}\right)\right)\Phi_{,\gamma}\right)\end{array}\right)}{F\cdot\Phi}\right)_{;\beta}\end{array}\right). \quad (499)$$

We may recognize the first term $\left(\frac{1}{2F}\aleph_{\alpha}^{\beta}\left[g_{\alpha\beta};F\right]\right)_{;\beta}$ as a possible "constant" for the third order polynomial, which results from a suitable function Φ from the rest in (496), which is to say from the term:

$$-\left(\frac{\left(\begin{array}{l}g^{\beta\mu}\Phi_{,\mu\alpha}-g^{\beta\mu}\Gamma^{\gamma}_{\mu\alpha}\Phi_{,\gamma}+\frac{F'}{2\cdot F}\left(g^{\beta\mu}\left(f_{,\mu}\cdot\Phi_{,\alpha}+f_{,\alpha}\cdot\Phi_{,\mu}\right)-\delta^{\beta}_{\alpha}f_{,\sigma}\cdot g^{\gamma\sigma}\Phi_{,\gamma}\right)\\ -\delta^{\beta}_{\alpha}\left(g^{\mu\nu}\Phi_{,\mu\nu}-\left(g^{\mu\nu}\Gamma^{\gamma}_{\mu\nu}+\frac{F'}{2\cdot F}\left(2\cdot f_{,\mu}\cdot g^{\mu\gamma}-n\cdot f_{,\sigma}\cdot g^{\gamma\sigma}\right)\right)\Phi_{,\gamma}\right)\end{array}\right)}{F\cdot\Phi}\right)_{;\beta}. \quad (500)$$

However, as recent investigations show (see [10, 15]), it is rather the Φ and the metric which provide the constant term, while the function F produces the third order polynomial with respect to parameters we might interpret as mass, thereby giving us three (!) solutions for these very parameters.

As our derivation was purely mathematical (we only used the Bianchi identity), we conclude that every system providing third order polynomials due to (500) (or any other of the many "intrinsic" Bianchi possibilities coming out from (486) as shown in [19]) has the potential to exist in three generations.

Thus, this should now help us to try and solve the 3-generation problem for elementary particles… and every other system, too.

12.4.3 Simplifying the Task via Intrinsic Solutions?

In order to keep things simple, we assume to have functions F and Φ as F[f] and Φ[f], concentrate on the scalar part of (486), and after forming the covariant derivative, we split the integrand as follows:

$$0=\left[R_{\alpha\beta}-\frac{g_{\alpha\beta}}{2}R-\frac{1}{2F}\left(\begin{array}{c}F_{,\alpha\beta}(n-2)+F_{,a}g^{ab}\left(g_{\beta b,\alpha}-g_{\beta\alpha,b}\right)\\-F_{,\alpha}g^{ab}g_{\beta b,a}-F_{,\beta}g^{ab}g_{\alpha b,a}\\+F_{,d}g^{cd}\frac{1}{2}n\left(\frac{2}{n}g_{\alpha c,\beta}-g_{\alpha c,\beta}-g_{\beta c,\alpha}+g_{\alpha\beta,c}\right)\\-\frac{1}{2F}\left(F_{,\alpha}\cdot F_{,\beta}(3n-6)\right)\end{array}\right)-\frac{\left(\partial_{\alpha}\Phi_{,\beta}-\Gamma^{*\gamma}_{\alpha\beta}\Phi_{,\gamma}\right)}{\Phi}\right]^{;\beta}$$

$$0=\left[\begin{array}{c}g_{\alpha}^{\beta}\frac{\left(\frac{F}{\sqrt{-gF^{n}}}\partial_{\mu}\left(\sqrt{-gF^{n}}\cdot\frac{g^{\mu\nu}}{F}\Phi_{,\nu}\right)\right)}{\Phi}\\-\frac{g_{\alpha}^{\beta}}{2}\left(\begin{array}{c}-\frac{1}{2F}\left(\begin{array}{c}2F_{,ij}(n-1)g^{ij}+2\Gamma^{a}_{ij}F_{,a}g^{ij}\\-F_{,i}g^{ab}g_{jb,a}g^{ij}-F_{,j}g^{ab}g_{ib,a}g^{ij}\\-n\Gamma^{d}_{ij}F_{,d}g^{ij}+\frac{n}{2}F_{,d}g^{cd}g_{ab,c}g^{ab}\end{array}\right)\\-\frac{F_{,i}\cdot F_{,j}}{4F^{2}}g^{ij}\left((n-7)(n-2)\right)\\+\frac{1}{F}\left(F_{,ab}g^{ab}+F_{,d}g^{cd}\frac{1}{2}g_{ab,c}g^{ab}\right)\end{array}\right)\end{array}\right]_{;\beta}\qquad(501)$$

or:

$$0=\left[R_{\alpha\beta}-\frac{g_{\alpha\beta}}{2}R-\frac{1}{2F}\left(\begin{array}{c}F_{,\alpha\beta}(n-2)+F_{,a}g^{ab}\left(g_{\beta b,\alpha}-g_{\beta\alpha,b}\right)\\ -F_{,\alpha}g^{ab}g_{\beta b,a}-F_{,\beta}g^{ab}g_{\alpha b,a}\\ +F_{,d}g^{cd}\frac{1}{2}n\left(\frac{2}{n}g_{\alpha c,\beta}-g_{\alpha c,\beta}-g_{\beta c,\alpha}+g_{\alpha\beta,c}\right)\\ -\frac{1}{2F}\left(F_{,\alpha}\cdot F_{,\beta}(3n-6)\right)\end{array}\right)-\frac{\left(\partial_{\alpha}\Phi_{,\beta}-\Gamma^{*\gamma}_{\alpha\beta}\Phi_{,\gamma}\right)}{\Phi}\right]^{;\beta}$$

$$0=\left[\begin{array}{c}g^{\beta}_{\alpha}\frac{\left(\frac{F}{\sqrt{-gF^{n}}}\partial_{\mu}\left(\sqrt{-gF^{n}}\cdot\frac{g^{\mu\nu}}{F}\Phi_{,\nu}\right)\right)}{\Phi}\\ -\frac{g^{\beta}_{\alpha}}{2}\left(\begin{array}{c}-\frac{1}{2F}\left(\begin{array}{c}2g^{\alpha\beta}F_{,\alpha\beta}(n-1)\\ +F_{,d}g^{cd}g^{\alpha\beta}\left((n-1)g_{\alpha\beta,c}-ng_{\alpha c,\beta}\right)\end{array}\right)\\ -\frac{F_{,i}\cdot F_{,j}}{4F^{2}}g^{ij}\left((n-7)(n-2)\right)\\ +\frac{1}{F}\left(F_{,ab}g^{ab}+F_{,d}g^{cd}\frac{1}{2}g_{ab,c}g^{ab}\right)\end{array}\right)\end{array}\right]_{;\beta}. \quad (502)$$

Using the results from [19] we can write:

$$0=\left[R_{\alpha\beta}-\frac{g_{\alpha\beta}}{2}R-\frac{1}{2F}\left(\begin{matrix}+F'\left(\begin{matrix}f_{,\alpha\beta}(n-2)f_{,a}g^{ab}\left(g_{\beta b,\alpha}-g_{\beta\alpha,b}\right)\\-f_{,\alpha}g^{ab}g_{\beta b,a}-F_{,\beta}g^{ab}g_{\alpha b,a}\\+f_{,d}g^{cd}\frac{1}{2}n\left(\frac{2}{n}g_{\alpha c,\beta}-g_{\alpha c,\beta}-g_{\beta c,\alpha}+g_{\alpha\beta,c}\right)\end{matrix}\right)\\f_{,\alpha}\cdot f_{,\beta}\left(F''(n-2)-\frac{(F')^2}{2F}(3n-6)-\frac{2F'\Phi'}{\Phi}\right)\end{matrix}\right)\\-\Phi'\frac{\left(\partial_\alpha f_{,\beta}-\Gamma^{\gamma}_{\alpha\beta}f_{,\gamma}\right)}{\Phi}\right]^{;\beta}$$

$$0=\overbrace{\left(R_{\alpha\beta}-\frac{g_{\alpha\beta}}{2}R\right)^{;\beta}}^{=0}-\left[\frac{1}{2F}\left(\begin{matrix}+F'\left(\begin{matrix}f_{,\alpha\beta}(n-2)f_{,a}g^{ab}\left(g_{\beta b,\alpha}-g_{\beta\alpha,b}\right)\\-f_{,\alpha}g^{ab}g_{\beta b,a}-F_{,\beta}g^{ab}g_{\alpha b,a}\\+f_{,d}g^{cd}\frac{1}{2}n\left(\frac{2}{n}g_{\alpha c,\beta}-g_{\alpha c,\beta}-g_{\beta c,\alpha}+g_{\alpha\beta,c}\right)\end{matrix}\right)\\f_{,\alpha}\cdot f_{,\beta}\left(F''(n-2)-\frac{(F')^2}{2F}(3n-6)-\frac{2F'\Phi'}{\Phi}\right)\end{matrix}\right)\\+\Phi'\frac{\left(\partial_\alpha f_{,\beta}-\Gamma^{\gamma}_{\alpha\beta}f_{,\gamma}\right)}{\Phi}\right]^{;\beta}$$

$$0=\left[\frac{1}{2F}\left(\begin{matrix}+F'\left(\begin{matrix}f_{,\alpha\beta}(n-2)f_{,a}g^{ab}\left(g_{\beta b,\alpha}-g_{\beta\alpha,b}\right)\\-f_{,\alpha}g^{ab}g_{\beta b,a}-F_{,\beta}g^{ab}g_{\alpha b,a}\\+f_{,d}g^{cd}\frac{1}{2}n\left(\frac{2}{n}g_{\alpha c,\beta}-g_{\alpha c,\beta}-g_{\beta c,\alpha}+g_{\alpha\beta,c}\right)\end{matrix}\right)\\f_{,\alpha}\cdot f_{,\beta}\left(F''(n-2)-\frac{(F')^2}{2F}(3n-6)-\frac{2F'\Phi'}{\Phi}\right)\end{matrix}\right)+\Phi'\frac{\left(\partial_\alpha f_{,\beta}-\Gamma^{\gamma}_{\alpha\beta}f_{,\gamma}\right)}{\Phi}\right]^{;\beta}, \quad (503)$$

$$0=\left[g^{\beta}_{\alpha}\frac{g^{ab}}{2F}\left(\begin{matrix}F'\left(\begin{matrix}2(n-2)f_{,ab}-\Gamma^{d}_{ab}f_{,d}(n-2)\\+f_{,d}g^{cd}\left(g_{ab,c}\frac{1}{2}(n-2)-2g_{cb,a}\right)\end{matrix}\right)\\+f_{,a}f_{,b}\left(\left(2\frac{F}{\Phi}\Phi''+2(n-1)F'\frac{\Phi'}{\Phi}+(n-2)\left(\frac{(F')^2}{2F}(n-7)+2F''\right)\right)\right)\\+4F\frac{\Phi'\left(f_{,ab}-\Gamma^{d}_{ab}f_{,d}\right)}{\Phi}\end{matrix}\right)\right]_{;\beta}. \quad (504)$$

In [19] we discussed the possibility to use the split-up as given above as an option to obtain intrinsic solutions to the integrand as follows:

$$0=\left(\begin{array}{c}R_{\alpha\beta}-\frac{g_{\alpha\beta}}{2}R-\frac{1}{2F}\left(\begin{array}{c}+F'\left(\begin{array}{c}f_{,\alpha\beta}(n-2)f_{,a}g^{ab}\left(g_{\beta b,\alpha}-g_{\beta\alpha,b}\right)\\-f_{,\alpha}g^{ab}g_{\beta b,a}-F_{,\beta}g^{ab}g_{\alpha b,a}\\+f_{,d}g^{cd}\frac{1}{2}n\left(\frac{2}{n}g_{\alpha c,\beta}-g_{\alpha c,\beta}-g_{\beta c,\alpha}+g_{\alpha\beta,c}\right)\end{array}\right)\\f_{,\alpha}\cdot f_{,\beta}\left(F''(n-2)-\frac{(F')^2}{2F}(3n-6)-\frac{2F'\Phi'}{\Phi}\right)\end{array}\right)\\-\Phi'\frac{\left(\partial_\alpha f_{,\beta}-\Gamma_{\alpha\beta}^{\gamma}f_{,\gamma}\right)}{\Phi}\end{array}\right), \tag{505}$$

$$0=g^{ab}\left(\begin{array}{c}F'\left(\begin{array}{c}2(n-2)f_{,ab}-\Gamma_{ab}^{d}f_{,d}(n-2)\\+f_{,d}g^{cd}\left(g_{ab,c}\frac{1}{2}(n-2)-2g_{cb,a}\right)\end{array}\right)\\+f_{,a}f_{,b}\left(\left(2\frac{F}{\Phi}\Phi''+2(n-1)F'\frac{\Phi'}{\Phi}+(n-2)\left(\frac{(F')^2}{2F}(n-7)+2F''\right)\right)\right)\\+4F\frac{\Phi'\left(f_{,ab}-\Gamma_{ab}^{d}f_{,d}\right)}{\Phi}\end{array}\right), \tag{506}$$

which is just (503) and (504) only without the covariant derivative.

The clever reader may already have guessed that we introduced the covariant derivative in order to obtain some kind of conservation law (just as in the classical General Theory of Relativity – e.g., [101]) and in aiming for some simplicity, we split up the complicated task of (485). For illustration, however, we here also want to briefly discuss the problems occurring with such intrinsic solution approaches. In order to get rid of the nonlinear operator terms, namely, we have to demand the following for the equations above:

$$0=F''(n-2)-\frac{(F')^2}{2F}(3n-6)-\frac{2F'\Phi'}{\Phi}, \tag{507}$$

$$0=2\frac{F}{\Phi}\Phi''+2(n-1)F'\frac{\Phi'}{\Phi}+(n-2)\left(\frac{(F')^2}{2F}(n-7)+2F''\right). \tag{508}$$

Unfortunately, we see ourselves unable to solve the above (coupled) differential equations for both functions F and Φ (only (507) can be solved in closed form with respect to Φ). So, with respect to find intrinsic solutions to (486), we might like to consider simpler forms, like, instead of the set-up (503) and (504), we could demand a set of intrinsic solutions to the general variational quest as follows:

$$0=-g_{\alpha\beta}\cdot\frac{1}{2}\cdot\left(\begin{array}{c}-\frac{g^{ab}}{2F}\left(\begin{array}{c}2F_{,ab}(n-2)-\Gamma^{d}_{ab}F_{,d}(n-2)\\+F_{,d}g^{cd}\left(g_{ab,c}\frac{1}{2}(n-2)-2g_{cb,a}\right)\end{array}\right)\\-\frac{F_{,i}\cdot F_{,j}}{4F^{2}}g^{ij}((n-7)(n-2))\end{array}\right)$$

$$0=R_{\alpha\beta}-\frac{g_{\alpha\beta}}{2}R+\left(\begin{array}{c}\frac{1}{2F}\left(\begin{array}{c}F_{,\alpha\beta}(n-2)+F_{,a}g^{ab}\left(g_{\beta b,\alpha}-g_{\beta\alpha,b}\right)\\-F_{,\alpha}g^{ab}g_{\beta b,a}-F_{,\beta}g^{ab}g_{\alpha b,a}\\+F_{,d}g^{cd}\frac{1}{2}n\left(\frac{2}{n}g_{\alpha c,\beta}-g_{\alpha c,\beta}-g_{\beta c,\alpha}+g_{\alpha\beta,c}\right)\\-\frac{1}{2F}\left(F_{,\alpha}\cdot F_{,\beta}(3n-6)\right)\end{array}\right)\\-\frac{\left(\partial_{\alpha}\Phi_{,\beta}-\Gamma^{*\gamma}_{\alpha\beta}\Phi_{,\gamma}-\frac{F\cdot g_{\alpha\beta}}{\sqrt{-gF^{n}}}\partial_{\mu}\left(\sqrt{-gF^{n}}\cdot\frac{g^{\mu\nu}}{F}\Phi_{,\nu}\right)\right)}{\Phi}\end{array}\right). \tag{509}$$

Apart from the fact that this time the linearity condition with vanishing operator $F_{,i}\cdot F_{,j}$ leads to a manageable differential equation for F only, when just concentrating on the scalar equation (the first one), we will also see further below, that this setting guarantees the correct asymptotic outcome for the F-scaled metric tensor.

We start with the first equation in (509) and obtain:

$$0=-g_{\alpha\beta}\cdot\frac{1}{2}\cdot\left(\begin{array}{c}-\frac{g^{ab}}{2F}\left(\begin{array}{c}2\left(F''f_{,a}\cdot f_{,b}+F'f_{,ab}\right)(n-2)-F'\Gamma^{d}_{ab}f_{,d}(n-2)\\+F'f_{,d}g^{cd}\left(g_{ab,c}\frac{1}{2}(n-2)-2g_{cb,a}\right)\end{array}\right)\\-(F')^{2}\frac{f_{,a}\cdot f_{,b}}{4F^{2}}g^{ab}((n-7)(n-2))\end{array}\right)$$
$$\Rightarrow 0=g_{\alpha\beta}\cdot\left(\begin{array}{c}\frac{g^{ab}F'}{2F}\left(\begin{array}{c}2f_{,ab}(n-2)-\Gamma^{d}_{ab}f_{,d}(n-2)\\+f_{,d}g^{cd}\left(g_{ab,c}\frac{1}{2}(n-2)-2g_{cb,a}\right)\end{array}\right)\\+\frac{f_{,a}\cdot f_{,b}}{2F}g^{ab}(n-2)\left(\frac{(F')^{2}}{2F}(n-7)+2F''\right)\end{array}\right). \tag{510}$$

As linearization condition we demand:

$$0=\frac{(F')^{2}}{2F}(n-7)+2F'', \tag{511}$$

which can be satisfied via the function F[f] as follows:

$$F[f]=\begin{cases} C_F\cdot(f+C_f)^{\frac{4}{n-3}} & n\neq 3 \\ C_F\cdot e^{f\cdot C_f} & n=3 \end{cases}. \tag{512}$$

Subsequently, equation (510) can be simplified and yields:

$$0=g_{\alpha\beta}\cdot\frac{g^{ab}F'}{F}\left(2f_{,ab}(n-2)-\Gamma^{d}_{ab}f_{,d}(n-2)+f_{,d}g^{cd}\left(g_{ab,c}\frac{1}{2}(n-2)-2g_{cb,a}\right)\right). \tag{513}$$

Interestingly, for metrics without shear elements:

$$g_{ij}=\begin{pmatrix} g_{00} & \cdots & 0 \\ \vdots & \ddots & \vdots \\ 0 & \cdots & g_{n-1n-1} \end{pmatrix}, \tag{514}$$

this converges to the ordinary Laplace operator, namely:

$$\begin{array}{c} 0=g_{\alpha\beta}\cdot\frac{g^{ab}F'}{F}\left(2f_{,ab}(n-2)-\Gamma^{d}_{ab}f_{,d}(n-2)+f_{,d}g^{cd}\left(g_{ab,c}\frac{1}{2}(n-2)-2g_{cb,a}\right)\right) \\ \xrightarrow{g_{ij}=\begin{pmatrix} g_{00} & \cdots & 0 \\ \vdots & \ddots & \vdots \\ 0 & \cdots & g_{n-1n-1} \end{pmatrix}} \\ 0=g_{\alpha\beta}\cdot\frac{g^{ab}F'}{F}\cdot 2(n-2)\Delta f \end{array}. \tag{515}$$

We see that the intrinsic approach gives us some manageable possibilities to partially solve (485), especially with respect to its scalar parts, which we have found to be connected with many quantum aspects [2, 4, 5, 7 – 23, 56]. Some critics may now argue that this equation is not of Klein-Gordon character as it does not contain any potential nor mass, but the third author has already shown that this problem is easily solved by adding additional dimensions carrying the right properties to produce masses and potentials (e.g., [2, 4, 7 – 24] and the section "Transition to the "Classical" Physics…" in this book).

With respect to our goal of finding proper conservation laws and still keeping things simple, we might resort to the "weak-gravity" condition and assume the influence of the true tensor part (503) small in comparison to the scalar equation (504) (after factorizing out the metric tensor $g_{\alpha\beta}$ only a scalar remains). And in fact, when performing the quite cumbersome and lengthy evaluations in (504), we obtain third order equations, which might give us an explanation for the 3-generation problem of elementary particles. However, as the split-up-method may be considered a little bit arbitrary, we here intend to seek for something more general when aiming for a scalar equation.

12.4.4 Simplification via Scalarizing the Integrand

Repetition: Regarding our problem of obtaining conservation laws, we might just take (485), form the covariant derivative (with respect to the unscaled metric) from the integrand as shown in the section "Why Every System Has the Potential to Come in Three Generations" in [21] and repeated here in the subsections above. We see that, because the contracted Bianchi identity erases the Ricci terms of the unscaled metric, we are left with terms containing F and Φ, providing us with the necessary conservation. The only problem with this: as (499) shows, we end up with some pretty complicated equations and we might like to look for some suitable simplifications. For this we once again apply the "weak gravity condition" in covariant:

$$\delta G^{\alpha\beta} = G^{\alpha\beta}\cdot\delta_0 + \overbrace{G^{ab}\delta^{\alpha\beta}_{ab}}^{\text{Gravity}} = \frac{1}{F}\cdot\left(g^{\alpha\beta}\cdot\delta_0 + \overbrace{g^{ab}\delta^{\alpha\beta}_{ab}}^{\text{Gravity}}\right) \xrightarrow{\forall\,\delta^{\alpha\beta}_{ab}\ll\delta_0} = \frac{g^{\alpha\beta}}{F}\cdot\delta_0 \tag{516}$$

or in mixed form:

$$\delta G^{\alpha}_{\beta} = G^{\alpha}_{\beta}\cdot\delta_0 + \overbrace{G^{b}_{a}\delta^{b\alpha}_{a\beta}}^{\text{Gravity}} \xrightarrow{\forall\,\delta^{b\alpha}_{a\beta}\ll\delta_0} = G^{\alpha}_{\beta}\cdot\delta_0 = g^{\alpha}_{\beta}\cdot\delta_0 = \delta^{\alpha}_{\beta}\cdot\delta_0\,, \tag{517}$$

and define:

$$I^{\beta}_{\alpha} \equiv R^{\beta}_{\alpha} - \frac{\delta^{\beta}_{\alpha}}{2}R + \frac{1}{2F}\not{\mathcal{R}}^{\beta}_{\alpha}\left[g_{\alpha\beta};F\right] - \frac{\left(G^{\beta\mu}\nabla_{\mu}\nabla_{\alpha} - \delta^{\beta}_{\alpha}\Delta_{G}\right)\Phi}{\Phi}, \tag{518}$$

again with:

$$\frac{1}{2F}\not{\mathcal{R}}^{\beta}_{\alpha}\left[g_{\alpha\beta};F\right] \equiv \frac{1}{2F}\left(\begin{array}{c} \frac{\delta^{\beta}_{\alpha}}{2}\left(\begin{array}{c}\left(\begin{array}{c} 2F_{,ij}(n-1)g^{ij} + 2\Gamma^{a}_{ij}F_{,a}g^{ij} \\ -F_{,i}g^{ab}g_{jb,a}g^{ij} - F_{,j}g^{ab}g_{ib,a}g^{ij} \\ -n\Gamma^{d}_{ij}F_{,d}g^{ij} + \frac{n}{2}F_{,d}g^{cd}g_{ab,c}g^{ab} \end{array}\right) \\ +\frac{F_{,i}\cdot F_{,j}}{2F}g^{ij}\left((n-6)(n-1)\right) \end{array}\right) \\ -g^{\beta\gamma}\left(\begin{array}{c} F_{,\alpha\gamma}(n-2) + F_{,ab}g_{\alpha\gamma}g^{ab} \\ +F_{,a}g^{ab}\left(g_{\gamma b,\alpha} - g_{\gamma\alpha,b}\right) - F_{,\alpha}g^{ab}g_{\gamma b,a} - F_{,\gamma}g^{ab}g_{\alpha b,a} \\ +F_{,d}g^{cd}\frac{1}{2}n\left(\begin{array}{c} \frac{2}{n}g_{\alpha c,\gamma} - g_{\alpha c,\gamma} - g_{\gamma c,\alpha} \\ +g_{\alpha\gamma,c} + \frac{1}{n}g_{\alpha\gamma}g_{ab,c}g^{ab} \end{array}\right) \\ -\frac{1}{2F}\left(F_{,\alpha}\cdot F_{,\gamma}(3n-6) + g_{\alpha\gamma}F_{,c}F_{,d}g^{cd}(4-n)\right) \end{array}\right) \end{array}\right), \tag{519}$$

or:

$$\frac{1}{2F}\aleph_{\alpha}^{\beta}\left[g_{\alpha\beta};F\right]\equiv\frac{1}{2F}\left(\begin{array}{c}\frac{\delta_{\alpha}^{\beta}}{2}\left(\left(\begin{array}{c}2g^{\alpha\beta}F_{,\alpha\beta}(n-1)\\+F_{,d}g^{cd}g^{\alpha\beta}\left((n-1)g_{\alpha\beta,c}-ng_{\alpha c,\beta}\right)\end{array}\right)\\+\frac{F_{,i}\cdot F_{,j}}{2F}g^{ij}\left((n-6)(n-1)\right)\end{array}\right)\\-g^{\beta\gamma}\left(\begin{array}{c}F_{,\alpha\gamma}(n-2)+F_{,ab}g_{\alpha\gamma}g^{ab}\\+F_{,a}g^{ab}\left(g_{\gamma b,\alpha}-g_{\gamma\alpha,b}\right)-F_{,\alpha}g^{ab}g_{\gamma b,a}-F_{,\gamma}g^{ab}g_{\alpha b,a}\\+F_{,d}g^{cd}\frac{1}{2}n\left(\begin{array}{c}\frac{2}{n}g_{\alpha c,\gamma}-g_{\alpha c,\gamma}-g_{\gamma c,\alpha}\\+g_{\alpha\gamma,c}+\frac{1}{n}g_{\alpha\gamma}g_{ab,c}g^{ab}\end{array}\right)\\-\frac{1}{2F}\left(F_{,\alpha}\cdot F_{,\gamma}(3n-6)+g_{\alpha\gamma}F_{,c}F_{,d}g^{cd}(4-n)\right)\end{array}\right)\end{array}\right). \tag{520}$$

Now we can evaluate the covariant derivative of the integrand of (485) as follows:

$$\begin{aligned}\left(I_{\alpha}^{\beta}\delta G_{\beta}^{\alpha}\right)_{;\gamma}&=\left(I_{\alpha}^{\beta}\left(G_{\beta}^{\alpha}\cdot\delta_{0}+\overbrace{G_{a}^{b}\delta_{\alpha\beta}^{b\alpha}}^{\text{Gravity}}\right)\right)_{;\gamma}\xrightarrow{\forall\,\delta_{\alpha\beta}^{b\alpha}\ll\delta_{0}}=\left(I_{\alpha}^{\beta}G_{\beta}^{\alpha}\cdot\delta_{0}\right)_{;\gamma}\\&=I_{\alpha;\gamma}^{\beta}G_{\beta}^{\alpha}\cdot\delta_{0}+\overbrace{I_{\alpha}^{\beta}G_{\beta;\gamma}^{\alpha}\cdot\delta_{0}}^{=0}+I_{\alpha}^{\beta}G_{\beta}^{\alpha}\cdot\delta_{0;\gamma}\\&=I_{\alpha;\gamma}^{\beta}G_{\beta}^{\alpha}\cdot\delta_{0}+I_{\alpha}^{\beta}G_{\beta}^{\alpha}\cdot\delta_{0;\gamma}\end{aligned}. \tag{521}$$

Assuming that the second term in the last line is small compared to the first, this simplifies to:

$$\left(I_{\alpha}^{\beta}\delta G_{\beta}^{\alpha}\right)_{;\gamma}\simeq I_{\alpha;\gamma}^{\beta}G_{\beta}^{\alpha}\cdot\delta_{0}+\overbrace{I_{\alpha}^{\beta}G_{\beta;\gamma}^{\alpha}\cdot\delta_{0}}^{=0}=\left(I_{\alpha}^{\beta}G_{\beta}^{\alpha}\right)_{;\gamma}\cdot\delta_{0}. \tag{522}$$

In consequence, the covariant derivative of the integrand (485) (using (518) and the equation above) reads:

$$\begin{aligned}\left(I_{\alpha}^{\beta}\delta G_{\beta}^{\alpha}\right)_{;\gamma}&\simeq\left(\left(R_{\alpha}^{\beta}-\frac{\delta_{\alpha}^{\beta}}{2}R+\frac{1}{2F}\aleph_{\alpha}^{\beta}\left[g_{\alpha\beta};F\right]-\frac{\left(G^{\beta\mu}\nabla_{\mu}\nabla_{\alpha}-\delta_{\alpha}^{\beta}\Delta_{G}\right)\Phi}{\Phi}\right)G_{\beta}^{\alpha}\right)_{;\gamma}\cdot\delta_{0}\\&=\left(\left(1-\frac{n}{2}\right)\left(R-\left(\begin{array}{c}\frac{1}{2F}\left(\begin{array}{c}2g^{\alpha\beta}F_{,\alpha\beta}(n-1)\\+F_{,d}g^{cd}g^{\alpha\beta}\left((n-1)g_{\alpha\beta,c}-ng_{\alpha c,\beta}\right)\end{array}\right)\\+\frac{F_{,i}\cdot F_{,j}}{4F^{2}}g^{ij}\left((n-6)(n-1)\right)\end{array}\right)\right)+(n-1)\frac{\Delta_{G}\Phi}{\Phi}\right)_{;\gamma}\cdot\delta_{0}\end{aligned}. \tag{523}$$

As shown before (e.g., [4]), we can get rid of the nonlinear differential operator with the approach F[f], via:

$$4FF''+F'\cdot F'(n-6)=0 \quad\Rightarrow\quad F[f]=\begin{cases} C_F\cdot(f+C_f)^{\frac{4}{n-2}} & n\neq 2 \\ C_F\cdot e^{f\cdot C_f} & n=2 \end{cases}, \tag{524}$$

making (523) to:

$$\begin{aligned}
\left(I_\alpha^\beta \delta G_\beta^\alpha\right)_{;\gamma} &\simeq \left(\left(R_\alpha^\beta-\frac{\delta_\alpha^\beta}{2}R+\frac{1}{2F}\mathfrak{G}_\alpha^\beta\left[g_{\alpha\beta};F\right]-\frac{\left(G^{\beta\mu}\nabla_\mu\nabla_\alpha-\delta_\alpha^\beta\Delta_G\right)\Phi}{\Phi}\right)G_\beta^\alpha\right)_{;\gamma}\cdot\delta_0 \\
&=\left(\left(1-\frac{n}{2}\right)\left(\begin{array}{c} R-\frac{g^{\alpha\beta}}{2F}F'\left(2f_{,\alpha\beta}(n-1)+f_{,d}g^{cd}\left(\begin{array}{c} g_{\alpha c,\beta}-g_{\beta\alpha,c}-g_{c\beta,\alpha} \\ +\frac{n}{2}\left(2g_{\alpha\beta,c}-g_{\alpha c,\beta}-g_{\beta c,\alpha}\right)\end{array}\right)\right) \\ -(n-1)\frac{f_{,\alpha}\cdot f_{,\beta}}{4F^2}g^{\alpha\beta}\overbrace{\left(4FF''+F'\cdot F'(n-6)\right)}^{=0}\end{array}\right) +(n-1)\frac{\Delta_G\Phi}{\Phi}\right)_{;\gamma}\cdot\delta_0 \\
&=\left(\left(1-\frac{n}{2}\right)\left(R-\frac{g^{\alpha\beta}}{2F}F'\left(2f_{,\alpha\beta}(n-1)+f_{,d}g^{cd}\left(\begin{array}{c} g_{\alpha c,\beta}-g_{\beta\alpha,c}-g_{c\beta,\alpha} \\ +\frac{n}{2}\left(2g_{\alpha\beta,c}-g_{\alpha c,\beta}-g_{\beta c,\alpha}\right)\end{array}\right)\right)\right) +(n-1)\frac{\Delta_G\Phi}{\Phi}\right)_{;\gamma}\cdot\delta_0
\end{aligned}. \tag{525}$$

As this should be a conservation law, the results should be zero and thus:

$$\begin{gathered}
\left(I_\alpha^\beta \delta G_\beta^\alpha\right)_{;\gamma}=0 \\
\Rightarrow \\
\left(\left(1-\frac{n}{2}\right)\left(R-\frac{g^{\alpha\beta}}{2F}F'\left(2f_{,\alpha\beta}(n-1)+f_{,d}g^{cd}\left(\begin{array}{c} g_{\alpha c,\beta}-g_{\beta\alpha,c}-g_{c\beta,\alpha} \\ +\frac{n}{2}\left(2g_{\alpha\beta,c}-g_{\alpha c,\beta}-g_{\beta c,\alpha}\right)\end{array}\right)\right)\right) +(n-1)\frac{\Delta_G\Phi}{\Phi}\right)_{;v}=0
\end{gathered}. \tag{526}$$

Now we assume that the function f[t,x,y,z]=f[t] shall code a Dirac particle at rest [27], which would be given due to:

$$f[t]=e^{\pm i\cdot\frac{m_R\cdot c^2}{\hbar}\cdot t}\cdot C_f=e^{\pm i\cdot\mu\cdot t}\cdot C_f, \tag{527}$$

with m_R giving the rest mass of the particle and $\hbar$ denoting the reduced Planck constant.

Further assuming an ordinary space-time in 4 dimensions with the metric tensor:

$$g_{\alpha\beta}=\begin{pmatrix} g_t[t] & 0 & 0 & 0 \\ 0 & 1 & 0 & 0 \\ 0 & 0 & 1 & 0 \\ 0 & 0 & 0 & 1 \end{pmatrix};\quad g_t[t]=e^{\pm i\cdot m\cdot t} \tag{528}$$

and an oscillating perturbation of the kernel of the Einstein-Hilbert action via the kernel scaling function:

$$\Phi[t]=e^{\pm i\cdot M\cdot t}\cdot C_{\Phi}, \tag{529}$$

and inserting all this into (526), results in the following algebraic equation:

$$\frac{3mM^2}{8}-\frac{m^2M}{8}-\frac{M^3}{4}+\frac{(2mM-m^2)}{4}\cdot\mu+\frac{1}{2}M\mu^2+\mu^3=0. \tag{530}$$

Thereby we have assumed only positive signs in the exponents for all functions f, Φ, and g_t. Equation (530) can be written as follows:

$$0=-C_0+C_1\cdot\mu-C_2\cdot\mu^2+\mu^3. \tag{531}$$

The general solution to a three-order polynomial could be given via the following product form:

$$\begin{gathered}(\mu-\mu_1)\cdot(\mu-\mu_2)\cdot(\mu-\mu_3)= \\ \mu^3-\mu^2\cdot(\mu_1+\mu_2+\mu_3)+\mu\cdot(\mu_1\mu_2+\mu_1\mu_3+\mu_2\mu_3)-\mu_1\mu_2\mu_3\end{gathered}. \tag{532}$$

Comparing the last line of (386) with (384) gives us:

$$\begin{aligned} C_2&=\mu_1+\mu_2+\mu_3=-\frac{M}{2} \\ C_1&=\mu_1\mu_2+\mu_1\mu_3+\mu_2\mu_3=\frac{(2mM-m^2)}{4} \\ C_0&=\mu_1\mu_2\mu_3=\frac{1}{4}\left(M^3-\frac{3mM^2}{2}+\frac{m^2M}{2}\right)\end{aligned}. \tag{533}$$

Unfortunately, we only have two parameters and thus, will not be able to find a metric setting for the explanation of the three generations of elementary particles. The reason for this may be seen in our by far too simple metric approach, which we have only chosen to explicitly keep things simple and stick to the classical (Dirac) particle at rest. However, keeping the lazy spirit, how about a slight extension to a 3-parameter approach with still purely time dependent metric components like:

$$g_{\alpha\beta} = \begin{pmatrix} g_t[t] & 0 & 0 & 0 \\ 0 & g_s[t] & 0 & 0 \\ 0 & 0 & g_s[t] & 0 \\ 0 & 0 & 0 & g_s[t] \end{pmatrix}; \quad g_s[t] = e^{\pm i \cdot v \cdot t} ? \qquad (534)$$

Now we result in the equation:

$$\begin{pmatrix} -\frac{m^2 M}{8} + \frac{3mM^2}{8} - \frac{M^3}{4} - \frac{m^2 v}{8} + \frac{mMv}{2} - \frac{3M^2 v}{8} + \frac{mv^2}{4} \\ -\frac{Mv^2}{4} - \frac{m^2 \mu}{4} + \frac{mM\mu}{2} + \frac{mv\mu}{2} + \frac{v^2 \mu}{2} + \frac{M\mu^2}{2} + \frac{3v\mu^2}{2} + \mu^3 \end{pmatrix} = 0. \qquad (535)$$

12.4.4.1 The Three Generations of Elementary Particles

The numerical solutions to the third order algebraic equation (535) for the charged leptons (electron μ=0.511 MeV/c², muon μ=105.7 MeV/c² and tau μ=1777 MeV/c²) would be (all in MeV/c²):

m → 1220.68, M → 1221.7, v → -548.507,

m → -1054.41, M → -1053.39, v → 209.857,

where we have given only the two real solutions, while there are 6 complex ones in total. The corresponding results for the neutrinos and the quarks are given in [21]. Thereby we found funny asymmetries of numerical character with respect to matter and antimatter particles [22].

12.4.5 Intermediate Conclusions with Respect to General Systems

We found a manageable way to extract a weak-gravity scalar from the quantum Einstein field equations. Subjecting it to the covariant derivative operation, we obtained conservation laws. Applying the latter to a simple metric with just time-dependent components directly resulted in a polynomial of third order in the mass-parameter for the object described by the metric. The polynomial gives three solutions for such masses and by feeding the known masses of the three elementary particle generations into the approach we were able to evaluate the metric settings for charged and uncharged leptons and all quarks [21]… reverse engineering, if one so will. Interpreting the masses in a more general way for other systems than just elementary particles, namely, as inertia, we might ask whether also intelligent systems, for which holds the same fundamental (because purely mathematical) law, have to exist in such three generations.

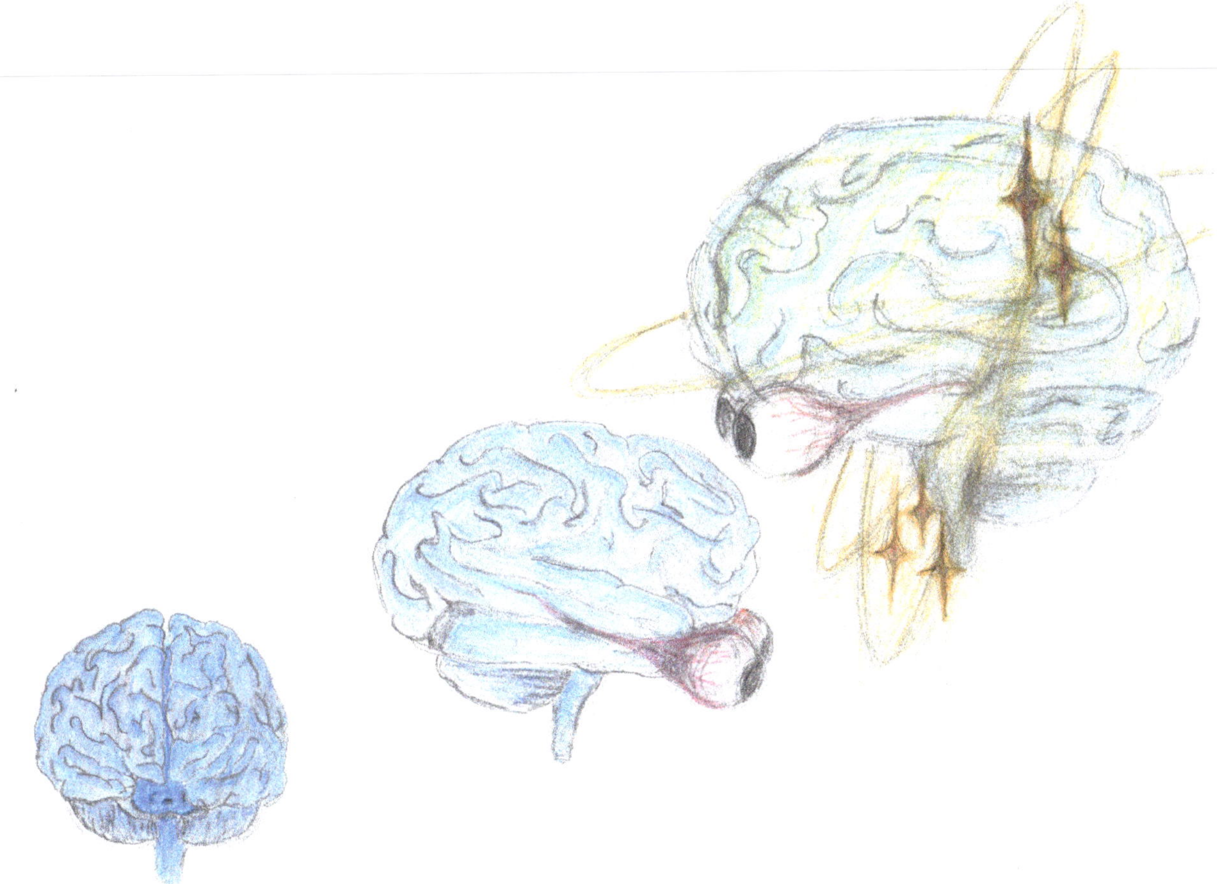

Fig. 36: The three generations of consciousness. [A] (see [102] for more)

So, for one thing, we should state that our finding with the quite fundamental third order differential equation based on the purely mathematical Bianchi identity automatically leads to the conclusion that the occurrence of three generations of matter forms must be a rather general aspect for any arbitrary system. This may help to understand certain three-fold appearances in socioeconomic, natural, psychological, medical, and many other fields. Thereby, we have the interesting aspect of apparently only one generation being truly stable.

And of course, being so fundamental, the 3-generation aspect should also be of importance for intelligent systems (figure 36).

12.4.6 Surfaces as Matter?

In section "Transition to the "Classical" Physics and Some Interesting Findings Along the Way" we discussed shell-like solutions to the quantum Einstein field equations which gave quite reasonable results when comparing their characteristics with elementary particles (of which we have considered the electron via "The Size of the Electron?" in the corresponding subsection). Thus, it should not come too much as a surprise that the behavior of the PSA in the second tower technology has a lot of apparent "particle characteristics".

12.4.7 About the Instability of the Second and Third Particle Generations

It was shown above in section "The Third Tower and the Three Generations of the Mind" that the existence of higher generations of particles requires the kernel of the Einstein-Hilbert action to be of the type (483). This action integral, however, also leads to surface terms. Assuming that nature tries to avoid such surface terms via the transformation of the variation in accordance with the technique elaborated in section "About Non-Vanishing Hilbert Surface Terms and the Scaling Trick" above, which is to say:

$$\begin{aligned}\delta\int_V d^n x\left(\sqrt{-\gamma}\cdot\Phi\cdot R\right)\rightarrow\delta\int_V d^n x\left(\sqrt{-g}\cdot F^{n/2}\cdot R^*\right)=\delta\int_V d^n x\left(\sqrt{-g}\cdot F^{n/2-1}\cdot R^*_{\alpha\beta}G^{\alpha\beta}\cdot F\right)\\ \Rightarrow\\ =\delta\int_V d^n x\left(\sqrt{-g}\cdot F^{n/2-1}\cdot R^*_{\alpha\beta}g^{\alpha\beta}\right)\end{aligned}\quad,\ (536)$$

we might just have found an explanation for the instability of higher generations elementary particles. Apparently, the universe's aspiration for a surface term free "environment" leads to the instability of solutions enforcing such terms, followed by their immediate removal via variational adjustments of the kind (536). In other words: the universe "heals" itself and removes "unpleasant intricacies", as which the surface terms apparently appear.

13 The Universe's Half Spin Deception

13.1 A Harmless Metric Object with Spin

We apply the following metric:

$$g^{8}_{\alpha\beta}=\begin{pmatrix}-c^2 & 0 & 0 & 0 & \cdots & 0\\ 0 & 1 & 0 & 0 & \cdots & 0\\ 0 & 0 & r^2 & 0 & \cdots & 0\\ 0 & 0 & 0 & r^2\cdot\sin^2\varphi_1 & \cdots & 0\\ \cdots & \cdots & \cdots & \cdots & \ddots & 0\\ 0 & 0 & 0 & 0 & 0 & g_{77}\end{pmatrix}$$
$$\times F\left[f\left[t,r,\vartheta,\varphi,\tau,\rho,\theta,\phi\right]\right];\quad F[f]=f^{\frac{2}{3}}$$
$$f\left[t,r,\vartheta,\varphi,\tau,\rho,\theta,\phi\right]=f_t[t]\cdot f_r[r]\cdot f_\vartheta[\vartheta]\cdot f_\varphi[\varphi]\cdot f_\tau[\tau]\cdot f_\rho[\rho]\cdot f_\theta[\theta]\cdot f_\phi[\phi];$$
$$g_{44}=c_4^2\cdot r^{-2};\quad g_{55}=1;\quad g_{66}=\rho^2;\quad g_{77}=\rho^2\cdot\sin[\theta]^2; \tag{537}$$

on our scalar field equation (38), in the form:

$$0=\left(R-\frac{g^{\alpha\beta}}{2F}F'\left(2f_{,\alpha\beta}(n-1)+f_{,d}g^{cd}\left(\begin{array}{c}g_{\alpha c,\beta}-g_{\beta\alpha,c}-g_{c\beta,\alpha}\\ +\frac{n}{2}\left(2g_{\alpha\beta,c}-g_{\alpha c,\beta}-g_{\beta c,\alpha}\right)\end{array}\right)\right)\right)\cdot\delta_0$$
$$\Rightarrow 0=R-\frac{g^{\alpha\beta}}{2F}F'\left(2f_{,\alpha\beta}(n-1)+f_{,d}g^{cd}\left(\begin{array}{c}g_{\alpha c,\beta}-g_{\beta\alpha,c}-g_{c\beta,\alpha}\\ +\frac{n}{2}\left(2g_{\alpha\beta,c}-g_{\alpha c,\beta}-g_{\beta c,\alpha}\right)\end{array}\right)\right) \tag{538}$$

Thereby we have used two spherical coordinate sets and can partially solve the task (538) when using the following partial solutions discussed elsewhere [9, 10] as follows:

$$f_\varphi[\varphi]=C_{\varphi-}\cdot e^{-i\cdot A\cdot\varphi}+C_{\varphi+}\cdot e^{+i\cdot A\cdot\varphi}, \tag{539}$$

$$f_\tau[\tau]=C_{\tau1}\cdot\cos\left[c_4\cdot A_2\cdot\tau\right]+C_{\tau2}\cdot\sin\left[c_4\cdot A_2\cdot\tau\right], \tag{540}$$

$$f_\vartheta[\vartheta]=C_{P\vartheta}\cdot P_L^A\left[\cos[\vartheta]\right]+C_{Q\vartheta}\cdot Q_L^A\left[\cos[\vartheta]\right], \tag{541}$$

$$f_\phi[\phi]=C_{\phi-}\cdot e^{-i\cdot B\cdot\phi}+C_{\phi+}\cdot e^{+i\cdot B\cdot\phi}, \tag{542}$$

$$f_t[t]=C_{t1}\cdot\cos\left[c\cdot E_t\cdot t\right]+C_{t2}\cdot\sin\left[c\cdot E_t\cdot t\right], \tag{543}$$

$$f_\theta[\theta]=C_{P\theta}\cdot P_\ell^B[\cos[\theta]]+C_{Q\theta}\cdot Q_\ell^B[\cos[\theta]], \tag{544}$$

with the associated Legendre polynomials $P_L^{A_1}, Q_L^{A_1}$. This gives us the following Ricci scalar R^* and subsequent differential equation for the r-ρ-dependency of f[…]:

$$R^*=0=\left(\frac{\ell\cdot(\ell+1)}{\rho^2}-\frac{2\cdot\partial}{\rho\cdot\partial\rho}-\frac{\partial^2}{\partial\rho^2}+\frac{L\cdot(L+1)}{r^2}+A_2^2\cdot r^2-\frac{\partial}{r\cdot\partial r}-\frac{\partial^2}{\partial r^2}-E_t^2\right)f. \tag{545}$$

As we can directly separate the two radii with the approach from (537), we obtain the two differential equations:

$$0=\left(\frac{\ell\cdot(\ell+1)}{\rho^2}-\frac{2\cdot\partial}{\rho\cdot\partial\rho}-\frac{\partial^2}{\partial\rho^2}\right)f_\rho=E_\rho^2\cdot f_\rho, \tag{546}$$

$$0=\left(\frac{L\cdot(L+1)}{r^2}+A_2^2\cdot r^2-\frac{\partial}{r\cdot\partial r}-\frac{\partial^2}{\partial r^2}-E_t^2+E_\rho^2\right)f_r. \tag{547}$$

While (546) gives us the spherical Bessel functions again via:

$$f_\rho[\rho]=C_j\cdot j_\ell\left[E_\rho\cdot\rho\right]+C_y\cdot y_\ell\left[E_\rho\cdot\rho\right], \tag{548}$$

we obtain the following solution for f_r:

$$\begin{aligned}f_r[r]&=\frac{2^{\frac{1}{2}\left(1+\sqrt{L\cdot(1+L)}\right)}e^{-\frac{1}{2}A_2\cdot r^2}\cdot\left(r^2\right)^{\frac{1}{2}\left(1+\sqrt{L\cdot(1+L)}\right)}}{r}\\&\quad\times\left(C_U\cdot U\left[-nn,1+\sqrt{L\cdot(1+L)},A_2\cdot r^2\right]+C_L\cdot L_{nn}^{\sqrt{L\cdot(1+L)}}\left[A_2\cdot r^2\right]\right).\\nn&=\frac{E_t^2-E_\rho^2-2\cdot A_2\cdot\left(1+\sqrt{L\cdot(1+L)}\right)}{4\cdot A_2}\end{aligned} \tag{549}$$

We recognize the hypergeometric function U[a,b,z] and the Laguerre polynomials $L_{nn}^{\sqrt{L\cdot(1+L)}}$ in the solutions above which we know from the Schrödinger hydrogen solution (e.g., [9]). The only reasonable solution (singularity free) can be found for the case of L=0 and then (549) simplifies to:

$$\begin{aligned}f_r[r]&=\sqrt{2}\cdot e^{-\frac{1}{2}A_2\cdot r^2}\cdot\left(C_U\cdot U\left[-nn,1,A_2\cdot r^2\right]+C_L\cdot L_{nn}^0\left[A_2\cdot r^2\right]\right)\\nn&=\frac{E_t^2-E_\rho^2-2\cdot A_2}{4\cdot A_2}=\frac{E_t^2-E_\rho^2}{4\cdot A_2}-\frac{1}{2}\end{aligned}. \tag{550}$$

Figure 37 shows the corresponding f_r-distribution for three settings of E_t.

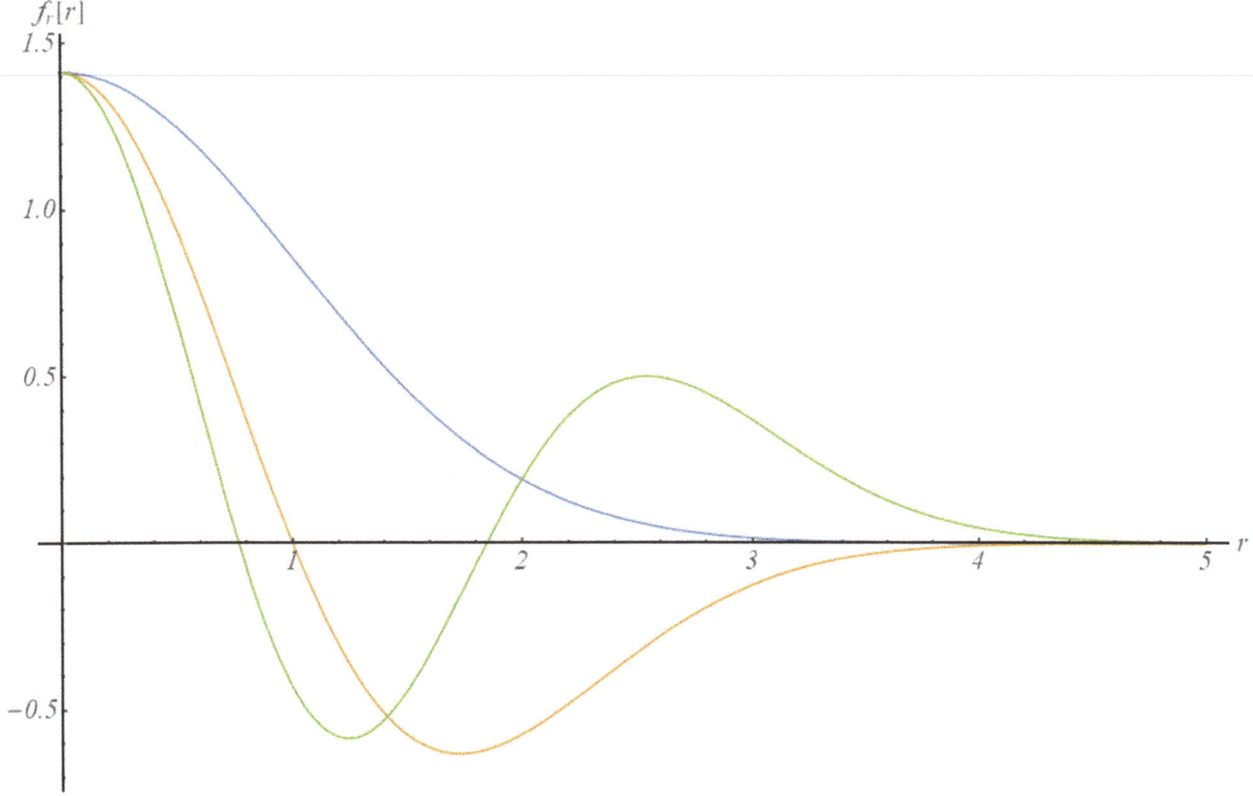

Fig. 37: f_r-distribution for solution (550) with setting A_2=1, E_ρ=±1, L=0, and E_t=$\sqrt{3},\sqrt{7},\sqrt{11}$, resulting in nn=0 (blue), 1 (yellow), 2 (green).

In order to also allow for L≠0-states, we change the metric (537) as follows:

$$g^{8}_{\alpha\beta}=\begin{pmatrix}-c^2 & 0 & 0 & 0 & \cdots & 0\\ 0 & 1 & 0 & 0 & \cdots & 0\\ 0 & 0 & r^2 & 0 & \cdots & 0\\ 0 & 0 & 0 & r^2\cdot\sin^2\varphi_1 & \cdots & 0\\ \cdots & \cdots & \cdots & \cdots & \ddots & 0\\ 0 & 0 & 0 & 0 & 0 & g_{77}\end{pmatrix} \tag{551}$$

$$\times F\left[f\left[t,r,\vartheta,\varphi,\tau,\rho,\theta,\phi\right]\right];\quad F[f]=f^{\frac{2}{3}}$$

$$f\left[t,r,\vartheta,\varphi,\tau,\rho,\theta,\phi\right]=f_t[t]\cdot f_r[r]\cdot f_\vartheta[\vartheta]\cdot f_\varphi[\varphi]\cdot f_\tau[\tau]\cdot f_\rho[\rho]\cdot f_\theta[\theta]\cdot f_\phi[\phi];$$

$$g_{44}=c_4^2\cdot r^{-2};\quad g_{55}=r^2;\quad g_{66}=r^2\cdot\rho^2;\quad g_{77}=r^2\cdot\rho^2\cdot\sin[\theta]^2;$$

and by keeping the partial solutions from (539) to (544), we end up with the following differential equation:

$$0=\left(\left(\frac{\ell\cdot(\ell+1)}{\rho^2}-\frac{2\cdot\partial}{\rho\cdot\partial\rho}-\frac{\partial^2}{\partial\rho^2}-\frac{18}{7}+L\cdot(L+1)\right)\cdot\frac{1}{r^2}+A_2^2\cdot r^2-\frac{4\partial}{r\cdot\partial r}-\frac{\partial^2}{\partial r^2}-E_t^2\right)f. \tag{552}$$

Again, we can directly separate the two radii and obtain two differential equations:

$$0=\left(\frac{\ell\cdot(\ell+1)}{\rho^2}-\frac{2\cdot\partial}{\rho\cdot\partial\rho}-\frac{\partial^2}{\partial\rho^2}\right)f_\rho=E_\rho^2\cdot f_\rho, \tag{553}$$

$$0=\left(\left(L\cdot(L+1)-\frac{18}{7}+E_\rho^2\right)\cdot\frac{1}{r^2}+A_2^2\cdot r^2-\frac{4\partial}{r\cdot\partial r}-\frac{\partial^2}{\partial r^2}-E_t^2\right)f_r. \tag{554}$$

Thereby, we already have applied solution (548) in (554). The solution for f_r reads:

$$f_r[r]=\frac{2^{\frac{1}{2}(1+L_x)}e^{-\frac{1}{2}A_2\cdot r^2}\cdot\left(r^2\right)^{\frac{1}{2}(1+L_x)}}{r^{5/2}}\times\left(C_U\cdot U\left[-nn,1+L_x,A_2\cdot r^2\right]+C_L\cdot L_{nn}^{L_x}\left[A_2\cdot r^2\right]\right). \tag{555}$$

$$nn=\frac{E_t^2-2\cdot A_2\cdot(1+L_x)}{4\cdot A_2};\quad L_x=\frac{1}{2}\cdot\sqrt{4\cdot\left(E_\rho^2+L\cdot(1+L)\right)-\frac{9}{7}}$$

Setting L=0, we find no singularity free solution unless $L_x\geq 2$. Subsequently, we have to demand $nn\geq 0$. Figure 38 shows the corresponding r-distribution for three settings of E_t.

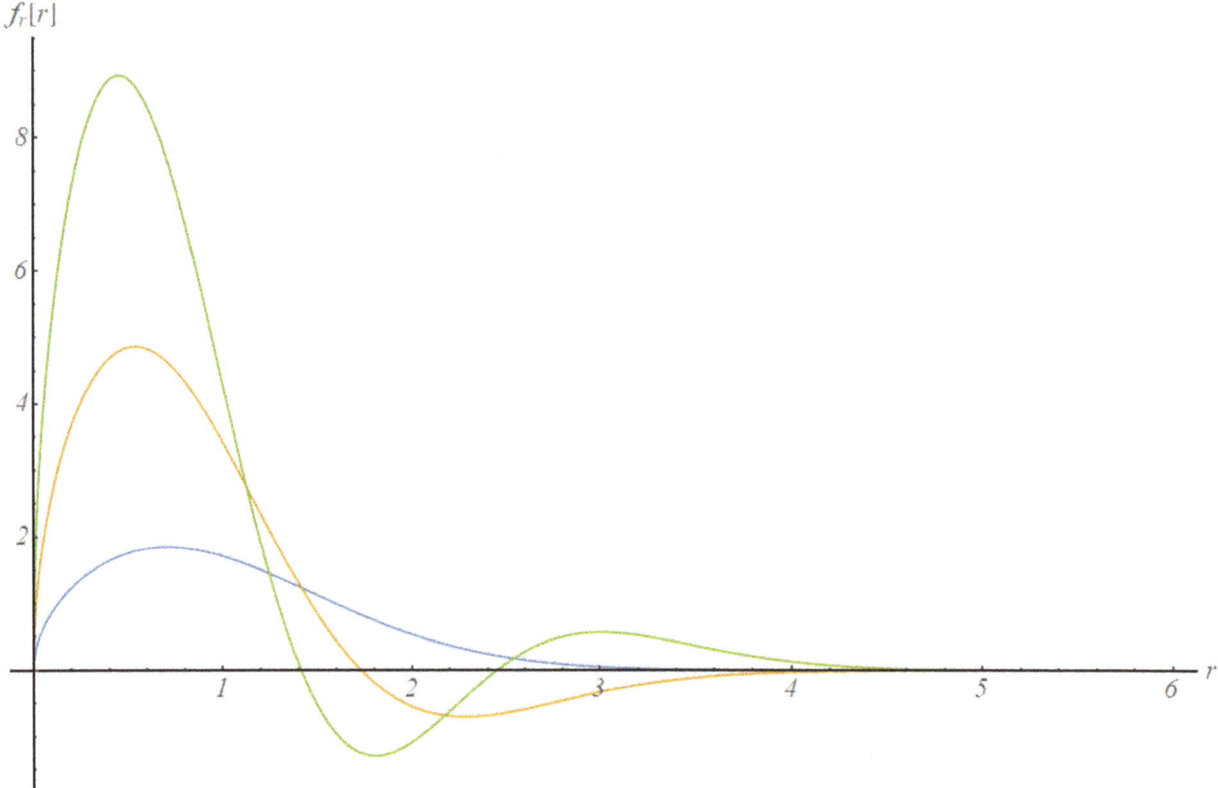

Fig. 38: f_r-distribution for solution (550) with setting $A_2=1$, $E_\rho=\pm\frac{11}{2\sqrt{7}}$, L=0, and $E_t=\sqrt{6},\sqrt{10},\sqrt{14}$, resulting in nn=0 (blue), 1 (yellow), 2 (green).

By choosing L=1/2, we can obtain reasonable (free of singularities) solutions for $L_x \geq 2$. Subsequently, we have to demand nn≥2. Figure 39 shows the corresponding r-distribution for three settings of E_t.

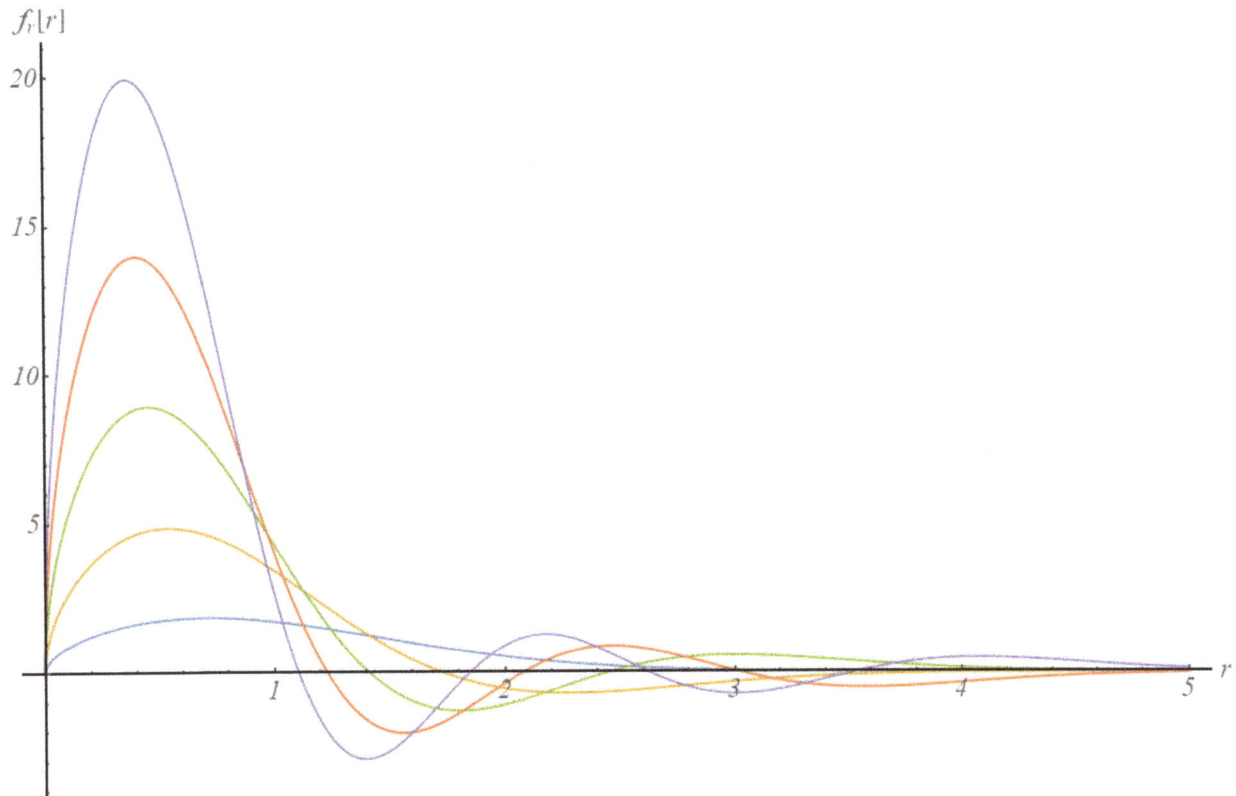

Fig. 39: f_r-distribution for solution (550) with setting $A_2=1$, $E_\rho=\pm\frac{5}{\sqrt{7}}$, L=1/2, and $(E_t)^2$=6,10,14,18,22, resulting in nn=0 (blue), 1 (yellow), 2 (green), 3 (red), 4 (violet).

13.2 Does Spin-½ Really Exist?

It is quite illustrative, if not to say shocking, to compare our finding with the corresponding half spin solution for the classical Schrödinger hydrogen case. Here L=1/2 would also lead to $L_x=2$ (c.f. equation (555)). This is quite interesting, because it leads to the question:

Is the spin-½ in the classical Quantum Theory only an illusion? Meaning, does it only occur to the observer who is unaware of certain internal dimensions bringing the L=1/2-about and then "pretending" a system with effective half spin states?

After all, the r-behavior we obtain is, thanks to $L_x=2=L^*(L+1)$ for the case L=1/2, definitively the one of a half spin object. In addition, we have the two states for E_ρ, being the states of two "spin-like energies" rather than spin orientations, but in connection with our metric (551) they contribute to the spin.

However, if spin-½ was only an illusion, which is to say an object pretending to have spin-½ by containing some kind of intrinsic sphere with always two states for the momentum-connected parameter E_ρ, should this approach not also – automatically – provide a metric understanding of the Pauli exclusion principle [77]?

Well, whatever the ominous additional radius ρ might be, we know that for odd integers L_ρ the spherical Bessel functions from (548) result in opposite signed outcomes for the two parameters $E_\rho=\pm X$. Thus, the total space-time curvature along ρ of two objects possessing opposite signed E_ρ leads to zero space-time and thus, probably a lower total energy. On the other hand, in the case of equally signed E_ρ the total curvature increases and leads to higher energy states. Nature does definitively prefer the state with lower total curvature and hence, the Pauli exclusion.

13.3 Our Finding and a Slightly Disturbing Question

It was shown that a scaled metric automatically provides Schrödinger and Schrödinger-like equations [24, 25, 26] directly out of the Einstein-Hilbert action when adding additional dimensions with certain radius dependencies. Thereby, we also found metric dependencies that produce confining potentials and bring about apparent half spin objects… but without the problematic 720-degree rotation abnormity of the classical spinors.

Our metric scale thereby takes on the classical quantum mechanical wave function. This just means that our space-time is distorted due to quantum processes via this wave function.

Question: Could it be that science was wrong about the half spin particles, which are in fact only complex geometrical structures of higher dimensionality, "pretending" to have half spin to a resolution-restricted observer who can only see the energy states $E_\rho=\pm X$?

13.4 Spin and the T1-T2 Activation Process

Having found that spin can effectively occur due to geometrical effects caused by additional degrees of freedom or dimensions, we should point out here that during the activation processes of our T1-T2-technologies, such degrees of freedom are being created. This insight is of importance with respect to the understanding and optimization of the technologies.

14 The Two Towers of Reality

Now we are ready to approach our two experiments with the fundamental tools derived above.

However, in order not to endanger the safety of our innovations, we here concentrate on the most essential theoretical basics without presenting their application, respectively, their explicit mathematical use regarding our key technologies. Instead, we pick the most important or most influential aspects from the towers T1 and T2 and make sure that those are derived in a fully falsifiable and mathematically rigorous manner.

14.1 The Key Aspects for the First Tower: Electromagnetic Interaction and Elasticity

As it was shown in table 1 that the activation is an essential phase for a successful deposition technology with subsequent good, if not to say perfect adhesivity of the coating, we here now derive the electromagnetism from the most fundamental starting point there is, the Einstein-Hilbert action. Along the way, as a by-product, we are going to obtain the Higgs field [56].

Fig. 40: Photonic activation is one of the key issues of the T1 deposition technology. The picture shows a localized (Gaussian), solenoidal photon passing along the z-axis. Absolute field values are shown. Due to the hollow shape of the structure, one easily recognizes the spin.

Another important field, which is of need when one intends to optimize the deposition technology towards a specific application is solid mechanics; especially the theory of elasticity. This, too, as we are going to demonstrate in this section can directly be derived from the Einstein-Hilbert action, thereby revealing astounding options for its generalization.

14.1.1 Electromagnetic Interaction (and Matter) via Space-Time Warping

To provide an easy introduction to the theoretical apparatus, we shall start with the Minkowski flat space metric tensor in Cartesian coordinates t, x, y, z as follows:

$$g_{\alpha\beta}^{\mathrm{flat}} = \begin{pmatrix} -c^2 & 0 & 0 & 0 \\ 0 & 1 & 0 & 0 \\ 0 & 0 & 1 & 0 \\ 0 & 0 & 0 & 1 \end{pmatrix}. \tag{556}$$

This metric solves the classical Einstein field equations which results from the Einstein-Hilbert action (6) (see [1, 3]) and corresponds to a Ricci scalar of R=0. Now we introduce the following special transformation:

$$G_{\alpha\beta} = g_{\alpha\beta}^{\mathrm{flat}} \cdot F\left[f\left[t,x,y,z\right]\right] = \begin{pmatrix} -c^2 & 0 & 0 & 0 \\ 0 & 1 & 0 & 0 \\ 0 & 0 & 1 & 0 \\ 0 & 0 & 0 & 1 \end{pmatrix} \cdot F\left[f\left[t,x,y,z\right]\right] \tag{557}$$

and evaluate the resulting new Ricci scalar R^*:

$$R^{*}=\frac{R}{F[f]}+\begin{pmatrix}-\Gamma_{\sigma\alpha}^{\mu}\Gamma_{\beta\mu}^{**\sigma}+\Gamma_{\alpha\beta}^{\sigma}\Gamma_{\sigma\mu}^{**\mu}-\Gamma_{\sigma\alpha}^{**\mu}\Gamma_{\beta\mu}^{\sigma}+\Gamma_{\alpha\beta}^{**\sigma}\Gamma_{\sigma\mu}^{\mu}\\+\Gamma_{\alpha\beta\,,\sigma}^{**\sigma}-\Gamma_{\beta\sigma\,,\alpha}^{**\sigma}-\Gamma_{\sigma\alpha}^{**\mu}\Gamma_{\beta\mu}^{**\sigma}+\Gamma_{\alpha\beta}^{**\sigma}\Gamma_{\sigma\mu}^{**\mu}\end{pmatrix}\frac{g^{\alpha\beta}}{F[f]}$$

$$\xrightarrow{R=0;\,g_{\alpha\beta}^{flat}}=\frac{1}{F[f]^{3}}\cdot\begin{pmatrix}\left(C_{N1}\cdot\left(\frac{\partial F[f]}{\partial f}\right)^{2}-C_{N2}\cdot F[f]\cdot\frac{\partial^{2}F[f]}{\partial f^{2}}\right)\cdot\left(\tilde{\nabla}_{g}f\right)^{2}\\-C_{N2}\cdot F[f]\cdot\frac{\partial F[f]}{\partial f}\cdot\Delta_{g}f\end{pmatrix}$$

$$=\frac{1}{F[f]^{3}}\cdot\left(\left(\frac{3}{2}\cdot\left(\frac{\partial F[f]}{\partial f}\right)^{2}-3\cdot F[f]\cdot\frac{\partial^{2}F[f]}{\partial f^{2}}\right)\cdot\left(\tilde{\nabla}_{g}f\right)^{2}-3\cdot F[f]\cdot\frac{\partial F[f]}{\partial f}\cdot\Delta_{g}f\right)\qquad .(558)$$

$$=\frac{3}{2\cdot F[f]^{3}}\cdot\begin{pmatrix}\left(\left(\frac{\partial F[f]}{\partial f}\right)^{2}-2\cdot F[f]\cdot\frac{\partial^{2}F[f]}{\partial f^{2}}\right)\cdot\left[-\frac{(\partial_{t}f)^{2}}{c^{2}}+(\partial_{x}f)^{2}+(\partial_{y}f)^{2}+(\partial_{z}f)^{2}\right]\\-2\cdot F[f]\cdot\frac{\partial F[f]}{\partial f}\cdot\left[-\frac{\partial_{t}^{2}}{c^{2}}+\partial_{x}^{2}+\partial_{y}^{2}+\partial_{z}^{2}\right]f\end{pmatrix}$$

Thereby we have used:

$$\Gamma_{\alpha\beta}^{*\gamma}=\frac{g^{\gamma\sigma}}{2\cdot F[f]}\left(\left[F[f]\cdot g_{\sigma\alpha}\right]_{,\beta}+\left[F[f]\cdot g_{\sigma\beta}\right]_{,\alpha}-\left[F[f]\cdot g_{\alpha\beta}\right]_{,\sigma}\right)$$
$$=\frac{g^{\gamma\sigma}}{2\cdot F[f]}\left(\left[F[f]\cdot g_{\sigma\alpha}\right]_{,\beta}+\left[F[f]\cdot g_{\sigma\beta}\right]_{,\alpha}-\left[F[f]\cdot g_{\alpha\beta}\right]_{,\sigma}\right)$$
$$=\frac{g^{\gamma\sigma}}{2}\left(g_{\sigma\alpha,\beta}+g_{\sigma\beta,\alpha}-g_{\alpha\beta,\sigma}\right)+\frac{g^{\gamma\sigma}}{2\cdot F[f]}\left(F[f]_{,\beta}\cdot g_{\sigma\alpha}+F[f]_{,\alpha}\cdot g_{\sigma\beta}-F[f]_{,\sigma}\cdot g_{\alpha\beta}\right).\qquad(559)$$
$$=\Gamma_{\alpha\beta}^{\gamma}+\frac{g^{\gamma\sigma}}{2\cdot F[f]}\left(F[f]_{,\beta}\cdot g_{\sigma\alpha}+F[f]_{,\alpha}\cdot g_{\sigma\beta}-F[f]_{,\sigma}\cdot g_{\alpha\beta}\right)$$
$$\equiv\Gamma_{\alpha\beta}^{\gamma}+\Gamma_{\alpha\beta}^{**\gamma}$$

The reader will find a more comprehensive elaboration about the motivation of our special transformation in the book [2] in section "3.3 The Ricci Scalar Quantization". However, it is quite entertaining to follow this simple trial started here and observe its amazing evolution into something rather unexpected (at least if taking its origin as a metric solution to the Einstein field equations) out of our "wrapper-transformation" (557).

The symbol $\tilde{\nabla}_{g}$ in (558) denotes a first order differential operator similar to the Nabla-operator in the metric $g_{\alpha\beta}$. The symbols C_{Ni} are standing for constants, which only depend on the number of dimensions. Their derivation was presented in [4] and its result can also be found in appendix B of [56]. Please note that the transformation (557) is just the simplest form of a general approach like:

$$G_{\alpha\beta}=F\left[f[t,x,y,z]\right]_{\alpha\beta}^{ij}g_{ij}\rightarrow G_{\alpha\beta}=F\left[f[t,x,y,z]\right]\cdot\delta_{\alpha}^{i}\delta_{\beta}^{j}g_{ij},\qquad(560)$$

whereby only the latter guarantees the maintained tensor character and thus, the covariance of the metric.

Without loss of generality we can now demand that the first term in parenthesis in (558) would be zero, which is to say, we choose our arbitrary function F such that we have:

$$C_{N1}\cdot\left(\frac{\partial F[f]}{\partial f}\right)^2-C_{N2}\cdot F[f]\cdot\frac{\partial^2 F[f]}{\partial f^2}=0. \tag{561}$$

For instance, in four dimensions this would always be the case for $F[f]=f^2$, giving us the Klein-Gordon-like equation:

$$\left[\Delta_g+\frac{R^*\cdot f^4}{f\cdot C_{N2}\cdot 2\cdot f}\right]f=\left[\Delta_g+\frac{R^*\cdot f^2}{2}\right]f=0;\quad \Delta_{g-\text{Coordinates}}=\Delta_g. \tag{562}$$

14.1.2 The Higgs Field

By setting the solution $F[f]=f^2$ into our transformation starting point (557), however, one would directly result in an all-scale quantum dominated metric solution for $G_{\alpha\beta}$. This is in total contrast to the daily experiences that quantum effects are only important in smaller scales. In order to overcome this problem, we evaluate the general solution of condition (561) in four dimensions and find:

$$F[f]=C_1\cdot f+\frac{C_1^2\cdot f^2}{4\cdot C_2}+C_2. \tag{563}$$

Thereby the constants C_i are arbitrary.

For entertainment and further motivation it should be pointed out here that the functional wrapper F[f] from (563) could also be adapted as follows:

$$F[f]=C_1\cdot f^2+\frac{C_1^2\cdot f^4}{4\cdot C_2}+C_2. \tag{564}$$

With arbitrary constants C_i, this assures the appearance of linear Laplace operator terms according to the condition (561) for the resulting Ricci scalar of the transformed metric in (558) as follows:

$$R^*=\frac{1}{F[f]^3}\cdot\left(-C_{N2}\cdot F[f]\cdot\frac{\partial F[f]}{\partial f}\cdot\Delta_g f^2\right)\xrightarrow{4D}R^*\cdot\frac{\left(2\cdot C_2+C_1\cdot f^2\right)^3}{8\cdot C_2\cdot C_1}=-3\cdot\Delta_g f^2. \tag{565}$$

Now we assume f to be a constant, which automatically leads to the simple equation:

$$R^*\cdot\frac{\left(2\cdot C_2+C_1\cdot f^2\right)^3}{8\cdot C_2\cdot C_1}=0. \tag{566}$$

Also assuming that the Ricci scalar curvature R^* shall be proportional to f^n, with an arbitrary exponent $n\geq 0$, we obtain the familiar trivial solution of $f_0=0$. However, from (566) we also obtain the non-trivial ground states:

$$f^2 = -\frac{2 \cdot C_2}{C_1}. \tag{567}$$

As the constants C_i are arbitrary, we could simply set them as follows:

$$2 \cdot C_2 = -\mu^2; \quad C_1 = 2 \cdot \lambda. \tag{568}$$

This gives us the additional ground state solutions directly in the classical Higgs field style [58], namely:

$$\left(f_{1,2}\right)^2 = \frac{\mu^2}{2 \cdot \lambda} \quad \Rightarrow \quad f_{1,2} = \pm \frac{\mu}{\sqrt{2 \cdot \lambda}}. \tag{569}$$

The measured value for the $f_{1,2}$ is known to be [59]:

$$\left|f_{1,2}\right| = \frac{\mu}{\sqrt{2 \cdot \lambda}} = \frac{246\,\text{GeV}}{\sqrt{2} \cdot c^2}. \tag{570}$$

Rewriting (566) with the use of our above definition for the C_i as:

$$R^* \cdot \frac{\left(-\mu^2 + 2 \cdot \lambda \cdot f^2\right)^3}{8 \cdot \mu^2 \cdot 2 \cdot \lambda} = f^n \cdot \frac{\left(2 \cdot \lambda \cdot f^2 - \mu^2\right)^3}{16 \cdot \mu^2 \cdot \lambda} = 3 \cdot \Delta_g f^2 \tag{571}$$

gives us the total Higgs-Ricci-curvature connection. It also, automatically, gives us a curvature value R^* at the ground state, which is:

$$R^* \sim \left|f_{1,2}\right|^n = \left(\frac{\mu}{\sqrt{2 \cdot \lambda}}\right)^n = \left(\frac{246\,\text{GeV}}{\sqrt{2} \cdot c^2}\right)^n. \tag{572}$$

Now, with the Higgs mass m_H known to be $125\text{GeV}/c^2$ and the fact that we have [59, 60]:

$$m_H = \sqrt{2 \cdot \mu^2} = \sqrt{4 \cdot \lambda \cdot \left(f_{1,2}\right)^2}, \tag{573}$$

we can obtain:

$$\lambda \simeq 0.13; \quad \mu \simeq 88.8\,\text{GeV}/c^2. \tag{574}$$

Please note that we obtain the simple case $F[f]=f^2$ with the condition:

$$C_1^2 = 2 \cdot C_2 \tag{575}$$

and the subsequent limit procedure $C_2 \rightarrow 0$ as follows:

$$\Rightarrow F[f] = \sqrt{2 \cdot C_2} \cdot f + f^2 + C_2 \xrightarrow[C_2 \to 0]{\lim} F[f] = f^2. \tag{576}$$

As such considerations can be generalized to any number of dimensions, arbitrary attributes, and systems, we can immediately state that also the systems of our innovative set-ups in T1 to T3 can produce a Higgs field, which provides an interaction force with the "things" being in there. As in the case of T3, these things are our thoughts; the field just gives the mass and moderates their forthcoming just as the real Higgs field does in our universal space-time with the elementary particles (not all of them, though). With respect to the key technologies T1 and T2 the finding clearly shows possibilities of states with a Higgs mechanism (on a higher level or order of interaction than the classical Higgs, of course) and those without. A cyclic process between the two states might just provide the explanation for some of the astounding properties of the innovations.

14.1.3 Back to the Derivation of the Electromagnetic Field

Using (563) instead of $F=f^2$ in (558) gives us:

$$R^* = \frac{1}{F[f]^3} \cdot \begin{pmatrix} \left(C_{N1} \cdot \left(\frac{\partial F[f]}{\partial f} \right)^2 - C_{N2} \cdot F[f] \cdot \frac{\partial^2 F[f]}{\partial f^2} \right) \cdot \left(\tilde{\nabla}_g f \right)^2 \\ -C_{N2} \cdot F[f] \cdot \frac{\partial F[f]}{\partial f} \cdot \Delta_g f \end{pmatrix}$$

$$= -C_{N2} \cdot \frac{\frac{\partial F[f]}{\partial f} \cdot \Delta_g f}{F[f]^2} \xrightarrow{4D} = -3 \cdot \frac{\left(C_1 + \frac{C_1^2 \cdot f}{2 \cdot C_2} \right)}{\left(C_1 \cdot f + \frac{C_1^2 \cdot f^2}{4 \cdot C_2} + C_2 \right)^2} \cdot \Delta_g f \qquad (577)$$

Immediately we see that in the many cases with $R^*=0$, we would obtain classical quantum Klein-Gordon equations. By the way: in the book [4] we have shown that the condition $R^*=0$ also directly follows from the Hamilton minimum principle and is not just a simple definition. Taylor expansion for small f (which is to say at f=0) for the general case in the second line in (577) also gives us equations leading to these classical equations. For instance, if only considering up to linear f as follows:

$$\frac{R^*}{3} \cdot \frac{\left(C_1 \cdot f + \frac{C_1^2 \cdot f^2}{4 \cdot C_2} + C_2 \right)^2}{C_1 + \frac{C_1^2 \cdot f}{2 \cdot C_2}} + \Delta_g f = 0 \xrightarrow{\text{small f}} \frac{R^*}{3} \cdot \left(\frac{C_2^2}{C_1} + \frac{3}{2} \cdot C_2 \cdot f + \ldots \right) + \Delta_g f = 0, \qquad (578)$$

and assuming constant R^*, we already have the typical structure of the Klein-Gordon equation plus a constant term $\frac{R^*}{3} \cdot \frac{C_2^2}{C_1}$. Such a constant would still not influence the principle structural character of our resulting quantum equations because a simple transformation of:

$$f^* = \frac{C_2^2}{C_1} + \frac{3}{2} \cdot C_2 \cdot f; \quad \Delta_g f^* = \Delta_g f \qquad (579)$$

does give us back the ordinary Klein-Gordon equation with:

$$\frac{R^*}{3} \cdot f^* + \Delta_g f^* = 0. \qquad (580)$$

The resulting transformation in four dimensions would give the following quantum metric:

$$G_{\alpha\beta} = g_{\alpha\beta} \cdot F[f] = g_{\alpha\beta} \cdot \left(C_1 \cdot f + \frac{C_1^2 \cdot f^2}{4 \cdot C_2} + C_2 \right). \qquad (581)$$

This clearly allows for the classical physics or Einstein situation with quantum effects becoming dominant only at small scales and thus, a dominating part C_2 at bigger scales. In result, one might separate as follows in four dimensions:

$$G_{\alpha\beta} = \overbrace{\overbrace{g_{\alpha\beta} \cdot C_2}^{\text{class. GTR}} + \overbrace{g_{\alpha\beta} \cdot \left(C_1 \cdot f \cdot \left(1 + \frac{C_1 \cdot f}{4 \cdot C_2} \right) \right)}^{\text{quantum metric}}}^{\text{Quantum Gravity in 4D}}. \tag{582}$$

Therefore it does not come as a surprise that we have already found so many cases where f has to be determined by classical quantum equations (see [2, 4] and references given there), respectively where our equations give similar or even equal solutions. Normalization of the quantum gravity metric $G_{\alpha\beta}$ requires division by C_2 and gives:

$$G_{\alpha\beta}^{\text{norm}} = \frac{G_{\alpha\beta}}{C_2} = \overbrace{\overbrace{g_{\alpha\beta}}^{\text{class. GTR}} + \overbrace{g_{\alpha\beta} \cdot \left(\frac{C_1}{C_2} \cdot f \cdot \left(1 + \frac{C_1 \cdot f}{4 \cdot C_2} \right) \right)}^{\text{quantum metric}}}^{\text{Quantum Gravity in 4D}}. \tag{583}$$

Now we apply the recipe outlined above to our simple flat space example in order to see how the transformation can bring about matter.

We saw that by taking the scaled metric (557), using rule (561) with the subsequent solution (563), we obtain the new Ricci scalar curvature:

$$R^* = -3 \cdot \frac{\left(C_1 + \frac{C_1^2 \cdot f}{2 \cdot C_2} \right)}{\left(C_1 \cdot f + \frac{C_1^2 \cdot f^2}{4 \cdot C_2} + C_2 \right)^2} \cdot \Delta_g f. \tag{584}$$

Now we demand that the Ricci scalar should be a conserved quantity and as we have R=0 with the flat space metric (556), we shall also demand $R^*=0$. This gives us the simple equation:

$$0 = \Delta_g f = \left[\partial_x^2 + \partial_y^2 + \partial_z^2 - \frac{\partial_t^2}{c^2} \right] f. \tag{585}$$

Aiming for more generality we extend (556) as follows (A, B, D are constants):

$$g_{\alpha\beta}^{\text{flat}} = \begin{pmatrix} -c^2 & 0 & 0 & 0 \\ 0 & A^2 & 0 & 0 \\ 0 & 0 & B^2 & 0 \\ 0 & 0 & 0 & D^2 \end{pmatrix}, \tag{586}$$

which makes (585) to:

$$0=\Delta_g f=\left[\frac{\partial_x^2}{A^2}+\frac{\partial_y^2}{B^2}+\frac{\partial_z^2}{D^2}-\frac{\partial_t^2}{c^2}\right]f. \tag{587}$$

Applying the separation approach f[t,x,y,z]=T[t]*X[x]*Y[y]*Z[z] gives us the solutions:

$$T[t]=C_{t1}\cdot\cos[c\cdot C_t\cdot t]+C_{t2}\cdot\sin[c\cdot C_t\cdot t], \tag{588}$$

$$X[x]=C_{x1}\cdot\cos[A\cdot C_x\cdot x]+C_{x2}\cdot\sin[A\cdot C_x\cdot x], \tag{589}$$

$$Y[y]=C_{y1}\cdot\cos[B\cdot C_y\cdot y]+C_{y2}\cdot\sin[B\cdot C_y\cdot y], \tag{590}$$

$$Z[z]=C_{z1}\cdot\cos[D\cdot C_z\cdot z]+C_{z2}\cdot\sin[D\cdot C_z\cdot z], \tag{591}$$

and the following characteristic equation:

$$0=C_x^2+C_y^2+C_z^2-C_t^2. \tag{592}$$

The matter coded with this solution for R^* can now directly be obtained via the classical Einstein field equations with matter, which reads:

$$R^{*\alpha\beta}-\frac{1}{2}R^*\cdot g^{\alpha\beta}+\Lambda\cdot g^{\alpha\beta}=-\kappa\cdot T^{\alpha\beta}. \tag{593}$$

As we had per demand R*=R=0 and as we also assume the cosmological constant to be equal to zero, we result in the following identity:

$$R^{*\alpha\beta}=-\kappa\cdot T^{\alpha\beta}. \tag{594}$$

Thus, in order to find out what kind of matter our transformation:

$$G_{\alpha\beta}=g_{\alpha\beta}^{\text{flat}}\cdot F[f[t,x,y,z]]=\begin{pmatrix}-c^2&0&0&0\\0&A^2&0&0\\0&0&B^2&0\\0&0&0&D^2\end{pmatrix}\cdot F[f[t,x,y,z]] \tag{595}$$

has created, we simply need to evaluate the corresponding Ricci tensor of the transformed/scaled metric $G_{\alpha\beta}$. In order to keep the presentation general[19] and as the evaluation of the derivatives is cumbersome, we give the Ricci tensor with the function f. But for simplicity we have set F[f] to $F[f]=(f+C_1)^2$

$$R^{*00}=\frac{(A\cdot B\cdot c)^2(f_{,z})^2+D^2\left(A^2c^2(f_{,y})^2+B^2\left(\begin{array}{c}c^2(f_{,x})^2+3A^2(f_{,t})^2\\-2A^2(C_1+f)f_{,t,t}\end{array}\right)\right)}{(A\cdot B\cdot D)^2c^4(C_1+f)^6}. \tag{596}$$

[19] This comes in handy the moment we want to exploit the additive character of our metric solutions. Thereby the additivity is assured by the means of condition (561). See further below in the main text!

$$R^{*11}=\frac{-(A\cdot B\cdot c)^2(f_{,z})^2+D^2\left(-A^2c^2(f_{,y})^2+B^2\begin{pmatrix}3c^2(f_{,x})^2\\-2c^2(C_1+f)f_{,x,x}\end{pmatrix}+A^2(f_{,t})^2\right)}{(c\cdot B\cdot D)^2A^4(C_1+f)^6}. \quad (597)$$

$$R^{*22}=\frac{-(A\cdot B\cdot c)^2(f_{,z})^2+D^2\begin{pmatrix}3A^2c^2(f_{,y})^2-2A^2c^2(C_1+f)f_{,y,y}\\-B^2c^2(f_{,x})^2+B^2A^2(f_{,t})^2\end{pmatrix}}{(c\cdot A\cdot D)^2B^4(C_1+f)^6}. \quad (598)$$

$$R^{*33}=\frac{3(A\cdot B\cdot c)^2(f_{,z})^2+D^2\begin{pmatrix}-A^2c^2(f_{,y})^2-B^2c^2(f_{,x})^2+B^2A^2(f_{,t})^2\\+2(C_1+f)\left(A^2c^2f_{,y,y}+B^2\left(c^2f_{,x,x}-A^2f_{,t,t}\right)\right)\end{pmatrix}}{(c\cdot A\cdot D)^2B^4(C_1+f)^6}. \quad (599)$$

$$\begin{aligned}R^{*01}&=\frac{2\left(-2f_{,x}f_{,t}+(C_1+f)f_{,t,x}\right)}{A^2c^2(C_1+f)^6};\quad R^{*02}=\frac{2\left(-2f_{,y}f_{,t}+(C_1+f)f_{,t,y}\right)}{B^2c^2(C_1+f)^6}\\R^{*03}&=\frac{2\left(-2f_{,z}f_{,t}+(C_1+f)f_{,t,z}\right)}{D^2c^2(C_1+f)^6};\quad R^{*12}=\frac{2\left(2f_{,y}f_{,x}-(C_1+f)f_{,x,y}\right)}{A^2B^2(C_1+f)^6}\\R^{*13}&=\frac{2\left(2f_{,z}f_{,x}-(C_1+f)f_{,x,z}\right)}{A^2D^2(C_1+f)^6};\quad R^{*23}=\frac{2\left(2f_{,y}f_{,z}-(C_1+f)f_{,z,y}\right)}{B^2D^2(C_1+f)^6}\end{aligned}. \quad (600)$$

Previously we had discussed corresponding solutions in spherical [61] and cylindrical [62] geometries and it is clear that we have obtained photons or photonic matter forms. What is more, with condition (561) we have automatically achieved additive character for all our solutions (588) to (592), allowing to apply simple integral transform methods in order to construct almost arbitrary photonic matter forms. The corresponding solution for a photon was given in [63] as a combined solution, consisting of two main parts f and F. Here we present the spatial distribution of the two functions.

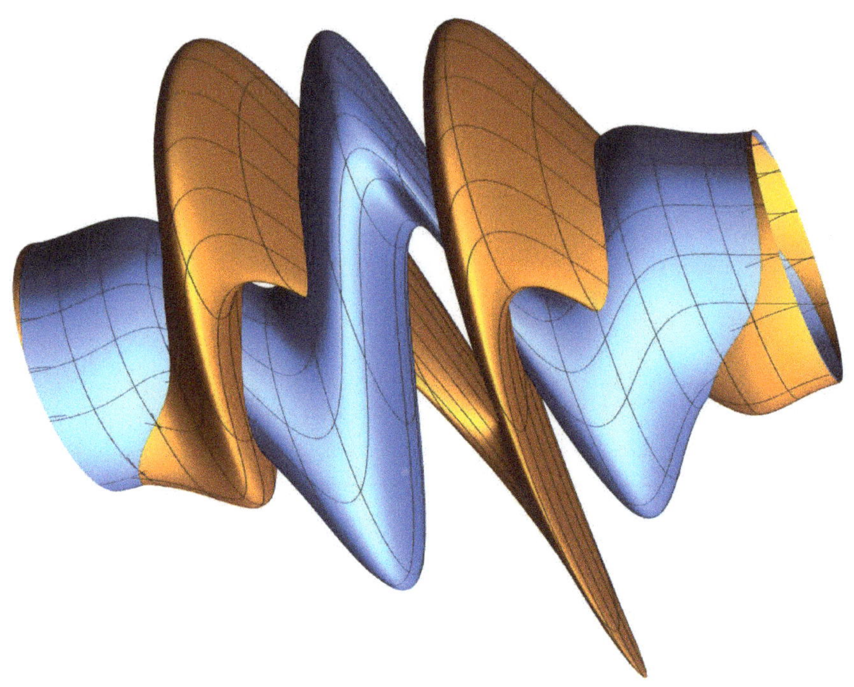

Fig. 41: Presentation of a photoninc solution for the wave function f as parametric plot with the real part (blue) and the imaginary part (yellow ochre). The moving-direction- or z-axis is shown as two tubes being bent and changed in radius according to the strength of the field.

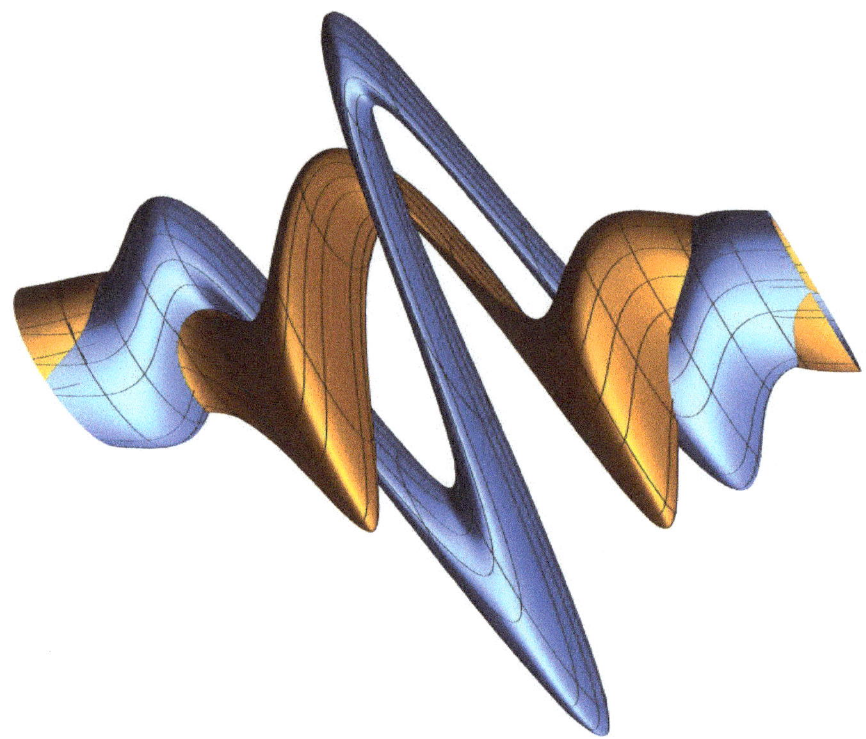

Fig. 42: Presentation of of a photoninc solution for the wave function F as parametric plot with the real part (blue) and the imaginary part (yellow ochre). The moving-direction- or z-axis of the photon is shown as two tubes being bent and changed in radius according to the strength of the field.

Without applying the technique of superposition of our solution (588) to (592), however, the subsequent photonic solution would be a plane wave structure with no changes along the x-axis. It was shown in [63] how such a structure can be localized also in lateral directions via standard integral transform methods and that the subsequent solutions do fulfill the Maxwell equations. The three figures below (figures 43, 44, and 45) illustrate the corresponding spatial deformation.

Thus, we can illustrate a localized photon by using an ensemble of 2-spheres or more easy tubes (understood as summed up spheres) and deform them in accordance to the electric and magnetic field, respectively the real and imaginary values of our solutions. At first, we consider the electric and magnetic part separately (two tubes, see figure 43) and discover that there is not much difference to our illustrations before (c.f. figures 41 and 42).

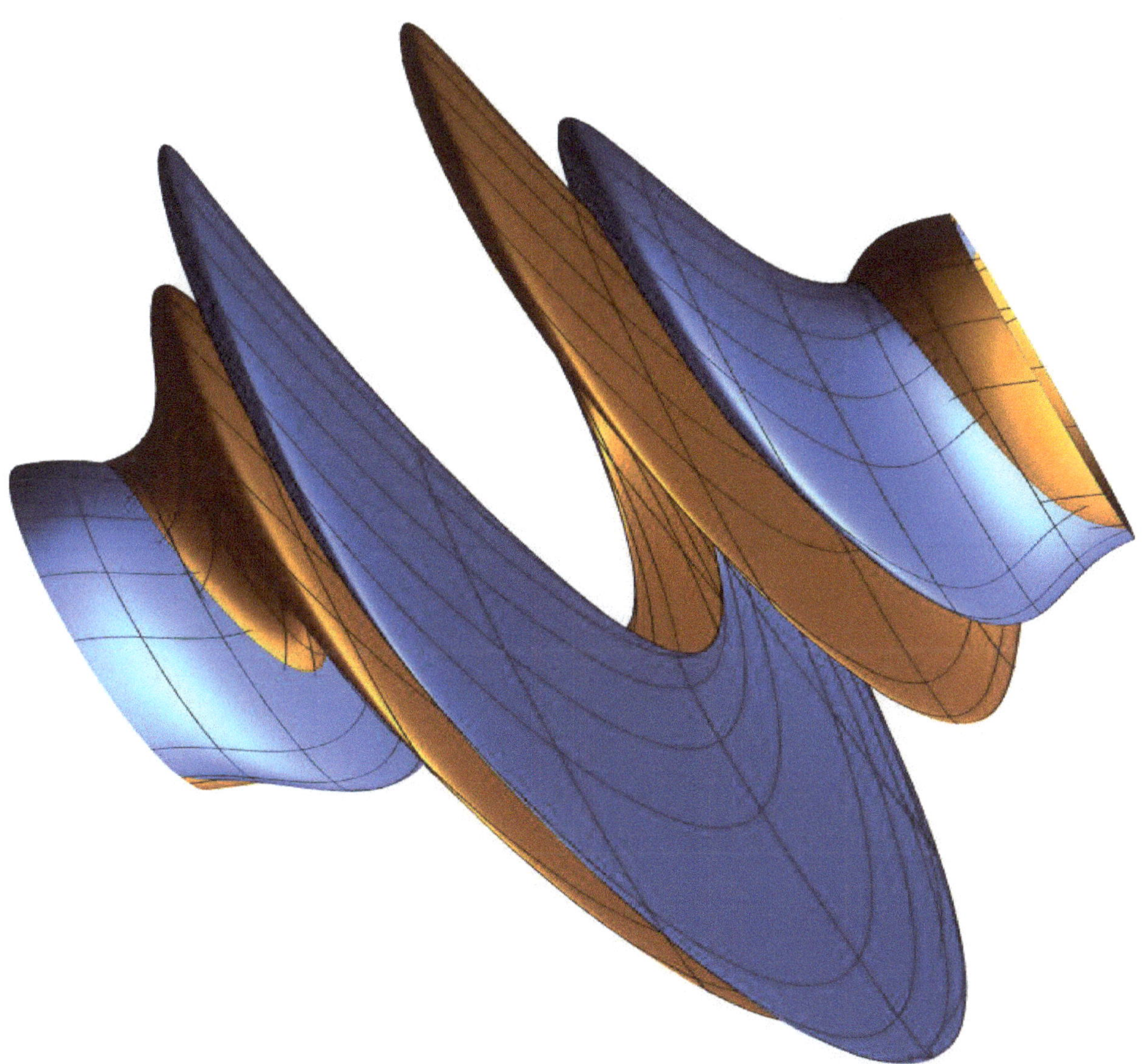

Fig. 43: Presentation of electric (yellow) and magnetic (blue) field. The moving-direction- or -axis is shown as two tubes being bent according to the strength of the fields.

Now however, with our clear assignment of the two fields to

a) deformation (bending) of the tube's axes in total in case of the electric and
b) deformation of the tube's wall, changing its diameter, in the case of the magnetic field

we can put all field information on just one tube (figure 44).

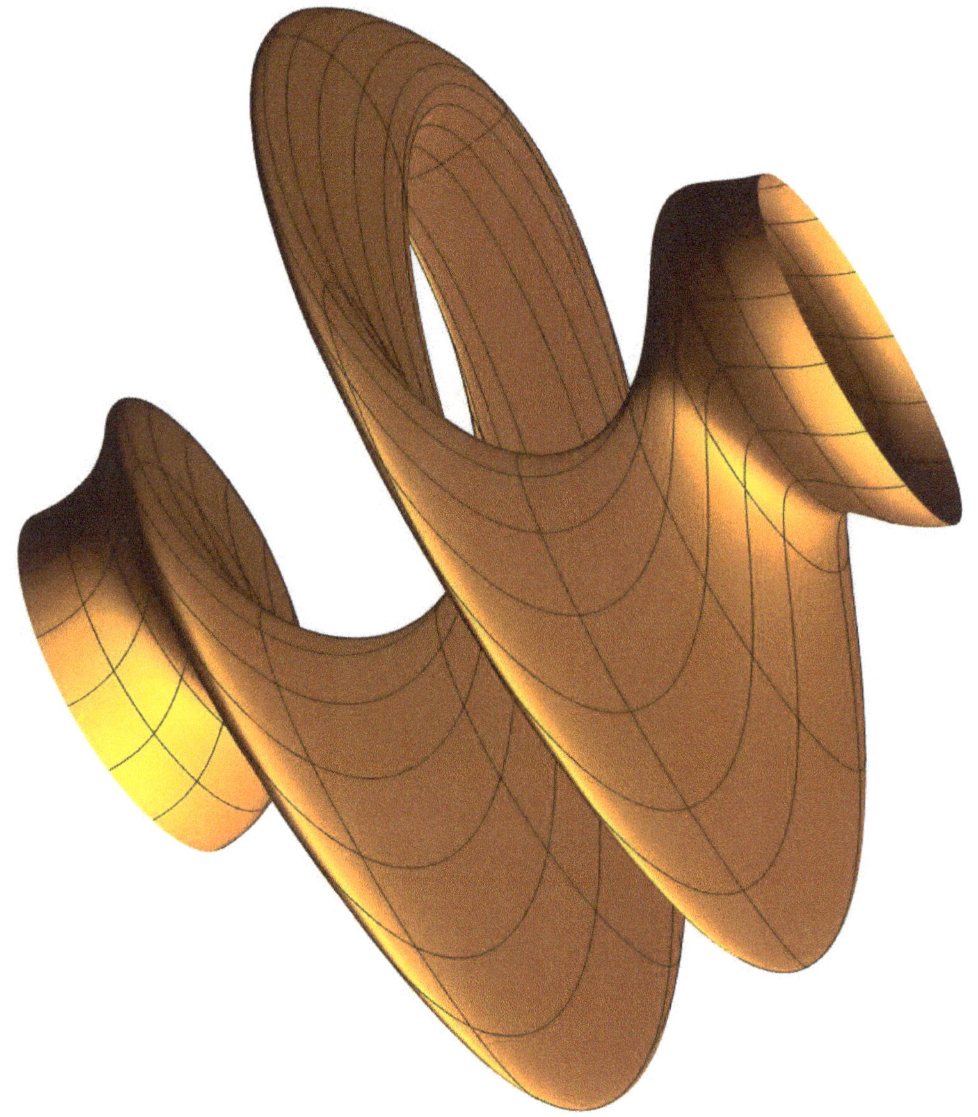

Fig. 44: Presentation of electric (axis-deformation) and magnetic (radius distortion) field of a photon. The moving-direction- or z-axis is shown as one tube being bent and distorted in radius according to the strength of the fields.

We recognize the double deformation of the "coordinate-tube" as x-y-deformation of its symmetry-axis (electric field) and its radius-shape deformation as result of the magnetic field along the z-axis (figure 44) for the passing photon. In figure 45 we have also coded the local energy density given as E^2+B^2 in the color of the distorted tube.

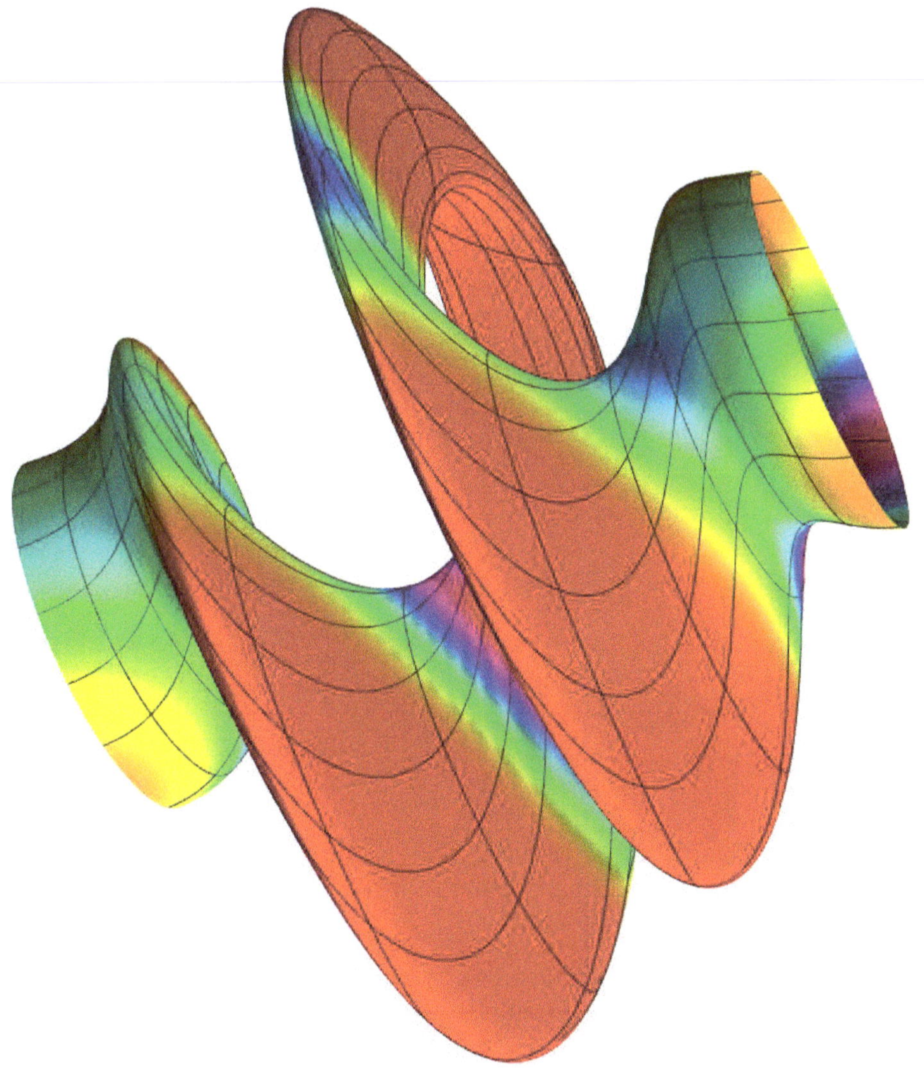

Fig. 45: Presentation of electric (axis-deformation) and magnetic (radius distortion) field of a photon. The z-axis is shown as one tube being bent and distorted in radius according to the strength of the fields. In addition the color is coding the energy density of the passing photon.

The connection with the energy momentum tensor for the electromagnetic field is achieved via (594). The electrostatic and magnetic vector fields E and B with components E_i and B_i are forming the energy momentum tensor in contravariant form as follows:

$$\left(T^{\alpha\beta}\right)=\begin{pmatrix}\frac{1}{2}\cdot(\mathbf{E}\cdot\mathbf{E}+\mathbf{B}\cdot\mathbf{B}) & (\mathbf{E}\times\mathbf{B})^{T}\\ \mathbf{E}\times\mathbf{B} & \frac{1}{2}\cdot(\mathbf{E}\cdot\mathbf{E}+\mathbf{B}\cdot\mathbf{B})\cdot\delta_{ik}-E_iE_k-B_iB_k\end{pmatrix}. \tag{601}$$

$$\delta_{ik}=\begin{pmatrix}1 & 0 & 0\\ 0 & 1 & 0\\ 0 & 0 & 1\end{pmatrix}$$

Please note that in SI-unites this tensor reads:

$$\left(T^{\alpha\beta}\right)=\begin{pmatrix}\dfrac{\varepsilon_0\cdot\mathbf{E}\cdot\mathbf{E}+\dfrac{\mathbf{B}\cdot\mathbf{B}}{\mu_0}}{2} & c\cdot\varepsilon_0\cdot(\mathbf{E}\times\mathbf{B})^T\\ c\cdot\varepsilon_0\cdot\mathbf{E}\times\mathbf{B} & \dfrac{\varepsilon_0\cdot\mathbf{E}\cdot\mathbf{E}+\dfrac{\mathbf{B}\cdot\mathbf{B}}{\mu_0}}{2}\cdot\delta_{ik}-\varepsilon_0\cdot E_iE_k-\dfrac{B_iB_k}{\mu_0}\end{pmatrix}. \tag{602}$$

The constants ε_0,μ_0 are the electric and the magnetic field constant, respectively.

For brevity, we shall leave it to the mathematically skilled and interested reader to work out the detailed connection of the two classical fields with our quantum metric distortion. For illustration, however, we just consider the energy density of the electromagnetic field given via the component T^{00} and thus, obtain:

$$\begin{aligned}R^{*00}&=\frac{(A\cdot B\cdot c)^2\left(f_{,z}\right)^2+D^2\left(A^2c^2\left(f_{,y}\right)^2+B^2\begin{pmatrix}c^2\left(f_{,x}\right)^2+3A^2\left(f_{,t}\right)^2\\-2A^2\left(C_1+f\right)f_{,t,t}\end{pmatrix}\right)}{(A\cdot B\cdot D)^2c^4\left(C_1+f\right)^6}\\&=-\kappa\cdot T^{00}=-\kappa\cdot\frac{\varepsilon_0\cdot\mathbf{E}\cdot\mathbf{E}+\dfrac{\mathbf{B}\cdot\mathbf{B}}{\mu_0}}{2}=-\frac{\kappa}{2}\cdot\left(\varepsilon_0\cdot\left(E_1^2+E_2^2+E_3^2\right)+\frac{B_1^2+B_2^2+B_3^2}{\mu_0}\right)\end{aligned}. \tag{603}$$

In fact, we find quadratic terms on both sides of the equation and it appears kind of attractive to seek the connections between the classical E- and B-fields and the quantum metric distortion in the following way:

$$\begin{aligned}\frac{\left(f_{,x}\right)^2+C_{tx}\left(3\left(f_{,t}\right)^2-2\left(C_1+f\right)f_{,t,t}\right)}{\left(C_1+f\right)^6}&=-\frac{\kappa_x}{2}\cdot\left(\varepsilon_0\cdot E_1^2+\frac{B_1^2}{\mu_0}\right)\\\frac{\left(f_{,y}\right)^2+C_{ty}\left(3\left(f_{,t}\right)^2-2\left(C_1+f\right)f_{,t,t}\right)}{\left(C_1+f\right)^6}&=-\frac{\kappa_y}{2}\cdot\left(\varepsilon_0\cdot E_2^2+\frac{B_2^2}{\mu_0}\right)\\\frac{\left(f_{,z}\right)^2+C_{tz}\left(3\left(f_{,t}\right)^2-2\left(C_1+f\right)f_{,t,t}\right)}{\left(C_1+f\right)^6}&=-\frac{\kappa_z}{2}\cdot\left(\varepsilon_0\cdot E_3^2+\frac{B_3^2}{\mu_0}\right)\end{aligned}. \tag{604}$$

Thus, obviously we have created photonic matter from a flat space vacuum solution to the Einstein field equations by the means of a simple metric transformation. The scalar Ricci curvature was thereby kept unchanged, which is to say $R=R^*$. The equations necessary to solve this were the classical quantum Klein-Gordon equations, which we directly obtain from the $R=R^*$-condition.

14.1.4 Solid Body Mechanics in the Metric Picture

It was shown in [2] that the typical equation of elasticity can also be derived when extending the classical variation in the following way:

$$\begin{aligned}
&\delta g_{\delta\gamma}=\delta\left(\mathbf{g}_\delta\cdot\mathbf{g}_\gamma\right)=\mathbf{g}_\delta\cdot\delta\mathbf{g}_\gamma+\delta\mathbf{g}_\delta\cdot\mathbf{g}_\gamma\\
&=\frac{\partial G^i\left[x_k\right]}{\partial x^\delta}\mathbf{e}_i\cdot\left(\frac{\partial^2 G^j\left[x_k\right]}{\partial x^\beta\partial x^\gamma}\mathbf{e}_j\right)\delta x^\beta+\left(\frac{\partial^2 G^i\left[x_k\right]}{\partial x^\chi\partial x^\delta}\mathbf{e}_i\right)\delta x^\chi\cdot\frac{\partial G^j\left[x_k\right]}{\partial x^\gamma}\mathbf{e}_j\\
&=\frac{\partial G^i\left[x_k\right]}{\partial x^\delta}\mathbf{e}_i\cdot\left(\frac{\partial^2 G^j\left[x_k\right]}{\partial x^\gamma\partial x^\gamma}\mathbf{e}_j\right)C^\gamma_\beta\delta x^\beta+\left(\frac{\partial^2 G^i\left[x_k\right]}{\partial x^\chi\partial x^\delta}\mathbf{e}_i\right)\delta x^\chi\cdot\frac{\partial G^j\left[x_k\right]}{\partial x^\gamma}\mathbf{e}_j\\
&=\frac{\partial G^i\left[x_k\right]}{\partial x^\delta}\mathbf{e}_i\cdot\left(\frac{\partial^2 G^j\left[x_k\right]}{\left(\partial x^\gamma\right)^2}\mathbf{e}_j\right)C^\gamma_\beta\delta x^\beta+\left(\frac{\partial^2 G^i\left[x_k\right]}{\partial x^\chi\partial x^\delta}\mathbf{e}_i\right)\delta x^\beta C^\chi_\beta\cdot C^\delta_\gamma\frac{\partial G^j\left[x_k\right]}{\partial x^\delta}\mathbf{e}_j
\end{aligned}\quad.\tag{605}$$

So far it is not clear to the authors how this could be connected to our scaled metric derivation from above, but it appears intriguing that elastic equations seem to pop up in connection with the base vector of the metric (general relativity) and the scale factor (quantum part). Before moving on and investigating this connection, however, we first need to repeat some essentials and try to work out what might be missing.

14.1.5 Alternative Way for Deriving the Equations of Elasticity from the Metric Origin out of the Einstein-Hilbert-Action

When observing (605), the experienced reader recognizes the structural elements of the basic equation of elasticity (e.g., [64], pp. 166 or [65]), which can be given for an isotropic material with the Poisson's ratio μ as:

$$\left(\overbrace{\left(1-2\cdot\mu\right)}^{\hat{=}a}\cdot\Delta G^j\left[x_k\right]+\left(\frac{\partial^2 G^j\left[x_k\right]}{\partial x^\gamma\partial x^\delta}\right)\right)\mathbf{e}_j=0\,.\tag{606}$$

A simple exchange of the dummy indices in (605) leads us to:

$$\begin{aligned}
&\delta g_{\delta\gamma}=\delta\left(\mathbf{g}_\delta\cdot\mathbf{g}_\gamma\right)=\mathbf{g}_\delta\cdot\delta\mathbf{g}_\gamma+\delta\mathbf{g}_\delta\cdot\mathbf{g}_\gamma\\
&=\frac{\partial G^i\left[x_k\right]}{\partial x^\delta}\mathbf{e}_i\cdot\left(\frac{\partial^2 G^j\left[x_k\right]}{\left(\partial x^\gamma\right)^2}\mathbf{e}_j\right)C^\gamma_\beta\delta x^\beta+\left(\frac{\partial^2 G^i\left[x_k\right]}{\partial x^\chi\partial x^\delta}\mathbf{e}_i\right)\delta x^\beta C^\chi_\beta\cdot C^\delta_\gamma\frac{\partial G^j\left[x_k\right]}{\partial x^\delta}\mathbf{e}_j\\
&=\frac{\partial G^j\left[x_k\right]}{\partial x^\delta}\mathbf{e}_j\cdot\left(\frac{\partial^2 G^i\left[x_k\right]}{\left(\partial x^\gamma\right)^2}\mathbf{e}_i\right)C^\gamma_\beta\delta x^\beta+\left(\frac{\partial^2 G^i\left[x_k\right]}{\partial x^\chi\partial x^\delta}\mathbf{e}_i\right)\delta x^\beta C^\chi_\beta\cdot C^\delta_\gamma\frac{\partial G^j\left[x_k\right]}{\partial x^\delta}\mathbf{e}_j\\
&=\left(\left(\frac{\partial^2 G^i\left[x_k\right]}{\left(\partial x^\gamma\right)^2}\mathbf{e}_i\right)C^\gamma_\beta+\left(\frac{\partial^2 G^i\left[x_k\right]}{\partial x^\chi\partial x^\delta}\mathbf{e}_i\right)C^\chi_\beta C^\delta_\gamma\right)\cdot\delta x^\beta\frac{\partial G^j\left[x_k\right]}{\partial x^\delta}\mathbf{e}_j
\end{aligned}\quad,\tag{607}$$

and assuming $C_{\beta}^{\gamma} = b \cdot \delta_{\beta}^{\gamma}, C_{\beta}^{\chi} C_{\gamma}^{\delta} = b^2 \cdot \delta_{\beta}^{\chi}\delta_{\gamma}^{\delta}, a = 1/b$ yields the isotropic equation:

$$\begin{aligned}
\delta g_{\delta\gamma} &= \delta(\mathbf{g}_\delta \cdot \mathbf{g}_\gamma) = \mathbf{g}_\delta \cdot \delta\mathbf{g}_\gamma + \delta\mathbf{g}_\delta \cdot \mathbf{g}_\gamma \\
&= \left(\left(\frac{\partial^2 G^i[x_k]}{(\partial x^\gamma)^2} \mathbf{e}_i \right) C_\beta^\gamma + \left(\frac{\partial^2 G^i[x_k]}{\partial x^\chi \partial x^\delta} \mathbf{e}_i \right) C_\beta^\chi C_\gamma^\delta \right) \cdot \delta x^\beta \frac{\partial G^j[x_k]}{\partial x^\delta} \mathbf{e}_j \\
&= b^2 \cdot \left(\frac{\Delta G^i[x_k]\mathbf{e}_i}{b} + \left(\frac{\partial^2 G^i[x_k]}{\partial x^\gamma \partial x^\delta} \mathbf{e}_i \right) \right) \cdot \delta x^\beta \frac{\partial G^j[x_k]}{\partial x^\delta} \mathbf{e}_j \\
&= b^2 \cdot \left(a \cdot \Delta G^i[x_k]\mathbf{e}_i + \left(\frac{\partial^2 G^i[x_k]}{\partial x^\gamma \partial x^\delta} \mathbf{e}_i \right) \right) \cdot \delta x^\beta \frac{\partial G^j[x_k]}{\partial x^\delta} \mathbf{e}_j
\end{aligned} \quad . \tag{608}$$

With the n-function-ansatz from [2] we can solve:

$$a \cdot \Delta G^j[x_k]\mathbf{e}_j + \left(\frac{\partial^2 G^j[x_k]}{\partial x^\gamma \partial x^\delta} \mathbf{e}_j \right) = \left(\overbrace{(1 - 2 \cdot \mu)}^{\triangleq a} \cdot \Delta G^j[x_k] + \left(\frac{\partial^2 G^j[x_k]}{\partial x^\gamma \partial x^\delta} \right) \right) \mathbf{e}_j = 0 \tag{609}$$

with α=-1-2*a.

As in connection with the general equation:

$$\begin{aligned}
\delta g_{\delta\gamma} &= \delta(\mathbf{g}_\delta \cdot \mathbf{g}_\gamma) = \mathbf{g}_\delta \cdot \delta\mathbf{g}_\gamma + \delta\mathbf{g}_\delta \cdot \mathbf{g}_\gamma \\
&= \frac{\partial G^i[x_k]}{\partial x^\delta} \mathbf{e}_i \cdot \left(\frac{\partial^2 G^j[x_k]}{\partial x^\alpha \partial x^\gamma} \mathbf{e}_j \right) \delta x^\alpha + \left(\frac{\partial^2 G^i[x_k]}{\partial x^\alpha \partial x^\delta} \mathbf{e}_i \right) \delta x^\alpha \cdot \frac{\partial G^j[x_k]}{\partial x^\gamma} \mathbf{e}_j \\
&= \frac{\partial G^i[x_k]}{\partial x^\delta} \mathbf{e}_i \cdot \sum_{\alpha=0}^{n-1} \left(\frac{\partial^2 G^j[x_k]}{\partial x^\alpha \partial x^\gamma} \mathbf{e}_j \right) \delta x^\alpha + \sum_{\alpha=0}^{n-1} \left(\frac{\partial^2 G^i[x_k]}{\partial x^\alpha \partial x^\delta} \mathbf{e}_i \right) \delta x^\alpha \cdot \frac{\partial G^j[x_k]}{\partial x^\gamma} \mathbf{e}_j \\
&= \frac{\partial G^i[x_k]}{\partial x^\delta} \mathbf{e}_i \cdot \sum_{\alpha=0}^{n-1} \left(\frac{\partial^2 G^j[x_k]}{\partial x^\alpha \partial x^\gamma} \mathbf{e}_j \right) \delta^\alpha + \sum_{\alpha=0}^{n-1} \left(\frac{\partial^2 G^i[x_k]}{\partial x^\alpha \partial x^\delta} \mathbf{e}_i \right) \delta^\alpha \cdot \frac{\partial G^j[x_k]}{\partial x^\gamma} \mathbf{e}_j \\
&\Rightarrow \sum_{\alpha=0}^{n-1} \left(\frac{\partial^2 G^j[x_k]}{\partial x^\alpha \partial x^\gamma} \mathbf{e}_j \right) \delta^\alpha = 0
\end{aligned} \tag{610}$$

from [32], it should be pointed out that with non-equal δx^α or δ^α for the various α, we could even derive anisotropic equations, respectively equations for anisotropic space-times.

It should be noted that, as shown before in [2] (see subsection "4.1. Matrix Option and Classical Dirac-Form"), one could also extract a Dirac-like equation from (608). This could be obtained via $\delta x^\beta \to \delta x^\delta$ leading to:

$$
\begin{aligned}
&\delta g_{\delta\gamma}=\delta\left(\mathbf{g}_{\delta}\cdot\mathbf{g}_{\gamma}\right)=\mathbf{g}_{\delta}\cdot\delta\mathbf{g}_{\gamma}+\delta\mathbf{g}_{\delta}\cdot\mathbf{g}_{\gamma}\\
&=b^{2}\cdot\left(a\cdot\Delta G^{i}\left[x_{k}\right]\mathbf{e}_{i}+\left(\frac{\partial^{2}G^{i}\left[x_{k}\right]}{\partial x^{\gamma}\partial x^{\delta}}\mathbf{e}_{i}\right)\right)\cdot\delta x^{\beta}\frac{\partial G^{j}\left[x_{k}\right]}{\partial x^{\delta}}\mathbf{e}_{j}\\
&=b^{2}\cdot\left(a\cdot\Delta G^{i}\left[x_{k}\right]\mathbf{e}_{i}+\left(\frac{\partial^{2}G^{i}\left[x_{k}\right]}{\partial x^{\gamma}\partial x^{\delta}}\mathbf{e}_{i}\right)\right)\cdot\delta x^{\delta}\frac{\partial G^{j}\left[x_{k}\right]}{\partial x^{\delta}}\mathbf{e}_{j}\\
&\Rightarrow\sum_{\delta=0}^{n-1}\delta x^{\delta}\frac{\partial G^{j}\left[x_{k}\right]}{\partial x^{\delta}}\mathbf{e}_{j}=0
\end{aligned}
\quad . \qquad (611)
$$

We have to point out that, even though these Dirac-like equations are anything but close to classical elasticity equations, they still give displacement distributions of spaces and might automatically include defects (like dislocations) which require quite some math-construction in the classical technical mechanics. Thus, our purely metric approach here may even find some use in practical applications of material science and elsewhere (see figures below, which are evaluated with a generalized version of our software FilmDoctor [108]).

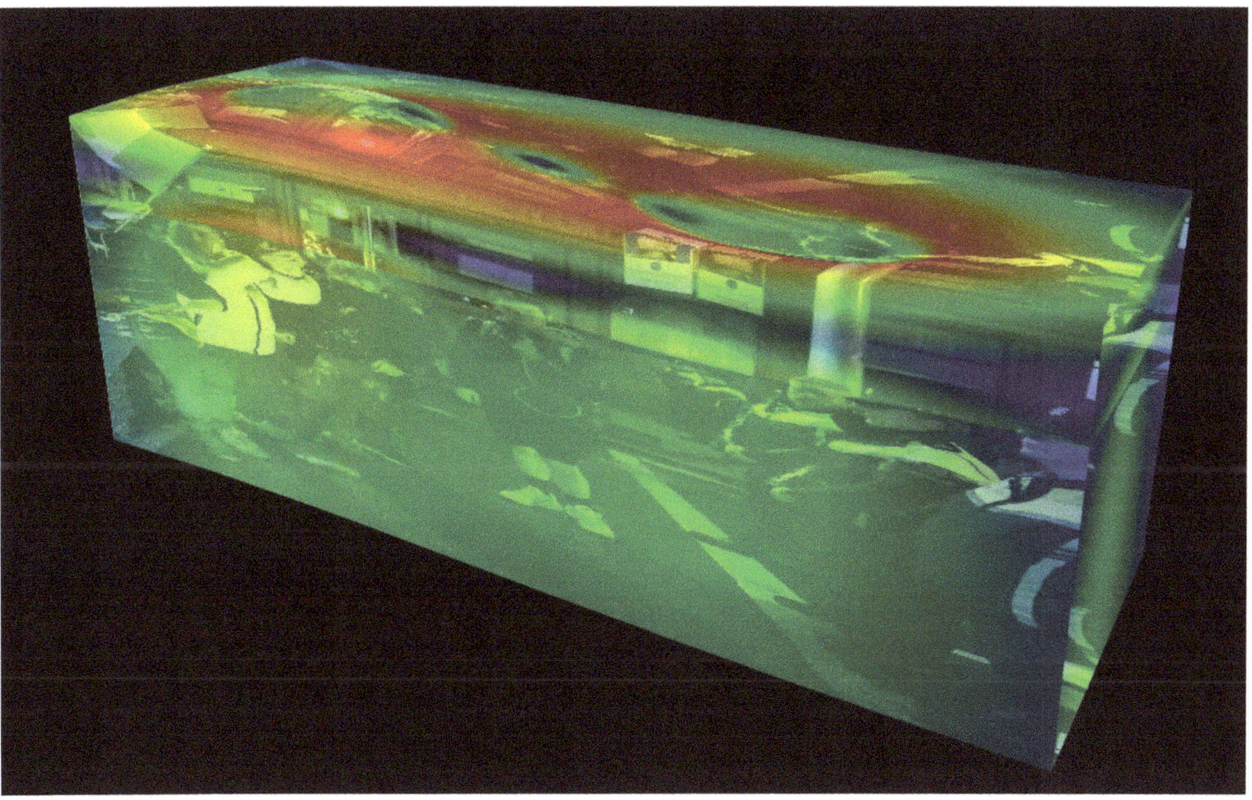

Fig. 46: Example for Socioeconomic Polarization – Picture 1 (see main text in this section and [108]).

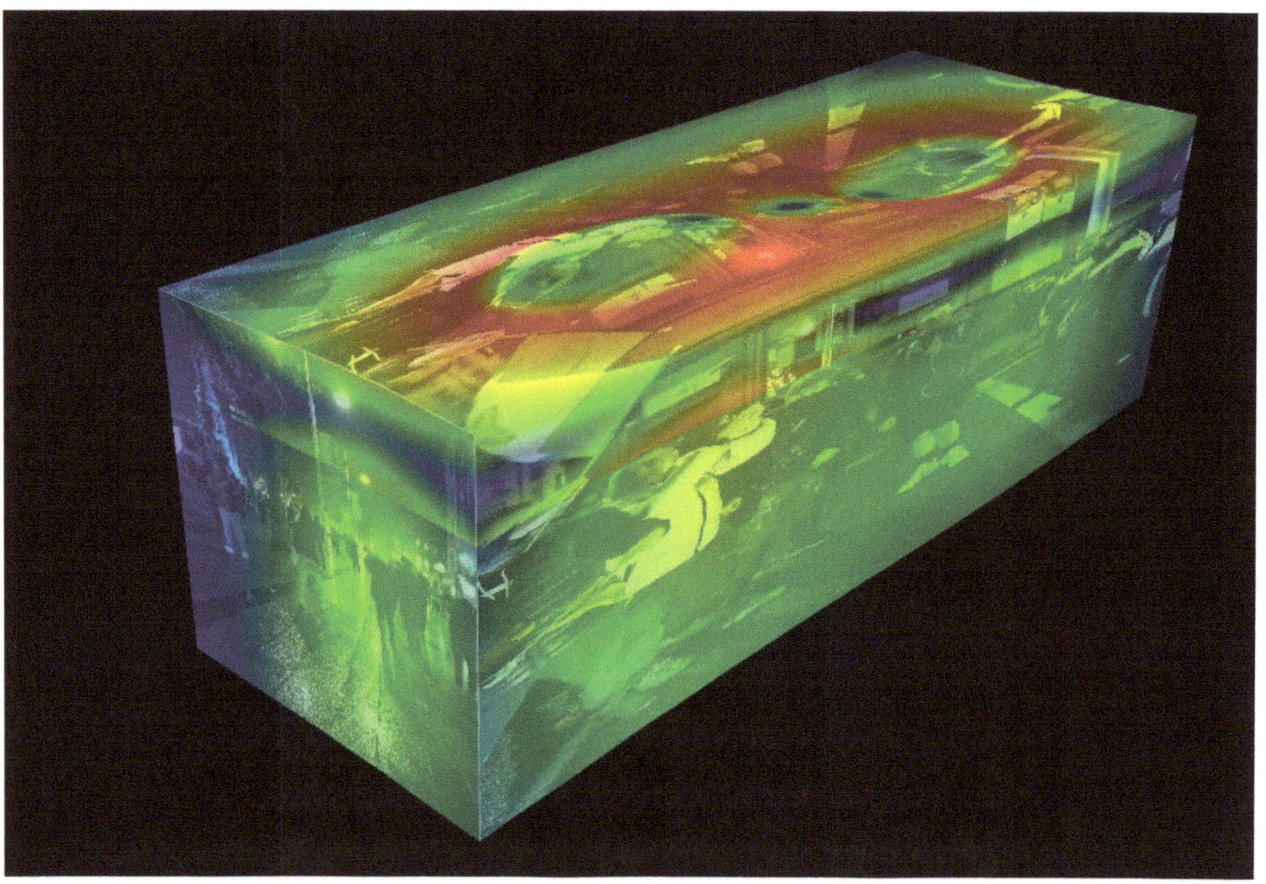

Fig. 47: Example for Socioeconomic Polarization – Picture 2 (see main text in this section and [108]).

14.2 The Key Aspects for the Second Tower

14.2.1 Metric Diffusion Path A: Evolution, Quantum Gravity Thermodynamics, Sigmoid-Curves, and Turbulences

14.2.1.1 Where Is Thermodynamics?

Having worked out the similarity of quantum gravity and the theory of elasticity in the last subsection has also allowed us to try and find other fundamental fields inside the already derived general equation(s). Such general considerations have been performed in [2]. Here we repeat the question about the quantum gravity origin of thermodynamics.

But why would we need a quantum gravity originated and thus, completely non-postulated thermodynamics for our innovations T1 and T2?

Well, the strangeness of the observations in connection with the two innovations, most prominent among the two, however, the second tower, forces us to also question certain dogma in classical or, let us better say "established" or "conventional" thermodynamics. In rederiving the latter from scratch, we want to make sure that we do not burden us with old bias, potentially compromising the whole approach.

It is well known from the thermodynamics of deformed bodies that the work of deformation can be described as the inner product of the stress tensor σ^{jk} with the deformation tensor u_{jk}:

$$W_D = \sigma^{jk} u_{jk} . \tag{612}$$

Knowing that the functions G^j from the subsection above are displacement fields and closely investigating them, we conclude that the variation of the metric tensor (611) is nothing else than the holistic equivalent of a deformation tensor and thus, that the integrand in the classical and the extended Einstein-Hilbert action is giving us nothing else but a very general form of deformation work.

If this is the case, however, and especially as we assume the Einstein-Hilbert action to "contain it all" (at least in its extended form), we must ask where we could find the equivalent to thermal energy? After all, sticking to the simpler classical action form, which reads:

$$\left(R^{\delta\gamma} - \frac{1}{2} R g^{\delta\gamma} + \Lambda g^{\delta\gamma} + \kappa T^{\delta\gamma} \right) \delta g_{\delta\gamma} , \tag{613}$$

none of the addends in parenthesis has the makings of a term T*dS or S*dT with S denoting the entropy, and T denoting temperature.

Obviously, we are missing something… at least as long as we consider thermodynamics as so fundamental that it has to show up in a truly fundamental theory.

Thus, with thermodynamics being missing inside the classical metric and the newer base-vector variation of the Einstein-Hilbert action, we have to look for a different way to obtain what we need.

14.2.1.2 The Base-Vector Variation

Our starting point shall be the usual tensor transformation rule for the covariant metric tensor:

$$g_{\delta\gamma} = \mathbf{g}_\delta \cdot \mathbf{g}_\gamma = \frac{\partial G^\alpha [x_k]}{\partial x^\delta} \frac{\partial G^\beta [x_k]}{\partial x^\gamma} g_{\alpha\beta} . \tag{614}$$

This, as we already saw above, leads to quite some varieties with respect to the apparently simple term $\delta g_{\mu\nu}$ (e.g., see [2]). The base vectors $\mathbf{g}_\delta$ to a certain metric are given as:

$$\mathbf{g}_\delta = \frac{\partial G^\alpha [x_k]}{\partial x^\delta} \mathbf{e}_\alpha , \tag{615}$$

where the functions $G^\alpha[\ldots]$ denote arbitrary functions of the coordinates x_k. Here the vectors $\mathbf{e}_\alpha$ shall denote the base vectors of a fundamental coordinate system of the right (in principle arbitrary) number of dimensions. Thus, we have the variation for $\delta g_{\delta\gamma}$ as follows:

$$\delta g_{\delta\gamma} = \delta(\mathbf{g}_\delta \cdot \mathbf{g}_\gamma) = \mathbf{g}_\delta \cdot \delta\mathbf{g}_\gamma + \delta\mathbf{g}_\delta \cdot \mathbf{g}_\gamma$$
$$= \frac{\partial G^\alpha[x_k]}{\partial x^\delta}\mathbf{e}_\alpha \cdot \delta\left(\frac{\partial G^\beta[x_k]}{\partial x^\gamma}\mathbf{e}_\beta\right) + \delta\left(\frac{\partial G^\alpha[x_k]}{\partial x^\delta}\mathbf{e}_\alpha\right) \cdot \frac{\partial G^\beta[x_k]}{\partial x^\gamma}\mathbf{e}_\beta . \tag{616}$$

Now we introduce two additional degrees of freedom, namely:

a) that neither the number of dimensions in which the base vectors exist and form a complete transformation (615) needs to be the same as the metric space they define,
b) nor that the variation δ is defined or fixed in any way.

In order to properly account for point a), we shall rewrite (614) and (615) as:

$$\delta g_{\delta\gamma} = \delta(\mathbf{g}_\delta \cdot \mathbf{g}_\gamma) = \mathbf{g}_\delta \cdot \delta\mathbf{g}_\gamma + \delta\mathbf{g}_\delta \cdot \mathbf{g}_\gamma$$
$$= \frac{\partial G^i[x_k]}{\partial x^\delta}\mathbf{e}_i \cdot \delta\left(\frac{\partial G^j[x_k]}{\partial x^\gamma}\mathbf{e}_j\right) + \delta\left(\frac{\partial G^i[x_k]}{\partial x^\delta}\mathbf{e}_i\right) \cdot \frac{\partial G^j[x_k]}{\partial x^\gamma}\mathbf{e}_j . \tag{617}$$
$$\mathbf{g}_\delta = \frac{\partial G^j[x_k]}{\partial x^\delta}\mathbf{e}_j$$

Thereby the Latin indices are running to a different (potentially higher) number of dimensions N than the Greek indices, which shall be defined for a space or space-time of n dimensions.

Now, having extended the options for the metric and its base vectors in such a way, which is to say by allowing them to have some kind of arbitrary intrinsic structure, we should ask ourselves whether this should not also be possible for our quantum scale factors F[f]. As already shown in connection with the solutions to the elastic equations, we may assume intrinsic structures and various quantum centers (see [2], section "Centers of Gravity and Quantum Centers"). The latter then automatically leads to ensembles of F[f]s and thus, to statistics and – necessarily – thermodynamics. In order to better understand this, we here repeat the corresponding considerations with respect to the metric or base vector variation from [2].

14.2.1.3 The Variation with Respect to Ensemble Parameters

Reconsideration of the classical Einstein-Hilbert action according to [2], thereby incorporating our degrees of freedom (617), leads to:

$$\delta_g W = 0 = \int_V d^n x \overbrace{\left(R^{\delta\gamma} - \frac{1}{2}Rg^{\delta\gamma} + \Lambda g^{\delta\gamma} + \kappa T^{\delta\gamma}\right)}^{\text{Relativity}}$$
$$\times\left(\underbrace{\frac{\partial G^i[x_k]}{\partial x^\delta}\mathbf{e}_i \cdot \delta\left(\frac{\partial G^j[x_k]}{\partial x^\gamma}\mathbf{e}_j\right) + \delta\left(\frac{\partial G^i[x_k]}{\partial x^\delta}\mathbf{e}_i\right) \cdot \frac{\partial G^j[x_k]}{\partial x^\gamma}\mathbf{e}_j}_{\text{Quantum}} + \underbrace{\frac{\partial g_{\delta\gamma}}{\partial n}\delta n}_{\text{ThD}}\right) . \tag{618}$$

We have assumed that we have a closed system and that therefore the amount of information, respectively the number of dimensions should not change, meaning $\frac{\partial g_{\alpha\beta}}{\partial n}\delta n=0$. What still could change, respectively be variated, however, would be the number of i-j-ensembles and the corresponding base-vectors. For instance, we could understand these ensembles as objects with different centers of gravity, like:

$$\frac{\partial G^{j}\left[x_{k}\right]}{\partial x^{\gamma}}\mathbf{e}_{j}=\frac{\partial G^{j}\left[x_{k}-\xi_{kj}\right]}{\partial x^{\gamma}}\mathbf{e}_{j}, \tag{619}$$

with the ξ_{kj} denoting the various centers of gravity. Now one could perform the variation with respect to exactly these coordinates and find (thereby simplifying via our condition of the closed system $\frac{\partial g_{\alpha\beta}}{\partial n}\delta n=0$):

$$X_{k}\equiv x_{k}-\xi_{k};\quad \delta_{g}W=\int_{V}d^{n}x\overbrace{\left(R^{\delta\gamma}-\frac{1}{2}Rg^{\delta\gamma}+\Lambda g^{\delta\gamma}+\kappa T^{\delta\gamma}\right)}^{\text{Relativity}}$$
$$\times\left(\begin{array}{c}\underbrace{\frac{\partial G^{i}[X_{k}]}{\partial x^{\delta}}\mathbf{e}_{i}\cdot\delta_{x}\left(\frac{\partial G^{j}[X_{k}]}{\partial x^{\gamma}}\mathbf{e}_{j}\right)+\delta_{x}\left(\frac{\partial G^{i}[X_{k}]}{\partial x^{\delta}}\mathbf{e}_{i}\right)\cdot\frac{\partial G^{j}[X_{k}]}{\partial x^{\gamma}}\mathbf{e}_{j}}_{\text{Quantum}}\\ \underbrace{+\frac{\partial G^{i}[X_{k}]}{\partial x^{\delta}}\mathbf{e}_{i}\cdot\delta_{\xi}\left(\frac{\partial G^{j}[X_{k}]}{\partial x^{\gamma}}\mathbf{e}_{j}\right)+\delta_{\xi}\left(\frac{\partial G^{i}[X_{k}]}{\partial x^{\delta}}\mathbf{e}_{i}\right)\cdot\frac{\partial G^{j}[X_{k}]}{\partial x^{\gamma}}\mathbf{e}_{j}}_{\text{"?"}\Rightarrow 2^{\text{nd}}\text{ law of Thermodynamics}}\end{array}\right). \tag{620}$$

We realize that the additional variation is just the positioning of the first derivatives of metric displacements and their vectors. In other words: This variation is about positioning the individual centers of gravity within the metric space-time and their vectors of changes against the various coordinates x. In our "Science Riddle 20" [66] we discussed the effect of chaotic distributions for huge ensembles. We came to the conclusion that states are statistically favored where the various displacements cancel each other out. This then leads to a disappearance of the term "?" in (620). Interestingly, however, such equal distributions of vector orientations and magnitudes also coincides with states with maximum classical entropy, where we exactly obtain ?=0 (c.f. figure 48). Thus, it is clear: The "?" must stand for thermodynamics and that it is a driving force for the second law of thermodynamics.

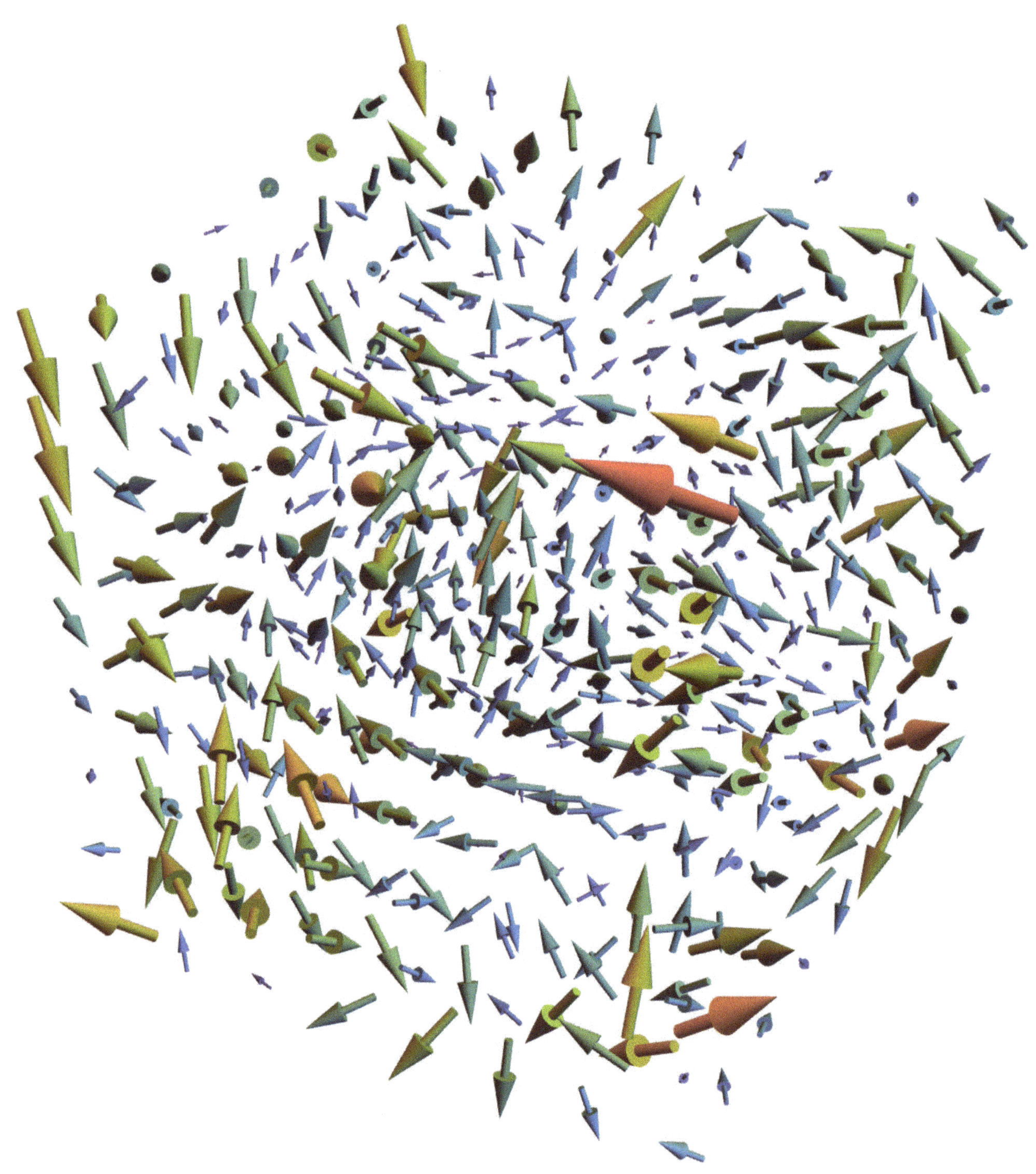

Fig. 48: Random distribution of displacement vector field for G^j in order to achieve a structural solution to (620). The solution also results in maximum entropy.

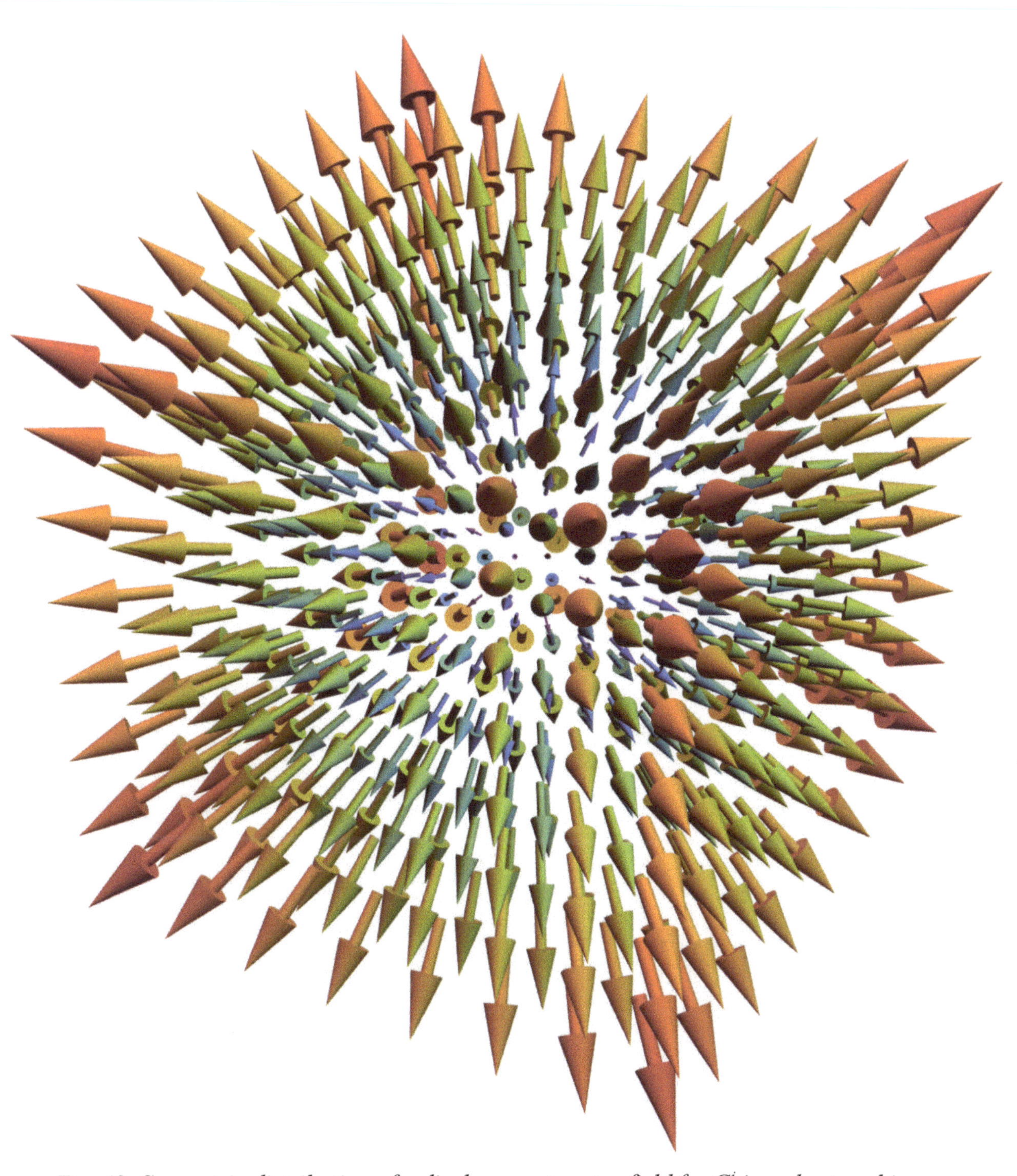

Fig. 49: Symmetric distribution of a displacement vector field for G^j in order to achieve a structural solution to (620). The solution corresponds to a non-extremal entropy.

However, there is also the chance to make this variational term to zero with suitable symmetrical structures of the ensembles also canceling each other out (figure 49 and 50). Then the term "?" would stand for a force of "self-organization" and/or self-structuring of the very system (or parts of it). The whole could also be achieved as a proper combination of the "?"- and the "quantum"-term which then has to give a resulting zero in sum.

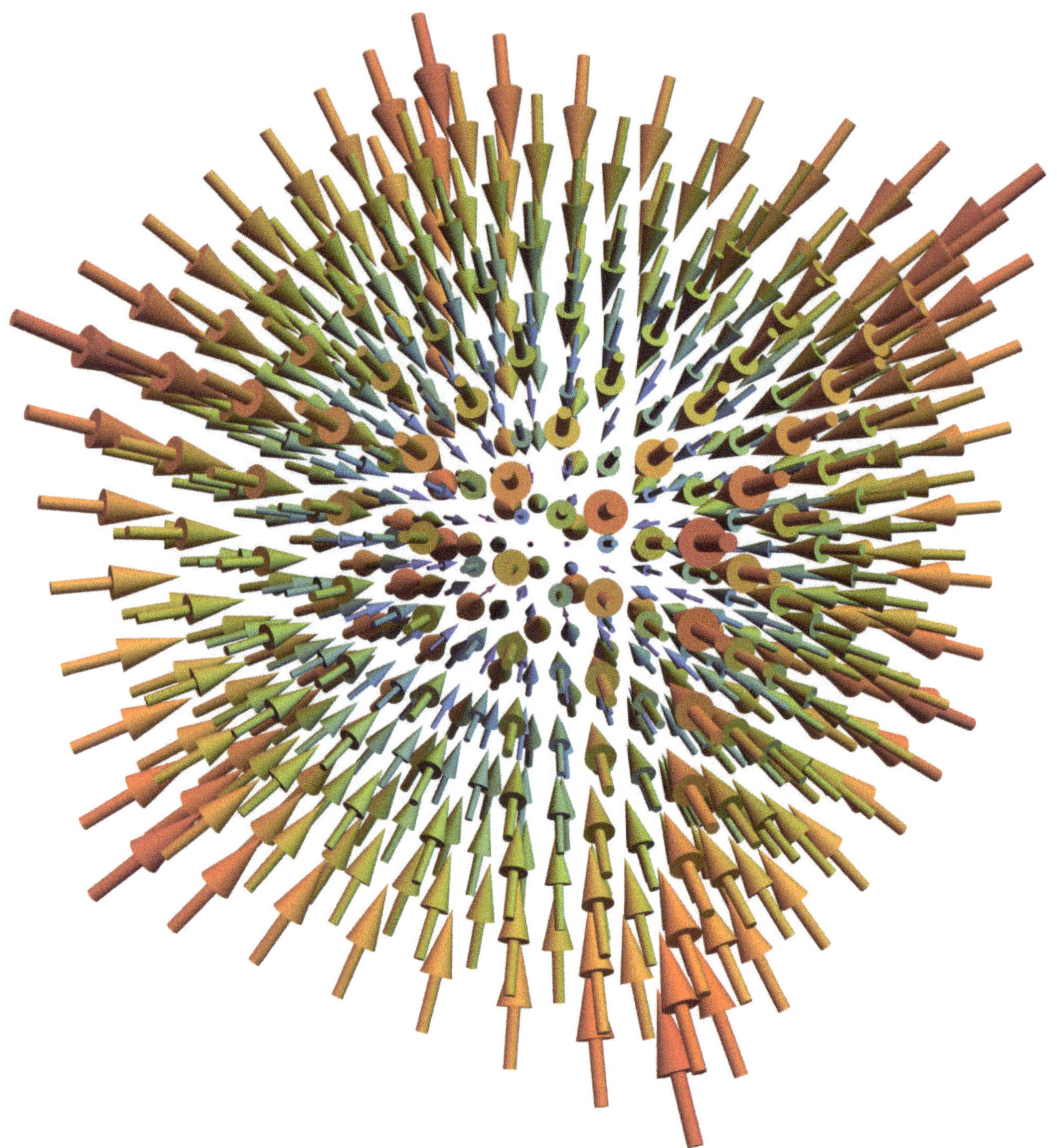

Fig. 50: Symmetric distribution of a displacement vector field for G^j in order to achieve a structural solution to (620). The solution corresponds to a non-extremal entropy.

In this whole context we also make out the parameter N (the number to which the indices i and j can run) as yet another quantity defining upon the form of enclosure of a given system. With n and N being fixed, we should speak about a truly closed system in the classical meaning.

From all that has been elaborated above we conclude that the term "?" does not only present the metric manifestation of the 2nd law of thermodynamics, but also gives the reason for the self-organization vector and evolution we observe within our universe as a rather omnipresent property.

With this we might want to rewrite (620) as follows:

$$X_k \equiv x_k - \xi_k; \quad \delta_g W = \int_V d^n x \overbrace{\left(R^{\delta\gamma} - \frac{1}{2} R g^{\delta\gamma} + \Lambda g^{\delta\gamma} + \kappa T^{\delta\gamma} \right)}^{\text{Relativity}} \times \left(\begin{array}{c} \underbrace{\frac{\partial G^i[X_k]}{\partial x^\delta} \mathbf{e}_i \cdot \delta_x \left(\frac{\partial G^j[X_k]}{\partial x^\gamma} \mathbf{e}_j \right) + \delta_x \left(\frac{\partial G^i[X_k]}{\partial x^\delta} \mathbf{e}_i \right) \cdot \frac{\partial G^j[X_k]}{\partial x^\gamma} \mathbf{e}_j}_{\text{Quantum}} \\ \underbrace{+ \frac{\partial G^i[X_k]}{\partial x^\delta} \mathbf{e}_i \cdot \delta_\xi \left(\frac{\partial G^j[X_k]}{\partial x^\gamma} \mathbf{e}_j \right) + \delta_\xi \left(\frac{\partial G^i[X_k]}{\partial x^\delta} \mathbf{e}_i \right) \cdot \frac{\partial G^j[X_k]}{\partial x^\gamma} \mathbf{e}_j}_{\text{Thermodynamics \& Evolution}} \end{array} \right). \tag{621}$$

It should be pointed out that the structure also allows an easy incorporation of interaction among the various centers of gravity (e.g., [66]).

In [3] we considered a variety of options with respect to the variation of ensemble parameters. To cover at least a few of these options before moving on to distinguish between centers of gravity and quantum centers (see below), we here repeat some of our earlier results.

14.2.1.3.1 Ordinary Derivative Variation and THE Ideal Gas

It should be noted that performing the variation as ordinary derivation and with the simple setting for X_k as set in (621), the variation of the quantum and the thermo-evolution term can cancel each other out, because we could perform it as follows:

$$\begin{aligned} \delta_x \left(\frac{\partial G^j[X_k]}{\partial x^\gamma} \mathbf{e}_j \right) &= \frac{\partial}{\partial x^\beta} \left(\frac{\partial G^j[X_k]}{\partial x^\gamma} \mathbf{e}_j \right) \delta x^\beta \\ \delta_\xi \left(\frac{\partial G^j[X_k]}{\partial x^\gamma} \mathbf{e}_j \right) &= \frac{\partial}{\partial \xi^\beta} \left(\frac{\partial G^j[X_k]}{\partial x^\gamma} \mathbf{e}_j \right) \delta \xi^\beta = \frac{\partial}{\partial x^\gamma} \left(\frac{\partial G^j[X_k]}{\partial \xi^\beta} \mathbf{e}_j \right) \delta \xi^\beta \end{aligned}. \tag{622}$$

Now we assume that we can perform the following simplification in the second line:

$$\frac{\partial}{\partial \xi^\beta} \left(\frac{\partial G^j[X_k]}{\partial x^\gamma} \mathbf{e}_j \right) \delta \xi^\beta = \left(\frac{\partial^2 G^j[X_k]}{\partial x^\gamma \partial \xi^\beta} \mathbf{e}_j \right) \delta \xi^\beta = \left(- \frac{\partial^2 G^j[X_k]}{\partial x^\gamma \partial x^\beta} \mathbf{e}_j \right) \delta x^\beta . \tag{623}$$

It has to be pointed out that this is not possible in cases of position dependent base vectors, for instance. However, in Cartesian coordinates we could also simplify the first line of (622) and then obtain in sum:

$$\begin{array}{c}\left(\frac{\partial^2 G^j[X_k]}{\partial x^\beta \partial x^\gamma}\mathbf{e}_j\right)\delta x^\beta \\ \Rightarrow \delta_x\left(\frac{\partial G^j[X_k]}{\partial x^\gamma}\mathbf{e}_j\right)+\delta_\xi\left(\frac{\partial G^j[X_k]}{\partial x^\gamma}\mathbf{e}_j\right)=0\end{array}, \tag{624}$$

which automatically gives us:

$$0=\left(\begin{array}{c}\underbrace{\frac{\partial G^i[X_k]}{\partial x^\delta}\mathbf{e}_i\cdot\delta_x\left(\frac{\partial G^j[X_k]}{\partial x^\gamma}\mathbf{e}_j\right)+\delta_x\left(\frac{\partial G^i[X_k]}{\partial x^\delta}\mathbf{e}_i\right)\cdot\frac{\partial G^j[X_k]}{\partial x^\gamma}\mathbf{e}_j}_{\text{Quantum}} \\ +\underbrace{\frac{\partial G^i[X_k]}{\partial x^\delta}\mathbf{e}_i\cdot\delta_\xi\left(\frac{\partial G^j[X_k]}{\partial x^\gamma}\mathbf{e}_j\right)+\delta_\xi\left(\frac{\partial G^i[X_k]}{\partial x^\delta}\mathbf{e}_i\right)\cdot\frac{\partial G^j[X_k]}{\partial x^\gamma}\mathbf{e}_j}_{\text{Thermodynamics \& Evolution}}\end{array}\right). \tag{625}$$

In order to avoid such a triviality, we either shall demand the variation with respect to ξ_k to be different, or we set X_k as follows:

$$X_k \equiv x_k - f_k[\xi_k]. \tag{626}$$

Obviously already the simplest linear form like $f[\xi_k]=m_k*\xi_k$ suffices to force both terms in (622) to give zero independently, which, having assumed the same structural variation with respect to ξ_k and x_k, only requires the solution of one. This only leaves statistical laws to take care about the distribution of the various i-j-gravity centers. There is no interaction.

As example we want to consider solutions to the following variational outcome (from [67]):

$$\sum_{\alpha=0}^{n-1}\frac{\partial^2 G^j[x_k]}{\partial x^\alpha \partial x^\gamma}\mathbf{e}_j = 0 = \sum_{\alpha=0}^{n-1}\frac{\partial^2 G^i[x_k]}{\partial x^\alpha \partial x^\delta}\mathbf{e}_i. \tag{627}$$

Even when introducing point solutions from [27], reading for a single center at $\xi_k=0$:

$$G^j[x_k]=\pm C_s\cdot\left(\begin{array}{c}\frac{t}{\left(x^2+y^2+z^2\right)^2\left(1+\frac{t^2}{x^2+y^2+z^2}\right)} \\ +\frac{\arctan\left[\frac{t}{\sqrt{x^2+y^2+z^2}}\right]}{\left(x^2+y^2+z^2\right)^{3/2}}\end{array}\right)\cdot\{t,x,y,z\}, \tag{628}$$

we still do not get anything but an identical solution already to the quantum term (regarding the evaluation in connection with the variation we have to refer to [67]). Thereby C_s stands for a suitable constant. We recognize the potential of a point charge as limit for t→∞. The resulting t→∞-limit would then read:

$$\lim_{t\to\infty} G^j[x_k] = \mp C_s \cdot \frac{\pi}{2} \cdot \frac{1}{\left(x^2+y^2+z^2\right)^{3/2}} \cdot \{0, x, y, z\}. \tag{629}$$

Assuming two equal centers placed at $\pm x_0$ would give us:

$$\lim_{t\to\infty} G^j[x_k] = \mp C_s \cdot \frac{\pi}{2} \cdot \frac{1}{\left((x \pm x_0)^2+y^2+z^2\right)^{3/2}} \cdot \{0, x \pm x_0, y, z\}, \tag{630}$$

already provides the fulfillment of the quantum term, which is guaranteed due to our solution (628) to (627) and thus, both terms in (622). This means that even the presence of "point charges" would not force our system to follow anything else but statistics, which leads to the 2nd law of thermodynamics and does neither provide any interaction nor evolutionary driving forces.

We might see this approach above therefore as a quantum gravity model for an ideal gas.

14.2.1.3.2 Combined Successive Variation

Things are getting significantly different the moment we allow the variation to be performed somehow simultaneously or in a successive manner, but then we have variations of second order and these are considered small of second order. So, here is an example:

$$\begin{aligned}
&\delta_\xi\left(\frac{\partial G^i[X_k]}{\partial x^\delta}\mathbf{e}_i \cdot \delta_x\left(\frac{\partial G^j[X_k]}{\partial x^\gamma}\mathbf{e}_j\right) + \delta_x\left(\frac{\partial G^i[X_k]}{\partial x^\delta}\mathbf{e}_i\right) \cdot \frac{\partial G^j[X_k]}{\partial x^\gamma}\mathbf{e}_j\right) \\
&= \delta_\xi\left(\frac{\partial G^i[X_k]}{\partial x^\delta}\mathbf{e}_i\right) \cdot \delta_x\left(\frac{\partial G^j[X_k]}{\partial x^\gamma}\mathbf{e}_j\right) + \delta_\xi\delta_x\left(\frac{\partial G^i[X_k]}{\partial x^\delta}\mathbf{e}_i\right) \cdot \frac{\partial G^j[X_k]}{\partial x^\gamma}\mathbf{e}_j \\
&+ \frac{\partial G^i[X_k]}{\partial x^\delta}\mathbf{e}_i \cdot \delta_\xi\delta_x\left(\frac{\partial G^j[X_k]}{\partial x^\gamma}\mathbf{e}_j\right) + \delta_x\left(\frac{\partial G^i[X_k]}{\partial x^\delta}\mathbf{e}_i\right) \cdot \delta_\xi\left(\frac{\partial G^j[X_k]}{\partial x^\gamma}\mathbf{e}_j\right)
\end{aligned} \tag{631}$$

Assuming all double variations to be small of higher order, we could simplify:

$$\begin{aligned}
&\delta_\xi\left(\frac{\partial G^i[X_k]}{\partial x^\delta}\mathbf{e}_i \cdot \delta_x\left(\frac{\partial G^j[X_k]}{\partial x^\gamma}\mathbf{e}_j\right) + \delta_x\left(\frac{\partial G^i[X_k]}{\partial x^\delta}\mathbf{e}_i\right) \cdot \frac{\partial G^j[X_k]}{\partial x^\gamma}\mathbf{e}_j\right) \\
&= \delta_\xi\left(\frac{\partial G^i[X_k]}{\partial x^\delta}\mathbf{e}_i\right) \cdot \delta_x\left(\frac{\partial G^j[X_k]}{\partial x^\gamma}\mathbf{e}_j\right) + \delta_x\left(\frac{\partial G^i[X_k]}{\partial x^\delta}\mathbf{e}_i\right) \cdot \delta_\xi\left(\frac{\partial G^j[X_k]}{\partial x^\gamma}\mathbf{e}_j\right)
\end{aligned} \tag{632}$$

Still, we have obtained products of variated terms that we might also consider small of second order and thus, for here and now we are not going to consider this path any further.

Thus, we conclude that our form of rather simple (even though extended) variation by the means of ordinary derivatives does not do the job and we have to investigate more general options.

14.2.1.3.3 Incorporating Interaction

It is evident that with the introduction of centers of gravity we should also consider the option for interaction among those various centers. This would simply be achieved by allowing the G^j to not only depend on the differential coordinates X_k as given in (620), but also to have dependencies as:

$$G^j[\ldots]=G^j\left[X_{kj}\right]\equiv G^j\left[X_{k0},X_{k1},\ldots,X_{ki},\ldots,X_{k(N-1)}\right];\quad X_{ki}\equiv x_k-\xi_{ki}\,. \tag{633}$$

This immediately renders the variation with respect to the relative or position coordinates extremely complex even in the smallest ensembles. But this is not so different from the usual complexity known from statistical mechanics. Here now we just face a quantum gravity interactive statistics, where the complete variation should read:

$$\delta_g W=\int_V d^n x\overbrace{\left(R^{\delta\gamma}-\frac{1}{2}Rg^{\delta\gamma}+\Lambda g^{\delta\gamma}+\kappa T^{\delta\gamma}\right)}^{\text{Relativity}}\times\left(\begin{array}{c}\underbrace{\frac{\partial G^i\left[X_{ki}\right]}{\partial x^\delta}\mathbf{e}_i\cdot\delta_x\left(\frac{\partial G^j\left[X_{kj}\right]}{\partial x^\gamma}\mathbf{e}_j\right)+\delta_x\left(\frac{\partial G^i\left[X_{ki}\right]}{\partial x^\delta}\mathbf{e}_i\right)\cdot\frac{\partial G^j\left[X_{kj}\right]}{\partial x^\gamma}\mathbf{e}_j}_{\text{Quantum}}\\+\underbrace{\sum_{i,j=0}^{N-1}\left(\frac{\partial G^i\left[X_{ki}\right]}{\partial x^\delta}\mathbf{e}_i\cdot\delta_{\xi j}\left(\frac{\partial G^j\left[X_{kj}\right]}{\partial x^\gamma}\mathbf{e}_j\right)+\delta_{\xi i}\left(\frac{\partial G^i\left[X_{ki}\right]}{\partial x^\delta}\mathbf{e}_i\right)\cdot\frac{\partial G^j\left[X_{kj}\right]}{\partial x^\gamma}\mathbf{e}_j\right)}_{2^{\text{nd}}\text{ law of Thermodynamics \& Interaction}}\end{array}\right). \tag{634}$$

Summing up all variations considered here (meaning without the scale factor) in just one world formula also for N- and n-variable (perhaps open) systems would now give us the following two forms [2]:

$$\delta_g W = \left(\overbrace{\left(\int_V d^n x \left([R - 2\kappa L_M - 2\Lambda] \cdot \frac{\partial \sqrt{-g}}{\partial n} + \sqrt{-g} \cdot \frac{\partial [R - 2\kappa L_M - 2\Lambda]}{\partial n} \right) \right) \delta n}^{\text{Thermodynamics / Exchange / Open Systems}} + \int_V d^n x \overbrace{\left(R^{\delta\gamma} - \frac{1}{2} R g^{\delta\gamma} + \Lambda g^{\delta\gamma} + \kappa T^{\delta\gamma} \right)}^{\text{Relativity}} \right.$$

$$\times \left(\begin{array}{c} \underbrace{\frac{\partial G^i [X_{ki}]}{\partial x^\delta} \mathbf{e}_i \cdot \delta_x \left(\frac{\partial G^j [X_{kj}]}{\partial x^\gamma} \mathbf{e}_j \right) + \delta_x \left(\frac{\partial G^i [X_{ki}]}{\partial x^\delta} \mathbf{e}_i \right) \cdot \frac{\partial G^j [X_{kj}]}{\partial x^\gamma} \mathbf{e}_j}_{\text{Quantum}} + \underbrace{\frac{\partial g_{\delta\gamma}}{\partial n} \delta n}_{\text{ThD}} \\ \underbrace{+ \delta_N \left(\sum_{i,j=0}^{N-1} \frac{\partial G^i [X_{ki}]}{\partial x^\delta} \mathbf{e}_i \cdot \delta_{\xi j} \left(\frac{\partial G^j [X_{kj}]}{\partial x^\gamma} \mathbf{e}_j \right) + \delta_{\xi i} \left(\frac{\partial G^i [X_{ki}]}{\partial x^\delta} \mathbf{e}_i \right) \cdot \frac{\partial G^j [X_{kj}]}{\partial x^\gamma} \mathbf{e}_j \right)}_{2^{nd} \text{ law of Thermodynamics \& Interaction}} \end{array} \right) \quad (635)$$

or somewhat simpler with the N-variation as separate addend:

$$\delta_g W = \left(\overbrace{\left(\int_V d^n x \left([R - 2\kappa L_M - 2\Lambda] \cdot \frac{\partial \sqrt{-g}}{\partial n} + \sqrt{-g} \cdot \frac{\partial [R - 2\kappa L_M - 2\Lambda]}{\partial n} \right) \right) \delta n}^{\text{Thermodynamics / Exchange / Open Systems}} + \int_V d^n x \overbrace{\left(R^{\delta\gamma} - \frac{1}{2} R g^{\delta\gamma} + \Lambda g^{\delta\gamma} + \kappa T^{\delta\gamma} \right)}^{\text{Relativity}} \right.$$

$$\times \left(\begin{array}{c} \underbrace{\frac{\partial G^i [X_{ki}]}{\partial x^\delta} \mathbf{e}_i \cdot \delta_x \left(\frac{\partial G^j [X_{kj}]}{\partial x^\gamma} \mathbf{e}_j \right) + \delta_x \left(\frac{\partial G^i [X_{ki}]}{\partial x^\delta} \mathbf{e}_i \right) \cdot \frac{\partial G^j [X_{kj}]}{\partial x^\gamma} \mathbf{e}_j}_{\text{Quantum}} + \underbrace{\frac{\partial g_{\delta\gamma}}{\partial n} \delta n}_{\text{ThD}} \\ \underbrace{+ \frac{\partial g_{\alpha\beta}}{\partial N} \delta_N + \sum_{i,j=0}^{N-1} \left(\begin{array}{c} \frac{\partial G^i [X_{ki}]}{\partial x^\delta} \mathbf{e}_i \cdot \delta_{\xi j} \left(\frac{\partial G^j [X_{kj}]}{\partial x^\gamma} \mathbf{e}_j \right) \\ + \delta_{\xi i} \left(\frac{\partial G^i [X_{ki}]}{\partial x^\delta} \mathbf{e}_i \right) \cdot \frac{\partial G^j [X_{kj}]}{\partial x^\gamma} \mathbf{e}_j \end{array} \right)}_{2^{nd} \text{ law of Thermodynamics \& Interaction}} \end{array} \right). \quad (636)$$

14.2.1.4 Thermodynamics of T2 or a Form-Based Thermodynamics

With the derivation of thermodynamics as a variational result from the Einstein-Hilbert action based in quantum and gravity centers, it can be deduced from the phenomenological characteristics of T2 that the process-relevant thermodynamics of the Polar Selective Agent, PSA, is determined by the individual positions intrinsic agent surfaces and their positions. Herein not only lays the fundamental understanding of the process but also its simulation and optimization with respect to a wider range of potential applications in many very different fields.

14.2.1.4.1 Derivation of the Diffusion Equation

As a very simple example we here want to repeat the essentials of the derivation of the diffusion equation from the fundamental equation (636) and demand the following boundary conditions to be fulfilled:

$$\begin{gathered}\int_V d^n x\left(\left[R-2\kappa L_M-2\Lambda\right]\cdot\frac{\partial\sqrt{-g}}{\partial n}+\sqrt{-g}\cdot\frac{\partial\left[R-2\kappa L_M-2\Lambda\right]}{\partial n}\right)\cdot\delta n=0\\ \frac{\partial g_{\alpha\beta}}{\partial n}\delta n=0\\ \frac{\partial G^i\left[X_{ki}\right]}{\partial x^\delta}\mathbf{e}_i\cdot\delta_x\left(\frac{\partial G^j\left[X_{kj}\right]}{\partial x^\gamma}\mathbf{e}_j\right)+\delta_x\left(\frac{\partial G^i\left[X_{ki}\right]}{\partial x^\delta}\mathbf{e}_i\right)\cdot\frac{\partial G^j\left[X_{kj}\right]}{\partial x^\gamma}\mathbf{e}_j=0\end{gathered}\quad,\qquad(637)$$

which results in the remaining equation:

$$\frac{\partial g_{\delta\gamma}}{\partial N}\delta_N+\sum_{i,j=0}^{N-1}\left(\frac{\partial G^i\left[X_{ki}\right]}{\partial x^\delta}\mathbf{e}_i\cdot\delta_{\xi j}\left(\frac{\partial G^j\left[X_{kj}\right]}{\partial x^\gamma}\mathbf{e}_j\right)+\delta_{\xi i}\left(\frac{\partial G^i\left[X_{ki}\right]}{\partial x^\delta}\mathbf{e}_i\right)\cdot\frac{\partial G^j\left[X_{kj}\right]}{\partial x^\gamma}\mathbf{e}_j\right)=0\,.(638)$$

For the reason of simplicity, we ignore any particle interaction and reduce (638) to:

$$\frac{\partial g_{\delta\gamma}}{\partial N}\delta_N+\frac{\partial G^i\left[X_k\right]}{\partial x^\delta}\mathbf{e}_i\cdot\delta_\xi\left(\frac{\partial G^j\left[X_k\right]}{\partial x^\gamma}\mathbf{e}_j\right)+\delta_\xi\left(\frac{\partial G^i\left[X_k\right]}{\partial x^\delta}\mathbf{e}_i\right)\cdot\frac{\partial G^j\left[X_k\right]}{\partial x^\gamma}\mathbf{e}_j=0,\qquad(639)$$

with the definition for X_k as given in (620). Similar to our evaluation from the section “Motivation” in [50, 68] (see also [2], subsection “2.2. Intelligent Zero Approaches – Just one Example”), we perform the variation of the second and third addend in (639) as follows:

$$\begin{aligned}G^j\left[X_k\right]\equiv G^j;\quad&\frac{\partial G^i}{\partial x^\delta}\mathbf{e}_i\cdot\delta_\xi\left(\frac{\partial G^j}{\partial x^\gamma}\mathbf{e}_j\right)+\delta_\xi\left(\frac{\partial G^i}{\partial x^\delta}\mathbf{e}_i\right)\cdot\frac{\partial G^j}{\partial x^\gamma}\mathbf{e}_j\\ &=-C_\delta^\gamma\frac{\partial G^i}{\partial x^\gamma}\mathbf{e}_i\cdot\left(\frac{\partial^2 G^j}{\partial x^\sigma\partial x^\gamma}\mathbf{e}_j\right)\delta^\sigma-\left(\frac{\partial^2 G^i}{\partial x^\sigma\partial x^\delta}\mathbf{e}_i\right)\delta^\sigma\cdot\frac{\partial G^j}{\partial x^\gamma}\mathbf{e}_j\end{aligned}.\qquad(640)$$

Exchanging the dummy indices and assuming $C_\delta^\gamma=b\cdot\delta_\delta^\gamma$ leads to:

$$=-\left(C^{\gamma}_{\delta}\left(\frac{\partial^2 G^i}{\partial x^\sigma \partial x^\gamma}\mathbf{e}_i\right)+\left(\frac{\partial^2 G^i}{\partial x^\sigma \partial x^\delta}\mathbf{e}_i\right)\right)\delta^\sigma\cdot\frac{\partial G^j}{\partial x^\gamma}\mathbf{e}_j$$

$$=-(b+1)\left(\frac{\partial^2 G^i}{\partial x^\sigma \partial x^\delta}\mathbf{e}_i\right)\delta^\sigma\cdot\frac{\partial G^j}{\partial x^\gamma}\mathbf{e}_j=-(b+1)C^{\delta}_{\sigma}\left(\frac{\partial^2 G^i}{\partial x^\sigma \partial x^\delta}\mathbf{e}_i\right)\delta^\sigma\cdot\frac{\partial G^j}{\partial x^\gamma}\mathbf{e}_j$$

$$=-(b+1)\cdot b\cdot\delta^{\delta}_{\sigma}\left(\frac{\partial^2 G^i}{\partial x^\sigma \partial x^\delta}\mathbf{e}_i\right)\delta^\sigma\cdot\frac{\partial G^j}{\partial x^\gamma}\mathbf{e}_j=-(b+1)\cdot b\left(\frac{\partial^2 G^i}{\left(\partial x^\sigma\right)^2}\mathbf{e}_i\right)\delta^\sigma\cdot\frac{\partial G^j}{\partial x^\gamma}\mathbf{e}_j \quad . \qquad (641)$$

$$=-(b+1)\cdot b\left(\frac{\partial^2 G^i\mathbf{e}_i}{\left(\partial x^\sigma\right)^2}\right)\delta^\sigma\cdot\frac{\partial G^j}{\partial x^\gamma}\mathbf{e}_j=-(b+1)\cdot b\cdot\Delta\left(G^i\mathbf{e}_i\right)\delta^\sigma\cdot\frac{\partial G^j}{\partial x^\gamma}\mathbf{e}_j$$

Now we assume the variation regarding N to result in zeros with respect to all spatial derivatives. The justification for this is the typical continuity equation approach where we demand that under the volume considered in our variational integral (636) any change of N does not change the derivatives with respect to the spatial coordinates, meaning:

$$\frac{\partial g_{\delta\gamma}}{\partial N}\delta_N=\frac{\partial\left(\frac{\partial G^j}{\partial x^\gamma}\mathbf{e}_j\frac{\partial G^i}{\partial x^\delta}\mathbf{e}_i\right)}{\partial N}\delta_N=0;\quad \delta,\gamma\neq 0\,. \qquad (642)$$

With respect to the time-coordinate, however, we can have very complex dependencies. The simplest would be a linear one like:

$$\frac{\partial\left(\frac{\partial G^j}{\partial x^0}\mathbf{e}_j\right)}{\partial N}\delta_N=D^{-1}\cdot\frac{\partial G^j}{\partial x^0}\mathbf{e}_j\cdot\delta^\delta. \qquad (643)$$

This gives us the following for the first addend:

$$\frac{\partial g_{\delta\gamma}}{\partial N}\delta_N=D^{-1}\cdot\left(\frac{\partial G^i}{\partial x^\delta}\mathbf{e}_i\cdot\frac{\partial G^j}{\partial x^0}\mathbf{e}_j+\frac{\partial G^i}{\partial x^0}\mathbf{e}_i\cdot\frac{\partial G^j}{\partial x^\gamma}\mathbf{e}_j\right)\cdot\delta^\delta. \qquad (644)$$

In connection with the result from (641), again using $C^{\gamma}_{\delta}=b\cdot\delta^{\gamma}_{\delta}$ and exchanging the dummy indices, we result in:

$$0=D^{-1}\cdot\left(\frac{\partial G^i}{\partial x^\delta}\mathbf{e}_i\cdot\frac{\partial G^j}{\partial x^0}\mathbf{e}_j+\frac{\partial G^i}{\partial x^0}\mathbf{e}_i\cdot\frac{\partial G^j}{\partial x^\gamma}\mathbf{e}_j\right)\cdot\delta^\sigma-(b+1)\cdot b\cdot\Delta\left(G^i\mathbf{e}_i\right)\delta^\sigma\cdot\frac{\partial G^j}{\partial x^\gamma}\mathbf{e}_j$$

$$=D^{-1}\cdot(b+1)\cdot\left(\frac{\partial G^i}{\partial x^0}\mathbf{e}_i\right)\cdot\frac{\partial G^j}{\partial x^\gamma}\mathbf{e}_j\cdot\delta^\sigma-(b+1)\cdot b\cdot\Delta\left(G^i\mathbf{e}_i\right)\delta^\sigma\cdot\frac{\partial G^j}{\partial x^\gamma}\mathbf{e}_j \quad , \qquad (645)$$

$$=(b+1)\cdot\left(\left(\frac{\partial G^i}{\partial x^0}\mathbf{e}_i\right)-D\cdot b\cdot\Delta\left(G^i\mathbf{e}_i\right)\right)\delta^\sigma\cdot\frac{\partial G^j}{\partial x^\gamma}\mathbf{e}_j$$

$$\Rightarrow\quad\left(\frac{\partial G^i}{\partial x^0}\mathbf{e}_i\right)=D\cdot b\cdot\Delta\left(G^i\mathbf{e}_i\right)$$

where we recognize the diffusion equation in its simplest (homogeneous) form. Along the way of our derivation, we can easily make out passages where we can incorporate anisotropy and inhomogeneity in very general manners.

14.2.2 Diffusion Equation Path B

14.2.2.1 Matter and Antimatter Diffusion Equation for Flat AND Curved Space-Times

One of the most important classical evolution equations is the diffusion equation. Its classical derivation (e.g., in form of the heat or concentration transfer equation in [69], chapter 25.5) only knows a scalar form. Here we will show that a fundamental derivation, thereby starting at the Einstein-Hilbert action, does not only show us the existence of quaternionic, respectively, Dirac-like diffusion equations, but also seems to demand corresponding matter and antimatter transfer processes. What this means with respect to practical applications will be discussed in this section… which is to say that we at least want to begin such a discussion here.

Before we can come to that, however, we have to repeat the decomposition of the results from (42) in the form:

$$0=R\cdot\left(C_f\pm f\right)\pm\frac{4\cdot(n-1)}{(n-2)}\left(\Delta_s f+\Delta_t f\right), \tag{646}$$

where – once again – we make use of condition:

$$4FF''+F'F'(n-6)=0\quad\Rightarrow\quad F=\left(C_f\pm f\right)^{\frac{4}{n-2}};\quad n>2. \tag{647}$$

To emphasize the relation to the Schrödinger equation we apply the Greek symbol Ψ for the function f. For simplicity we factor out the term 4/(n-2) and we also assume a Minkowski-like metric time component. Now and as already shown in [7], section "First Order Evolution Equations from the Einstein-Hilbert Action – Option A: The Schrödinger-Evolution-Equation", we can write (646) as follows:

$$0=\left(\Psi-C_{f0}\right)\cdot R+\frac{4}{(n-2)}\cdot(1-n)\cdot\left(\Delta_{n-1}\Psi-\frac{\partial^2\Psi}{c^2\cdot\partial t^2}\right)\quad n>2. \tag{648}$$

In the further derivation we ignore the n=2-case, where we would have the starting point:

$$0=R-C_{f0}\cdot\left(\Delta_{n-1}\Psi-\frac{\partial^2\Psi}{c^2\cdot\partial t^2}\right)\quad n=2\ \Rightarrow 0=R-C_{f0}\cdot\left(\frac{\partial^2}{\partial x^2}\Psi-\frac{\partial^2\Psi}{c^2\cdot\partial t^2}\right). \tag{649}$$

Now we Taylor-expand the curvature R, thereby assuming it to depend on the function Ψ:

$$0=\left(\Psi-C_{f0}\right)\cdot\left(R_{c1}\cdot\Psi+R_{c2}\cdot\Psi^2+\ldots\right)+\frac{4}{(n-2)}\cdot(1-n)\cdot\left(\Delta_{n-1}\Psi-\frac{\partial^2\Psi}{c^2\cdot\partial t^2}\right). \tag{650}$$

With Ψ being our diffusion function, this setting simply assumes that the curvature of our system depends on the diffusion processes taking place in it. Further assuming that all Ψ^k with k≥2 are small of higher order, we can simplify the last expression to:

$$0=\left(\Psi-C_{f0}\right)\cdot\mathrm{R_c}\cdot\Psi+\frac{4}{(\mathrm{n}-2)}\cdot(1-\mathrm{n})\cdot\left(\Delta_{\mathrm{n}-1}\Psi-\frac{\partial^2\Psi}{\mathrm{c}^2\cdot\partial\mathrm{t}^2}\right)$$

$$=\overbrace{\Psi\cdot\mathrm{R_c}\cdot\Psi}^{\approx 0}-C_{f0}\cdot\mathrm{R_c}\cdot\Psi+\frac{4}{(\mathrm{n}-2)}\cdot(1-\mathrm{n})\cdot\left(\Delta_{\mathrm{n}-1}\Psi-\frac{\partial^2\Psi}{\mathrm{c}^2\cdot\partial\mathrm{t}^2}\right), \tag{651}$$

$$=-C_{f0}\cdot\mathrm{R_c}\cdot\Psi+\frac{4}{(\mathrm{n}-2)}\cdot(1-\mathrm{n})\cdot\left(\Delta_{\mathrm{n}-1}\Psi-\frac{\partial^2\Psi}{\mathrm{c}^2\cdot\partial\mathrm{t}^2}\right)$$

$$\xrightarrow{C_{f0}\cdot\mathrm{R_c}=C_{\mathrm{R}}}$$

$$\Rightarrow\quad 0=\left[\frac{4}{(\mathrm{n}-2)}\cdot(1-\mathrm{n})\cdot\left(\Delta_{\mathrm{n}-1}-\frac{\partial^2}{\mathrm{c}^2\cdot\partial\mathrm{t}^2}\right)-C_{\mathrm{R}}\right]\Psi\cdot \tag{652}$$

We see that in contrast to the classical diffusion equation, where no constant term is present, we have the C_R, which came from the Ricci curvature. Thus, what we have just derived is nothing else but the source for a diffusion equation in a mildly curved space-time. We will consider the situation in strong curvatures at the end of this section.

Here now we proceed with the classical assumptions, where there is no curvature and so we set $C_R=0$. Then we can separate the time derivative and reformulate everything as follows:

$$\xrightarrow{\left(\Delta_{\mathrm{n}-1}-\frac{\partial^2}{\mathrm{c}^2\cdot\partial\mathrm{t}^2}\right)-\frac{4}{(\mathrm{n}-2)}\frac{C_{\mathrm{R}}=0}{(1-\mathrm{n})}=\mathrm{C1}\cdot\frac{\partial_{\mathrm{t}}^2}{\mathrm{c}^2}+\Delta_{\mathrm{n}-1}}$$

$$\Rightarrow 0=\Delta\Psi=\left(\mathrm{C1}\cdot\frac{\partial_{\mathrm{t}}^2}{\mathrm{c}^2}+\underbrace{\frac{1}{\sqrt{\mathrm{g}}}\partial_\alpha\sqrt{\mathrm{g}}\cdot\mathrm{g}^{\alpha\beta}\partial_\beta}_{\{\mathrm{n}-1\}\mathrm{Dim}-\Delta-\mathrm{Operator}}\right)\Psi\cdot \tag{653}$$

Please note that our Greek indices α and β are running only from 1 to n-1 and not – as usual – from 0 to n-1, because the time or 0-component has been separated. Now we can rigidly follow the receipt from [70], section "First Order Evolution Equations from the Einstein-Hilbert Action – Option A: The Schrödinger-Evolution-Equation", introduce a function $\Psi=\Phi+\mathrm{X}$ and demand the following additional condition:

$$\partial_{\mathrm{t}}\Psi=\mathrm{c}^2\cdot\mathrm{C2}\cdot(\Phi-\mathrm{X}). \tag{654}$$

Together with (653) we obtain:

$$0=\mathrm{C1}\cdot\mathrm{C2}\cdot\partial_{\mathrm{t}}(\Phi-\mathrm{X})+\frac{1}{\sqrt{\mathrm{g}}}\partial_\alpha\sqrt{\mathrm{g}}\cdot\mathrm{g}^{\alpha\beta}\partial_\beta(\Phi+\mathrm{X}). \tag{655}$$

The following two equations summed up would result in (655):

$$\begin{aligned} 0 &= C1\cdot C2\cdot\partial_t\Phi+\frac{1}{\sqrt{g}}\partial_\alpha\sqrt{g}\cdot g^{\alpha\beta}\partial_\beta\Phi \\ 0 &= \frac{1}{\sqrt{g}}\partial_\alpha\sqrt{g}\cdot g^{\alpha\beta}\partial_\beta X - C1\cdot C2\cdot\partial_t X \end{aligned}. \tag{656}$$

The attentive reader already spots the first order evolution equation, which we recognize easily when reshaping the last equations as follows:

$$\begin{aligned} C1\cdot C2\cdot\partial_t\Phi &= -\frac{1}{\sqrt{g}}\partial_\alpha\sqrt{g}\cdot g^{\alpha\beta}\partial_\beta\Phi \\ C1\cdot C2\cdot\partial_t X &= +\frac{1}{\sqrt{g}}\partial_\alpha\sqrt{g}\cdot g^{\alpha\beta}\partial_\beta X \end{aligned}, \tag{657}$$

and sees the similarities to the classical diffusion equation. Comparing with the original diffusion equation with diffusion coefficient D as given in the form below (e.g., [69]):

$$\left[\partial_t - D\cdot\Delta_{diff}\right]\Psi = 0, \tag{658}$$

does – once again – give us the matter and antimatter solutions. By having kept the symmetry and the subsequent Laplace operator general, we automatically also cover for complex diffusion equations with convection and other generalizations.

Setting the constants C1*C2=1/D:

$$C1\cdot C2=\frac{1}{D}\quad\Rightarrow\quad\begin{cases}\partial_t\Phi=-\frac{D}{\sqrt{g}}\partial_\alpha\sqrt{g}\cdot g^{\alpha\beta}\partial_\beta\Phi\\ \partial_t X=+\frac{D}{\sqrt{g}}\partial_\alpha\sqrt{g}\cdot g^{\alpha\beta}\partial_\beta X\end{cases}, \tag{659}$$

shows us that only the second equation with the function X would give us the classical dependency. We here see it as the ordinary matter diffusion. In consequence we have to state that the alternative equation, with the diffusion function Φ giving us an "unnatural" diffusion direction, is potentially violating the second law of thermodynamics.

We should therefore ignore it… or should we not?

Well, diffusion is a process we usually associate with huge numbers of particles. Thus, it would definitively make sense to exclude an "antimatter" diffusion against the statistically governing of the second law of thermodynamics. However, in smaller systems, with rather non-statistical behavior of the constituents, we definitely cannot exclude ordering processes which run against the "usual", which is to say entropy-defined, direction. In such cases we would then have to except the existence of two diffusion processes, namely one for X and one for Φ, with the latter becoming recessive in statistically governed systems.

14.2.2.1.1 About the Matter- Antimatter Asymmetry

That being said, one might even suggest that the observation of the matter- antimatter asymmetry in our universe [71] is nothing else but a statistical effect. So far it is not clear to the authors how one could mathematically put this idea into action, but the existence of matter and antimatter on one hand, plus their diffusion equation analogues on the other, and the statistical or entropy-related exclusion of transport processes in certain directions, just strongly motivates the association of certain forms of matter with the dynamics of sub-particles or sub-constituents. Seeing an elementary particle like the electron not as one entity moving through time, but as an ensemble of sub-particles in a dynamic process, it is clear why the corresponding antiparticle, the positron, then, which is to say in our sub-constituents-picture, cannot be stable. Its internal "diffusion" or "transport" process simply moves into the wrong – statistically or entropically wrong – time-direction. Of course, all matter, sporting matter, and antimatter options has then to be considered as some dynamic "thing", which actually is "doing something within the flow of time". One rather simple option here could just be the time-vector defining a movement parallel or antiparallel to the general universal expansion or the latter's acceleration [72].

14.2.2.1.2 What Is Life?... and Is the Polar Selective Agent Alive?

In the same context and almost as a by-product one might just answer the question about "what is life?" in a completely different way. Usually, the question is answered via the concept of entropy, thereby assuming that living systems decrease their entropy or at least maintain it due to energy consumption and interaction with an "entropy-absorbing" / entropy increasing environment [73]. This just perfectly agrees with our result from above when considering the antimatter diffusion process. Thus, with our findings here, we might also say that life behaves antimatter like. Then, however, antimatter has not truly disappeared during the creation and evolution process of the universe, but is just present as a structural manifestation within the matter itself.

With respect to our second question in the subsection headline, we could simply state that our Polar Selective Agent from the innovation T2 most definitively is not alive, because there are more criteria to be fulfilled than to follow an anti-diffusion equation; but, nevertheless, T2 has quite some potential to become alive, though. After all, it is an old question why the famous Miller experiment from 1952 (figure 51) could never produce true life, but only certain amino acids.

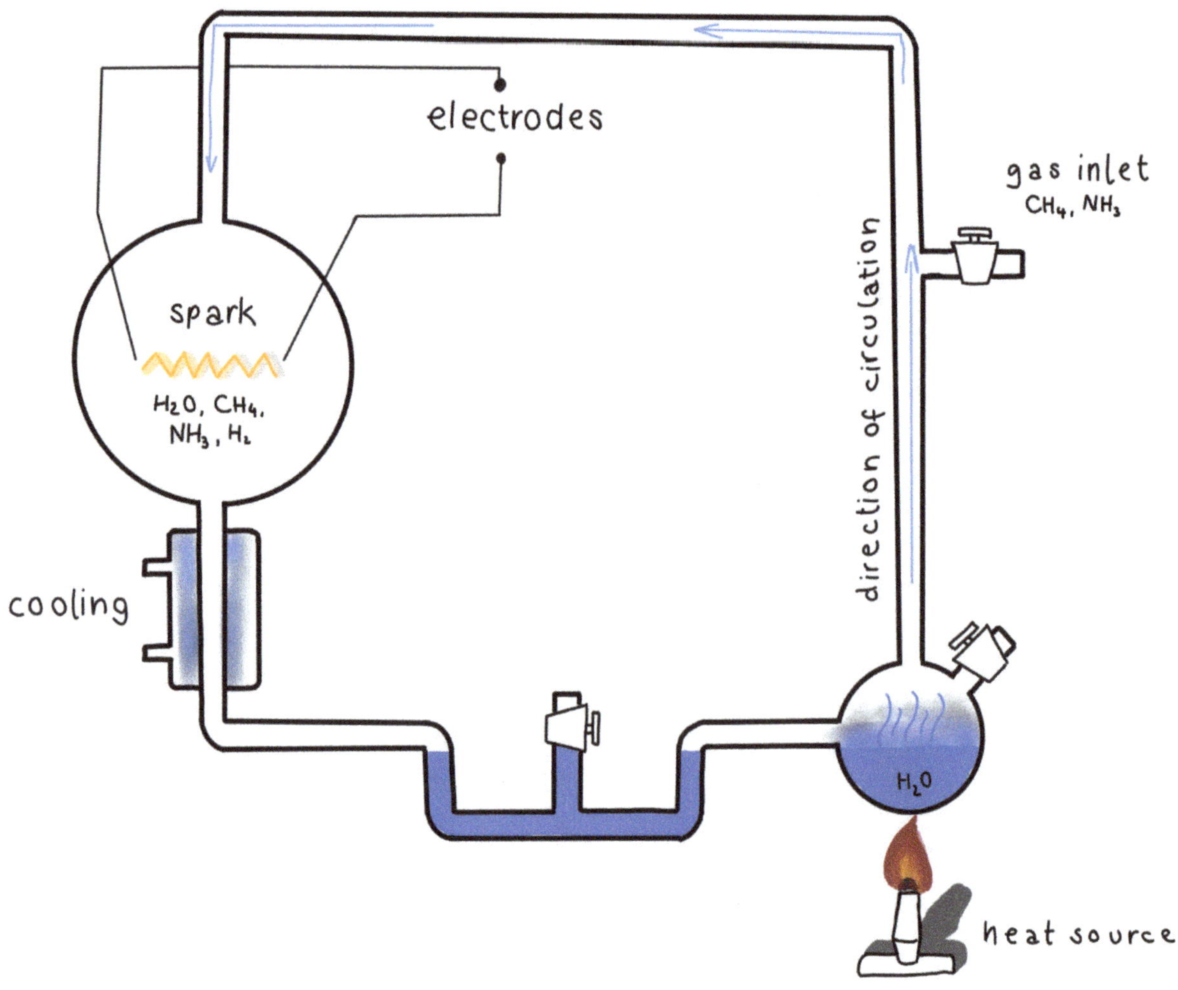

Fig. 51: Miller or Miller-Urey experiment from 1952. [A]

In order to work out the full potential of our innovation, we better make sure that we also cover the remotest application.

14.2.2.1.2.1 In the Beginning Was the Molecule

(from [94])

According to the current state of knowledge, organic life emerged from a purely molecular phase in which chemical interactions took place under the influence of the most diverse forms of energy (thermally through intensive volcanic activity of the early Earth, electrically through atmospheric discharges in the phase of primordial soup formation, electromagnetically through solar radiation and particulate through particle radiation from space and the solar wind), from which the first organic molecules and molecule ensembles capable of life and reproduction emerged in several steps (figure 52).

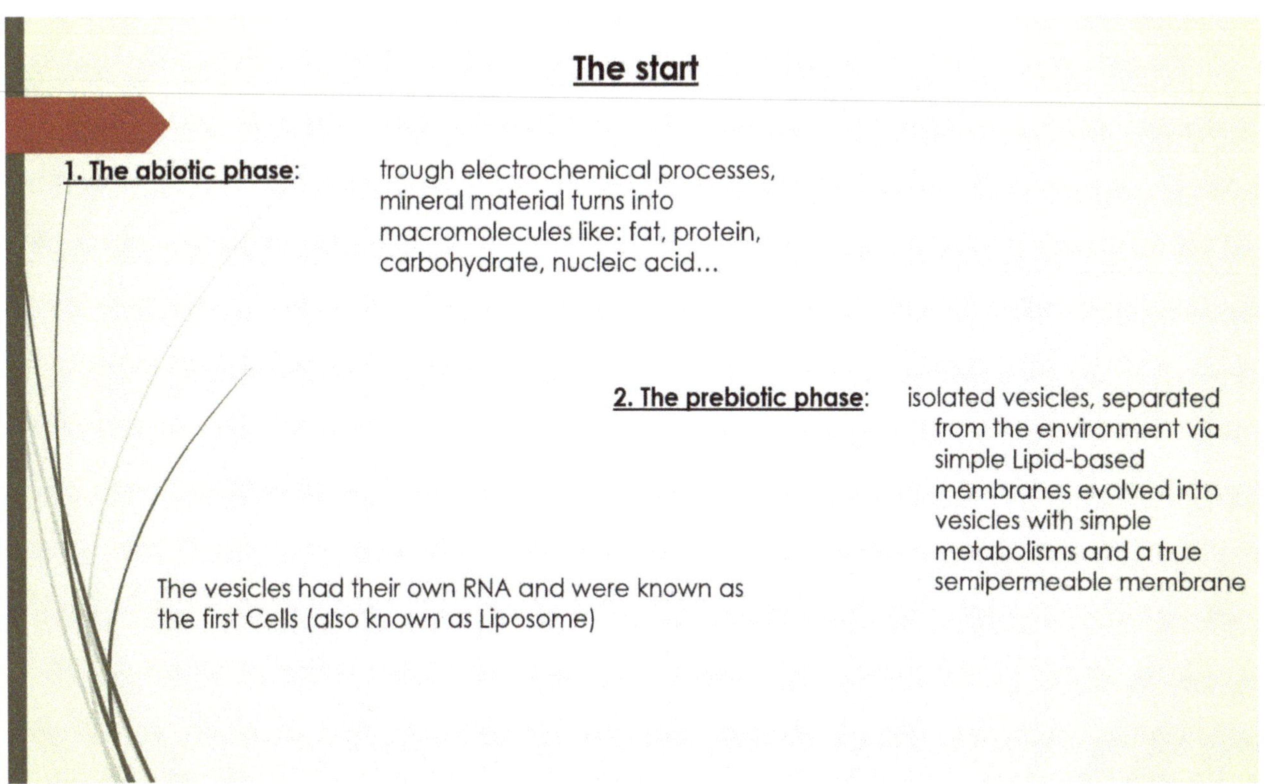

Fig. 52: The beginning of the origin of life with the abiotic phase and the prebiotic phase.

14.2.2.1.2.2 Abiotic Phase

As the Miller or Miller-Urey experiment of 1953 (figure 51) showed, a mixture of amino acids, the organic basic building blocks of life, could be produced from completely inorganic starting materials under the influence of energy.

14.2.2.1.2.3 Prebiotic Phase

The first molecular ensembles surrounded themselves with a lipid shell through so-called self-organizing processes and thus, formed the first liposomes.

14.2.2.1.2.4 But Something Was Always Missing

Even though the Miller experiment showed that the simplest ingredients of life, the amino acids could be produced in an abiotic process, it was never possible to obtain living systems. Something was always missing.

What if the missing piece was a Polar Selective Agent, which, as some kind of chemo-physical selector (or creator), assuring just the right components coming together, could provide this missing link to truly make life crawl out of Stanley Miller's reactor?

We, which is to say the authors, are going to further investigate this possibility.

14.2.2.1.3 Curved Space Diffusion Equations

Even though we concentrate on crucial aspects for the second tower key technology in this section, we sometimes find tools which are useful for both T1 and T2. The mathematical description of diffusion processes in curved space-times is such a tool. As already said above, in connection with the first and the second tower, we are in need of an unbiased description of diffusion processes anyway (see subsections "The Origin of Evolution" and "Metric Derivation of the Diffusion Equation" in main section "Creation and Separation – The Two Towers of Key Technology"). The fundamental derivation of the latter has not only given us the ordinary and the anti-diffusion, but will also provide us with diffusion equations in curved spaces. This is essential for the understanding of the two processes T1 and T2. Regarding T1, it helps us with the understanding why our deposition process has so much less problems on curved surfaces, sharp edges, and corners. With respect to T2 we will be able to describe diffusion through internal (but curved) interfaces inside a liquid agent.

Now we want to repeat the derivation of the diffusion equations from the previous section, but this time we will perform the evaluation for curved spaces. We also aim for a metric adjustment leading us to perfect Laplace operators as demonstrated in [7], section "Dissolving an Apparent Conflict". Our starting point shall be the following equation, resulting from the variation of the Einstein-Hilbert action with respect to the scaling wrapper function F[f] (regarding the full evaluation we refer to the appendix C in this book):

$$0=\left[\begin{array}{c}\frac{\sqrt{-g}}{2}(n+2)FF'\Delta f\\ +F^2\left(\begin{array}{c}\left(\sqrt{-g}\cdot g^{\alpha\beta}\right)_{,\alpha\beta}-\frac{n}{2}\cdot\frac{\sqrt{-g}\cdot R}{(n-1)}\\ -\frac{1}{2(n-1)}\left(\sqrt{-g}\cdot g^{\alpha\beta}\cdot g^{ij}\left(\begin{array}{c}g_{i\beta,j}-g_{ij,\beta}-g_{\beta j,i}\\ +\frac{n}{2}\left(2g_{ij,\beta}-g_{i\beta,j}-g_{j\beta,i}\right)\end{array}\right)\right)_{,\alpha}\end{array}\right)\end{array}\right], \tag{660}$$

where in fact we had begun with the Einstein-Hilbert action, but had slightly altered its variation. That this is by no means an arbitrary alteration but just a technical trick of immense power and flexibility in treating more general forms of the Einstein-Hilbert action was shown in [8] (also presented here in section "Uncertainties One Cannot Get Rid Of"). The results was:

$$0=\frac{\sqrt{-g}}{2}(n+2)\Delta f+\frac{F}{F'}\left(\begin{array}{c}\left(\sqrt{-g}\cdot g^{\alpha\beta}\right)_{,\alpha\beta}-\frac{n}{2}\cdot\frac{\sqrt{-g}\cdot R}{(n-1)}\\ -\frac{1}{2(n-1)}\left(\sqrt{-g}\cdot g^{\alpha\beta}\cdot g^{ij}\left(\begin{array}{c}g_{i\beta,j}-g_{ij,\beta}-g_{\beta j,i}\\ +\frac{n}{2}\left(2g_{ij,\beta}-g_{i\beta,j}-g_{j\beta,i}\right)\end{array}\right)\right)_{,\alpha}\end{array}\right). \tag{661}$$

Inserting condition[20]:

$$0=\frac{1}{2}(n-4)(n+2)(F')^2+2FF", \tag{662}$$

with the solution (we recognize the special situation in the case of n=4, where the simple linear setting of F[f]=f satisfies (662)):

$$\begin{matrix} F=\left(f+C_f\right)^{\frac{4}{n\cdot(n-2)-4}}; \quad n\neq 4 \\ F=f+C_f; \quad n=4 \end{matrix}, \tag{663}$$

we obtain from (661):

$$0=\left(\begin{matrix}\frac{\sqrt{-g}}{2}(n+2)\Delta f+\frac{f+C_f}{4}(n\cdot(n-2)-4) \\ \times\left(\begin{matrix}\left(\sqrt{-g}\cdot g^{\alpha\beta}\right)_{,\alpha\beta}-\frac{n}{2}\cdot\frac{\sqrt{-g}\cdot R}{(n-1)} \\ -\frac{1}{2(n-1)}\left(\sqrt{-g}\cdot g^{\alpha\beta}\cdot g^{ij}\left(\begin{matrix}g_{i\beta,j}-g_{ij,\beta}-g_{\beta j,i} \\ +\frac{n}{2}\left(2g_{ij,\beta}-g_{i\beta,j}-g_{j\beta,i}\right)\end{matrix}\right)\right)_{,\alpha}\end{matrix}\right)\end{matrix}\right). \tag{664}$$

As before in the case R=0, we now substitute as follows:

$$\partial_t f=c^2\cdot C2\cdot(\Phi-X). \tag{665}$$

Together with (664), thereby assuming that we can separate the n-Laplace operator into its time derivative (being Minkowski-like) and the spatial part, we obtain:

$$0=\left(\begin{matrix}C1\cdot C2\cdot\partial_t(\Phi-X)+\frac{1}{\sqrt{g}}\partial_A\sqrt{g}\cdot g^{AB}\partial_B(\Phi+X)+\frac{f+C_f}{2}\frac{(n\cdot(n-2)-4)}{\sqrt{-g}\cdot(n+2)} \\ \times\left(\begin{matrix}\left(\sqrt{-g}\cdot g^{\alpha\beta}\right)_{,\alpha\beta}-\frac{n}{2}\cdot\frac{\sqrt{-g}\cdot R}{(n-1)} \\ -\frac{1}{2(n-1)}\left(\sqrt{-g}\cdot g^{\alpha\beta}\cdot g^{ij}\left(\begin{matrix}g_{i\beta,j}-g_{ij,\beta}-g_{\beta j,i} \\ +\frac{n}{2}\left(2g_{ij,\beta}-g_{i\beta,j}-g_{j\beta,i}\right)\end{matrix}\right)\right)_{,\alpha}\end{matrix}\right)\end{matrix}\right). \tag{666}$$

---

[20] This condition is of need in order to get rid of the nonlinear differential operator in the original variational result. Please see appendix C for the full evaluation.

Please note that the capital indices A and B run from 1 to n-1, while the other indices run from 0 to n-1. The following two equations summed up would result in (666):

$$0=\left(\begin{array}{c}\frac{1}{D}\cdot\partial_t\Phi+\frac{1}{\sqrt{g}}\partial_A\sqrt{g}\cdot g^{AB}\partial_B\Phi+\frac{f+C_f}{4}\frac{(n\cdot(n-2)-4)}{\sqrt{-g}\cdot(n+2)}\\ \times\left(\begin{array}{c}\left(\sqrt{-g}\cdot g^{\alpha\beta}\right)_{,\alpha\beta}-\frac{n}{2}\cdot\frac{\sqrt{-g}\cdot R}{(n-1)}\\ -\frac{1}{2(n-1)}\left(\sqrt{-g}\cdot g^{\alpha\beta}\cdot g^{ij}\left(\begin{array}{c}g_{i\beta,j}-g_{ij,\beta}-g_{\beta j,i}\\ +\frac{n}{2}\left(2g_{ij,\beta}-g_{i\beta,j}-g_{j\beta,i}\right)\end{array}\right)\right)_{,\alpha}\end{array}\right)\end{array}\right). \quad (667)$$

$$0=\left(\begin{array}{c}\frac{1}{\sqrt{g}}\partial_A\sqrt{g}\cdot g^{AB}\partial_B X-\frac{1}{D}\cdot\partial_t X+\frac{f+C_f}{4}\frac{(n\cdot(n-2)-4)}{\sqrt{-g}\cdot(n+2)}\\ \times\left(\begin{array}{c}\left(\sqrt{-g}\cdot g^{\alpha\beta}\right)_{,\alpha\beta}-\frac{n}{2}\cdot\frac{\sqrt{-g}\cdot R}{(n-1)}\\ -\frac{1}{2(n-1)}\left(\sqrt{-g}\cdot g^{\alpha\beta}\cdot g^{ij}\left(\begin{array}{c}g_{i\beta,j}-g_{ij,\beta}-g_{\beta j,i}\\ +\frac{n}{2}\left(2g_{ij,\beta}-g_{i\beta,j}-g_{j\beta,i}\right)\end{array}\right)\right)_{,\alpha}\end{array}\right)\end{array}\right). \quad (668)$$

As before we have used the setting for C1*C2 from (659). Assuming a perfectly flat space in cartesian coordinates, we would just get back our diffusion equations from above (Eq. (659)). Thus, the equations (667) and (668) are giving us the curved space-time matter and antimatter diffusion equations.

14.2.2.1.3.1 A Note about Possible Applications of the Curved-Space-Time Diffusion Equations

Some more applied thinking readers may understand the equations (667) and (668) only to be of use in cases of strong gravitational fields. But this is not justified as any form of geometrical or other – generalized dimensional – restriction of the diffusion system (including intrinsic or extrinsic boundaries) could effectively be described by suitable metrics and corresponding curvatures. We just remind the reader of the coating of edges and corners, regarding T1, or the diffusion through intrinsic barriers of curved shape in connection with T2. This way (667) and (668) are providing a very general tool-box for the handling of a great variety of complex diffusion or related problems in systems of arbitrary numbers of dimensions.

In addition, it should be noted that (668) also presents an evolution equation… in a curved space-time.

14.2.2.1.4 Diffusion Equations of Dirac Character

When considering T2 for the cleaning of tar sands and assuming a Polar Selective Agent of fermionic character (like turbulences excluding each other – similar or even equal to the Pauli principle [77]), we cannot apply the second order diffusion equations as derived above, but need to find a diffusion equation of Dirac style.

Using:

$$\begin{array}{c}\delta W=0=\delta\int\limits_V d^n x\left(\sqrt{g\cdot\Phi^n}\cdot R^*\right)\\ \xrightarrow{g_{\alpha\beta}\cdot\Phi=\gamma_{\alpha\beta}\cdot F=\gamma_{\alpha\beta}\cdot F[f];\;\; F^{\frac{n}{2}+1}=\Phi^{\frac{n}{2}}}\\ 0=\delta_f\int\limits_V d^n x\left(\sqrt{-\gamma}\cdot F^{\frac{n}{2}+1}R^*\right)\Rightarrow 0=\delta_f\int\limits_V d^n x\left(\sqrt{-\gamma\cdot F^n}\cdot F\cdot R^*\right)\end{array},\tag{669}$$

the resulting variational task reads:

$$\frac{\delta W}{\delta f}=0=\delta_f\int\limits_{V_G} d^n x\sqrt{-g\cdot F^n}\left(\begin{array}{c}R-\dfrac{1}{2F}\left(\begin{array}{c}2F_{,\alpha\beta}(n-1)g^{\alpha\beta}+2\Gamma^a_{\alpha\beta}F_{,a}g^{\alpha\beta}\\ -F_{,\alpha}g^{ab}g_{\beta b,a}g^{\alpha\beta}-F_{,\beta}g^{ab}g_{\alpha b,a}g^{\alpha\beta}\\ -n\Gamma^d_{\alpha\beta}F_{,d}g^{\alpha\beta}+\dfrac{n}{2}F_{,d}g^{cd}g_{\alpha\beta,c}g^{\alpha\beta}\end{array}\right)\\ -\dfrac{F_{,\alpha}\cdot F_{,\beta}}{4F^2}g^{\alpha\beta}\left((n-6)(n-1)\right)\end{array}\right)\tag{670}$$

and, when following the derivation shown in appendix C of this book, we obtain the scalar equation:

$$0=\left[\begin{array}{c}\dfrac{\sqrt{-g}}{2}\left(\dfrac{f_{,\alpha}f_{,\beta}}{2}g^{\alpha\beta}\left(\dfrac{1}{2}(n-4)(n+2)(F')^2+2FF''\right)+\boxed{(n+2)FF'\Delta f}\right)\\ +F^2\left(\begin{array}{c}\left(\sqrt{-g}\cdot g^{\alpha\beta}\right)_{,\alpha\beta}-\dfrac{\sqrt{-g}\cdot R}{(n-1)}\\ -\dfrac{1}{2(n-1)}\left(\sqrt{-g}\cdot g^{\alpha\beta}\cdot g^{ij}\left(\begin{array}{c}g_{i\beta,j}-g_{ij,\beta}-g_{\beta j,i}\\ +\dfrac{n}{2}\left(2g_{ij,\beta}-g_{i\beta,j}-g_{j\beta,i}\right)\end{array}\right)\right)_{,\alpha}\end{array}\right)\end{array}\right].\tag{671}$$

Applying the condition:

$$0=\frac{1}{2}\left(\frac{1}{2}(n-4)(n+2)(F')^2+2FF''\right)\pm(n+2)F\frac{F'}{f}$$
$$\Rightarrow\begin{cases}+\Rightarrow\quad F[f]=C_2\cdot f^{-\frac{4(1+n)}{(n-2)n-4}}\left(4+2n-n^2+4f^{1+n}(1+n)C_1\right)^{\frac{4}{(n-2)n-4}},\\ -\Rightarrow\quad F[f]=C_2\cdot\left(f^{3+n}\left((n-2)n-4\right)+4(3+n)C_1\right)^{\frac{4}{(n-2)n-4}}\end{cases}\tag{672}$$

yields:

$$0=\left[\begin{array}{c}\pm\frac{f_{,\alpha}f_{,\beta}}{f}g^{\alpha\beta}+\Delta f\\ +\frac{2\cdot F}{\sqrt{-g}\,(n+2)F'}\left(\begin{array}{c}\left(\sqrt{-g}\cdot g^{\alpha\beta}\right)_{,\alpha\beta}-\frac{\sqrt{-g}\cdot R}{(n-1)}\\ -\frac{1}{2(n-1)}\left(\sqrt{-g}\cdot g^{\alpha\beta}\cdot g^{ij}\left(\begin{array}{c}g_{i\beta,j}-g_{ij,\beta}-g_{\beta j,i}\\ +\frac{n}{2}\left(2g_{ij,\beta}-g_{i\beta,j}-g_{j\beta,i}\right)\end{array}\right)\right)_{,\alpha}\end{array}\right)\end{array}\right]$$
$$=\left[\begin{array}{c}\pm\frac{f_{,\alpha}f_{,\beta}}{f}g^{\alpha\beta}+\Delta f+\frac{2}{\sqrt{-g}\,(n+2)}\left\{\begin{array}{c}\frac{f(4-(n-2)n)}{4(1+n)}+f^{2+n}C_1\\ \frac{f((n-2)n-4)}{4(3+n)}+\frac{C_1}{f^{2+n}}\end{array}\right\}\\ \times\left(\begin{array}{c}\left(\sqrt{-g}\cdot g^{\alpha\beta}\right)_{,\alpha\beta}-\frac{\sqrt{-g}\cdot R}{(n-1)}\\ -\frac{1}{2(n-1)}\left(\sqrt{-g}\cdot g^{\alpha\beta}\cdot g^{ij}\left(\begin{array}{c}g_{i\beta,j}-g_{ij,\beta}-g_{\beta j,i}\\ +\frac{n}{2}\left(2g_{ij,\beta}-g_{i\beta,j}-g_{j\beta,i}\right)\end{array}\right)\right)_{,\alpha}\end{array}\right)\end{array}\right],\tag{673}$$

respectively:

$$0=\left\{\begin{array}{l}\left[\begin{array}{l}\frac{f_{,\alpha}f_{,\beta}}{f}g^{\alpha\beta}+\Delta f+\frac{2}{\sqrt{-g}\,(n+2)}\left(\frac{f\left(4-(n-2)n\right)}{4(1+n)}+f^{2+n}C_1\right)\\ \times\left(\begin{array}{c}\left(\sqrt{-g}\cdot g^{\alpha\beta}\right)_{,\alpha\beta}-\frac{\sqrt{-g}\cdot R}{(n-1)}\\ -\frac{1}{2(n-1)}\left(\sqrt{-g}\cdot g^{\alpha\beta}\cdot g^{ij}\left(\begin{array}{c}g_{i\beta,j}-g_{ij,\beta}-g_{\beta j,i}\\ +\frac{n}{2}\left(2g_{ij,\beta}-g_{i\beta,j}-g_{j\beta,i}\right)\end{array}\right)\right)_{,\alpha}\end{array}\right)\end{array}\right]\\ \left[\begin{array}{l}-\frac{f_{,\alpha}f_{,\beta}}{f}g^{\alpha\beta}+\Delta f+\frac{2}{\sqrt{-g}\,(n+2)}\left(\frac{f\left((n-2)n-4\right)}{4(3+n)}+\frac{C_1}{f^{2+n}}\right)\\ \times\left(\begin{array}{c}\left(\sqrt{-g}\cdot g^{\alpha\beta}\right)_{,\alpha\beta}-\frac{\sqrt{-g}\cdot R}{(n-1)}\\ -\frac{1}{2(n-1)}\left(\sqrt{-g}\cdot g^{\alpha\beta}\cdot g^{ij}\left(\begin{array}{c}g_{i\beta,j}-g_{ij,\beta}-g_{\beta j,i}\\ +\frac{n}{2}\left(2g_{ij,\beta}-g_{i\beta,j}-g_{j\beta,i}\right)\end{array}\right)\right)_{,\alpha}\end{array}\right)\end{array}\right]\end{array}\right. . \quad (674)$$

Setting $C_1=0$, we obtain:

$$0=\left\{\begin{array}{l}\left[\begin{array}{l}f_{,\alpha}f_{,\beta}g^{\alpha\beta}+f\cdot\Delta f+f^2\frac{\left(4-(n-2)n\right)}{2\cdot\sqrt{-g}\,(n+2)(1+n)}\\ \times\left(\begin{array}{c}\left(\sqrt{-g}\cdot g^{\alpha\beta}\right)_{,\alpha\beta}-\frac{\sqrt{-g}\cdot R}{(n-1)}\\ -\frac{1}{2(n-1)}\left(\sqrt{-g}\cdot g^{\alpha\beta}\cdot g^{ij}\left(\begin{array}{c}g_{i\beta,j}-g_{ij,\beta}-g_{\beta j,i}\\ +\frac{n}{2}\left(2g_{ij,\beta}-g_{i\beta,j}-g_{j\beta,i}\right)\end{array}\right)\right)_{,\alpha}\end{array}\right)\end{array}\right]\\ \left[\begin{array}{l}-f_{,\alpha}f_{,\beta}g^{\alpha\beta}+f\cdot\Delta f+f^2\frac{(n-2)n-4}{2\cdot\sqrt{-g}\,(n+2)(3+n)}\\ \times\left(\begin{array}{c}\left(\sqrt{-g}\cdot g^{\alpha\beta}\right)_{,\alpha\beta}-\frac{\sqrt{-g}\cdot R}{(n-1)}\\ -\frac{1}{2(n-1)}\left(\sqrt{-g}\cdot g^{\alpha\beta}\cdot g^{ij}\left(\begin{array}{c}g_{i\beta,j}-g_{ij,\beta}-g_{\beta j,i}\\ +\frac{n}{2}\left(2g_{ij,\beta}-g_{i\beta,j}-g_{j\beta,i}\right)\end{array}\right)\right)_{,\alpha}\end{array}\right)\end{array}\right]\end{array}\right. \quad (675)$$

and thus, have the starting point for the Dirac quaternion decomposition, respectively, factorization as described in [70], section "First Order Evolution Equations from the Einstein-Hilbert Action – Option B revisited: The Dirac-Evolution Equation for Curved Space-Times".

For completeness only we here also want to give a general setting for the condition (672), reading:

$$\begin{aligned}&\frac{1}{2}\left(\frac{1}{2}(n-4)(n+2)(F')^2+2FF''\right)=F\cdot F'\cdot\frac{H'}{H};\quad H=H[f]\\&\Rightarrow\quad F=\left(C_f-\int_1^f H[\phi]\cdot\frac{n\cdot(2-n)+4}{4}\cdot d\phi\right)^{\frac{4}{(n-2)n-4}}\cdot C_{f1}\end{aligned}\qquad(676)$$

We leave it to the reader to play with the options of various functions H[f].

By defining:

$$\begin{aligned}V_+&=\frac{(4-(n-2)n)}{2\cdot\sqrt{-g}\,(n+2)(1+n)}\left(\begin{array}{c}\left(\sqrt{-g}\cdot g^{\alpha\beta}\right)_{,\alpha\beta}-\frac{\sqrt{-g}\cdot R}{(n-1)}\\-\frac{1}{2(n-1)}\left(\sqrt{-g}\cdot g^{\alpha\beta}\cdot g^{ij}\left(\begin{array}{c}g_{i\beta,j}-g_{ij,\beta}-g_{\beta j,i}\\+\frac{n}{2}\left(2g_{ij,\beta}-g_{i\beta,j}-g_{j\beta,i}\right)\end{array}\right)\right)_{,\alpha}\end{array}\right)\\V_-&=\frac{(n-2)n-4}{2\cdot\sqrt{-g}\,(n+2)(3+n)}\left(\begin{array}{c}\left(\sqrt{-g}\cdot g^{\alpha\beta}\right)_{,\alpha\beta}-\frac{\sqrt{-g}\cdot R}{(n-1)}\\-\frac{1}{2(n-1)}\left(\sqrt{-g}\cdot g^{\alpha\beta}\cdot g^{ij}\left(\begin{array}{c}g_{i\beta,j}-g_{ij,\beta}-g_{\beta j,i}\\+\frac{n}{2}\left(2g_{ij,\beta}-g_{i\beta,j}-g_{j\beta,i}\right)\end{array}\right)\right)_{,\alpha}\end{array}\right)\end{aligned}\qquad(677)$$

we can significantly abbreviate (675) as follows:

$$0=\begin{cases}f_{,\alpha}f_{,\beta}g^{\alpha\beta}+f\cdot\Delta f+f^2\cdot V_+\\-f_{,\alpha}f_{,\beta}g^{\alpha\beta}+f\cdot\Delta f+f^2\cdot V_-\end{cases}.\qquad(678)$$

Now we could demand the Laplace operator term in (678) to have an eigenvalue M^2:

$$\Delta f_i=M^2\cdot f_i\qquad(679)$$

and further assume to have a set f_i of possible solutions leading us to:

$$0=\begin{cases}f_{i,\alpha}f_{i,\beta}g^{\alpha\beta}+f_i^2\cdot\left(V_++M^2\right)\\-f_{i,\alpha}f_{i,\beta}g^{\alpha\beta}+f_i^2\cdot\left(V_-+M^2\right)\end{cases}.\qquad(680)$$

Please note that by setting:

$$0=\begin{cases} f_{i,\alpha}f_{i,\beta}g^{\alpha\beta}+f_i^2\cdot\left(V_++M^2\right)=f_{i,\alpha}f_{i,\beta}g^{\alpha\beta}+f_i^2\cdot\overbrace{\left(V_++M^2\right)}^{\mu_+\cdot\mu_+} \\ -f_{i,\alpha}f_{i,\beta}g^{\alpha\beta}+f_i^2\cdot\left(V_-+M^2\right)=-f_{i,\alpha}f_{i,\beta}g^{\alpha\beta}+f_i^2\cdot\underbrace{\left(V_-+M^2\right)}_{-\mu_-\cdot\mu_-} \end{cases}$$

$$\Rightarrow \qquad\qquad (681)$$

$$0=f_{i,\alpha}f_{i,\beta}g^{\alpha\beta}+\begin{Bmatrix}\mu_+\cdot\mu_+ \\ \mu_-\cdot\mu_-\end{Bmatrix}\cdot f_i^2$$

we would once again have the perfect starting point for the Dirac factorization, which is even leading us to the classical form:

$$0=f_{i,\beta}\gamma^\beta-i\cdot\beta\cdot m_\pm\cdot f_i, \qquad (682)$$

only with the m-letter now containing much more than just mass. One may also state that we now know what – metrically – is behind the mass term in Dirac's classical equation.

Thereby we have used the definition:

$$\begin{Bmatrix}\mu_+\cdot\mu_+ \\ \mu_-\cdot\mu_-\end{Bmatrix}\cdot I=\beta^2\cdot m_\pm^2 \qquad (683)$$

and the generalized β-matrix with $\beta=\begin{pmatrix}1 &&&& \\ & 1 &&& \\ && \ddots && \\ &&& -1 & \\ &&&& -1\end{pmatrix}$.

Now we apply the known relation between the Dirac matrices and the metric tensor, reading:

$$g^{\alpha\beta}\cdot I=\frac{\gamma^\alpha\gamma^\beta+\gamma^\beta\gamma^\alpha}{2}, \qquad (684)$$

and can follow the path from the sections above by putting (682) into the following form:

$$0=\left(f_{i,\alpha}f_{i,\beta}g^{\alpha\beta}+m_{\pm}\cdot m_{\pm}\cdot f_i^2\right)\cdot I$$

$$\Rightarrow \quad = \quad \Rightarrow f_{i,\alpha}f_{i,\beta}\left(\frac{\gamma^{\alpha}\gamma^{\beta}+\gamma^{\beta}\gamma^{\alpha}}{2}\right)+m_{\pm}\cdot m_{\pm}\cdot f_i^2$$

$$\Rightarrow\Rightarrow\Rightarrow$$

$$\begin{gathered}
A)\Rightarrow f_{i,\alpha}f_{i,\beta}\frac{\gamma^{\alpha}\gamma^{\beta}}{2}+\frac{1}{2}\cdot m_{\pm}\cdot m_{\pm}\cdot f_i^2=0\\
\Rightarrow f_{i,\alpha}f_{i,\beta}\gamma^{\alpha}\gamma^{\beta}+m_{\pm}\cdot m_{\pm}\cdot f_i^2=0\\
\Rightarrow\left(f_{i,\alpha}\gamma^{\alpha}+i\cdot m_{\pm}\cdot f_i\right)\left(f_{i,\beta}\gamma^{\beta}-i\cdot m_{\pm}\cdot f_i\right)=0\\
B)\Rightarrow f_{i,\alpha}f_{i,\beta}\frac{\gamma^{\beta}\gamma^{\alpha}}{2}+\frac{1}{2}\cdot m_{\pm}\cdot m_{\pm}\cdot f_i^2=0\\
\Rightarrow f_{i,\alpha}f_{i,\beta}\gamma^{\beta}\gamma^{\alpha}+m_{\pm}\cdot m_{\pm}\cdot f_i^2=0\\
\Rightarrow\left(f_{i,\alpha}\gamma^{\alpha}+i\cdot m_{\pm}\cdot f_i\right)\left(f_{i,\beta}\gamma^{\beta}-i\cdot m_{\pm}\cdot f_i\right)=0
\end{gathered} \quad . \qquad (685)$$

The subsequent curved Dirac equations are:

$$\begin{gathered}
f_{i,\alpha}\gamma^{\alpha}+i\cdot m_{\pm}\cdot f_i=0\\
f_{i,\beta}\gamma^{\beta}-i\cdot m_{\pm}\cdot f_i=0
\end{gathered}. \qquad (686)$$

We realize that the expression $m_{\pm}$, which gives us the equivalent of the classical mass, is not a simple scalar anymore, but has become a complex function of the space-time curvature and dimension.

Separating now the time derivative ∂_0 gives us the most general innate Dirac-type evolution equations of any arbitrary system:

$$\begin{gathered}
f_{i,0}\gamma^{0}+f_{i,\alpha}\gamma^{\alpha}+i\cdot m_{\pm}\cdot f_i=0;\quad \alpha=1,\dots,n-1\\
f_{i,0}\gamma^{0}+f_{i,\beta}\gamma^{\beta}-i\cdot m_{\pm}\cdot f_i=0;\quad \beta=1,\dots,n-1
\end{gathered}. \qquad (687)$$

These are also – automatically – the system's most general diffusion equations, where the metric and subsequently also the corresponding gamma matrices (c.f. identity (684)) assure the correction relation between time and spatial derivatives. That being said it is clear that our gamma matrices here cannot be perfectly equal to the classical Dirac matrices for n=4 or in [75] for general n. To give an example (simplest case): if we assume the mass-term to play no role and thus, setting $m_{\pm}=0$, such a diffusion Dirac equation should then look like:

$$f_{i,0}\gamma^{0}+\frac{1}{\sqrt{D}}\cdot f_{i,\alpha}\gamma^{\alpha}=0;\quad \alpha=1,\dots,n-1\cdot \qquad (688)$$

Now, the gamma matrices in (688) would in fact just be the classical ones and we are left with a Dirac equation of the neutrino type… only that here it is a very general diffusion equation.

14.2.2.2 S-Curve or Sigmoid Evolution

Often it is assumed that S- or Sigmoid-curve behavior itself is just the result of an evolution equation and in many cases this is perfectly true (just consider typical diffusion processes). However, in this section we will show that S-curve behavior already comes directly out of the Einstein field equations as a very basic solution for all metrics where we have pairwise dimensional entanglement and certain symmetries.

Especially the deposition technique T1 as a growth process will profit from a thorough understanding of the functional dependencies coming with "change" in general. However, these functions are also connected with one of the diffusion equations we have covered in the subsection before. Diffusion, on the other hand, is strongly connected with separation processes and as we have shown in this book that a first principle-based derivation of this equation leads to an ordinary and an anti-diffusion, providing some insight into the strange behavior observable in the separation experiment T2, we definitely should also cover the metric origin of S-curves in here.

Thereby, most interestingly, no extension of the classical theory or quantization as introduced and applied in the previous sections of this book is needed. As we will see, it totally suffices to functionally combine two orthogonal dimensions and the outcome are S-curves.

Other authors (e.g., [76]) tended to express quite some amazement about the omnipresence of S-curves.

Thus, in this subsection it will be shown that sigmoid-curves are – if starting from an Einstein or metric perspective – of most principal character and occur in so many different fields in basic physics, biology, chemistry, economy, social science, medicine (see pandemic dynamics, for instance), propagation processes, etc. [76], because they are one of the fundamental solutions of the Einstein field equations [3]. We will demonstrate that sigmoid-curve-solutions are always obtained the moment orthogonal properties of arbitrary mathematical problems, which is to say orthogonal dimensions of arbitrary spaces, are entangled in spaces or space-times (in general these are just systems formulated in a Riemann apparatus) with cosmological constants.

14.2.2.2.1 About S- or Sigmoid Curves

Humans have the unfortunate habit to take basic experiences for granted that they do not ask for the reasons behind them and thereby – even if it would bring them great insight – they often miss to discover fundamental laws, which are holding together our universe.

Failing to run through a wall can be seen as such an experience. It usually proves to be quite difficult, if not to say painful. This is one of the earliest things every human child experiences in its life and one might see it as the physical manifestation of the Pauli-Exclusion-Principle [77] and the electrostatic repulsion. Where there is one body, there cannot be another one. That is just it, what children realize when finding out that they cannot crawl through solid objects. After such incidents, usually (politicians excluded), human beings know that their own head and the wall do quite some fighting against any attempt to be merged or "interfused".

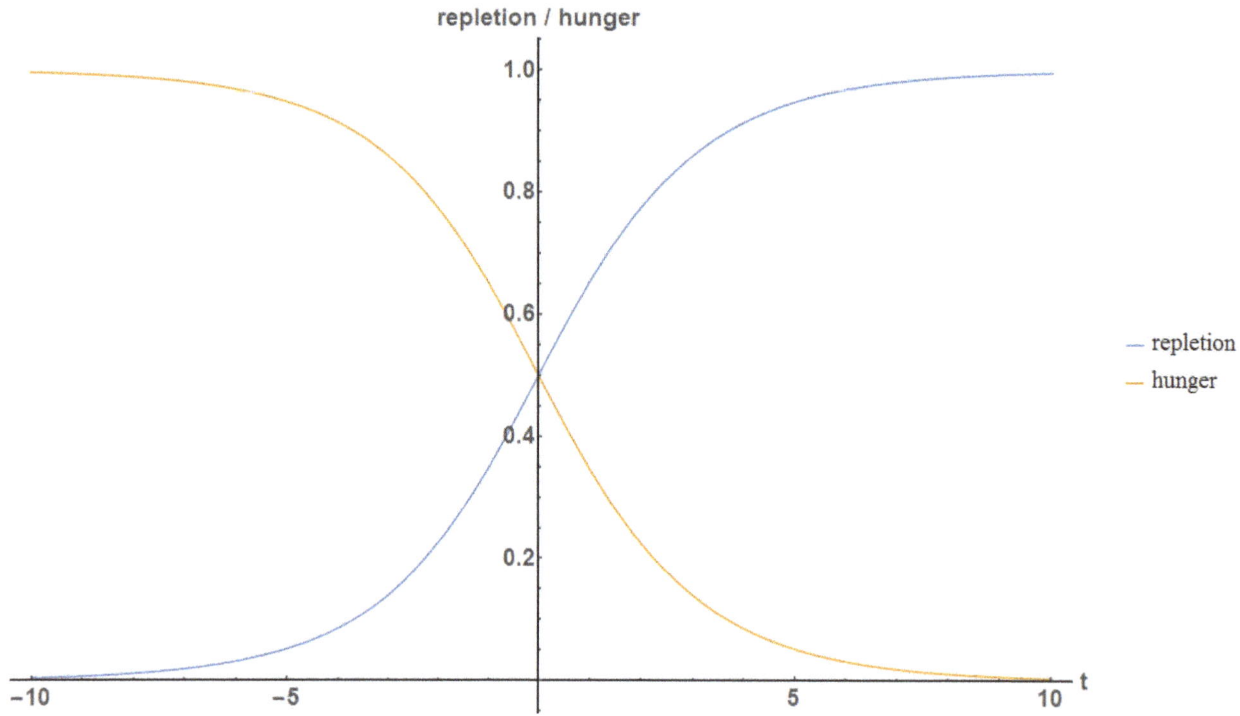

Fig. 53: Sigmoid-dependencies in baby-feeding. The figure shows the "coevolution" of hunger and repletion of a child being fed by its mother in its time dependency (t).

An almost similar basic experience can be made out for situations where two apparently independent properties (or dimensions) are interlinked or connected in such a way that one of the properties governs the metric evolution of the other. In quantum physics such a state is called entangled. It should be pointed out, however, that such "entanglement" is by no means restricted to quantum applications, but always occurs when pairs of orthogonal quantities are "forced" to interact in a simultaneous kind of way. Sticking to small humans, just imagine a baby being fed by its mother. At first, the child tastes what is offered, then, after realizing that the thing is very good, it accelerates its consummation until repletion kicks in and the child finally stops eating (hopefully satisfied and quiet).

Here, the entangled dimensions are time and hunger. Their entanglement does not only lead to a very busy mother, but also to a certain time-dependency of the parameter hunger. This dependency is a so-called Sigmoid curve. An example is given in figure 53. Sometimes these types of dependencies are also just named S-curves.

After we were able to extract the Dirac equation (see theory section above, [70], section "First Order Evolution Equations from the Einstein-Hilbert Action – Option B: The Dirac-Evolution Equation") and in connection with it also the Pauli-Exclusion-Principle [77] directly from the Einstein-Hilbert action [1] and the subsequent Einstein field equations [3] even in an obviously quantized form (c.f. equation (29) and corresponding discussion in one of our earlier publications [2, 4]), thereby explaining the very basic wall-experience mentioned above, we here intend to find out why so many and very different fields follow the same fundamental behavior, namely the Sigmoid or S-curve-dependency [76].

14.2.2.2.2 Einstein Field Equation Solutions Are Already Quantized or Just the Ground States of Their Quantum Gravity Forms

As shown in [2, 4] and partially also the sections above, the most basic assumption of n properties leading to n-dimensional spaces or space-times and these spaces following a minimum principle, we automatically end up in the Einstein-Hilbert action [1] (either in its classical form with linear Ricci scalar R or an extended form with the Ricci curvature being substituted by a function of the Ricci scalar f(R) as investigated in [6] and in [70]):

$$\delta_g W = 0 = \delta_g \int_V d^n x \left(\sqrt{-g}\left[R - 2\kappa L_M\right]\right). \tag{689}$$

The solution of this variational problem directly leads us to the Einstein field equations:

$$R^{\alpha\beta} - \frac{1}{2} R g^{\alpha\beta} + \Lambda g^{\alpha\beta} = -\kappa T^{\alpha\beta}. \tag{690}$$

Here we have: $R^{\alpha\beta}$, $T^{\alpha\beta}$ the Ricci- and the energy momentum tensor, respectively, while the parameters Λ and κ are constants (usually called cosmological and coupling constant, respectively). These are the well-known Einstein field equations in n dimensions with the indices α and β running either from 1 to n or – as usually done in here – from 0 to n-1. The theory behind it is called General Theory of Relativity (GTR). Please note that Einstein and Hilbert had artificially added the energy momentum tensor, respectively its Lagrangian density L_M, because they thought it a necessity in order to obtain matter solutions. Obviously, it did never bother people that many solutions to the vacuum field equations with $L_M = T^{\alpha\beta} = 0$, like e.g., the Schwarzschild [78, 79] or the Kerr-solution [80, 81], do not need such terms and still describe matter of a certain kind. The excuse always was that matter in these metrics is concentrated in the singularity in the center of the system. We have already shown here that this assertion may not be the only explanation, nor would it be the best one (see "The Evolution of Matter" in here or chapter "About the Dimensional Size of Systems" and subsection "Hawking Radiation of a Black Hole" in [70] and corresponding sections in our book [56]).

Thus, we assume the introduction of L_M and $T^{\alpha\beta}$ to be a mistake and suggest instead that all matter forms can be extracted from the vacuum field equations in either its classical:

$$R^{\alpha\beta} - \frac{1}{2} R g^{\alpha\beta} + \Lambda g^{\alpha\beta} = 0 \tag{691}$$

or its extended form (c.f. equations (29) and (359)).

Considering the Hamilton principle and the subsequent Einstein-Hilbert action [1] and Einstein field equations [3] (potentially in its generalized form as e.g., given in (29); see also [4]), the most basic physical law there is, these equations must also contain Quantum Theory. Apparently, as demonstrated above, it is only a question of finding the right solutions or – even simpler – the right transformations to actually find these equations. We suspect the technique of metric scalar generalization and Ricci scalar multiplication as given in the sections above to provide the key or at least help us along the way.

14.2.2.2.3 The Most Principal Origin of Sigmoid-Dependencies

From the many solutions presented in our earlier publications (e.g., [2, 4]), we here are only interested in the ones of pairwise coupling of dimensions.

It was shown in [83] that there are at least two principal ways to combine ("entangle") orthogonal dimensions of metric spaces (see also [84] for deeper and [85] a bit more entertaining reading).

For convenience, we here repeat the main part of the derivation.

Setting the following metric for the four coordinates t, x, y, z:

$$g_{\alpha\beta} = \begin{pmatrix} H\cdot f[t] & 0 & 0 & 0 \\ 0 & A\cdot f[t] & 0 & 0 \\ 0 & 0 & B\cdot g[y] & 0 \\ 0 & 0 & 0 & D\cdot g[y] \end{pmatrix} \tag{692}$$

into the Einstein field equations (689) with the momentum-energy tensor set to zero, gives solutions if the functions f[t] and g[y] satisfy the following ordinary differential equations:

$$\begin{aligned} 0 &= 2\cdot H\cdot\Lambda\cdot f^3 + f\cdot\partial_t^2 f - (\partial_t f)^2 \\ 0 &= 2\cdot B\cdot\Lambda\cdot g^3 + g\cdot\partial_y^2 g - (\partial_y g)^2 \end{aligned}. \tag{693}$$

Immediately we see that for Λ=0 we obtain an exponential solution for f and g of the form:

$$f[t] = C_1\cdot e^{C_2\cdot t};\quad g[y] = C_3\cdot e^{C_4\cdot y}. \tag{694}$$

Generalization as done in [83] is possible as follows:

$$g_{\alpha\beta} = \begin{pmatrix} H\cdot g_t'[t]^2\cdot f & 0 & 0 & 0 \\ 0 & A\cdot g_x'[x]^2\cdot f & 0 & 0 \\ 0 & 0 & B\cdot g_y'[y]^2\cdot g & 0 \\ 0 & 0 & 0 & D\cdot g_z'[z]^2\cdot g \end{pmatrix}. \tag{695}$$
$$f = f\left[g_t[t]\right] = C_1\cdot e^{C_2\cdot g_t[t]};\quad g = g\left[g_y[y]\right] = C_3\cdot e^{C_4\cdot g_y[y]}$$

It was shown in [86] how this solution can be extended to an arbitrary number of even dimensions.

While for Λ=0 we obtained the typical exponential Dirac-like solution as we have often seen in connection with quantized field equations (c.f. references given in [83]), we get a very interesting solution for the metric (692) in the case of $\Lambda \neq 0$, namely:

$$f[t]=\frac{C_1\cdot\left(1-\tanh\left[\pm\frac{\sqrt{C_1}}{2}(t+C_2)\right]^2\right)}{4\cdot H\cdot\Lambda};\quad g[y]=\frac{C_3\cdot\left(1-\tanh\left[\pm\frac{\sqrt{C_3}}{2}(y+C_4)\right]^2\right)}{4\cdot B\cdot\Lambda}. \tag{696}$$

As it directly follows from the transformation rules for tensors, we can immediately expand the solution (696) to a more general form:

$$g_{\alpha\beta}=\begin{pmatrix} H\cdot g_t'[t]^2\cdot f & 0 & 0 & 0\\ 0 & A\cdot g_x'[x]^2\cdot f & 0 & 0\\ 0 & 0 & B\cdot g_y'[y]^2\cdot g & 0\\ 0 & 0 & 0 & D\cdot g_z'[z]^2\cdot g \end{pmatrix} \tag{697}$$

where now we have to satisfy the following differential equations:

$$\begin{aligned} 0&=2\cdot H\cdot\Lambda\cdot f^3+f\cdot\partial_{g_t}^2 f-\left(\partial_{g_t} f\right)^2\\ 0&=2\cdot B\cdot\Lambda\cdot g^3+g\cdot\partial_{g_y}^2 g-\left(\partial_{g_y} g\right)^2, \end{aligned} \tag{698}$$

which are as easily solvable as the ones above, only that this time we have the functions $g_t[t]$ and $g_y[y]$ instead of the coordinates t and y themselves. Now we obtain:

$$\begin{aligned} f[t]&=\frac{C_1\cdot\left(1-\tanh\left[\pm\frac{\sqrt{C_1}}{2}(g_t[t]+C_2)\right]^2\right)}{4\cdot H\cdot\Lambda}=\frac{C_1\cdot\cosh\left[\pm\frac{\sqrt{C_1}}{2}(g_t[t]+C_2)\right]^{-2}}{4\cdot H\cdot\Lambda}\\ g[y]&=\frac{C_3\cdot\left(1-\tanh\left[\pm\frac{\sqrt{C_3}}{2}(g_y[y]+C_4)\right]^2\right)}{4\cdot B\cdot\Lambda}=\frac{C_3\cdot\cosh\left[\pm\frac{\sqrt{C_3}}{2}(g_y[y]+C_4)\right]^{-2}}{4\cdot B\cdot\Lambda} \end{aligned}. \tag{699}$$

It was shown in [87] how the solution can be extended to an arbitrary even number of dimensions. Odd numbers of dimensions, on the other hand, were considered in [89, 90, 92].

Regarding the discussion of these solutions, we have to refer to [84, 85, 86, 88, 89, 90].

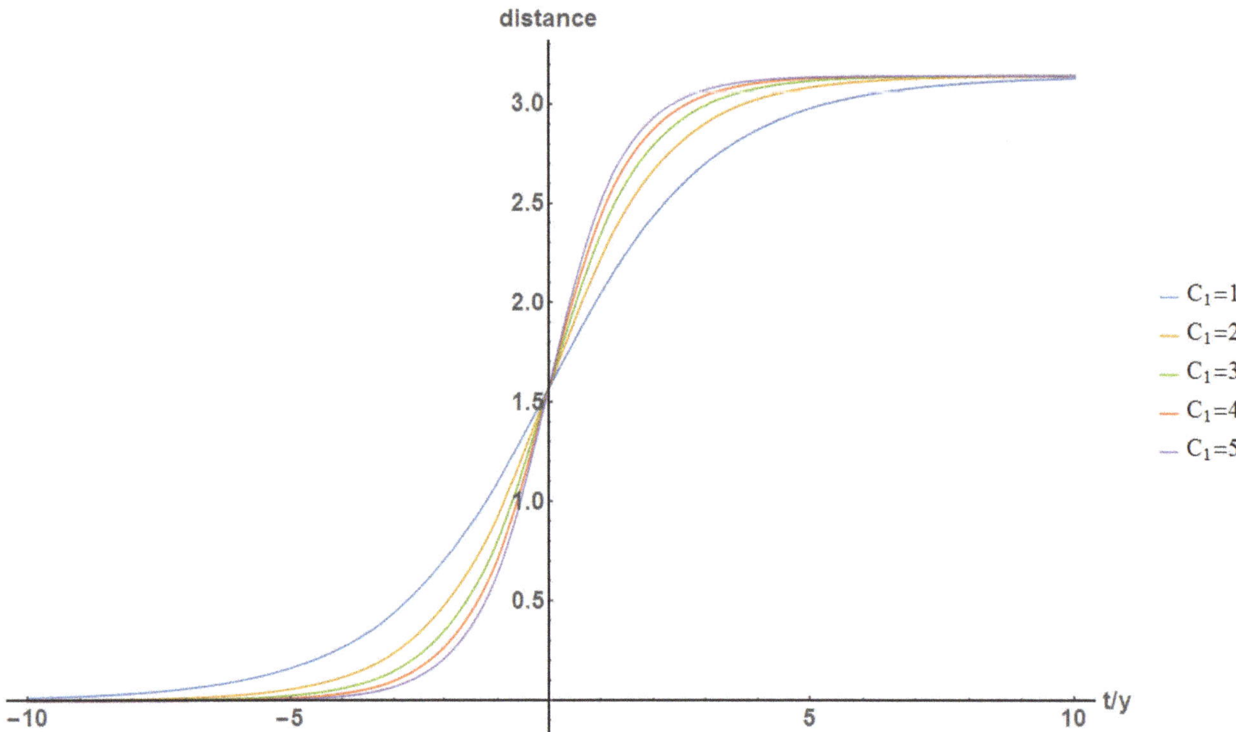

Fig. 54: Sigmoid-curve behavior as principal result of distances of entangled dimensions in metric spaces.

Before proceeding, we emphasize that we have obtained the solutions above from the Einstein field equations under the assumption of a pairwise coupling or entanglement of two dimensions as given in (697). The same kind of solution is always obtained when a coupling of dimensions is demanded in such a way. This was shown in [87].

It should be noted that the metric structure also allows the construction of the most basic and general Turing machine [88, 90, 92]. Thereby the classical and the quantum computer appears just as degenerated forms of such a generalized machine, which we had simply named Einstein Quantum Computer.

Now we evaluate the length of distances an arbitrary (fictive or real) object could travel along such coupled (entangled) dimensions. This evaluation is simple. Taking our example from above, it requires the solution of the following integrals:

$$d_t[\tau] = \int_{-\infty}^{\tau} \sqrt{H \cdot g_t'[t]^2 \cdot f} \cdot dt; \quad d_y[Y] = \int_{-\infty}^{Y} \sqrt{B \cdot g_y'[y]^2 \cdot g} \cdot dy. \tag{700}$$

It totally suffices to give the solution for just one of these integrals, which reads:

$$d_t[\tau] = \left[\frac{2\arctan\left[\tanh\left[\frac{1}{4}\sqrt{C_1}\left(C_2 + g_t[t]\right)\right]\right]}{\sqrt{\Lambda}} \right]_{-\infty}^{\tau}. \tag{701}$$

In the simple case of $g_t[t]=t$ and C_1 real positive, the lower limit evaluates to $(-\pi/2)$. A set of examples is given in figure 54 (with $C_2=0$ and $\Lambda=1$). The typical completely antisymmetric S-curve-shape is obtained by subtraction $\pi/2$ (see figure 55).

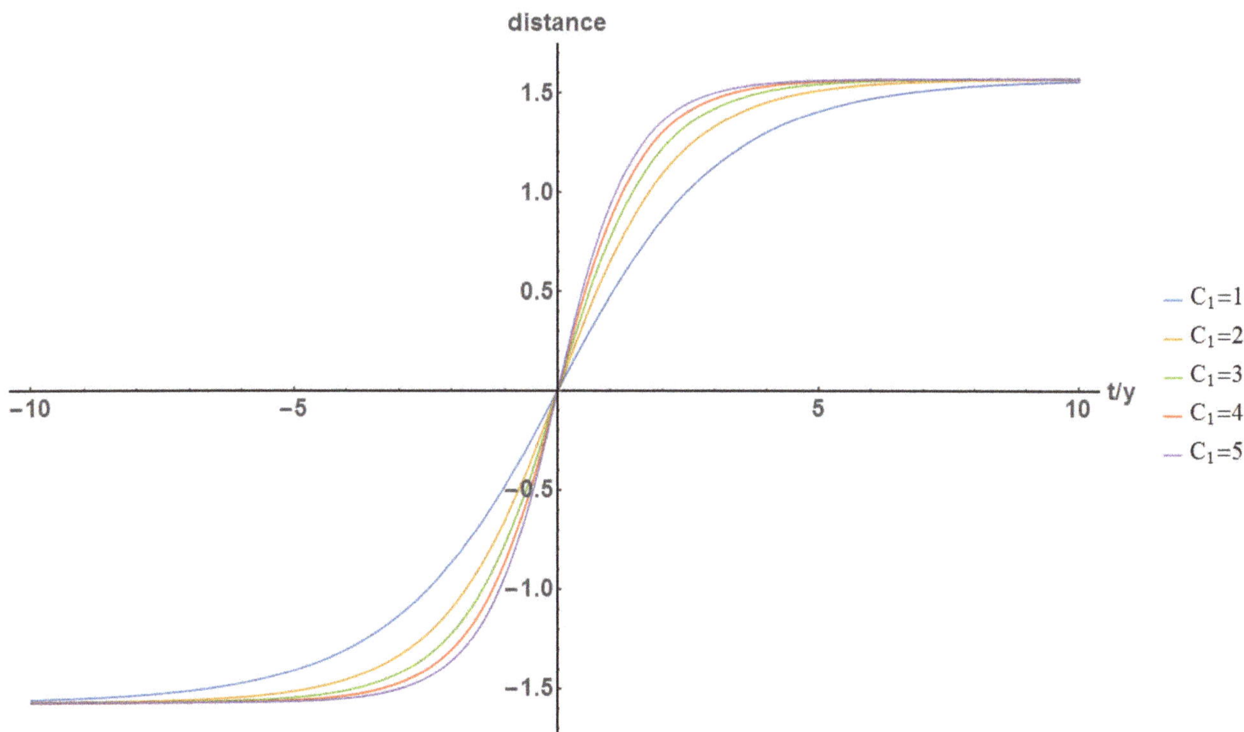

Fig. 55: Sigmoid-curve behavior as principal result of distances of entangled dimensions in metric spaces. In contrast to figure 54 we here subtracted the upper limit of our line integral (700) simply in order to obtain the classical symmetric appearance of S-curves.

Applying the invariant-theory onto our metric and assuming, that such invariants play the role of certain global parameters like mass (c.f. section "Theoretical Prerequisites" in [93]), we can easily evaluate quite a few of such parameters. E.g., taking our parameter and function set from above and considering our baby-feeding-example from the introduction again, we result in the following integral for the state of repletion:

$$d_t[\tau]=\left[\frac{2\arctan\left[\tanh\left[\frac{1}{4}\cdot t\right]\right]}{\sqrt{\Lambda}}\right]_{-\infty}^{\tau}+\frac{1}{2};\quad \text{with}\quad \Lambda=\pi^2. \tag{702}$$

Considering just one feeding-process, it is obvious that this is also equivalent to the amount of food the baby has consumed in a certain time τ. Thus, quite obviously, this could be seen as the mass (or – depending on the baby's state of education and "housekeeping" abilities – mess) of the system.

Please note that also in the case of $\Lambda=0$, Sigmoid-functions can be observed. Simply setting $g_t[t]=t^2$ and $g_y[y]=y^2$ in (695) (with $C_2=C_4=-1$) gives the desired result. The resulting S-curves then are described by the Gaussian error functions, but in contrast to the case $\Lambda\neq0$ they do only appear in the dimension being entangled with t and y. The line-integrals for t and y appear as Gaussian bell-curves, which is to say, they are of particle-like character. This has been illustrated in figure 56 for the dimensions t and x. Obviously, the perfect symmetric S-behavior for the entangled pair of dimensions is only possible in connection with a non-zero cosmologic constant Λ.

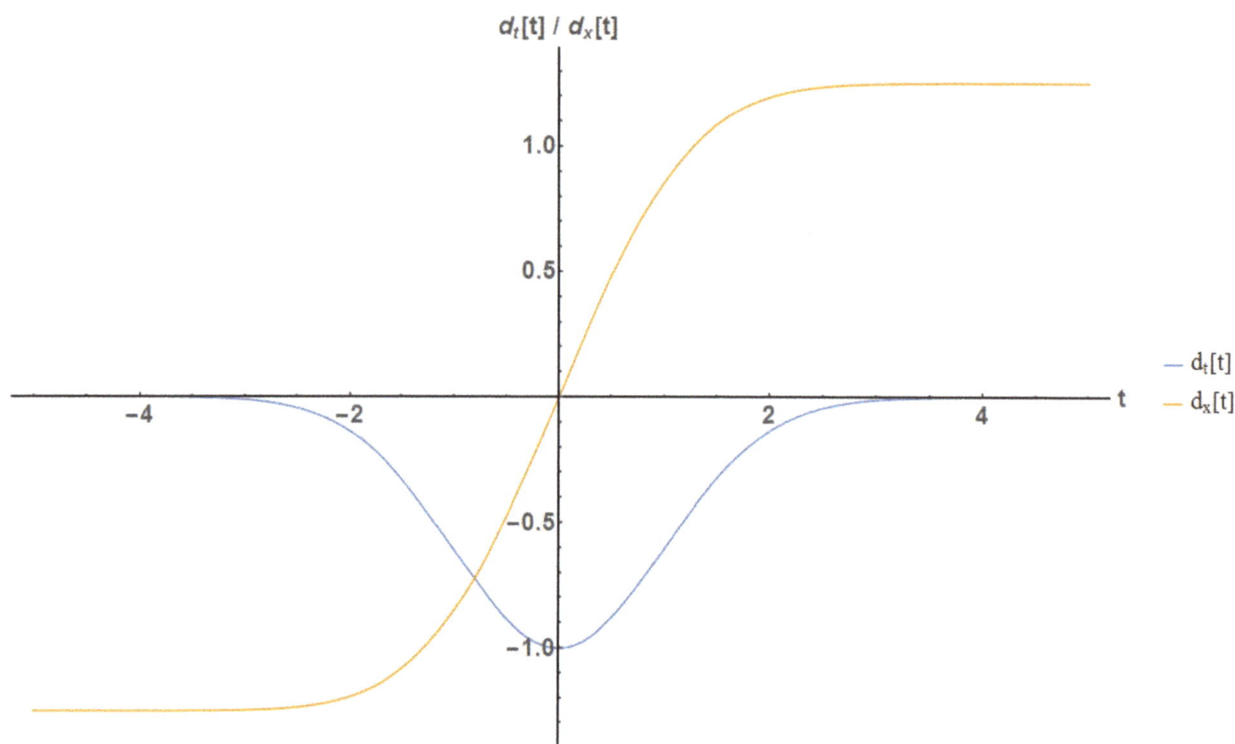

Fig. 56: Evaluation of line length in t and x as function of t for a case with Λ=0 (see text).

There are also options for combined dependencies of the form f[t, x] and g[y, z], which have been considered in certain issues of the third author's series of "Einstein had it…" (following issue LIII).

14.2.2.2.4 Evolution in an S-Curve Metric Space-Time

It was shown that Sigmoid dependencies are very basic metric solutions to the Einstein field equations and occur when there is an entanglement of orthogonal dimensions in completely arbitrary systems. Thus, such solutions are of very fundamental character. We assume that it is for this reason that Sigmoid curves are so omnipresent in natural processes and that we recognize S-curve behavior in so many principally different fields. In the subsections above we already achieved to undergird the omnipresence of S-curves by solving the classical vacuum Einstein field equations for cases of pairwise entangled properties (dimensions) and showing that the results are metrics with S-curve properties.

Now we want to generalize our considerations by extending the investigation towards scaled S-curve-metrics to analyze the global evolutionary potential of such basic solutions.

In order to do so we simply perturb the kernel of the Einstein-Hilbert action (395) just as shown in section "Uncertainties One Cannot Get Rid Of", leading us to the following field equation with respect to the perturbation functions f (also representing, as stated at the end of section "Uncertainties One Cannot Get Rid Of", the system's wave functions):

$$0=\left(\begin{array}{c}R_{\alpha\beta}-\frac{\gamma_{\alpha\beta}}{2}R+\Lambda\cdot\gamma_{\alpha\beta}+\frac{\gamma_{\alpha\beta}}{2}\left(\begin{array}{c}\frac{1}{2F}\left(\begin{array}{c}2F_{,ij}(n-1)\gamma^{ij}+2\Gamma_{ij}^{a}F_{,a}\gamma^{ij}\\ -F_{,i}\gamma^{ab}\gamma_{jb,a}\gamma^{ij}-F_{,j}\gamma^{ab}\gamma_{ib,a}\gamma^{ij}\\ -n\Gamma_{ij}^{d}F_{,d}\gamma^{ij}+\frac{n}{2}F_{,d}\gamma^{cd}\gamma_{ab,c}\gamma^{ab}\end{array}\right)\\ +\frac{F_{,i}\cdot F_{,j}}{4F^{2}}\gamma^{ij}\left((n-6)(n-1)\right)\end{array}\right)\\ -\frac{1}{2F}\left(\begin{array}{c}F_{,\alpha\beta}(n-2)+F_{,ab}\gamma_{\alpha\beta}\gamma^{ab}\\ +F_{,a}\gamma^{ab}\left(\gamma_{\beta b,\alpha}-\gamma_{\beta\alpha,b}\right)-F_{,\alpha}\gamma^{ab}\gamma_{\beta b,a}-F_{,\beta}\gamma^{ab}\gamma_{\alpha b,a}\\ +F_{,d}\gamma^{cd}\frac{1}{2}n\left(\frac{2}{n}\gamma_{\alpha c,\beta}-\gamma_{\alpha c,\beta}-\gamma_{\beta c,\alpha}+\gamma_{\alpha\beta,c}+\frac{1}{n}\gamma_{\alpha\beta}\gamma_{ab,c}\gamma^{ab}\right)\\ -\frac{1}{2F}\left(F_{,\alpha}\cdot F_{,\beta}(3n-6)+\gamma_{\alpha\beta}F_{,c}F_{,d}\gamma^{cd}(4-n)\right)\end{array}\right)\end{array}\right). \qquad (703)$$

Inserting the metric solution from above (e.g., (692)) the equations simplify to:

$$0=\left(\begin{array}{c}\frac{\gamma_{\alpha\beta}}{2}\left(\begin{array}{c}\frac{1}{2F}\left(\begin{array}{c}2F_{,ij}(n-1)\gamma^{ij}+2\Gamma_{ij}^{a}F_{,a}\gamma^{ij}\\ -F_{,i}\gamma^{ab}\gamma_{jb,a}\gamma^{ij}-F_{,j}\gamma^{ab}\gamma_{ib,a}\gamma^{ij}\\ -n\Gamma_{ij}^{d}F_{,d}\gamma^{ij}+\frac{n}{2}F_{,d}\gamma^{cd}\gamma_{ab,c}\gamma^{ab}\end{array}\right)\\ +\frac{F_{,i}\cdot F_{,j}}{4F^{2}}\gamma^{ij}\left((n-6)(n-1)\right)\end{array}\right)\\ -\frac{1}{2F}\left(\begin{array}{c}F_{,\alpha\beta}(n-2)+F_{,ab}\gamma_{\alpha\beta}\gamma^{ab}\\ +F_{,a}\gamma^{ab}\left(\gamma_{\beta b,\alpha}-\gamma_{\beta\alpha,b}\right)-F_{,\alpha}\gamma^{ab}\gamma_{\beta b,a}-F_{,\beta}\gamma^{ab}\gamma_{\alpha b,a}\\ +F_{,d}\gamma^{cd}\frac{1}{2}n\left(\frac{2}{n}\gamma_{\alpha c,\beta}-\gamma_{\alpha c,\beta}-\gamma_{\beta c,\alpha}+\gamma_{\alpha\beta,c}+\frac{1}{n}\gamma_{\alpha\beta}\gamma_{ab,c}\gamma^{ab}\right)\\ -\frac{1}{2F}\left(F_{,\alpha}\cdot F_{,\beta}(3n-6)+\gamma_{\alpha\beta}F_{,c}F_{,d}\gamma^{cd}(4-n)\right)\end{array}\right)\end{array}\right). \qquad (704)$$

We realize that with the metric tensor already being fixed, we may not be able to find a suitable (non-trivial) solution to the function F[f]. That is why we either would need to go back to (703) and solve the equation with respect to the metric tensor and the wave function, or we just assume to have the "weak gravity" condition (33) fulfilled. This latter approach would give us a scalar equation as follows:

$$0=\left(\begin{array}{c}\frac{g^{\alpha\beta}}{2F}F'\left(2f_{,\alpha\beta}(n-1)+f_{,d}g^{cd}\left(\begin{array}{c}g_{\alpha c,\beta}-g_{\beta\alpha,c}-g_{c\beta,\alpha}\\+\frac{n}{2}\left(2g_{\alpha\beta,c}-g_{\alpha c,\beta}-g_{\beta c,\alpha}\right)\end{array}\right)\right)\\+(n-1)\frac{f_{,\alpha}\cdot f_{,\beta}}{4F^{2}}g^{\alpha\beta}\left(4FF''+F'\cdot F'(n-6)\right)\end{array}\right), \tag{705}$$

which we could always fulfill for F[f] in a non-trivial manner. Most interestingly, this leads to the Klein-Gordon, Schrödinger, and Dirac type evolution equations already considered in this book with the underlying metric being also of evolutionary (S-curve) character.

Alternatively, we might just apply our variation with respect to the scaling function f (c.f. section "Dissolving an Apparent Conflict" in [70]) where we obtain the following scalar equation:

$$0=\left[\begin{array}{c}\frac{\sqrt{-\gamma\cdot F^{n}}}{2}\left(\frac{f_{,\alpha}f_{,\beta}}{2}\gamma^{\alpha\beta}\left(\frac{1}{2}(n-4)(n+2)(F')^{2}+2FF''\right)+(n+2)FF'\Delta_{\gamma}f\right)\\+F^{2}\left(\begin{array}{c}\left(\sqrt{-\gamma\cdot F^{n}}\cdot\gamma^{\alpha\beta}\right)_{,\alpha\beta}-\frac{n}{2}\cdot\frac{\sqrt{-\gamma\cdot F^{n}}\cdot R}{(n-1)}\\-\frac{1}{2(n-1)}\left(\sqrt{-\gamma\cdot F^{n}}\cdot\gamma^{\alpha\beta}\cdot\gamma^{ij}\left(\begin{array}{c}\gamma_{i\beta,j}-\gamma_{ij,\beta}-\gamma_{\beta j,i}\\+\frac{n}{2}\left(2\gamma_{ij,\beta}-\gamma_{i\beta,j}-\gamma_{j\beta,i}\right)\end{array}\right)\right)_{,\alpha}\end{array}\right)\end{array}\right]. \tag{706}$$

$$\text{with: }\quad \Delta_{\gamma}f=\frac{1}{\sqrt{\gamma}}\partial_{\alpha}\sqrt{\gamma}\cdot\gamma^{\alpha\beta}\partial_{\beta}f$$

This is totally equivalent to the "weak gravity" condition, but has the advantage of directly providing us with a perfect Laplace operator.

Both such equations (705) and (706) can be developed to the various evolution equations being derived in this book by using the techniques given in the corresponding sections above and will result in equations of Klein-Gordon, Schrödinger, and Dirac type with either matter or antimatter properties. The interesting aspect with the S-curves metrics, being considered in the subsection above, is that even the trivial solution F=const already has evolution potential, because the unscaled metric $g_{\alpha\beta}$ (or, as used in the equations above $\gamma_{\alpha\beta}$) provides it.

However, just in order to give a simple example here, when extending the metric (692) with the innate functions (696) as follows (thereby exchanging f[t] with h[t] in order to have the symbol f for the global scaling or wave function again):

$$g_{\alpha\beta} = F\left[f\left[t,x,y,z\right]\right]\cdot\begin{pmatrix} H\cdot h[t] & 0 & 0 & 0 \\ 0 & A\cdot h[t] & 0 & 0 \\ 0 & 0 & B\cdot g[y] & 0 \\ 0 & 0 & 0 & D\cdot g[y] \end{pmatrix}, \tag{707}$$

and either demanding (705) with the setting for $F[f]=(C_f+f)^2$ or (706) with $F[f]=(C_f+f)$, we find that a set of wave-like solutions of the kind:

$$f[t,x,y,z] = \begin{Bmatrix} f_0\left[\frac{\sqrt{-H\cdot A}}{A}\cdot t+x, \frac{\sqrt{-B\cdot D}}{D}\cdot y+z\right] \\ f_1\left[\frac{\sqrt{-H\cdot A}}{A}\cdot t+x, \frac{B}{\sqrt{-B\cdot D}}\cdot y+z\right] \\ f_2\left[\frac{H}{\sqrt{-H\cdot A}}\cdot t+x, \frac{\sqrt{B\cdot D}}{D}\cdot y+z\right] \\ f_3\left[\frac{H}{\sqrt{-H\cdot A}}\cdot t+x, \frac{B}{\sqrt{-B\cdot D}}\cdot y+z\right] \end{Bmatrix} \tag{708}$$

would perfectly do the job. Waves, however, are typical quantum mechanical features and so we conclude that the evolutionary potential of our S-curve metric provides degrees of freedom in form of volumetric waves. For illustration we set A=1, $H=-1/c^2$ and assume h to be only a function of t and x, which gives us:

$$f[t,x,y,z] \Longrightarrow f[t,x] = \{f_0[c\cdot t+x], f_1[-c\cdot t+x]\} \tag{709}$$

and makes the metric (707) to:

$$g_{\alpha\beta} = F\left[C_f + \begin{Bmatrix} f_0[c\cdot t+x] \\ f_1[-c\cdot t+x] \end{Bmatrix}\right]\cdot\begin{pmatrix} -\frac{h[t]}{c^2} & 0 & 0 & 0 \\ 0 & h[t] & 0 & 0 \\ 0 & 0 & B\cdot g[y] & 0 \\ 0 & 0 & 0 & D\cdot g[y] \end{pmatrix}. \tag{710}$$

With the setting for $F[f]=(C_f+f)^2$ this metric would have a vanishing Ricci curvature everywhere, but still, even though very simple and of low number of dimensions, provides a rather active dynamics. The metric could be illustrated as two sigmoid entanglements of two pairs of dimensions being wrapped in a cloud of t-x waves.

As for $F[f]=(C_f+f)^2$ the governing quantum or evolution equation (705) becomes even linear, because we have:

$$0=\begin{pmatrix}\frac{g^{\alpha\beta}}{2F}F'\left(2f_{,\alpha\beta}(n-1)+f_{,d}g^{cd}\begin{pmatrix}g_{\alpha c,\beta}-g_{\beta\alpha,c}-g_{c\beta,\alpha}\\+\frac{n}{2}\left(2g_{\alpha\beta,c}-g_{\alpha c,\beta}-g_{\beta c,\alpha}\right)\end{pmatrix}\right)\\+(n-1)\frac{f_{,\alpha}\cdot f_{,\beta}}{4F^2}g^{\alpha\beta}\underbrace{\left(4FF''+F'\cdot F'(n-6)\right)}_{=0}\end{pmatrix},\qquad(711)$$

$$\Rightarrow 0=g^{\alpha\beta}\left(2f_{,\alpha\beta}(n-1)+f_{,d}g^{cd}\begin{pmatrix}g_{\alpha c,\beta}-g_{\beta\alpha,c}-g_{c\beta,\alpha}\\+\frac{n}{2}\left(2g_{\alpha\beta,c}-g_{\alpha c,\beta}-g_{\beta c,\alpha}\right)\end{pmatrix}\right)$$

the solutions for f would be additive and could be superposed, leading to arbitrarily complex wave function structures.

15 The Third Tower of Infinite Spiritual Dimensionality

Fig. 57: "Inside my head", a six-year-old describes what he "feels" and "sees" when closing his eyes. [D]

15.1 Why Does Consciousness Need Spirituality?

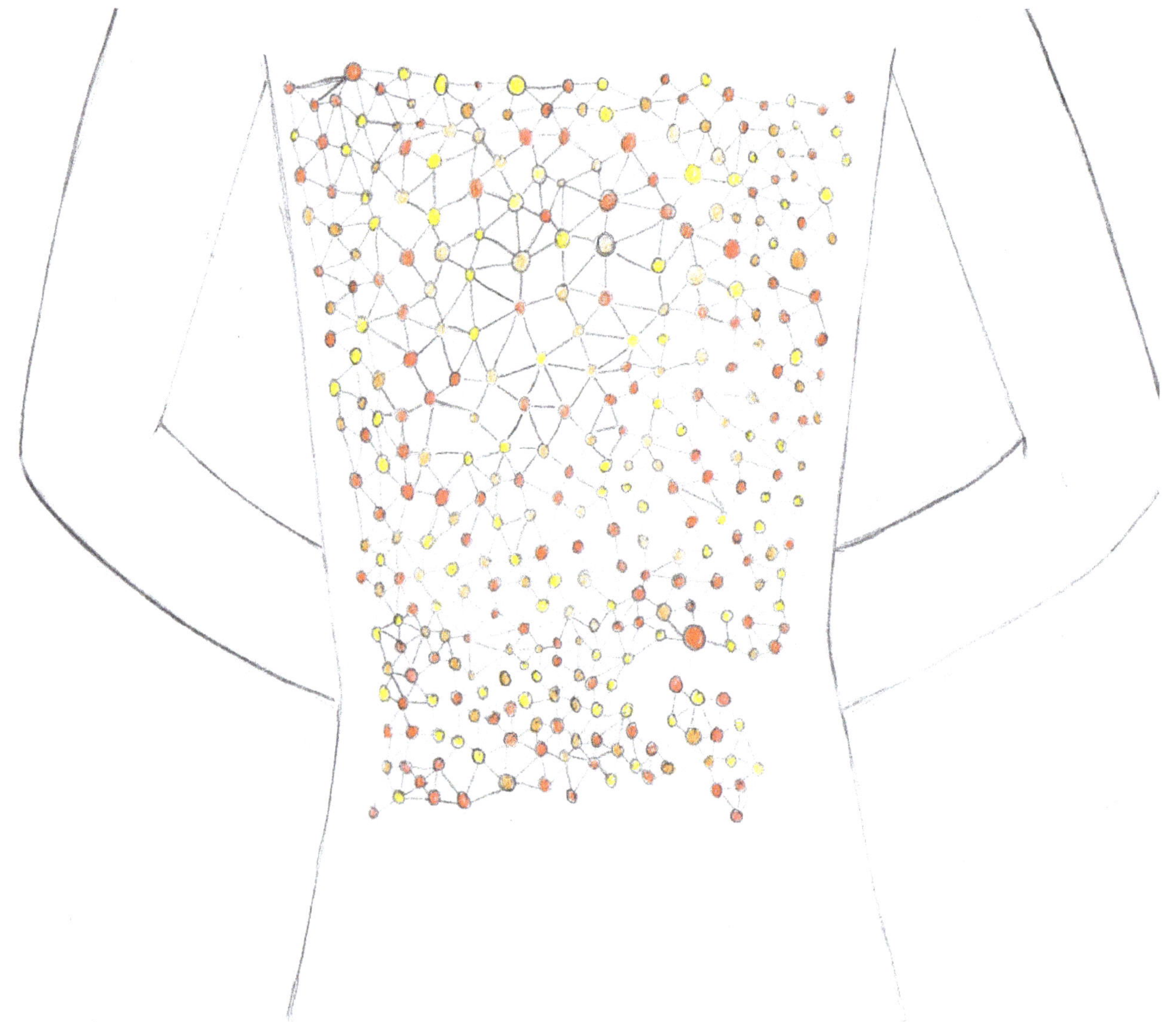

Fig. 58: The spiritual body. [A]

As most human beings see themselves not only as intelligent, but also as spiritual entities (figure 58), we find ourselves in the position of needing to also somehow incorporate the aspect of spirituality into our holistic consideration of the human mind and mathematical psychology.

Thus, in this chapter we are going to investigate how evolution needed to invent spirituality in order to save consciousness. It will be shown that, due to Darwin's evolutionary laws, the latter is not possible without the first.

15.1.1 The Evolutionary Dilemma – The Self-Conscious Bottleneck Situation

There was a moment when evolution "had the amazing idea" to develop consciousness, which, as we hinted in our derivation above, seems to be "not more" than the ability to actually and actively steer certain parametric quantum gravity solutions for the system "mind" by the system (the mind) itself. Thereby, evolution realized, which is to say the process of mutation and selection "discovered", that life forms equipped with consciousness actually weren't the fittest. As a remark, however, it needs to be pointed out that consciousness probably was not the thing evolution had aimed for. It only wanted to have cleverness. Through cleverness, evolution has produced species that are able to be innovative, use tools, perhaps even create some themselves.

Through cleverness, evolution could bring about entities that are able to help evolution with the ever more complicated inventions necessary to really make the difference. Consciousness, on the other hand, merely came as a by-product nobody really had aimed or even wished for. One may say that consciousness was the true original sin. A sin, by the way, nobody should be blamed for, but evolution. After all, if we all had stayed dumb like earthworms, the word "sin" would not even exist. The third author even is of the opinion or rather sees this as a possibility, that a calculating or perhaps "thinking" universe reproduced consciousness, which already existed in a metric form on an extremely low scale, by the means of biological substrates. The "unconscious goal" was to bring an intelligent vector into the apparently chaotic short scale processes of evolutionary progress and regress (figure 59). Consciousness might have just been the outcome of a search process for a vector stabilizing evolutionary force in order to bring a bit more order into the semi-chaos of purely self-organizing-based structural formation.

Fig. 59: To evolve or not to evolve. [A]

The main part of this little piece of text about the connection of the evolution of consciousness and spirituality (co-evolution even) is to be found in the book "The World Formula: A Late Recognition of David Hilbert's Stroke of Genius" from the third author [2] (Section 17.5.3, starting page 556). There one finds – among other things – this interesting story about a godlike entity the author had characterized as a "young nothingness".

The end of the story reads as follows:

On first sight it was good to have some kind of awareness or self-awareness, but within the cruel reality of life on earth it often became a disadvantage to "feel" respectively to consciously recognize what was going on and what life or being consciously alive really meant.

So, evolution had to learn it the hard way, that it could create sufficiently fit life forms, who also have consciousness only when adding yet another piece to that recipe and this was spirituality. Only spirituality allowed these self-aware and feeling life forms to put all those difficult or impossible to understand, often cruel or even unbearable experiences to a place in those bigger and bigger growing brains where it (meaning all the mess, cruelty, and madness around) would hurt a bit less (figure 60).

Fig. 60: Trust. [A]

Of course, in order to properly investigate this problem, one requires not only a completely new and most holistic understanding about evolution, but also about the psyche of sufficiently cognitively capable species, which not necessarily started on this planet with the early human beings [7, 8, 55, 70, 74, and this book]. We should also point out that evolution had, next to spirituality, also another option to play with and to repair the damage, respectively, to mitigate the danger emanating from the conscious self-awareness. Interestingly, this option has most generally to do with our way of digesting information, mulling things over and thereby creating innovative ideas. We will show in this chapter further below how this "individual innovation center" can also protect us from the harms of a conscious state. Along the way we will learn about the negative effect being connected with this intrinsic "thought cooking process", the Dunning and Kruger mechanism [207, 208]. We might even suspect that the two centers, the spiritual and the innovative one, are effectively the same, because both need to deal with thoughts as finite pieces of information, consequently resulting in a quantum statistic and the corresponding Planck radiation law.

But now we want to go on with our little story about how evolution had to invent spirituality.

Of course, evolution never intended to make this storage for spiritual thought bigger than absolutely necessary. After all, evolution only wanted to create fitter life forms and had no interest whatsoever in bringing dim and passive believers into being. Thus, the organ of spirituality only ever was thought as a temporary storage room for everything which could not be explained by the conscious (quantum gravity equation solving and individual parameter-set-solution creating) mind straight away.

Unfortunately, however, it turned out that explanations for some things took longer than the typical life span of such life forms. In fact, there were problems where even the typical lifespan of a whole species would not suffice[21]. Not even the strongest Dunning-Kruger effect would in such cases suffice to fool the individual onto Mount Stupid (see section below) long enough so that it would not suffer from the consequences of the conscious mental stress fields. Thus, it happened that there was always something in these spiritual storages. They were never empty and the bigger brains, keen to also get these unexplainable things explained, were extremely susceptible to all sorts of prophets who promised to have solutions, answers, guidelines, and – apparent – orientation. In other words: our hunger for "meaning" made us extremely vulnerable regarding promises of salvation.

And so, the epoch of the shamans, priests, imams, Gretas, and CO2-smellers, esoterics, astrologists, politicians, political scientists, soul-experts, climate-apostles, philosophers, and coffee shop owners had begun.

But we do not want you to get us wrong. We are far from criticizing evolution for its invention. On the contrary, because with spirituality, evolution gave us something else…

It gave us hope (figure 61) …

[21] Maybe this universe as a whole – if a thinking entity – will come to the conclusion that it will not be able to answer all the questions it has in its own life span and then has to think about a way to recreate itself with some of the earlier results being implanted into the new being. Maybe this has already happened "before."

Fig. 61: "Thank you, brother!" [A]

15.2 How Do We Emit Confidence or the Mathematically Derived Dunning-Kruger Effect

Current politics is the incarnation of stupidity and ideology in one systemic disease favoring the dumbest and most parasitic elements.

How can we avoid climbing "Mount Stupid"?

There are two dominating processes in which we emit confidence. Of course, there are also other options, often only derivatives of the two dominant ones, but we leave these apparently recessive processes for later. Those two dominating ways can be – briefly and dramatically simplified – described as follows:

a) We absorb information and give it away (show or mirror it in our confidence level to the outside world) in dependence on how well we have digested it. In other words, we have a perfectly monotonic behavior between knowledge and confidence emittance.
b) We internally reflect upon a certain topic and radiate to the outside world pieces of our inside "confidence" in dependence on our inner state of "excitement" and / or self-assurance.

Thereby, the outside observer can only detect the sum[22] of the confidence emittance of the two sources a) and b) and does not truly see the inner confidence level. This is hidden inside the individual (one may call it individual censorship or confidence horizon).

We therefore will prefer to use the expression "confidence emittance" instead of just the word "confidence" as the two might be severely different.

While we can assume that source a) should be approximately a linear dependency, which is to say that our "confidence emittance" increases linearly with the properly absorbed and digested information, we have to perform quite some contemplation on point b). This is our more or less innovative source, where new ideas could come from and where we mull things over before making up our minds or even uttering an opinion. It is only fair to consider this source as some kind of cavity, in which bits of information are being processed, meaning they are permanently internally exchanged, respectively, intrinsically emitted and absorbed. This, however, is nothing else but the description of a black body or cavity radiation system. As the derivation of the corresponding radiation law is textbook knowledge, we will not present it here but only refer to corresponding literature [96, 97]. When we now take into account that all information has to be discrete so that b) results in a quantum system or quantum-dominated object that automatically produces a Planck-like radiation, we can evaluate the connection between information being absorbed and "confidence emittance" for an individual.

---

[22] It has to be pointed out that the word "sum" does not restrict the addends to positive signs. While the monotonic knowledge-confidence-emittance as described in source a) in fact is positive definite, the sources of type b) can have negative signs. There are enough psychological diseases and clinical studies backing this conjecture up, but in this book we will only concentrate on the ordinary Dunning and Kruger effect and not its antipode.

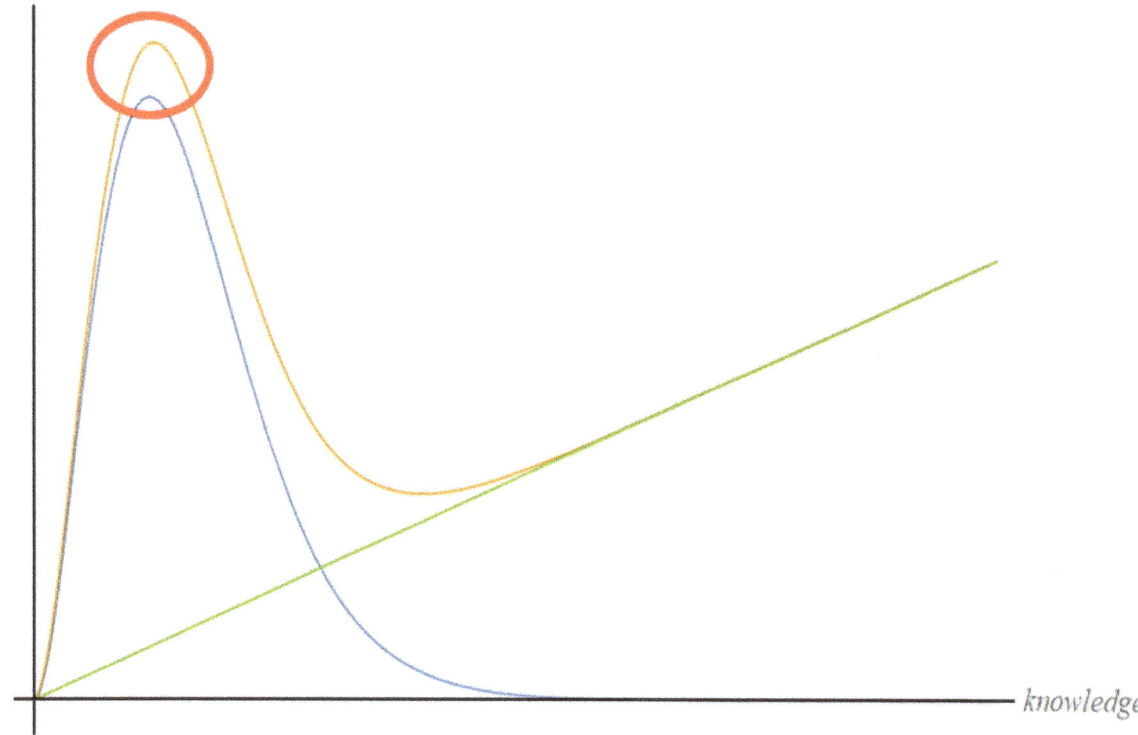

Fig. 62: Individual confidence emittance to the outside world as sum of source a) (green), source b) (blue), and the sum of the two sources (yellow). The red ellipse marks the so-called Mount Stupid (see text).

We immediately recognize the typical Dunning-Kruger dependency [207, 208] with the well-known "Mount Stupid" (local maximum of the yellow curve) at relatively low knowledge levels (see red ellipse in figure 62). As far as the authors know, this is the first time that this effect has been connected with the quantum statistics and has been mathematically derived.

15.2.1 The Other Temperature

But having a mathematical description of this effect now at hand, we can try to find out more. For here and now, we are only interested in the influence of the inner state of "excitement" and / or self-assurance. The dependency is shown in figure 63 where we clearly see how with growing self-assurance the quantum-dominated b)-source becomes ever more dominant at lower amounts of absorbed information or knowledge, and consequently the mount-stupid-effect increases.

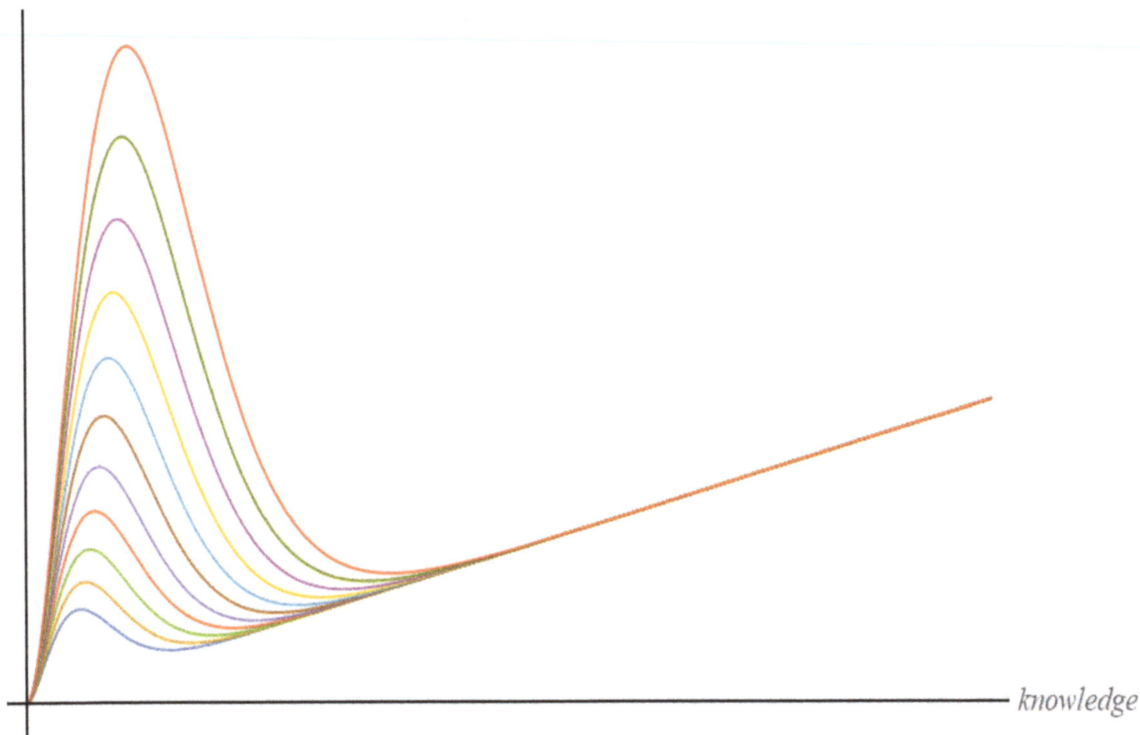

Fig. 63: Individual confidence emittance to the outside world as sum of sources a) and b) for a variety of inner states of "excitement" and / or self-assurance. Thereby, from dark blue to red, the Mount-Stupid-effect increases with growing self-assurance.

Most interestingly, we find that what in Planck's radiation law plays the role of the temperature, becomes the intrinsic self-assurance in connection with the confidence emittance. This makes intuitive sense, because self-assurance could be seen as comparable with some kind of inner excitement. Seeing the latter as an average of the "kinetic" energy of our inner thoughts, we immediately recognize the classical temperature definition and hence, the congruence between self-assurance / self-excitement and physical temperature. Unfortunately, the inner b)-confidence is not coupled to any real knowledge, but only the result of more or less isolated inner reflection (literally coming out of a bubble). It can therefore – just as the Dunning-Kruger effect illustrates – lead to a confidence emittance which is more an embarrassment than reality-based wisdom. Hence, the expression "Mount Stupid."

15.2.2 How to Avoid Climbing Mount Stupid

After sending the calculations and figure 62 to a few friends, including even some experts on the field, one of them pointed out to the third author that he had hoped to obtain hints about how to get over the mountain of stupidity, i.e., avoid or just "tunnel through" it, and asked what ways and means there are of not making this mistake that is obviously inherent in us.

Fig. 64: Mount Stupid effect in action. [A]

Well, first of all we should realize from which time in the evolutionary development of mankind our more recent brain structures originate. They were constructed and formed when we roamed the countryside as hunters, gatherers, and fishermen in at best medium-sized groups and the group already ensured – via its integral collection of experiences and knowledge – that Mount Stupid could not become too high. Those who did not learn this quickly became victims of the omnipresent dangers: they landed under the mighty hooves of the mammoth, in the clutches of a cave lion, poisoned themselves on plants or mushrooms, froze to death like Ötzi somewhere in the mountains, or simply died of an avoidable disease or injury without the chance of reproduction. It was simply the experience and the group interaction enforcing severe caution with the imaginary knowledge that made people quickly realize that they had to be careful in unknown areas of knowledge in order not to fall into the trap of being too stupid to realize that one is too stupid.

At the same time, however, the center of innovation was an advantage for those who had ideas over those who were not quite so imaginative and – consequently – creative. One could assume that this is why Neanderthals were inferior to homo sapiens sapiens[23], even if they were physically the more robust species. So, despite the Mount Stupid disadvantage, the innovation center was an advantage and was therefore retained by evolution.

How could evolution have foreseen that there would one day be politicians, Tuna-Gretas and CO2-smellers, media scribblers and fact-checkers who, in their protected bubbles ("black body radiation caves"), would never be exposed to the experience of the Dunning-Kruger effect and its negative outcomes? Evolution could not have known that there would one day be self-contained areas of gigantic groups of individuals in which not only individual mountains of stupidity would be able to grow, but whole continents of total stultification would sprout from the ocean floor into the sky.

Real experts, i.e., those who have already experienced Mount Stupid at some point because they have worked their way beyond it in a certain field, therefore hardly ever show the Dunning-Kruger effect. The same holds true for sensibly educated people who have not been kept in a bubble of well-being or a protected area and who have therefore automatically learned the pitfalls of the competence conceit through the "corrective of external reason"... who, yes, here the word is appropriate, have adopted nothing other than modesty. This also explains why there are so many mount-stupid-climbers today and – here is an irony for you – this automatically increases and intensifies dangerous and self-destructive mass formation processes as described by the leading psychologist M. Desmet [216] (see also [56] and [217]). Such processes, however, are the driving forces for the fulfillment of the Fermi paradox [12] in its worst aspect. They – the processes, ultimately being caused by the very structures the growing brains "wanted" to overcome the consciousness bottleneck problem with – create the new bottleneck for evolution which, most ironically, evolution wanted to overcome

[23] Something the authors definitively intend to calculate through in one of their next little projects [70].

when solving the consciousness problem by the spiritual-innovation center (c.f. subsection above under headline "The Evolutionary Dilemma – The Self-Conscious Bottleneck Situation"). These mount-stupid-climbers are all people (politicians, mainstream media flacks, and fact-checkers above all) who have been and are being bred matrix-like into spheres and to whom one is not allowed to say anything, least of all the truth about their own stupidity. In climate or Covid "debates" alone, it has happened to the authors all the time that we could not get through with facts and evidence (e.g., introduction sections of [2, 4] and – MOST IMPORTANT – [218]) only because our counterpart, sitting in his bubble, seriously considered the presentation of facts and evidence as "unfair and inappropriate". Such people (and only too often they are "academics", i.e., students who have been made stupid on purpose) thus refuse precisely to take in that very knowledge which could bring them to the realization that they do not know nearly enough to justify their big headedness, which is to say, to cavort so arrogantly and self-confidently in a certain field, as they are doing right now as actually stupid know-nothings, but not being aware of this fact.

So, back to our question:

How can we avoid becoming victims of the Dunning-Kruger effect?

Well, we just have to be aware of it and get as much supervisory authority and constructive criticism as we can get… to each and every topic. A tiny bit of humility helps along the way.

15.2.3 Summary about the Metric Dunning and Kruger Effect

It was found that, by assuming that the internal thinking process of the human mind consists of discrete units (quantizable pieces which we might call "thoughts"), we can easily derive the Dunning and Kruger effect as a necessary characteristic of human behavior. The opposite, the anti-Dunning-Kruger effect can be explained similarly, but in this chapter was only mentioned as a theoretical possibility which also might explain certain clinical observations. The same holds for all other thought-related processes (fear, individual assessment of objects and other individuals and – oh yes – all attraction and repulsion creating potential or feelings, like antipathy, affection, if not so say, hate and love) where the inner reflections of quantized pieces of information play a role.

15.2.4 The "Other Temperature" also in T1 and T2

It should be noted here that the generalization of the Planck radiation law with respect to generalized photonic concepts [2, 4, 205, 213] (see also section "The Key Aspects for the First Tower: Electromagnetic Interaction and Elasticity") provides a quite comfortable way for a thorough and comprehensive parameter identification of various activation agents and PSAs via simple black body like experimental setups[24] with just the right – PSA and activation agent related – wavelengths selections, resonance frequencies, and quite general order-parameter variations (not only temperature). As elaborated in various sections above, the corresponding analysis imbibes nothing else but generalized Fourier techniques and their holistic understanding as sequences of separation and creation applications.

[24] For IP protection reasons this aspect will not be further elaborated in this RASA strategy book.

16 Can One Man Change the World without Showing His Hand?

And what has this to do with quantum gravity and material science?

Fig. 65: A man changes the world without showing his hand. [C]

We start with the material science aspect of the 2-fold question from above.

Thereby the question was brought forth to one of the authors, Norbert Schwarzer, whether a single and most localized, which is to say almost point-like, defect could influence an ensemble of atoms or molecules within a component in such a way that it (the defect) dramatically influences the properties of the component, but does not show itself as the culprit for the observed parameter change.

Schwarzer had the answer ready, because he had already dealt with this question in connection with a slightly different and rather more theoretical application, namely the physical riddle about dark matter and energy [219]. The answer is mathematically clear: It is possible that a certain defect creates a parameter change for a – compared to its own size – huge area inside any material structure without showing itself. Thereby the source only needs to radiate its influence in a certain attitude which leads to a suitable collection of plane waves, which are themselves just superposed signals coming from the defect in the following form:

$$\begin{aligned} e^{i \cdot C_{jt} \cdot z} &= \sum_{l=0}^{\infty} (2 \cdot l + 1) \cdot i^{l} \cdot j_{jk}\left[C_{jt} \cdot r\right] \cdot P_{l}\left[\cos[\vartheta]\right]; \\ k = k_{l} &= \frac{\sqrt{1 + 4 \cdot l \cdot (l+1)} - 1}{2} = 0,1,2,3,\ldots \text{ for } l = 0,1,2,3,\ldots \end{aligned} \quad . \tag{712}$$

An outside observer, however, would only see the plane waves and cannot make out the source.

From this automatically follows that now engineers, who cannot find the source for a certain global failure mechanism in a who-knows-what for an application, have just the perfect excuse.

Only measurements of very local character would allow to move further forward in such cases. In mechanics, for instance, this could be nanoindentation and highly sensitive scratch and tribo-tests, potentially combined with the so-called "calo method" for higher depth sensitivity [220, 221].

Now we are ready to also answer the question in the title of this chapter, which reads:

"Can One Man Change the World without Showing His Hand?"

Well, as it is shown in [56], in bigger socioeconomic systems (societies), single individuals or even groups of the latter are radiating their influence to the environment (rest of the society) in multidimensional sphere-like waves. Thus, the mathematical structures of any form of communication or information-transmittance are quite similar to the ones shown in the equation above and hence, are equipped with the same potential to be superposed to globally appearing (and acting) as apparent plane waves. Clearly it follows that a global influencer does not necessarily need to show his face (or hand) in a potentially evil game, changing the world really badly. Unfortunately, we cannot think of any example where something like this has been done in the past or is currently being attempted, can we…

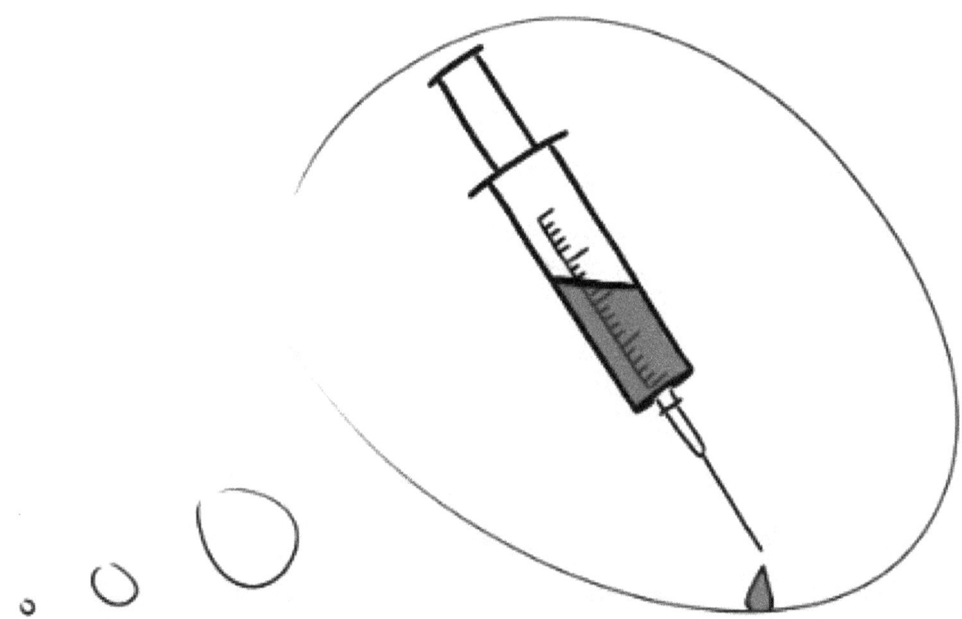

… but at least we now know that there is such a possibility.

17 Planck's Principle – On the Evolution of Science

17.1 "Science Progresses One Funeral at a Time" or Why Almost Nobody Has Heard About Our Quantum Gravity Theory?

Perhaps we find the answer in the so-called Planck's principle.

Fig. 66: Science progresses one funeral at a time. [A]

According to Wikipedia[25] this sentence was formulated by Max Planck as follows [209]:

"*A new scientific truth does not triumph by convincing its opponents and making them see the light, but rather because its opponents eventually die and a new generation grows up that is familiar with it ...*
An important scientific innovation rarely makes its way by gradually winning over and converting its opponents: it rarely happens that Saul becomes Paul. What does happen is that its opponents gradually die out, and that the growing generation is familiarized with the ideas from the beginning: another instance of the fact that the future lies with the youth.

— Max Planck, Scientific autobiography, 1950, p. 33, 97

Colloquially, this is often paraphrased as "Science progresses one funeral at a time"."

It is quite obvious that this principle can be derived from a generalized evolutionary concept [70], because as all living creatures, also the "old farts" in science and their protégés are fighting for resources and only – if at all – secondarily for the scientific truth.

In other words, which is to say in the words of the third author of this book:

"… the true Fausts are very rare… and to make this absolutely clear, I do not consider myself a Dr. Faustus myself. I am also fighting for resources, just as my fellow next. Hence, you better check everything I say or write, prove my starting point, falsify my derivations, and double-check my conclusions, PLEASE!"

[25] https://en.wikipedia.org/wiki/Planck%27s_principle

18 Global Summary

Even if the authors should torpedo their status as holistic "scientific entrepreneurs" with this last summary statement, they – as honest persons and true scientists – cannot help but notice that the key technologies, herein named the towers T1, T2, and T3 for their stunning character and appearance, obviously have far more potential than mortals like ourselves can be aware of at this point in time… and space. In unearthing this full potential and setting it free via the most general and holistic approach there can currently be seen, the authors see one of the many contributions they could bring into the community of ALL human beings [222].

19 Appendix A (by W. Wismann and D. Martin): Using *eSurface*® Technology to Enhance & Improve PCB Manufacturing Capabilities

19.1 Background

Printed Circuit Board (PCB) evolution over the last decade has been challenged with manufacturing constraints keeping up with the demand for supporting smaller, lighter, more feature rich products, all while lowering costs. These new products often demand incorporation of more components, device packaging with smaller interconnect and finer pitches, and demanding more layers and associated interconnect from layer to layer (vias). As the semiconductor industry struggles with Moore's Law[26], the PCB must keep up with the semiconductor's ASIC and processor packaging features often stretching the manufacturing capabilities of historical construction processes. Some of these demands push standard volumetric constraints to utilize significantly finer features and must be designed to accommodate more layers forcing a larger quantity of vias. Typically, this also forces the use of thinner dielectric materials, which can complicate sequentially layered construction.

Advances such as Laser microvia technology, Laser / LED Direct Imaging, Phototool resolution, and advanced Dry Films, have somewhat enabled progress over some of these challenges, but the inherent subtractive (etching) process still has basic principles of physics to overcome, and does not fully support the ultra fine feature requirements of advancing circuitry in PCBs.

Over the last three decades a recognizably more efficient solution using Additive processing has been desired, repeatedly introduced, but failed to meet requisite industry standards especially in adhesion / peel requirements and resistance to solvents (RTS). Historically, the offered solutions have been topical agents, metal laden inks, or adhesive based approaches.

eSurface® technology using a global pioneer patented process is unique in that the typically tried topical solutions are abandon and ***eSurface***® employs an *"in"* the surface technology creating an elemental covalent bond at available tangent points through use of its chemistry, referred to as "Covaler". This complex coordinated chemistry is a photo imageable activated catalyst capable of utilizing industry prevalent UV (or other wavelengths) imaging spectrum to trigger the deposition and bonding of industry standard catalyst to the substrate without epoxies, resins, or adhesives. ***eSurface***® has selected Pd to be compatible with current conditioning / plating practices, but can switch out this element to other metals if desired. The ***eSurface***® process can be used in a myriad of ways incorporating conventional PCB practices or opening up doors to completely new methods of metalizing materials, planar or with multi-dimensional features. ***eSurface***® goes beyond PCB manufacturing and has far reaching application potentials in Semiconductor, Solar /Photovoltaic, RAD Hard /EMP, EMI / EMF, Wearables, etc.

The ***eSurface***® process lends itself to use in the PCB environment through fully-additive methods as well as semi-additive processing with significant benefit as described herein.

---

[26] https://www.pcworld.com/article/426997/moores-law-at-50-the-past-and-future.html

19.2 Fully-Additive

19.2.1 Fully-Additive: General Overview

The use of fully-additive technology has been sought for several decades and recognized as a more efficient method to producing PCBs. Applying Cu only to the requisite areas of circuitry as opposed to etching away the predominant Cu mass of the panel has obvious advantages. Historically adhesion of additive Cu was the typical inhibitor precluding this process method adoption. ***eSurface***® solved this issue with superior bond, more often surpassing clad peel strengths. The current demands of finer traces and features, higher density circuitry, and smaller via structures has pushed print and etch technology beyond efficient or reliable manufacturing capabilities. The ***eSurface***® fully-additive process[27] drastically reduces process steps and associated variables often present with etchants and sequential plating operations.

19.2.2 Fully-Additive: Key Advantages

A) Utilizes far less process steps and smaller facility footprint improving cycle time and cost structures
B) Reduces chemistries needed through process step elimination
C) Reduces specific Cu clad thickness inventory improving material management
D) Reduces or eliminates etch artifacts such as trapezoidal traces, undercut, overhang / entrapment, and scalloped trace edges
E) Improved adhesion
F) Recoverable processing of panel (prior to drill, panel outer layer Cu can be etched / restarted without compromising or scrapping already completed work

While the PCB industry is working on the complete ecosphere to better support a complete Fully Additive processes, ***eSurface***® has successfully accomplished and demonstrated this industry changing goal by using Covaler and currently achievable processing techniques.

The ***eSurface***® technology is a huge step toward commercializing Fully Additive processing of PCBs as a key enabler to a new era of manufacturing techniques. Advanced Cu chemistries are being further developed to improve electroless Cu deposition composition that will bring Fully Additive practices to fruition. Globally a multitude of market sectors are looking to advance manufacturing capabilities and additive attributes are a common modality.

---

[27] https://www.youtube.com/watch?v=qoCWU2djmKc

19.3 Semi-Additive

19.3.1 Semi-Additive: General Overview

Semi-Additive Processing (SAP) has gained popularity over the last several years and is present in many PCB manufacturing facilities as a method to accomplish advanced or finer features. The objective is to minimize the base or "floor" Cu to reduce time in etchants, which results in a less propensity to distort traces. The traditional print and etch technology is modified by starting with a thinner layer of Cu than the standard (1/4oz –1oz) foils. This is conventionally accomplished by procuring thin foil clad materials, chemically milling of thicker Cu clad foils, or using alternate activation /Cu deposition. These typically have either a significant cost impact and / or not usable in outer layers due to poor adhesion. ***eSurface***® accomplishes this in a different way by activating the entire surface by flood exposing the entire Covaler applied panel, then putting a 1 – 1.5 micron "Seed" layer of electroless Cu.

19.3.2 Semi-Additive: Key Advantages

The semi-additive processes outlined herein simply provide a more efficient way to replace the typical clad Cu foils with a much thinner layer of Cu that exhibits a far superior bond when ***eSurface***® is properly employed. The following images are general processing steps and not intended to depict the complete processing cycle used.

Advantages in using this ***eSurface***® variant to conventional semi-additive processes are:

A) Standard depositions are typically an order of magnitude thinner than thinnest Cu foils (1micron Vs 10micron)
B) Less expensive alternative than procuring thin foils
C) More robust & easier to handle than thin foils
D) Better controlled Cu thickness than costly chemical milling of foils
E) Typically, superior adhesion compared to Cu clad /foils
F) Better controls in holding tolerances compared to etching of thicker Cu
G) Recoverable Process (prior to drill, panel outer layer Cu can be etched / restarted without compromising or scrapping already completed work)

19.3.3 Semi-Additive: Summary

The unique highly minimized Cu seed layer applied in the ***eSurface***® Semi-Additive Process provides a substantially thinner "floor" of Cu to etch away. This eliminates the desired fine features and traces from being affected by the time in etchants. This drastically reduced etch-time, affords more even control, and significantly minimizes lateral (side) etching, which typically forms trapezoidal or scalloped trace walls. The typical undercut or over etching is potentially eliminated through this process. Additionally, the traces retained have an elemental covalent bond with the substrate material and typically has far superior bond to that of clad Cu foils, low tooth foils, and alternate additive plating techniques.

19.4 *eSurface®* Technology Overview

The ***eSurface®*** process consists of a few simple steps; Application & Dry, Exposure, and Developing of the Covaler. These processes must be applied in a clean, controlled environment with a filtered light spectrum under 450nm.

Application utilizes two pieces of recommended equipment; the Covaler Applicator and Dryer / Oven, typically configured in-line with no handling between operations. The ***eSurface®*** Applicator is the most critical piece of equipment in process as it applies a controlled coating of Covaler that typically is only angstroms thick (depending on material porosity), and is very efficient due to its closed loop design.

Exposure utilizes imaging systems with a spectrum of UV between 355nm – 402nm, (dosing and spectrum typically like those used in standard dry film processing). For semi additive processing, where the entire panel is exposed, this operation can ideally be configured with an in-line UV exposure unit to further minimize handling and better support HVM with less operator intervention. Fully Additive uses “pattern” imaging (DI, or photo tool) and can also be configured in-line if such equipment is available.

Developing is a process of eliminating unexposed or residual Covaler (post imaging) through a controlled impingement clean rinse, using filtered or mild DI water, then dry. The ***eSurface®*** Developer equipment is designed specifically for conveyor processing and can also be configured to accept outputs from inline UV exposure platforms.

It is very important to understand that once the Covaler process is completed through Developing, there is NO residual left on the substrate. Only the desired metallization where UV exposure was applied will be present.

19.5 *eSurface®* Processing Overview

The processing methods provided herein are general processing steps and intended to provide a high-level overview with the understanding there are many other potential variants which can be explored combining current in-house practices and new novel methods. Please refer to the “***eSurface®*** Introduction and Training Guide”[28] where examples of suggested detailed process instructions can be found using proven methods and practices.

---

[28] Please request a copy of the confidential training guide: info@esurface.com

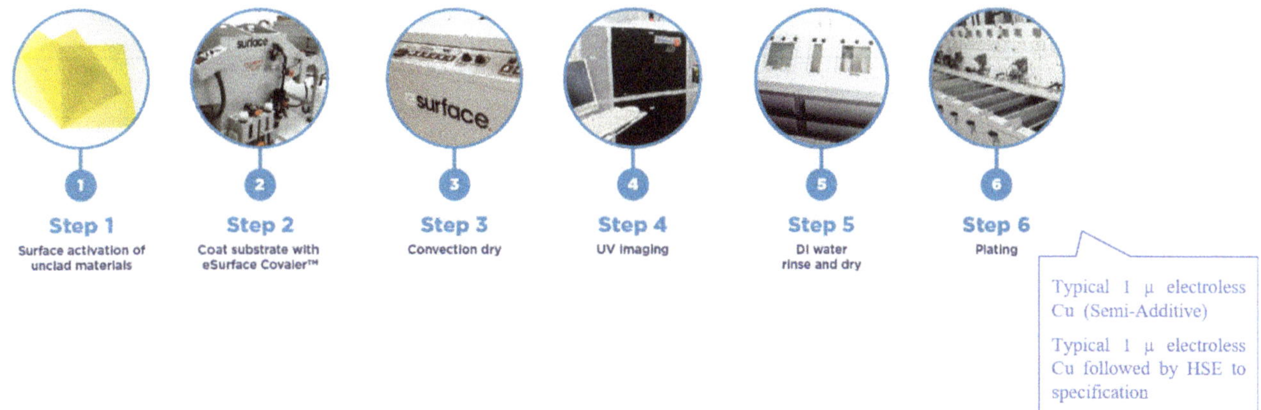

19.6 Feature Advantages

Through proper application, trace lines and spacing or features can see significant improvement through the reduction or elimination of etchant time. In ***eSurface***® SAP the typical Cu requiring etch is an order of magnitude thinner than economically available thin Cu clad materials, and 35X thinner than standard 1oz Cu foil. This is significant when understanding the effect of etchants and the lateral or side etch associated with etching down to the laminate base to create isolation (space between traces or features).

For fully additive, the etching process is eliminated as deposition of Cu starts at the bare base material and is built up only in the requisite areas.

Whether utilizing SAP or FAP, the basic principles of ***eSurface***® activation are the same and accomplished with pattern imaging for FAP, and panel (flood) imaging for SAP.

The ***eSurface***® process improvement of linear (vertical) trace geometry affords closer spacing, closer impedance matching, signal integrity, better ability to hold trace tolerances, and highly improves signal integrity.

With better signal integrity and less loss associated with undercut, scalloped edges, and other negative attributes associated with etching. Power budgets can be lowered, less heat is expended, and better volumetric efficiency is gained. The volumetric efficiency increases opportunity to design more circuitry into the same space, allowing more features on the same layer, thus reducing the number of layers and associated vias to make interconnect.

The ***eSurface***® use and effects on RF circuitry can often improve signal attributes to a 5x gain in bandwidth filtering and improve high-speed signal transmission (please request for "RF Test"[29]).

[29] Email request to obtain proprietary test results: info@esurface.com

19.7 Adhesion

Adhesion of the Cu to the substrate is an important element in PCB fabrication. Adhesion is measured in "peel strength" through industry standard practices. This test actually is designed for measuring the point of delamination (where the Cu foil releases from the substrate which is bonded through thermally pressed resin). This failure point is quite different in most ***eSurface***® applications as the novel elemental covalent bond provides a far more aggressive adhesion to the substrate and typically a cohesive failure occurs when tested (actual tearing of substrate). The cohesive failure typically retains elements of the substrate on the Cu being torn from the substrate and not merely delaminated. This is an important key factor when comparing Cu clad foils to ***eSurface***® processed applications. In fine trace applications, the Cu clad foils (even with acceptable adhesion) are dependent on the thin resin interface and typically roll off the board (very susceptible to peeling off when physically touched). This often precludes the use of fine traces on outer layers of the PCB where components are mounted, and subjected to sheer forces and soldering operations. ***eSurface***® with its superior bond, can enable the use of such fine features with more robust capability.

20 Appendix B (by W. Wismann and D. Martin): The Separation and Removal of Target Elements within a Host Media Through – The Enhancement of Polar Selectivity

If you want to have clean energy, how about cleaning the agents before using them?

20.1 Synopsis

By this innovation's definition, the host media is the feed stream, liquid stream, or wherever the target element is resident. The target element, compound or composite material may or may not be undesirable impurities or contamination. A particulate catalyst impregnated sorbent comprising of a metal, alkali, or alkali earth metal, metal oxide, a bimetallic combination (combination of metals), as the catalyst, impregnated into a carbon particulate, silica, or an alumina, a zeolite, a perlite form, or any structurally sound, porous sorbent, is provided for the separation and removal of the targeted elements, compounds, or composite materials from a liquid feed stream by a process which comprises the feed streams to be fractioned, and entered into a oxidation vessel or column, for a micro-bubbled oxidation reaction and then into a reactor or series of reactors and contacting the sorbent impregnated catalyst for the separation of the oxidize target element or compound by an enhanced selectivity to the catalyst, allowing the feed stream without the target element to pass through the reactor. A polar organic or inorganic solvent is passed through the reactor to wash the catalyst impregnated sorbent and then the solvent is separated from the target elements or compounds by distillation. The catalyst impregnated sorbent is then dried by heat or hot nitrogen which then regenerates the catalyst impregnated sorbent for reuse.

20.2 Exposition of the Innovation

20.2.1 Theme of the Innovation

This innovation relates to the separation and removal of target elements, compounds or composite material, which may or may not be undesirable impurities or contamination, from feed streams or fluid streams of hydrocarbons, carbonaceous liquids, and aqueous based solutions. In another aspect this innovation relates to sorbent compositions impregnated with a catalyst suitable for use in the targeting of the elements or compounds from the fluid streams by the means of an enhanced selectivity. A further aspect of this innovation relates to a process for the production of the catalyst impregnated sorbents for use in the removal of these targeted elements or compounds from the fluid streams.

20.2.2 Background of the Innovation

The present innovation relates to a process for the treatment of contaminated water, oil, hydrocarbons, and liquid carbonaceous fluid streams, and more particularly, a process for the selective removal of metals, nitrogen compounds, hydrogen sulfide and other sulfur species, aromatics, polynuclear aromatics, and specific hydrocarbon species, with the use of a particulate catalyst impregnated sorbent after oxidation of selective target element or compound.

Over the years various approaches have been used to separate specific hydrocarbon liquids from water and other polar solutions. In particular there exists a need for an efficient and economical process to remove oil or petroleum products from bodies of water. Moreover, these methods should permit the cost-effective recycling of at least a portion of the recovered hydrocarbon contaminant.

A steady demand for petroleum products among the industrialized nations has ensured a high volume of oil transfer with a corresponding release of material into the environment. The detrimental economic and ecological impact associated with such releases have generated several techniques for limiting the spread of the oil. However, such techniques are often labor intensive, expensive and provide less than optimal results.

In order to avoid the costly and often intractable problems associated with separating liquids from liquids, absorbents have often been used to facilitate the removal of hydrocarbons from aqueous solutions. These sorbent materials absorb the oil. Accordingly, effective sorbents should primarily absorb the contaminating hydrocarbon and not the water. In other words, the best materials for these applications are both oleophilic and hydrocarbon resistant. Among the absorbent materials which have been proposed for removing oil from water are wood chips, sawdust, certain clays, polymeric substances, cellulosic materials, and many others. Currently most of the sorbents used for such tasks are not salvageable or reusable and are intended to be destroyed or discarded along with the sorbed oil. Significantly one of the major drawbacks to the use of these materials is the prohibitive cost inherent in the preparation and utilization of a non-recyclable sorbent. In an effort to circumvent some of the difficulties inherent in the use of sorbent materials, hydrocarbon resistant sorbent materials have been applied to bodies of water contaminated with hydrocarbons. Hydrocarbon resistant sorbent materials differ from sorbent materials in that they distribute any associated liquid hydrocarbons in a film over the surface of the adsorbing particle. In contrast, absorption involves the uptake of the liquid hydrocarbon into the body of the solid sorbent and is closely related to the porous structure of the sorbent material.

Commonly used sorbent materials include fine sands, clays, solid inorganic compounds, polymers and treated natural fibers. For example, peat fibers, coconut husk, cotton fibers, jute, or wool may be coated with hydrocarbon resistant materials such as rubber or paraffin to provide a floating sorbent. However, such coated fiber sorbents generally involve labor-intensive fabrication techniques and relatively sophisticated production facilities. Further such sorbents are usually non-recyclable, often making them prohibitively expensive to employ. Similarly, other sorbents such as polymers may be too expensive for routine utilization and, if not easily biodegradable, may actually increase the adverse environmental impact.

More specifically, U.S. Pat. No. 3,891,574 discloses a sphere of carbon, which consists of a porous shell enclosing an empty space. The carbon particle is obtained by coating a core material with the carbon and subsequently removing the core through thermal decomposition. This process reportedly forms a heliosphere having a bulk density of approximately 275 g/l which has active surfaces on both the interior and exterior surface. Among other uses, the disclosure teaches that the resultant material may be used to adsorb crude oil. Yet, in addition to being fairly heavy, the material has a relatively low loading capacity of approximately 1.5 times its weight after extended exposure to the oil.

Accordingly, it is an object of the present innovation to provide a method for separating hydrocarbon liquids from water or polar liquid bodies through the use of a reusable, relatively low cost, sorbent. In addition to those materials previously mentioned, it is this innovation's disclosure that a particulate catalyst impregnated carbon may provide an efficient sorbent of liquid hydrocarbons in an aqueous environment. The resultant particulate catalyst impregnated carbon is usually in the form of small crystallites having dimensions considerably smaller than those observed in natural graphite. Adsorption properties of such particulate catalyst impregnated carbon materials are generally related to the amount of inner surface area and the selectivity of the catalyst impregnated.

Contaminated waters generally have a high dissolved iron (ferrous and/or ferric) content and other dissolved main group and/or transition metals (non-ferrous and non-ferric), such as, for example, copper, zinc, aluminum, manganese, silver, lead, cadmium, gold, nickel, arsenic, and other contaminated industrial wastewaters. The composition of waters contaminated with metals can vary substantially depending on the source or origin of the water.

One of the major problems facing the mining and mineral processing industry is the disposal and management of sulfide containing tailings. In certain instances, tailings containing pyrite, marcasite and pyrrhotite create particular problems because they oxidize readily due to weathering to form contaminated acid mine drainage. The rate of oxidation depends on the sulfide content, morphology, bacterial activity, ferric ion concentration, and oxygen availability. The acid mine drainage contains a high concentration of iron and other dissolved metals and has an excessively acidic pH. Virtually all heavy metal wastewaters from mine runoff to pickling liquors and rinses, galvanizing wastes, plating wastes, and hardening wastes must be treated to remove metals which contaminate the waste before the waste can be discharged into streams and other bodies of water.

Two principal methods have been utilized to treat acid mine drainage. Preventive controls include attempts to remove the sulfides, control the bacterial activity, control oxygen diffusion, coat the sulfide particles and agglomerate the tailings. Treatment techniques include neutralization, precipitation of hydroxides, and precipitation with sulfides, adsorption and removal. In the treatment and recovery of heavy metal ions from acid mine drainage, the conventional approach has been lime neutralization to precipitate the metal hydroxide. The precipitated hydroxides are difficult to filter. The metal hydroxides are not chemically stable and they are contained and will have to be decontaminated in the future. The sulfide precipitation technique, which utilizes sulfides as a precipitating agent, produces metal sulfides which are more stable as compared to the hydroxides. The metal sulfides are difficult to filter from solution. Furthermore, under certain circumstances, when an excess of sodium sulfide is used as a precipitating agent, a hazardous gas, H.sub.2 S, is often produced during the precipitation. A closed reactor vessel with secure venting would be required to minimize safety risks. The consumption of sulfides and other sulfur-containing compounds is excessive in the prior art processes due to the oxygen-sensitive nature of sulfide. Metal precipitates involving sulfur-containing organic compounds are easier to filter than inorganic sulfides and have been more widely used for wastewater treatment in recent times. However, when a waste stream contains a very large amount of metal to be treated, it is sometimes not economically feasible to employ these organic precipitates.

In general, the prior art processes are disadvantageous because they are non-selective, bulk precipitation processes, and they require high doses of ferrous ion at a high pH and at high temperatures (60-70 degrees C) for excessively long aging times to achieve successful oxidation and the formation of ferrite. Another disadvantage of the prior art co-precipitation process is the requirement of a protracted aging time, e.g., two to three days to permit ferrite product to acquire magnetic properties so that the magnetic ferrite particles can be separated from non-magnetic ferrite particles by a magnetic separator.

It can be seen that it would be beneficial to improve the processes for the removal of metals from contaminated water. It would be desirable to selectively remove metals or their oxides or salts from the wastewaters. It can also be seen that it would be advantageous to develop a process to treat contaminated waters, for example, acid mine drainage and mineral industries wastewaters, to provide a final product or products which can be conveniently removed from the bulk solution and effectively recovered and which provide a treated water which meets quality standards for discharge into the environment. Accordingly, the present innovation is directed to a process, which substantially overcomes one or more of the problems due to the limitations and disadvantages of the prior art.

By the process of the present innovation, most metals, which are present in contaminated water, can be selectively removed and recovered by oxidation. One skilled in the art using the process of the present innovation can oxidize most metals in contaminated water and for each metal to be removed from the contaminated water wherein the oxidation-reduction potential of the contaminated water is maintained at a positive potential.

In the field of liquid and wastewater treatment, filtration has long been a major method of removing suspended solids from liquid streams. In wastewater treatment applications in particular, the presence of suspended solids materials is frequently a major process problem and filtration has commonly been employed as a means of reducing and/or removing suspended solids from fluid streams. In such applications, down flow and up flow sand filters and dual or mixed media filters have been widely employed and have in general been shown to be cost effective and efficient in use. Nonetheless, work by practitioners in the field has shown that sand and mixed media filters are in general effective in removing suspended solids, but only under limited solids loading conditions. In general, solids concentrations of the liquid stream entering the filter must be below about 100-200 milligrams/liter. At suspended solids concentration values above this level, the filtration bed is susceptible to clogging and high-pressure drop across the bed.

For removal of contaminants, such as toxic organic chemical species, from liquid, it has been common to employ sorbent beds through which the liquid containing adsorbable contaminants is passed for removal of such contaminants. In particular, both carbon and synthetic resins have been widely employed as an adsorbent medium due its high selectivity for many organic and inorganic contaminants of liquid streams. Both of these are available in forms, which can be packed in columns, so the water can be passed through the medium without any need for subsequent solid/liquid separation steps. However, granular carbon cannot be regenerated on-site, and synthetic resins are often fouled by particulate matter. In the field of wastewater treatment, such carbons are generally quite large in size, involving 30-60 minute adsorbent bed liquid retention times. The liquid undergoing treatment requires prior filtration in order to avoid clogging of the sorbent bed, and the sorbent usage rate has to be substantial before on-site regeneration renders the sorbent costs economical, e.g., carbon usage must generally be greater than about 500 pounds per day before on-site regeneration renders the carbon sorbent costs reasonable on a per-pound basis.

To overcome the requirement of prior filtration up-stream of the sorbent bed, the prior art has proposed to operate the adsorbent bed in an expanded or fluidized bed-operating mode. Such mode of operation has utility primarily for treatment of liquids with low solids levels, e.g., less than 120 milligrams suspended solids per liter, while higher solids content liquid streams are likely to still require filtration, and even in low solids content liquid treatment operation where expanded or fluidized bed operation is possible, the liquid effluent from the sorbent bed will still contain substantial levels of suspended solids.

The prior art has further proposed to employ powdered adsorbents for removal of adsorbable contaminants from liquid in wastewater treatment by the active sludge process, by adding the powdered sorbent, such as carbon, directly to the wastewater treatment aeration tank. In such modifications, the powdered carbon or other sorbent is mixed with the biological active sludge solids and, as a result, both the sorbent and the biological solids have to be dewatered and regenerated together, which is disadvantageous from the standpoint of operating system complexity and treatment cost. Furthermore, such modification results in only slight polishing, i.e., removal of adsorbable contaminants, from the treated liquid.

Accordingly, it is an object of the innovation to provide an improved, integrated process for treatment of suspended solids and adsorbable contaminant containing liquid, by the use of an absorbent particulate catalyst impregnated sorbent. Other objects and advantages of the innovation will be apparent from the ensuing disclosure and appended claims.

Another aspect of the prior art in relation to the present innovation is the structural formation. A microfilter is generally a microporous structure composed of either crystalline aluminosilicate, chemically similar to clays and feldspars and belonging to a class of materials known as zeolites, or crystalline aluminophosphates derived from mixtures containing an organic amine or quaternary ammonium salt, or crystalline silicoaluminophosphates which are made by hydrothermal crystallization from a reaction mixture comprising reactive sources of silica, alumina and phosphate. Microfilters have a variety of uses. They can be used to dry gases and liquids; for selective molecular separation based on size and polar properties; as ion-exchangers; as catalysts in cracking, hydrocracking, disproportionation, alkylation, isomerization, oxidation, and conversion of oxygenates to hydrocarbons, particularly alcohol and di-alkyl ether to olefins; as chemical carriers; in gas chromatography; and in the petroleum industry to remove normal paraffins from distillates.

Microfilters are manufactured by reacting a mixture of several chemical components. One of the components used in the reaction process is a template, although more than one template can be used. The templates are used to form channels or tunnel-like structures (also called a microporous structure) within the composition. When the template is removed, an open microporous structure is left behind in which chemical compositions can enter, as long as the chemical compositions are small enough to be able to fit inside the tunnels. Thus, a microfilter acts to sieve or screen out large molecules from entering a molecular pore structure.

Microfilters are particularly desirable for use as catalytic agents. The microfilter can act as catalysts, and have catalytic sites within their microporous structures. Once the template is removed, a chemical feedstock that is small enough to enter into the tunnels can come into contact with a catalytic site, react to form a product, and the product can leave the microfilter through any number of the tunnels or pores as long as the product has not become too large to pass through the structure. The pore sizes typically range from around 2 to 10 angstroms in many catalytic microfilters.

Although finished sorbent microfilter particles are generally harder than the components, they are prone to damage due to physical stresses encountered during the manufacture of the finished sorbent particles or during the use of the finished sorbent particles in a reaction system. This damage tends to physically wear down or break apart (i.e., attrit) the sorbent particle until it is too small to efficiently recapture for reuse. The attritted particle is then discarded as waste from the system in which it is used. In the manufacture of finished sorbent particles, there may also be produced particles that are too small for subsequent use in a reaction system. For example, through misoperation of equipment or transient operations at the beginning or end of one cycle of a batch sorbent manufacturing operation, clumps or sheets of the microfilter or composite material may form on the walls or floors of equipment. The clumps are then discarded as a loss in the sorbent manufacturing process. The discarding of sorbent attrition particles or undersized clumps is problematic from an economic standpoint. Therefore, methods for effectively recovering and reusing these attrition particles and clumps are highly desired.

In order to limit losses of the microfilter-containing attrition particles and/or clumps during manufacture or during use, this innovation provides a particulate catalyst impregnated sorbent composition which comprises of a catalyst impregnated into a defined sized carbon structure or other composition, calcinated with a binder.

An additional fluid stream that has undesirable impurities or contaminants that need removal is that of a hydrocarbon based fluid stream, or a carbonaceous liquid feed stream. The contaminants include metals of vanadium, nickel, iron, and compounds of nitrogen, sulfur, and other aromatics.

One such process which has been proposed for the removal of the above mentioned contaminants from a hydrocarbon fluid stream is called hydrodesulfurization. While hydrodesulfurization of a hydrocarbon fluid stream can remove these undesirable compounds, it can result in the saturation of most, if not all, of the olefins contained in the gasoline. This saturation of olefins greatly affects the octane number (both the research and motor octane number) by lowering it. These olefins are saturated due to, in part, the hydrodesulfurization conditions required to remove thiophenic compounds (such as, for example, thiophene, benzothiophene, alkyl thiophenes, alkylbenzothiphenes and alkyl dibenzothiophenes), which are some of the most difficult compounds to removed. Additionally, the hydrodesulfurization conditions required to remove thiophenic compounds can also saturate aromatics.

In view of the ever-increasing need to be able to produce cleaner automotive fuel, a variety of processes have been proposed for achieving industry compliance with the Federal mandates. As a result of the lack of success in providing successful and economically feasible process for the reduction of the contaminant levels in cracked-gasolines, diesel fuels, kerosene, naphtha, vacuum distillates, fuel oils, and other hydrocarbon fluid streams or products, it is apparent that there is still a need for a better process.

Consequently, there is a need for a process wherein sulfur can be removed without hydrogenation of aromatics is achieved so as to provide a more economical process for the treatment of the hydrocarbon fluid streams.

It is therefore an object of the present innovation to provide a novel particulate catalyst impregnated sorbent for the removal of said contaminants from fluid streams of cracked-gasoline, diesel fuels, kerosene, naphtha, vacuum distillates, fuel oils, and other hydrocarbon fluid streams or products.

Another object of this innovation is to provide a process for the production of the novel particulate catalyst impregnated sorbent which is useful in the separation and removal of the contaminants in such fluid streams.

Another object of this innovation is to provide a process system for the removal of sulfur-containing compounds from cracked-gasoline, diesel fuels, kerosene, naphtha, vacuum distillates, fuel oils, and other hydrocarbon fluid streams or products, which minimize saturation of olefins and aromatics therein.

Other aspects, objects, and the several advantages of this innovation will be apparent from the following description of the innovation and the appended claims.

20.2.3 Summation of the Innovation

The present innovation overcomes the problems that are outlined above and advances the art by providing improved method of separation and removal of target elements and compounds, and provides for the novel particulate structured catalyst impregnated sorbent. The process also uses oxidation; fractional control of the fluid feeds or feed streams, continuous adsorption, to improve process efficiencies in the removal of impurities and contaminants from liquid flow streams. These advances in catalyst/sorbent construction, adsorption art pertain to increased the yield of adsorption treated product; improved quality in the treated product, and reduction of the utilities (low temperature and pressure) that are required to process a given liquid flow stream through use of a superior regeneration processes and apparatus. Additional advantages include extending process utility to a wide variety of liquid feed stream other than aqueous based or hydrocarbons that were not amenable to prior processes.

The novel approach and the description disclosed herein pertain to a process and system for use in treating a liquid stream to remove impurities or contaminants, where the oxidized impurities or contaminants have a greater selectivity for the catalyst within the porous sorbent particulates than do other components of the liquid.

From the foregoing, it can be seen that it would be advantageous to improve the processes for the removal of metals from contaminated water. It would be desirable to selectively remove metals or their oxides or salts from the wastewaters. It can also be seen that it would be advantageous to develop a process to treat contaminated waters, for example, acid mine drainage and mineral industries wastewaters, to provide a final product or products which can be conveniently removed from the bulk solution and effectively recovered, a product or products which are saleable and/or have immediate applications and which provide a treated water which meets quality standards for discharge into the environment.

These and other objects are achieved by the process of the present innovation, which provides a method by which hydrocarbon liquids may be separated and removed from water or other polar solutions. In particular, the present innovation is successful in overcoming the problems associated with the prior art methods for removing petroleum-based products from aqueous solutions. The materials used in the process may be recycled repeatedly thereby reducing the costs and the amount of material necessary for the effective separation of the contaminating liquid. Moreover, the hydrocarbon liquid recovered using this method may be processed and employed as originally intended thus eliminating waste disposal problems.

It is an object of this innovation to separate and remove the target elements and compounds (impurities or contaminants) by oxidizing the target element or compounds. The oxidizing gas is formed into micron size bubbles and is passed through the fluid stream or feed stream. The micron size bubbles of oxidizing gas are dispersed into the fluid stream or feed stream, whereupon the target elements or compounds contained within the fluid stream or feed stream are efficiently oxidized into oxides. Due to the micron size of the bubbles, the surface area of the oxidizing gas is greatly increased, thereby greatly increasing the efficiency of the oxidation reaction.

The first objective of the process is to oxidize the target elements or compounds (impurities or contaminants). A second objective is to allow simultaneous separation of metals, sulfur compounds, nitrogen-containing and aromatic hydrocarbons from the fluid stream so that a desired combination of residual aromatics and low sulfur and nitrogen content can be obtained. The operating conditions are relatively mild throughout the separation and removal process. Pressures are near ambient and temperatures are less than 80-90 degrees C throughout the separation and removal process.

The liquid flow stream then enters a reactor containing a fixed bed of a particulate catalyst impregnated sorbent and contacts the porous sorbent particulates with sufficient overall residence time for the target element or compound (impurity or contaminant) absorbance to the sorbent and slight covalent bonding to the catalyst which separates and removes the target elements or compounds in the liquid stream to produce both a purified liquid stream having a reduced impurity or contamination concentration and an impurity-bonded catalyst impregnated sorbent. The purified liquid stream is generally discharged from the reactor as a treated product with excellent characteristics. For example, with diesel feedstock, the treated product may be expected to be clear, colorless, free from any objectionable odors, have a very low ppm of sulfur compounds such as mercaptans, thiophenes, benzothiophenes and dibenzothiophenes, reduced nitrogen compounds, lower aromatics and poly nuclear aromatics, a smaller percentage of metals (vanadium, nickel, and iron), and an improved cetane index or octane quality.

In another embodiment of this innovation, the particulate catalyst impregnated sorbent used in the reactor has a composition which comprises bonding together a catalyst within a sorbent particle. The sorbent for the catalyst can be carbon, silica, or an alumina, a zeolite, a perlite form, or any structurally sound, porous sorbent. The catalyst can be a metal, alkali, or alkali earth metal, metal oxide, a bimetallic combination (combination of metals). For example, a sorbent comprising of a zeolite on which a specific type of catalytic metal, such as silver, is supported thereon through ion exchange exhibits excellent absorptivity of sulfur compounds at or in the vicinity of normal temperatures.

Following the separation of the target element or compound onto the particulate catalyst impregnated sorbent, the sorbent can be washed and reactivated through a flushing of a polar organic or inorganic solvent. This flushing or washing of the particulate catalyst impregnated sorbent, removes any residue of the target elements or compound. Drying the particulate catalyst impregnated sorbent with heat or hot nitrogen regenerates the absorbance ability of the particulate catalyst impregnated sorbent and it is ready for reuse. The organic or inorganic polar solvent with the target element or compounds can be separated by distillation.

While the present innovation is very useful in the separation and removal of liquid hydrocarbons such as oil from aqueous solutions, it is applicable to the separation of any target element or compound (impurity or contamination) from a hydrocarbon liquid. Examples of hydrocarbon liquids which may be separated from a target element or compound using the present innovation include, but are not limited to, gasoline, diesel, naphtha, kerosene, vacuum distillates, fuel oil, crude oil, paraffinic oils, xylenes, toluene, styrene, alkylbenzenes, naphthas, liquid organic polymers, vegetable oils, and the like. Examples of target elements or compounds within those hydrocarbon liquids can include, but are not limited to metals (vanadium, nickel, iron), nitrogen compounds, sulfur compounds (mercaptans, thiophenes, benzothiophenes, dibenzothiophenes), aromatics, poly nuclear aromatics, and the like.

The process design can be modified to accommodate a variety of hydrocarbon feed streams; however, the boiling range of the feed will to a large extent determine the suitability of the specific solvent or solvent combination because of the need to recover the solvent for recycle. Several process design variations and economic optimizations are distinguished to the present innovation. For example, depending on the final product specifications and feedstock quality the first oxidation step may be designed to remove a lighter fraction or larger amount of thiophenes and aromatic compounds, leaving a second fraction to be oxidized and separated downstream. The design optimization is a trade-off between the slower space velocity within the reactors after oxidation and the cost of the two fraction separation and an increased feed stream flow, including the solvent recovery and recycle.

In accordance with the present innovation, water, any polar aqueous fluid stream, hydrocarbons, or any carbonaceous liquid stream, containing heavy metals, sulfur compounds, impurities or contaminant can be treated to reduce the concentration of the target element or compounds by oxidizing it first then contacting the impurity latent water with a particulate catalyst impregnated sorbent structure. In accordance with the present innovation, the particulate catalyst impregnated sorbent can be provided in accordance with the present innovation, preferably using the method described below. The resulting particulate catalyst impregnated sorbent is particularly useful and effective in removing oxidized target elements or compounds from water or polar aqueous fluid streams when the sorbent structure is contacted with the fluid stream or feed stream containing the impurities under conditions normally associated with fixed bed adsorption techniques. The use of such structures under such conditions results in an unexpectedly high removal rate of the oxidized target elements and compounds from fluid stream or feed stream in accordance with the present innovation.

In the following detailed description, several embodiments of methods for the preparation of the particulate catalyst impregnated sorbent are described in more detail, as well as methods of using the medium sorbent to separate and remove the target elements or contaminants in accordance with the present innovation. Applicants have observed that the distinctions between the prior art discussed above and the present innovation can have dramatic and unpredictable effects on industry, the enhanced economical impact of the present innovation's implementation, and the present innovation's simplicity in comparison to other alternative methods or processes.

20.2.4 Detailed Description of the Innovation

The process of the present innovation is applicable to the removal of any target element or compound that can be oxidized from a substantially polar environment, water, other aqueous liquids, hydrocarbons, hydrocarbon by-products, or liquid carbonaceous substances. Use of particulate catalyst impregnated carbons, zeolite, perlite, or other structurally sound porous sorbent, in the disclosed process advantageously allows the oxidation, separation and removal, recovery and subsequent treatment of the separated target element or compounds, as well as the regeneration of the particulate catalyst impregnated sorbent. Accordingly, the present innovation is highly suitable for separation and removal or purification applications in manufacturing or commercially. It is the intent of this process patent to show utility, novelty, and efficiency, with the use, and combination of use, of the components of this innovation, the sub systems, and the process system as a whole.

In this section of the "Detailed Description of the Innovation", a brief summary of the sorbent composition and the making thereof will be introduced in the beginning with a more detailed description as a separate section. This is followed by the process description, design explanations and clarifications, and notes on specific water or hydrocarbon process differentiations.

This adsorption oxidation and the sorbent absorption process are characterized by the advantage that they can operate to remove the target elements and compounds at ambient temperatures and needs no extraordinary refrigeration or heating in order to accomplish this task.

Liquid phase absorption differs from gas phase adsorption in that diffusion is at least two orders of magnitude slower in the liquid phase than in the gas phase. Diffusion of components in a liquid phase requires additional residence time. Impurities absorb on the solid absorbent because the attraction of the absorbent surface is stronger than the attractive force that keeps the impurities in the surrounding fluid. Liquid absorption may be defined as a type of adhesion that, in a thermodynamically preferred sense, occurs in the internal surfaces of a solid having an absorbable impurity in the liquid medium. This preference results in a relatively increased concentration of absorbable impurities entering the absorbent particle pores and being drawn to the catalyst center by either the physisorption due to electronegative attraction or Van der Waals forces, or chemisorption, that is due to chemical or valence forces.

The election to remove more or less target elements and compounds in the separation and removal oxidation/absorption sections of the process is an economic decision that depends on the relative cost of the two operations. The absorbers can be on a single bed or on several separate beds or reactors, which has many variation variables. For example the trade-off between the speed of flow or space velocity and the costs of two separation steps, including the solvent recovery and recycle steps. These process steps can be performed equally well as batch or continuous flow operations as dictated by the economic requirements. The absorbers can be regenerated by circulating an amount of a polar organic or inorganic solvent at temperatures of between about 30 degrees C and 90 degrees C, through the beds and recycling the target elements and compound-rich effluent back to the solvent separation or distillation section after cooling.

The sorbent compositions of the present innovation, which are useful in the separation and removal of the target elements and compounds in the process of the present innovation, can be prepared by a method comprising of:

(a) mixing a support component preferably comprising of; a zeolite, or expanded perlite and alumina, or a carbon particulate structure, so as to form a mixture selected from the group consisting of a wet mix, a dough, a paste, a slurry, and the like and combinations thereof;
(b) particulating, preferably spray-drying, the mixture to form particulates, granules, spheres, micro-spheres, and the like and combinations thereof, preferably a micro-sphere particulate;
(c) drying the particulate under a drying condition as disclosed herein to form a dried particulate;
(d) calcining the dried particulate under a calcining condition as disclosed herein to form a calcined particulate;
(e) incorporating, preferably impregnating, the calcined particulate with a catalyst component thereof to form a particulate impregnated catalyst within the support component;
(f) drying the promoted particulate under a drying condition as disclosed herein to form a dried, particulate impregnated catalyst sorbent;
(g) calcining the dried, promoted particulate under a calcining condition as disclosed herein to form a calcined, particulate impregnated catalyst sorbent;
(h) reducing the calcined, promoted particulate with a suitable reducing agent so as to produce a sorbent composition having a reduced-valence catalyst component content therein, and wherein the reduced-valence catalyst component content is present in an amount effective for the removal of a specified target element or compound from a hydrocarbon-containing fluid, water, any polar aqueous fluid stream, or any carbonaceous liquid stream, when contacted with a sorbent composition(s) of the present innovation according to a process(es) of the present innovation.

Optimum adsorbent size and residence time for a particular adsorbent is a matter for empirical study under actual process conditions.

Although not wanting to be limited to a particular mechanism, it appears that these target elements and compounds undergo a catalytic conversion on the sorbent during oxidation resulting in the formation of substances having an increased molecular weight. For example, mercaptans are oxidized to sulfides and/or polysulfides. These higher molecular weight sulfur compounds are then absorbed by these particulate catalyst impregnated sorbents. The physical absorption of these sulfur compounds on these sorbents is increased, due to their higher molecular weight. Because the absorption of the sulfur compounds on the particulate catalyst impregnated sorbents of the present innovation is a two-stage process, i.e., first catalytic oxidation conversion of sulfur contaminated compounds, followed by physical absorption of the catalytically converted products, these sorbents which are the subject of the present innovation are termed "particulate catalyst impregnated sorbent".

The process for separation and removal of target elements and compounds consists of:

(a) Fractionation: in the host media target elements and compounds may reside in different concentrations or species within different boiling point fractions. This may be an integral function to the efficiency of the separation and removal steps.
(b) Oxidation: the target elements or compounds are best selected for separation from the host media by the oxidation of such target elements or compounds. The higher the efficiency of oxidation, the lower of the level of target elements or compounds will remain in the host media.
(c) Separation: the host media with the target elements or compounds is then entered into a reactor that contains a particulate catalyst impregnated sorbent, as described herein. The particulate catalyst impregnated sorbent absorbs the target elements or compounds by the enhanced selectivity of the sorbent for the target elements and compounds, whereby separating them for the host media.
(d) Removal: the target elements and compounds are removed from the particulate catalyst impregnated sorbent, by the washing of the sorbent with a polar organic or inorganic solvent. The reactors are washed or flushed with the solvent until the desired level of the target elements or compounds are eliminated from the particulate catalyst impregnated sorbent.
(e) Regeneration: the particulate catalyst impregnated sorbent is then regenerated by the drying with heat or hot nitrogen. The gaseous solvent is vacuum pumped from the reactors, cooled and recondensed. The solvent is then returned to the solvent holding tank for reuse. If hot nitrogen is used, it is collected and reused.
(f) Recycle: the washed target elements or compounds with the solvent flush are separated from the solvent by distillation. The separated solvent is then returned to the solvent holding tank for reuse.
(g) Refinement: the separated target elements or compounds can be reintroduced to the process system at a much higher concentration to be refined to a lower volume by separating any, and theoretically all, of the remaining host media. If the host media is a hydrocarbon feed stream or fuel, the target elements or compounds may be used as a fuel to any of the steps of this innovation process.

The separation and removal steps of the present innovation is carried out under a set of conditions that includes (1) total pressure being consistently throughout the process at ambient pressure, or a small amount of pump pressure caused by the movement of the host media through the reactors, or a small amount of vacuum or draw upon the system during regeneration, (2) the temperature being maintained at a low operating level of 70 degrees C to 80 degrees C while oxidizing and separating, or 40 degrees C to 140 degrees C during removal, regeneration, and recycle steps, depending upon the solvent and/or solvents used, (3) hourly space velocity to be determined by the specific target element or compound levels desired. Preferably it is 2.0 to 3.0 at the minimum flow and up to 10.0 to 15.0 maximum flows. These conditions are such that the particulate catalyst impregnated sorbent composition can separate and remove the target elements and compounds from the host media.

Such passage of suspended solids and adsorbable contaminants-containing influent liquid through the fixed sorbent bed is continued until the sorbent bed is at least partially loaded with deposited solids and absorbed contaminants, whereupon deposited solids and absorbate loaded sorbent is washed or flushed with a polar solvent to regenerate same, prior to again re-initiating normal on-stream separation and absorption operation.

The enhanced adsorption oxidation is accompanied by evolution of heat because the adsorbate molecules are stabilized on the adsorbent surface. For limited quantities of impurities in the fresh feed, temperature increase of the fluid is limited by the amount of adsorbable impurities that are typically present, i.e. the sensible heat of the other liquid components offsets the heat evolution due to impurities, which will cause only a small increase in the process temperature. The operating conditions during this process step are: temperature between about 20 and 90 degrees C, pressure is ambient and only increased by the process pump pressure. For example, the partition coefficient for thiophene-dioxides between the hydrocarbon and sorbent phases at 20 degrees C is double the value at 70 degrees C. The optimum oxidation temperature depends upon the adsorbent composition, the thiophene and thiophene-oxides molecular composition and molecular weight distribution, and the feed hydrocarbon composition, notably the aromatics content and composition. It is therefore necessary to optimize the extraction processes for each individual feedstock based on experimental oxidation and separation data. The oxidation can be performed in a packed or trayed column with or without intermittent mixing or induced pulsation; however, any suitable single or multi-stage fixed bed contacting equipment can be used. Other values outside these ranges are also possible.

The regeneration of the sorbent fixed bed is accomplished by draining the column and passing a heated gas, preferably nitrogen, upward through the packed sorbent bed to remove the solvent from the sorbent. The temperature of the gas used to cause this regeneration may be from about 120 degrees C to about 140 degrees C with a flow rate of about 15 ml/min and an outlet pressure gauge reading of about 8 psi.

The evaporated solvent could be channeled or vacuum pumped into a cooling column, recondensed, and then collected in its holding tank for reuse.

An alternative method for the regeneration for the sorbent fixed bed is to use a secondary solvent with a high dissolution coefficient and low vapor point to dissolve the first solvent from the sorbent. The two solvent mixtures are then sent to a distillation unit to separate, which recycles both for reuse. The reactor columns are then heated to 40 degree C, with the use of an insulated blanket or heating element. The evaporated secondary solvent could be channeled or vacuum pumped into a cooling column, recondensed, and then collected in its holding tank for reuse.

The method of this innovation preferably contemplates the creation of columnar fixed bed reactors packed with the material, tailored as set forth above, operating in parallel such that when one packed vessel reaches its absorption capacity, it can be taken off-line for regeneration when the second one is brought on-line.

Traditional engineering principles may be used to design the size and dimensions of the vessels to be packed with the particulate catalyst impregnated sorbent material to match the flow rate of the process system involved and yet achieve proper residence time for treatment. Because of the highly porous nature of the sorbents, the beds may be relatively tall.

The separation reactor column can be operated over a range of temperatures, solvent compositions, and flows of different space velocities to accommodate various feed compositions and product specifications.

The process can be conducted in a single reaction zone or a number of reaction zones arranged in series or in parallel, and contains significant advances in the use of adsorption over the system shown and described in U.S. Pat. No. 5,730,860.

The distillation column can be designed and operated so that the solvent recovered at the bottom meets recycle solvent specifications. It therefore does not need to be pure solvent but may contain small amounts or trace amounts of other compounds.

Note: In a zeolite column some of the thermal energy from the heated air is absorbed by the silane-zeolite bonds and regenerates their strength to restabilize the bed. The packed bed is cooled, refilled with ambient water, and put back on line for subsequent use when the capacity of the companion is reached, necessitating regeneration.

For cases where the feedstock quality results in a small build-up of residual hydrocarbons, a small side stream taken from the recycle solvent stream may be used as fuel for the process.

Thus, a method for enhanced adsorption oxidation and subsequent separation and removal of target elements and compounds using particulate catalyst impregnated sorbents has been discovered.

20.3 Notes on Oxidation

The oxidation step of the process method comprises the induction of an oxidizing gas by the utilization of air, oxygen, ozone, hydrogen peroxide or other well-known oxidation techniques. The oxidizing gas used in the detailed description of this innovation is air; however, other oxidizing gases may be utilized in the practice of the innovation, if desired.

The combined feed mixture is then heated to the desired reaction temperature in the fixed bed reactor column. For this step, the oxidation reactor can be a packed column, one or more reactors in series, or a similar configuration that provides adequate surface contact with the oxidation gas, minimum back-mixing, and the required residence time of 15-30 minutes to reach the desired oxidation conversion.

The air or oxidizing gas bubbles are dispersed into the oxidation column whereupon the target elements or compounds contained within the feed stream are immediately oxidized, thereby forming mono- and di-oxides. The micron size of the bubbles of the air or oxidizing gas greatly increases the surface area of the oxidizing gas/feed stream interface thereby substantially increasing the efficiency of the oxidizing reaction. The oxidant-to-target element ratio can vary according to the nature and reactivity of the compounds, the product specification, the operating temperature, and the selected catalyst. Depending on the amount and type of oxidant used, the reaction mixture may be in one or two liquid phases. The oxidation process reaction temperature should be maintained at 80 degrees C, converting the target elements or compounds quantitatively to oxidize compounds. Upon completion of the oxidation reaction, the feed stream continues to a fixed bed reactor column for adsorption of the oxidized target elements or compounds.

It is contemplated that all of the air or oxidizing gas will be consumed by the oxidizing reaction. If not, excess air or oxidizing gas will be reinitiated into the oxidation column of the feed stream or vented.

Furthermore, the particulate catalyst sorbents of the innovation are especially suitable for the oxidation reactions of organic molecules due to their superior adsorption qualities, which maintains an adequate retention time for the oxidation reaction. Using the particulate catalyst sorbents of the innovation, the reaction can be carried out either in the liquid phase by passing the mixture of the oxidized target elements or compounds over the particulate catalyst sorbents in a fixed-bed reactor.

20.4 Notes on Solvent Used in This Innovation

Properties of solvent mixtures, used in this innovation are ease of recovery; recycle ability, chemical stability, low cost, low toxicity, and strength of polarity. Solvent mixtures should also be essentially inert to reactions with the liquid feed streams.

With regard to this innovation, suitable polar organic and inorganic solvents include aromatics, halogenated aromatics, organo-chlorinated compounds, ketones and alcohols. An example of which would be, but not limited to, toluene, dichlorobenzene, dichloromethane, dichloroethane, acetone, ethanol, and methanol could be used.

The solvent separation phase or extraction, contain approximately 80 to 90% solvent and 10 to 20% target elements and compounds. The solvent can be removed from the target elements and compounds by distillation. The process design choice is determined by economics and also depends on the specifications of the original feedstock, design capacity of the plant, solvent selection, boiling points and phase density differences, as will be apparent to workers skilled in the art.

It is also an advantage for the solvent to be compatible with the oxidation reaction. In addition, if it is needed, one or more of the solvent components in the process can participate as an intermediary reactant in the oxidation reaction. An example of such a compound is acetic acid and its homologues, which form peroxy-acid intermediary oxidants with hydrogen peroxide. High extraction selectivity, strength of polarity, dissolution coefficients, safety considerations, process compatibility and low cost tend to dominate the solvent selection for industrial application.

Several process design variations and economic optimizations for hydrocarbon feed stocks are readily apparent to the process designer skilled in the art. For example, depending on the final product specifications and feedstock quality, the first separation process step may be designed to remove a lighter boiling point fraction of the compounds, leaving the rest to be oxidized and separated downstream. The design optimization is a trade-off between the speed of flow or hourly space velocity and the costs of two separations including the solvent recovery and recycle steps.

The process design can be modified to accommodate a variety of liquid feed streams. However, the boiling range of the feed largely determines the suitability of any specific solvent combination because of the need to recover the solvent for recycling. Another example of optimization is the possibility of processing the two separations together for solvent recovery. Doing this reduces complexity by eliminating several pieces of equipment.

After solvent recovery, depending on the character of the hydrocarbon feedstock, the recovered extract will consist of approximately 10 to 25% sulfur-containing compounds and 10 to 30% aliphatic compounds, with the balance being aromatic compounds.

20.5 Notes on Water Treatment with This Innovation

For the general practice of the innovation wherein the solids and adsorbable contaminants containing liquid is water, as for example municipal water purification or the treatment of municipal and/or industrial wastewaters, a particularly preferred particulate adsorbent is carbon, which may for example be in the form of powder, particulate, or micro-sphere. Alternatively, in other water treatment applications, the particulate adsorbent may suitably comprise a zeolite, an expanded perlite, or other structurally sound porous sorbent.

Generally, the waters which are treated by the process of the present innovation are contaminated with iron metal, iron oxides, and/or iron in the form of ferrous and/or ferric salts, as well as the non-ferrous and non-ferric metals, metal oxides and/or metal salts of copper, zinc, aluminum, manganese, silver, lead, cadmium, gold, nickel, arsenic, and the like, as well as the lanthanide and actinide metals. Thus, as used herein, the oxidation of a metal in any of the stages includes the separation of the metal and its oxides and salts. In accordance with the present innovation, the contaminated water must contain metals from which an oxide can be produced. Accordingly, the contaminated water must contain any combination of Zn, Mn, Mg, Cu, Ni, Cd, Pb, and the like and/or the oxides and salts thereof, so that said metals can be oxidized to be separated. In most instances in the case of wastewater treatment, these metals are the so-called "heavy metals", i.e., those transition metals which are toxic and will cause environmental harm if discharged into rivers, lakes or other natural water resources. However, the metals, which may be recovered from the separation from the host media, are not limited to the "heavy metals".

In the process of metal separation stage (a), at least one oxidizing agent is added to the contaminated water to increase the oxidation-reduction potential of the water to a more positive potential to convert the metal, i.e., iron, if present in the water, to the ferric ion form wherein iron will precipitate from the water. The metal precipitate is removed from the water during separation by the sorbent. In other instances, it may be desirable to selectively remove and recover a metal or metals which have value, e.g., a metal or metals that can be sold to a smelter, from the contaminated waters.

Support material used to produce adsorbent media for water treatment typically ranges from micron size, a few tenths of a millimeter in diameter, to several millimeters in diameter. In general, larger media have the advantage that they provide less resistance to water flow and clog less easily; however, they have the disadvantage that they have less surface area per unit volume, so less coating can be packed into a given size column. An additional limitation of some large media in the context of the present innovation is that the shear force needed to fluidize a bed increases as the size of the medium increases (for a given medium density). Depending on the specific support material, the increased shear forces for larger particles may cause the coating to be sheared off the surface.

The process of the present innovation is applicable to the removal of any hydrocarbon liquid from a substantially polar environment or water. Use of the particulate catalyst impregnated sorbent in the disclosed process, advantageously allows the recovery and subsequent treatment of the separated hydrocarbon liquid as well as the regeneration of the particulate catalyst impregnated sorbent. Accordingly, the present innovation is highly suitable for separation or purification applications in manufacturing or laboratory settings. Yet, based on the lack of equivalent alternatives, one of the most significant applications of the present innovation is the removal of contaminating hydrocarbons from bodies of water. Consequently, while the following exemplary embodiments are discussed within such a framework, this should in no way limit the applications of the present innovation.

The flow rate of contaminated water through a catalytic zeolite sorbent may be from about 0.05 to about 0.2 bed volumes per minute with a preferred flow rate of the contaminated water of about 0.1 bed volumes per minute, comparable with normal on-line cleanup rates. Empirical data based upon the BTEX loading of the contaminated water and the flow rate given the tailoring of the catalytic zeolite sorbent would lead, with simple experimentation, to a determination of the length of time it should take to use the capacity of a particular catalytic zeolite sorbent bed before breakthrough of the benzene or alkyl benzenes. Analysis of the effluent is also possible which can lead to automated control of the process.

20.6 Notes on Hydrocarbon Impurity Separation with This Innovation

The method of improving the quality of hydrocarbons comprised in the present innovation may be used either as the sole process for treating the hydrocarbons or in combination with existing hydrotreating techniques. In particular, when the method of the present innovation is used subsequently to hydrotreating, the aromatic compounds which remain after the hydrotreating process are removed thereby increasing the cetane rating of the diesel fuel.

For example, the hydrocarbon feed impurities may include heteroatom compounds, such as those containing nitrogen, oxygen, and sulfur. Sulfur-containing compounds include, for example, aliphatic, naphthenic, and aromatic mercaptans, sulfides (i.e., hydrogen sulfide, carbonyl sulfide), di- (i.e., carbon disulfide) and polysulfides, thiophenes and their higher homologs and analogs, benzothiophenes, and dibenzothiophenes. Other impurities may consist of nitrogen containing compounds including, but not limited to, nitrites and pyridines, and oxygen-containing compounds. For hydrocarbon feeds, heteroatom impurities have a polar atom which facilitates preferential adsorption.

Hydrocarbon mixtures, feed stocks, feed streams, by-products, heavy hydrocarbons and carbonaceous liquid feeds, applicable for this innovation, but not limited to, could be represented by the terms, "gasoline", "cracked gasoline", "diesel fuel", "heavy bottoms". The term "gasoline" denotes a mixture of hydrocarbons boiling in the range of from about 38 degrees C to about 205 degrees C, or any fraction thereof. Examples of suitable gasoline include, but are not limited to, hydrocarbon streams in refineries such as naphtha, straight-run naphtha, coker naphtha, catalytic gasoline, visbreaker naphtha, alkylate, isomerate, reformate, and the like and combinations thereof. The term "cracked-gasoline" denotes a mixture of hydrocarbons boiling in the range of from about 38 degrees C to about 205 degrees C, or any fraction thereof, that are products from either thermal or catalytic processes that crack larger hydrocarbon molecules into smaller molecules. Examples of suitable thermal processes include, but are not limited to, coking, thermal cracking, visbreaking, and the like and combinations thereof. Examples of suitable catalytic cracking processes include, but are not limited to, fluid catalytic cracking, heavy oil cracking, and the like and combinations thereof.

Thus, examples of suitable cracked-gasoline include, but are not limited to, coker gasoline, thermally cracked gasoline, visbreaker gasoline, fluid catalytically cracked gasoline, heavy oil cracked gasoline, and the like and combinations thereof. The term "diesel fuel" as employed herein is intended to mean a fluid composed of a mixture of hydrocarbons boiling from about 150 degrees C to approximately 370 degrees C or any fraction thereof. Such hydrocarbon streams include light cycle oil, kerosene, jet fuel, straight-run diesel and hydrotreated diesel. The term "heavy bottoms" as employed herein is intended to mean a fluid composed of a mixture of hydrocarbons boiling from about 275 degrees C and higher, or any fraction thereof. Such hydrocarbon streams include fuel oils, bunker C, bitumen, mazoot, and residuals.

In some instances, the cracked-gasoline may be fractionated and/or hydrotreated prior to the removal of its sulfur compounds when used as a hydrocarbon-containing fluid in a process of the present innovation.

A particulate catalyst impregnated sorbent composition having a reduced-valence catalyst component of the present innovation is a composition that has the ability to react chemically and/or physically with sulfur compound species. It is also preferable that the sorbent composition removes diolefins and other gum-forming compounds from cracked-gasoline.

20.7 Particulate Catalyst Impregnated Sorbent Construction Pertaining to This Innovation

20.7.1 Sorbent Support Structures

The carrier for the catalyst can be carbon, a zeolite, a perlite form, or any structurally sound, porous sorbent. The catalyst is impregnated inside the carrier, within a designed structure. The surface area of the carrier, the pore size of the carrier and the density of pores surrounding the catalyst, toward the center of the carrier's concentricity, are all equally important in the design of the structure, of the carrier. The surface area would typically range from approximately 500 to 2500 m.sup2/g. The size of the pores should range from 5 to 100 angstroms with the density increasing toward the center of the carrier. The optimized carrier would have the largest surface area (2500 m.sup2/g) with the largest pore size toward the outer area (100 angstroms) and maintaining this size toward the center of the carrier, where the catalyst is bonded, increasing the density of structure toward the center. This is the challenge of the optimized carrier, to push the parameters to the extreme, without losing the integrity of the structure. The catalyst could be a metal, alkali, or alkali earth metal, metal oxide, a bimetallic combination (combination of metals).

Among the amorphous carbon materials which are suitable for the particulate support of the catalysts described within this innovation, are products known as "onion carbon" or micro-sphere carbon structures. It has onion-like structures with a diameter of 100 angstroms or more, which are embedded firmly in the surrounding carbon, are susceptible catalyst impregnation sites or surface sites.

In general, catalyst impregnated sorbent carbon compounds are formed by inserting extra atoms or molecules into a host structure, without disrupting the chemical bonds of the host material. Carbon atoms in graphite are located at the points of a hexagonal lattice and kept in place through relatively strong covalent bonds. In contrast the hexagonal lattices are displaced relative to each other and held in place by weaker Van Der Waals forces. The lower bond energy between the lattices and packing defects make the graphite particles susceptible to the insertion of catalyst materials. For the production of particulate catalyst impregnated carbon sorbent, graphite powder is heat treated in the presence of a gaseous or liquid agent. The pressure at which actual impregnation begins is dependent on the polarity and the structural disorder of the graphite. Expansion to well over 100 times their original particle size may be realized.

A zeolite, which may be used in this innovation, has a substitution of sodium cations in a synthetic faujasite structure with a catalyst, preferably Zn, Ag, Cu, Ni, Co, and Sn, or a combination thereof, results in a 1.5-3.0 times increase in the adsorption capacity of the synthetic faujasites for certain target elements and compounds. This innovation will also note that these particulate catalyst sorbents of synthetic faujasites display enhanced adsorption capacity even at low concentrations of the target elements and compounds, i.e., below 1 ppm. This high capacity for removal of target elements and compounds results in an enhanced level of purification for feed streams. Zeolites suitable for use herein are those having a relatively high silica-to-alumina ratio and having a pore size greater than about 10 angstroms in diameter. Included among the synthetic zeolites are zeolites X, Y, L, ZK-4, ZK-5, E, H, J, M, Q, T, Z, alpha and beta, ZSM-types and omega. Preferred are the faujasites, particularly zeolite Y and zeolite X, more preferably those having a pore size greater than 10 angstroms in diameter. For example, Zeolite Y and X have a pore size of approximately 13 angstroms which is large enough to allow target elements and compound molecules to enter, and larger converted (oxidized) target elements and compound molecules to leave.

It is an embodiment of this innovation that an acceptable range of ion exchange of catalyst ions in the faujasite structure is about 50-75%. The transformation of target elements and compound contaminants is less efficient where substitution levels are below about 50% or above 75%. Therefore, catalyst forms of zeolite faujasites with ion exchange levels of from about 50% to about 75% possess a superior capacity for adsorbing certain target elements and compounds and provide a significant level of adsorption of these compounds from liquid streams.

The balance of the ions in the faujasite structure are preferably alkali and/or alkaline earth metals from about 25 to about 50% (equiv.) alkali and/or alkaline earth metals. Preferably, the alkali and/or alkaline earth metals are selected from sodium, potassium, calcium, and magnesium.

It is also asserted that these catalysts adsorb certain target elements and compounds reversibly. In contrast to transition metal oxides of other zeolites, such as zinc oxide and manganese oxide, the respective Zn, Ag, Cu, or Ni, Co, Sn, or a combination thereof, faujasite X or Y zeolites adsorb significant quantities of certain target elements and compounds by means of physisorption. These catalysts or bimetallic combinations can desorb these target elements and compounds by washing with a polar organic or inorganic solvent. Therefore, it has been discovered that these particulate catalyst impregnated sorbents can serve as regeneratable adsorbents with enhanced target element and compound adsorption capacity.

The term "perlite" as used herein is the petrographic term for a siliceous volcanic rock, which naturally occurs in certain regions throughout the world. The distinguishing feature, which sets it apart from other volcanic minerals, is its ability to expand four to twenty times its original volume when heated to certain temperatures. When heated above 815 to 875 degrees C, crushed perlite expands due to the presence of combined water with the crude perlite rock. The combined water vaporizes during the heating process and creates countless tiny bubbles in the heat softened glassy particles. It is these minuscule glass sealed bubbles, which account for its lightweight. Perlite can be manufactured to weigh as little as 2.0 lbs per cubic foot.

The perlite, which may be used in this innovation, will generally be present in the sorbent support composition in an amount in the range of about 15 to about 30 weight percent. In the manufacture of this type of a sorbent composition of the present innovation, the support component is generally prepared by combining the components of the support component, zinc tin oxide, perlite, and alumina in appropriate proportions by any suitable method or manner which provides for the intimate mixing of such components to thereby provide a substantially homogeneous mixture comprising zinc tin oxide, perlite, and alumina. Any suitable means for mixing the components of the support component can be used to achieve the desired dispersion of such components. Examples of suitable mixing means include, but are not limited to, mixing tumblers, stationary shells or troughs, batch or continuous mixers, and impact mixers.

The resulting product is an adsorbent medium that is capable of adsorbing or the absorption of target elements or compounds and filtering particulate material from water polar solutions, hydrocarbons, hydrocarbon by-products or other carbonaceous liquid substances. This innovation provides an attrition resistant particulate catalyst impregnated sorbent that is stable under basic conditions, which are generally encountered when adsorbent materials are regenerated after adsorbing or the absorption of target elements or compounds.

Smaller adsorbent particles beneficially enhance the heat transfer and mass transfer for a given feed stream at otherwise constant conditions. Smaller particles are more difficult to break than larger particles because smaller particles tend to have fewer faults, flaws or discontinuities. Porous particles of a given size are more resilient than non-porous particles of similar size and less prone to fracture.

A disadvantage of smaller particles may be that a smaller cross section flow is required for otherwise constant conditions, such as type of liquid feed, inlet temperature, and adsorbent replenishment rate. This requirement is offset by the fact that a larger diameter provides a larger adsorbent authory for a given bed height. Higher bed expansions can help offset this disadvantage of smaller particles.

Without the unique physical characteristics of the particulate catalyst impregnated sorbents, the numerous advantages of the present innovation would be impossible to obtain. Other materials fail to combine the low bulk density of the carbon catalyst impregnated sorbent, with its high adsorption capacity, cost efficient fabrication from readily available materials and high absorption capacity. Due to the previous use of carbon in other applications, several inexpensive fabrication methods have been developed to produce material, which may be employed effectively in the present innovation. Most of these fabrication processes depend on the use of inexpensive carbon as a starting material.

20.7.2 Construction or Manufacture of Particulate Catalyst Impregnated Sorbent

The catalyst component(s) may be incorporated onto, or within, the particulate sorbent (preferably spray-dried), calcined support component by any suitable means or method(s) for incorporating the catalyst component(s) onto, or within, a substrate material, such as the dried and calcined particulates, which results in the formation of a catalyst impregnated sorbent composition which can then be dried under a drying condition and calcined under a calcining condition to thereby provide dried, calcined, promoted particulate catalyst impregnated sorbent. The dried, calcined, promoted particulates can then be subjected to reduction with a reducing agent, preferably hydrogen, to thereby provide a sorbent composition of the present innovation. Examples of means for incorporating the catalyst component include impregnating, soaking or spraying, and combinations thereof.

20.7.3 Catalyst Metals

The elemental metals, metal oxides or metal-containing compounds of the selected bimetallic catalyst can be added to the particulate mixture by impregnation of the mixture with a solution, either aqueous or organic that contains the selected elemental metal, metal oxide or metal containing compounds.

Metal particles containing 10 to 1000 atoms are able to form the slight covalent bonds characteristic of the catalysts of the innovation. With larger metal particles, the strength of the bonds decreases. As the size of the metal particles increases, the binding strength finally decreases to a value, which, for the size of the metal particles, is no longer significantly greater than the binding strengths of conventional carbon supports.

The term "catalyst" as used herein is intended to mean a catalyst composite derived from one or more metals, metal oxides or metal oxide precursors wherein the metal is selected from the group consisting of cobalt, nickel, iron, manganese, zinc, copper, molybdenum, silver, tin, vanadium, tungsten, and antimony and wherein the bimetallic catalyst composite is in a substantially reduced valence state and wherein such catalyst is present in an effective amount to permit the removal of target elements and compounds from the host media. Low-valence metals are understood in the context of this innovation to mean metals in the valet, monovalent and divalent states.

Bimetallic catalyst combinations have unique physical and chemical properties that are important to the chemistry of the inventive sorbent composition described herein. These bimetallic catalyst combinations are formed by the direct change and inter-exchange of the solute metal for the solvent metal atoms in the crystal structure. There are three basic criteria that favor the formation of bimetallic catalyst combinations: (1) the electronegativities of the two components are similar; (2) the atomic radii of the two elements are within 20 percent of each other; and (3) the crystal structures of the two pure phases are the same. The catalyst metals (as the elemental metal or metal oxide) employed in the inventive sorbent composition preferably meet at least two of the three criteria set forth above.

Preferably, the reduced-valence catalyst component is reduced nickel, cobalt, silver, copper, tin, zinc, or a bimetallic combination. Such amounts of reduced-valence catalyst component are generally in the range from about 15 to about 40 weight percent of the total weight of the sorbent composition. When the catalyst component comprises a bimetallic catalyst component, the bimetallic catalyst component should comprise a ratio of the two metals forming such bimetallic catalyst component in the range of from about 10:1 to about 1:10. In a presently preferred embodiment of the present innovation, the catalyst component is a bimetallic catalyst component comprising silver and copper in a weight ratio of about 1:1, nickel and cobalt in a weight ratio of about 1:1, or a zinc and tin in a weight ratio of about 2:1.

For the impregnation of the particulates it can be desirable to use an aqueous solution of a catalyst component. Dissolving a metal-containing compound, in the form of a metal salt, such as, a metal chloride, a metal nitrate, a metal sulfate, and the combinations in a solvent, such as, water, alcohols, esters, ethers, ketones, and combinations is a preferred impregnating solution comprised in the aqueous solution formed, thereof. Preferably, the weight ratio of metal catalyst component to the aqueous medium of such aqueous solution can be in the range of from 2.0:1 to 2.5:1.

Once the catalyst component has been impregnated in the particulate sorbent, or calcined base support sorbent, the desired reduced-valence particulate catalyst impregnated sorbent is prepared by drying the resulting composition under a drying condition, followed by calcining to provide dried, calcined particulate catalyst sorbent, which are subjected to reduction with a suitable reducing agent, preferably hydrogen so as to produce a composition having a substantially reduced-valence catalyst component content, preferably a substantially zero content, with such zero valence catalyst component being present in an amount sufficient to permit the removal of target elements or compounds from a host media fluid.

Bimetallic catalyst compounds may be added to the support components prior to drying and calcining, or by impregnating the dried and calcined support particulates with a solution either aqueous or organic that contains the elemental metals, metal oxides or metal-containing compounds of the selected catalyst group after the first drying and calcining.

A preferred method of incorporating is impregnating using any standard wetness impregnation technique (i.e., essentially completely filling the pores of a substrate material with a solution of the incorporating elements) for impregnating a substrate. The impregnating solution, comprising of the desirable concentration of a catalyst component, impregnates the particulate sorbent support, which can then be subjected to drying and calcining followed by reduction with a reducing agent such as hydrogen. The impregnating solution can be any aqueous solution and amounts of such solution which suitably provides for the impregnation of the particulates of support component to give an amount of catalyst component that provides, after reduction with a reducing agent, a reduced catalyst component content sufficient to permit the removal of target elements and compounds from a host media, when such fluid is treated in accordance with the process of the present innovation.

In preparing the spray-dried sorbent material, a catalyst component can be added to the spray-dried sorbent material as a component of the original mixture, or they can be added after the original mixture has initially been spray dried and calcined. If a catalyst component is added to the spray-dried sorbent material after it has been spray dried and calcined, the spray-dried sorbent material should be dried and calcined a second time. The spray-dried sorbent material is preferably dried a second time at a temperature generally in the range of from about 90 degrees C to about 300 degrees C. The time period for conducting the drying a second time is generally in the range of from about 1.5 hours to 4 hours. Such drying a second time is generally carried out at a pressure preferably about atmospheric. This spray-dried sorbent material is then calcined, preferably in an oxidizing atmosphere such as in the presence of oxygen or air, under a calcining condition. Any drying methods(s) known to one skilled in the art such as, for example, air drying, heat drying, and the like and combinations thereof can be used.

Generally, a calcining condition can include a temperature in the range of from 480 degrees C to about 780 degrees C. Such calcining condition can also include a pressure, generally in the range of from about 7 pounds per square inch absolute (psia) to about 150 psia, and a time period in the range of from about 2 hours to 15 hours.

Impregnation of the particulate support can be achieved by use of solutions of the selected metal, which is formed of the metal per se, metal oxide or a precursor for the same. Such impregnation can be carried out in separate steps whereby the particulate support is dried or dried and calcined prior to the addition of the second metal component to the support.

Following impregnation of the particulate compositions with the appropriate bimetallic catalyst, the resulting impregnated particulate is then subjected to drying and calcination under the conditions noted supra prior to subjection of the calcined particulate to reduction with a reducing agent, preferably hydrogen.

Once the bimetallic catalyst has been incorporated in the particulate support, the desired reduced valence of the metals is achieved by drying the resulting composition followed by calcination and thereafter subjecting the resulting calcined composition to reduction with a suitable reducing agent, preferably hydrogen, so as to produce a composition having a substantially reduced valence metals content which is present in an amount to permit the removal of certain target elements and compounds from a host media.

If desired the components of the bimetallic catalyst can be added to the support individually rather than by co-impregnation.

The resulting product is an adsorbent medium that is capable of adsorbing or the absorption of target elements or compounds and filtering particulate material from water polar solutions, hydrocarbons, hydrocarbon by-products or other carbonaceous liquid substances. This innovation provides an attrition resistant particulate catalyst impregnated sorbent that is stable under basic conditions, which are generally encountered when adsorbent materials are regenerated after adsorbing or the absorption of target elements or compounds.

20.8 Notes on Carbon

A preferred method of obtaining the formless or unstructured carbon is to vaporize pure graphite in an enclosed apparatus containing at least two electrodes and having an inert atmosphere to produce a vaporized composition. Rapidly cooling the vapor to deposit the composition on a surface of the apparatus and/or the electrodes and then remove any impurities contained in the composition by way of solvent extraction.

The superior catalytic properties of the metal-containing catalysts described in the innovation are attributed to the fact that all the catalytically active metal is anchored at the surface of the particulate catalyst sorbent used in the oxidation adsorption part of the process of this current innovation, and therefore is very readily accessible for the components of the reaction being oxidized. In the case of the particulate catalyst impregnated sorbent (carbon) compounds, by contrast, a significant proportion of the catalytically active metal is in the interior of the macroscopic particles of the particulate catalyst impregnated sorbent (carbon) compound, and therefore improves the selectivity for the components of the reaction being catalyzed. The superior stability of the metal-containing catalysts of the innovation compared to a simple carbon sorbent is attributed to the slight covalent bonds formed with the metal or bimetallic combinations.

A particularly interesting aspect of the innovation is the qualitative change, which can be achieved in the catalytic properties of the bound metal. It was possible to demonstrate that the structural properties of the metal particles contained in the systems described in the innovation differed markedly from those of simple carbon sorbent systems. These structural differences are attributed to qualitatively different attractive interactions between the metal particles and the carbon support system in question. The structural differences in the metal particles are not, however, exclusively of geometrical nature; it is assumed that there are also differences in the electronic structure, as a result of which the active centers at the surface of the metal, which are crucial for the heterogeneous catalysis of a reaction.

An additional subject of the innovation is a method of producing the new catalysts. According to the method of the innovation, carbon is vaporized in a non-oxidizing atmosphere by means of an electric arc struck between at least two graphite electrodes in a vacuum apparatus, during which process one

a) works with a.c. or d.c. under a pressure of 100 Pa or less in a vacuum apparatus the walls of which are cooled, the product being deposited on the cooled walls, or
b) works with d.c. under a pressure of 1 to 100 kPa and arc lengths of 0.1 to 20 mm, the product accumulating on the electrode connected to the negative pole of the power supply, or
c) works with a.c. under a pressure of 1 to 100 Pa and arc lengths of 0.1 to 20 mm, the product accumulating on the carbon electrodes.

Optionally, according to one embodiment of the present innovation, thereafter the product of a), b) or c) may be reacted with a metal, or metal oxide, low-valence compound or combination of catalytically active metals, as pre-suggested.

The graphite used should be as pure as possible. It is preferable to work in a noble-gas atmosphere, most preferably of helium, argon or a mixture of helium and argon.

If the formless or unstructured carbon is produced according to procedure a), it is expedient to cool the walls of the vacuum apparatus with water. However, other cooling methods or coolants can be used in a similar manner. For preparing the formless or unstructured carbon it is also of advantage to work with two carbon electrodes, since this is how a commercially available apparatus is usually equipped. However, for the method of the innovation it is also possible to use modified electric-arc equipment with more than two graphite electrodes.

The reaction of the carbon with the metal compound is preferably performed in the absence of air in a suspension of the supporting carbon in a solvent in which the metal compound is soluble. It is of advantage to work at an elevated temperature, preferably at the reflux temperature of the solvent. Under these conditions, the reaction is generally allowed to proceed for between 15 and 25 hours. If the support material contains any impurities, these are removed prior to the reaction, preferably by extraction with a suitable organic solvent.

The general temperature range for the reaction of the carbon with the metal is between the solidification point of the solvent and its boiling point. The boiling point of the solvent can be raised according to standard practice by applying pressure.

The fabrication of the particulate catalyst impregnated sorbent (carbon) used in the present innovation involves the thermal shock of previously formed impregnated carbon compounds. The process of oxidizing and expanding carbon has been previously described in the literature, and is known by one competent in the art of carbon particulate manufacturing. The volume of the oxidizing solution used is not critical as long as it is sufficient to suspend the particulate mass and ensure effective impregnation. Large industrial processing may require relatively greater volumes or extended mixing times. While not necessary, the temperature may be elevated between 50 degrees C and 100 degrees C to increase the rate of oxidation. After the desired particulate catalyst impregnated sorbent is formed, the particles are thoroughly rinsed in water and then rapidly heated to approximately 1000 degrees C. This heating, which results in further expansion, is generally performed in an electrical furnace allowing substantial amounts of expanded particles to be produced.

Depending on the fabrication process employed, the bulk densities of the particulate catalyst impregnated sorbent may vary. Such properties are important in that they correspond to the absorption capacity of the particulate catalyst impregnated sorbent. More specifically the lower the bulk density, the higher the specific surface area will be, and, therefore, the higher the absorption capacity for target elements and compounds. Like other particulate catalyst impregnated compounds, as herein mentioned, these carbons impregnated sorbent particles are resistant to temperature, aging, and most corrosive media.

Without the unique physical characteristics of the particulate catalyst impregnated sorbents, the numerous advantages of the present innovation would be impossible to obtain. Other materials fail to combine the low bulk density of the carbon catalyst impregnated sorbent, with its high adsorption capacity, cost efficient fabrication from readily available materials and high absorption capacity. Due to the previous use of carbon in other applications, several inexpensive fabrication methods have been developed to produce material, which may be employed effectively in the present innovation. Most of these fabrication processes depend on the use of inexpensive carbon as a starting material.

20.9 Notes on Zeolite

For the use of a zeolite as the adsorbent of the innovation, one or more metals selected from silver, copper, zinc, tin, cobalt, and nickel are supported on a zeolite according to an ion exchange technique. More particularly, compounds of silver, copper and the like are dissolved in water to provide an aqueous solution, followed by ion exchange with use of the solution. The compounds of the metals should be ion-exchanged with a cation in the zeolite, and thus, should be made of a metal compound capable of being dissolved in water and existing as a metal ion in the aqueous solution. This aqueous solution is brought into contact with a zeolite according to a general ion exchange procedure including (1) an agitation procedure, (2) an impregnation procedure or (3) a flow procedure, thereby causing the cations in the zeolite to be exchanged with these metal ions. Thereafter, the zeolite is washed such as with water and dried to obtain an adsorbent of the innovation. Although the zeolite may be calcinated after drying, the calcination is not always necessary.

The zeolite powder produced is then admixed with a binder to produce a final adsorbent-catalyst product. The binder can be chosen from conventional mineral or synthetic materials, such as clays (kaolinite, bentonite, montmorillonite, attapulgite, smectite, etc.), silica, alumina, alumina hydrate, alumina trihydrate, alumosilicates, cements, etc. The mixture is then mixed and worked into itself and with 18-35% water to form a paste, which is then aggregated to form shaped articles of conventional shapes such as particulates, spheres, micro-spheres, etc. The product is then washed with deionized water to remove excess ions, dried, and calcined at a temperature from about 250 to about 550 degrees C. The adsorbent-catalysts, according to the present innovation, can provide improved and more reliable protection of the catalysts in large-scale commercial processes.

Zeolites typically have silica-to-alumina mole ratios of at least about 2, and have uniform pore diameters from about 3 to 15 angstroms. They also generally contain alkali metal cations, such as sodium and/or potassium and/or alkaline earth metal cations, such as magnesium and/or calcium. In order to increase the catalytic activity of the zeolite, it may be desirable to decrease the alkali metal content of the crystalline zeolite to less than about 0.5 wt. %. The alkali metal content reduction, as is known in the art, may be conducted by exchange with one or more cations selected from the Groups IIB through VIII of the Periodic Table of Elements.

20.10 Notes on Perlite

The perlite sorbent compositions of the present innovation, which are useful in the separation and removal process of the present innovation, can be prepared by a process comprising:

(a) mixing a support component, so as to form a mixture selected from the group consisting of a wet mix, a dough, a paste, a slurry, and the like and combinations thereof;
(b) particulating, preferably spray-drying, the mixture to form particulates selected from the group consisting of granules, spheres, micro-spheres, and the like and combinations thereof, preferably micro-spheres;
(c) drying the particulate under a drying condition as disclosed herein to form a dried particulate;
(d) calcining the dried particulate under a calcining condition as disclosed herein to form a calcined particulate;
(e) incorporating, preferably impregnating, the calcined particulate with a catalyst component thereof to form a particulate impregnated catalyst sorbent;
(f) drying the particulate sorbent under a drying condition as disclosed herein to form a dried, particulate catalyst impregnated sorbent;
(g) calcining the dried, particulate sorbent under a calcining condition as disclosed herein to form a calcined, particulate catalyst impregnated sorbent; and
(h) reducing the calcined, particulate catalyst impregnated sorbent with a suitable reducing agent so as to produce a sorbent composition having a reduced-valence catalyst component content therein, and wherein the reduced-valence catalyst component content is present in an amount

effective for the removal of target elements and compounds from a host media when such target elements and compound-containing fluid is contacted with a sorbent composition(s) of the present innovation according to a process(es) of the present innovation.

When using perlite for the selected sorbent, the alumina component of the base support can be any suitable compound of alumina that has cement-like properties, which can help to bind the particulate composition together. Presently preferred is alumina, preferably peptized alumina, however colloidal alumina solutions and generally those alumina compounds produced by the dehydration of alumina hydrates may be used.

The components of the support component are mixed to provide a resulting mixture which can be in a form selected from the group consisting of wet mix, dough, paste, slurry, and the like. Such resulting mixture can then be shaped to form a particulate selected from the group consisting of a granule, a sphere, or a micro-sphere. For example, if the resulting mixture is in the form of a wet mix, the wet mix can be densified, dried under a drying condition as disclosed herein, calcined under a calcining condition as disclosed herein, and thereafter shaped, or particulated, through the granulation of the densified, dried, calcined mix to form micro-spheres. Also, for example, when the mixture of the components of the support component results in a form of a mixture, which is either in a dough state or paste state, such mixture can then be shaped, to form a particulate. The resulting particulates are then dried under a drying condition as disclosed herein and then calcined under a calcining condition as disclosed herein. More preferably, when the mix is in the form of a slurry, the particulation of such slurry is achieved by spray drying the slurry to form micro-spheres thereof having a size in the range of from about 10 to about 500 microns. Such micro-spheres are then subjected to drying under a drying condition as disclosed herein and calcining under a calcining condition as disclosed herein.

When the particulation is achieved by preferably spray drying, a dispersant component may be utilized and can be any suitable compound that helps to promote the spray drying ability of the mix which is preferably in the form of a slurry. In particular, these components are useful in preventing deposition, precipitation, settling, agglomerating, adhering, and caking of solid particles in a fluid medium.

The silica used in the preparation of such sorbent compositions may be either in the form of silica or in the form of one or more silicon-containing compounds. Any suitable type of silica may be employed in the sorbent compositions of the present innovation. Examples of suitable types of silica include diatomite, silicalite, silica colloid, flame-hydrolyzed silica, hydrolyzed silica, silica gel and precipitated silica, with diatomite being presently preferred. In addition, silicon compounds that are convertible to silica such as silicic acid, sodium silicate and ammonium silicate can also be employed. Preferably, the silica is in the form of diatomite. The silica will generally be present in the sorbent composition in an amount in the range of from about 20 weight percent to about 60 weight percent, when the weight percents are expressed in terms of the silica based upon the total weight of the sorbent composition.

The alumina will generally be present in the sorbent composition in an amount in the range of from about 5.0 weight percent to about 20 weight percent, when such weight percents are expressed in terms of the weight of the alumina compared with the total weight of the sorbent system.

Zeolite and Perlite sorbent selections and substitutions are optimum and preferred, when the adsorption properties in the oxidation step of the process require the specific qualities or adsorptive properties designed by either the target elements or compounds, or the host media.

20.11 Conclusion

It is apparent there is provided in accordance with this innovation a sorbent, an improved method for preparing such sorbent and an on-line method whereby target elements and compounds can be successfully removed from contaminated water streams, hydrocarbon fluid streams or feed streams, and other carbonaceous liquids. It is further noted that the objects, aims, and advantages of this innovation as set forth above are accomplished by the described innovation.

Accordingly, the disclosed embodiments and instrumentalities are not exhaustive of all options or mannerisms for practicing the disclosed principles of the innovation. While the innovation has been described in connection with specific embodiments, it is evident and understood that the innovation is not limited to the embodiments disclosed, but is capable of numerous rearrangements, many alternatives, modifications, variations, and substitutions of parts and elements, thereof and would be apparent and could be made in the process, without departing from the spirit or scope of the innovation, to those skilled in the art after having been instructed by the foregoing description. Accordingly, it is intended to embrace all such alternatives, modifications and variations of this innovation provided they come within the scope of this thesis, as set forth, and their equivalents. The author hereby states his intention to rely upon the Doctrine of Equivalents in protecting the full scope and spirit of the innovation.

21 Appendix C (by N. Schwarzer): Derivation of the Functional Derivative of the Scaled and Adjusted Einstein-Hilbert Action with Respect to the Function f

The functional derivative:

$$\delta W=\frac{\delta W}{\delta f[\mathbf{y}]}=0=\frac{\delta}{\delta f}\int_V d^n x\left(\sqrt{g\cdot F^n}\cdot F\cdot R^*\right)$$
$$=\lim_{\in\to 0}\frac{1}{\in}\left(\begin{array}{c}\int_V d^n x\left(\sqrt{g\cdot F\left[f[\mathbf{x}]+\in\cdot\delta[\mathbf{x}-\mathbf{y}]\right]^n}\cdot R^{**}\left[f[\mathbf{x}]+\in\cdot\delta[\mathbf{x}-\mathbf{y}]\right]\right)\\ -\int_V d^n x\left(\sqrt{g\cdot F[f[\mathbf{x}]]^n}\cdot R^{**}[f[\mathbf{x}]]\right)\end{array}\right) \tag{713}$$

with:

$$R^{**}=F\cdot R^*=R-\frac{g^{\alpha\beta}}{2F}\left(2F_{,\alpha\beta}(n-1)+F_{,d}g^{cd}\left(\begin{array}{c}\left(g_{\beta c,\alpha}+g_{\alpha c,\beta}-g_{\beta\alpha,c}\right)-2g_{c\beta,\alpha}\\ +\frac{n}{2}\left(\left(-g_{\alpha c,\beta}-g_{\beta c,\alpha}+g_{\alpha\beta,c}\right)+g_{\alpha\beta,c}\right)\end{array}\right)\right)$$
$$+\frac{1}{4F^2}\left(F_{,\alpha}\cdot F_{,\beta}(3n-6)+g_{\alpha\beta}F_{,c}F_{,d}g^{cd}(4-n)\right)g^{\alpha\beta}$$
$$=R-\frac{g^{\alpha\beta}}{2F}\left(2F_{,\alpha\beta}(n-1)+F_{,d}g^{cd}\left(\begin{array}{c}\left(g_{\beta c,\alpha}+g_{\alpha c,\beta}-g_{\beta\alpha,c}\right)-2g_{c\beta,\alpha}\\ +\frac{n}{2}\left(\left(-g_{\alpha c,\beta}-g_{\beta c,\alpha}+g_{\alpha\beta,c}\right)+g_{\alpha\beta,c}\right)\end{array}\right)\right)$$
$$+\frac{1}{4F^2}\left(F_{,\alpha}\cdot F_{,\beta}(3n-6)+g_{\alpha\beta}F_{,c}F_{,d}g^{cd}(4-n)\right)g^{\alpha\beta}$$
$$=R-\frac{g^{\alpha\beta}}{2F}F'\left(2f_{,\alpha\beta}(n-1)+f_{,d}g^{cd}\left(\begin{array}{c}g_{\alpha c,\beta}-g_{\beta\alpha,c}-g_{c\beta,\alpha}\\ +\frac{n}{2}\left(2g_{\alpha\beta,c}-g_{\alpha c,\beta}-g_{\beta c,\alpha}\right)\end{array}\right)\right)$$
$$-(n-1)\frac{f_{,\alpha}\cdot f_{,\beta}}{4F^2}g^{\alpha\beta}\left(4FF''+F'\cdot F'(n-6)\right) \tag{714}$$

gives:

$$\frac{\delta W}{\delta f}=0=\frac{\delta}{\delta f}\int_V d^n x\left(\sqrt{g\cdot F^n}\cdot R^{**}\right)$$

$$=-\left[\begin{array}{l}\left((n-1)\cdot\sqrt{g}\cdot F^{\frac{n}{2}-1}g^{\alpha\beta}\cdot F'\right)_{,\alpha\beta}-\sqrt{g}\cdot R\frac{dF^{\frac{n}{2}}}{df}\\ -\frac{1}{2}\left(\sqrt{g}\cdot F^{\frac{n}{2}-1}g^{\alpha\beta}\cdot F'g^{cd}\left(\begin{array}{c}g_{\alpha c,\beta}-g_{\beta\alpha,c}-g_{c\beta,\alpha}\\ +\frac{n}{2}\left(2g_{\alpha\beta,c}-g_{\alpha c,\beta}-g_{\beta c,\alpha}\right)\end{array}\right)\right)_{,d}\\ +\frac{\sqrt{g}\cdot g^{\alpha\beta}}{2}\left(2f_{,\alpha\beta}(n-1)+f_{,d}g^{cd}\left(\begin{array}{c}g_{\alpha c,\beta}-g_{\beta\alpha,c}-g_{c\beta,\alpha}\\ +\frac{n}{2}\left(2g_{\alpha\beta,c}-g_{\alpha c,\beta}-g_{\beta c,\alpha}\right)\end{array}\right)\right)\frac{d\left[F^{\frac{n}{2}-1}F'\right]}{df}\\ +\frac{(n-1)}{2}\left(\begin{array}{c}-\left(f_{,\alpha}\cdot\sqrt{g}\cdot g^{\alpha\beta}F^{\frac{n}{2}-2}(4FF''+F'\cdot F'(n-6))\right)_{,\beta}\\ +\frac{f_{,\alpha}\cdot f_{,\beta}}{2}\sqrt{g}\cdot g^{\alpha\beta}\frac{d\left[F^{\frac{n}{2}-2}(4FF''+F'\cdot F'(n-6))\right]}{df}\end{array}\right)\end{array}\right]_{y}. \quad (715)$$

Thereby, just for brevity, we assume in this appendix that g denotes the absolute value of the determinant of the metric tensor $g_{\alpha\beta}$.

We can significantly simplify the outcome in the case of metrics consisting only of constant components:

$$\frac{\delta W}{\delta f}=0=\frac{\delta}{\delta f}\int_V d^n x\left(\sqrt{g}\cdot F^n\cdot R^{**}\right)$$

$$=-\left[\begin{array}{c}(n-1)\cdot\sqrt{g}\cdot g^{\alpha\beta}\cdot\left(\left(F^{\frac{n}{2}-1}\cdot F'\right)_{,\alpha\beta}+f_{,\alpha\beta}\frac{d\left[F^{\frac{n}{2}-1}F'\right]}{df}\right)-\sqrt{g}\cdot R\frac{dF^{\frac{n}{2}}}{df}\\+\frac{(n-1)}{2}\left(\begin{array}{c}-\left(f_{,\alpha}\cdot\sqrt{g}\cdot g^{\alpha\beta}F^{\frac{n}{2}-2}\left(4FF''+F'\cdot F'(n-6)\right)\right)_{,\beta}\\+\frac{f_{,\alpha}\cdot f_{,\beta}}{2}\sqrt{g}\cdot g^{\alpha\beta}\frac{d\left[F^{\frac{n}{2}-2}\left(4FF''+F'\cdot F'(n-6)\right)\right]}{df}\end{array}\right)\end{array}\right]_y$$

$$=-\left[\begin{array}{c}(n-1)\cdot\sqrt{g}\cdot g^{\alpha\beta}\cdot\left(\begin{array}{c}\left(\left(F^{\frac{n}{2}-1}\cdot F'\right)_{,\alpha\beta}+f_{,\alpha\beta}\frac{d\left[F^{\frac{n}{2}-1}F'\right]}{df}\right)\\+\frac{1}{2}\left(\begin{array}{c}-\left(f_{,\alpha}\cdot F^{\frac{n}{2}-2}\left(4FF''+F'\cdot F'(n-6)\right)\right)_{,\beta}\\+\frac{f_{,\alpha}\cdot f_{,\beta}}{2}\frac{d\left[F^{\frac{n}{2}-2}\left(4FF''+F'\cdot F'(n-6)\right)\right]}{df}\end{array}\right)\end{array}\right)\\-\sqrt{g}\cdot R\frac{dF^{\frac{n}{2}}}{df}\end{array}\right]_y\cdot\qquad(716)$$

Further simplification gives:

$$\frac{\delta W}{\delta f}=0=\frac{\delta}{\delta f}\int_V d^n x\left(\sqrt{g}\cdot F^n\cdot R^{**}\right)$$

$$=-\left[(n-1)\cdot\sqrt{g}\cdot g^{\alpha\beta}\cdot\begin{pmatrix}\frac{f_{,\alpha}\cdot f_{,\beta}}{4}\frac{d\left[F^{\frac{n}{2}-2}\left(4FF''+F'\cdot F'(n-6)\right)\right]}{df}\\-\frac{1}{2}\left(f_{,\alpha}\cdot F^{\frac{n}{2}-2}\left(4FF''+F'\cdot F'(n-6)\right)\right)_{,\beta}\\+\left(F^{\frac{n}{2}-1}\cdot F'\right)_{,\alpha\beta}+f_{,\alpha\beta}\frac{d\left[F^{\frac{n}{2}-1}F'\right]}{df}\end{pmatrix}-R\frac{dF^{\frac{n}{2}}}{df}\right]_y$$

$$=-\left[\begin{matrix}(n-1)\cdot\sqrt{g}\cdot g^{\alpha\beta}\cdot\begin{pmatrix}\frac{f_{,\alpha}\cdot f_{,\beta}}{4}\frac{d\left[F^{\frac{n}{2}-2}\left(4FF''+F'\cdot F'(n-6)\right)\right]}{df}\\-\frac{1}{2}\begin{pmatrix}f_{,\alpha\beta}\cdot F^{\frac{n}{2}-2}\left(4FF''+F'\cdot F'(n-6)\right)\\+f_{,\alpha}f_{,\beta}\cdot\frac{d\left[F^{\frac{n}{2}-2}\left(4FF''+F'\cdot F'(n-6)\right)\right]}{df}\end{pmatrix}\\+\left(\left(\left(\frac{n}{2}-1\right)F^{\frac{n}{2}-2}\cdot F'F'+F^{\frac{n}{2}-1}\cdot F''\right)f_{,\alpha}\right)_{,\beta}\\+f_{,\alpha\beta}\frac{d\left[F^{\frac{n}{2}-1}F'\right]}{df}\end{pmatrix}\\-\sqrt{g}\cdot R\frac{dF^{\frac{n}{2}}}{df}\end{matrix}\right]_y. \quad (717)$$

$$\frac{\delta W}{\delta f}=0=\frac{\delta}{\delta f}\int_V d^n x\left(\sqrt{g\cdot F^n}\cdot R^{**}\right)$$

$$=-\left[\begin{array}{c}(n-1)\cdot\sqrt{g}\cdot g^{\alpha\beta}\cdot\left(\begin{array}{c}\frac{f_{,\alpha}\cdot f_{,\beta}}{4}\frac{d\left[F^{\frac{n}{2}-2}\left(4FF''+F'\cdot F'(n-6)\right)\right]}{df}\\ -\frac{1}{2}\left(\begin{array}{c}f_{,\alpha\beta}\cdot F^{\frac{n}{2}-2}\left(4FF''+F'\cdot F'(n-6)\right)\\ +f_{,\alpha}f_{,\beta}\cdot\frac{d\left[F^{\frac{n}{2}-2}\left(4FF''+F'\cdot F'(n-6)\right)\right]}{df}\end{array}\right)\\ +\left(\left(\begin{array}{c}\left(\frac{n}{2}-1\right)\left(\left(\frac{n}{2}-2\right)F^{\frac{n}{2}-3}(F')^3+2F^{\frac{n}{2}-2}F'F''\right)\\ +F^{\frac{n}{2}-1}\cdot F'''\end{array}\right)f_{,\alpha}f_{,\beta}\right)\\ \left(\left(\frac{n}{2}-1\right)F^{\frac{n}{2}-2}\cdot F'F'+F^{\frac{n}{2}-1}\cdot F''\right)f_{,\alpha\beta}\\ +f_{,\alpha\beta}\frac{d\left[F^{\frac{n}{2}-1}F'\right]}{df}\end{array}\right)\\ -R\frac{dF^{\frac{n}{2}}}{df}\end{array}\right]_{y}. \quad (718)$$

For brevity we ignore the R-term (just for the moment) and evaluate as follows:

$$\frac{\delta W}{\delta f}=0=\frac{\delta}{\delta f}\int_V d^n x\left(\sqrt{g}\cdot F^n\cdot R^{**}\right)=-(n-1)\cdot\sqrt{g}\cdot g^{\alpha\beta}\Big|_y \times\left[\begin{array}{c}\frac{f_{,\alpha}\cdot f_{,\beta}}{4}\frac{d\left[F^{\frac{n}{2}-2}(4FF''+F'\cdot F'(n-6))\right]}{df}\\ -\frac{1}{2}\left(\begin{array}{c}f_{,\alpha\beta}\cdot F^{\frac{n}{2}-2}(4FF''+F'\cdot F'(n-6))\\ +f_{,\alpha}f_{,\beta}\cdot\frac{d\left[F^{\frac{n}{2}-2}(4FF''+F'\cdot F'(n-6))\right]}{df}\end{array}\right)\\ +\left(\left(\frac{n}{2}-1\right)\left(\left(\frac{n}{2}-2\right)F^{\frac{n}{2}-3}(F')^3+2F^{\frac{n}{2}-2}F'F''\right)+F^{\frac{n}{2}-1}\cdot F'''\right)f_{,\alpha}f_{,\beta}\\ F^{\frac{n}{2}-2}\cdot\left(\left(\frac{n}{2}-1\right)F'F'+F\cdot F''\right)f_{,\alpha\beta}\\ +f_{,\alpha\beta}\left(\left(\frac{n}{2}-1\right)F^{\frac{n}{2}-2}F'F'+F^{\frac{n}{2}-1}\cdot F''\right)\end{array}\right]_y, \tag{719}$$

$$\frac{\delta W}{\delta f}=0=\frac{\delta}{\delta f}\int_V d^n x\left(\sqrt{g}\cdot F^n\cdot R^{**}\right)=-(n-1)\cdot\sqrt{g}\cdot g^{\alpha\beta}\Big|_y \times\left[\begin{array}{c}\frac{f_{,\alpha}\cdot f_{,\beta}}{4}\frac{d\left[F^{\frac{n}{2}-2}(4FF''+F'\cdot F'(n-6))\right]}{df}\\ -\frac{1}{2}\left(\begin{array}{c}f_{,\alpha\beta}\cdot F^{\frac{n}{2}-2}(4FF''+F'\cdot F'(n-6))\\ +f_{,\alpha}f_{,\beta}\cdot\frac{d\left[F^{\frac{n}{2}-2}(4FF''+F'\cdot F'(n-6))\right]}{df}\end{array}\right)\\ +F^{\frac{n}{2}-3}f_{,\alpha}f_{,\beta}\left(\left(\frac{n}{2}-1\right)\left(\left(\frac{n}{2}-2\right)(F')^3+2FF'F''\right)+F^2\cdot F'''\right)\\ +2f_{,\alpha\beta}F^{\frac{n}{2}-2}\left(\left(\frac{n}{2}-1\right)F'F'+F\cdot F''\right)\end{array}\right]_y, \tag{720}$$

$$\frac{\delta W}{\delta f}=0=\frac{\delta}{\delta f}\int_V d^n x\left(\sqrt{g\cdot F^n}\cdot R^*\right)=-(n-1)\cdot\sqrt{g}\cdot g^{\alpha\beta}\Big|_y$$

$$\times\left[\begin{array}{c}-\frac{f_{,\alpha}\cdot f_{,\beta}}{4}\frac{d\left[F^{\frac{n}{2}-2}\left(4FF''+F'\cdot F'(n-6)\right)\right]}{df}\\ -\frac{1}{2}\left(f_{,\alpha\beta}\cdot F^{\frac{n}{2}-2}\left(4FF''+F'\cdot F'(n-6)\right)\right)\\ +F^{\frac{n}{2}-3}f_{,\alpha}f_{,\beta}\left(\left(\frac{n}{2}-1\right)\left(\left(\frac{n}{2}-2\right)(F')^3+2FF'F''\right)+F^2\cdot F'''\right)\\ +2f_{,\alpha\beta}F^{\frac{n}{2}-2}\left(\left(\frac{n}{2}-1\right)F'F'+F\cdot F''\right)\end{array}\right]_y$$

$$=-(n-1)\cdot\sqrt{g}\cdot g^{\alpha\beta}\Big|_y$$

$$\times\left[\begin{array}{c}-\frac{f_{,\alpha}\cdot f_{,\beta}}{4}F^{\frac{n}{2}-3}\left(\begin{array}{c}F\left(4(F'F''+FF''')+2F'\cdot F''(n-6)\right)\\ +\left(\frac{n}{2}-2\right)F'\left(4FF''+F'\cdot F'(n-6)\right)\end{array}\right)\\ -\frac{1}{2}\left(f_{,\alpha\beta}\cdot F^{\frac{n}{2}-2}\left(4FF''+F'\cdot F'(n-6)\right)\right)\\ +F^{\frac{n}{2}-3}f_{,\alpha}f_{,\beta}\left(\left(\frac{n}{2}-1\right)\left(\left(\frac{n}{2}-2\right)(F')^3+2FF'F''\right)+F^2\cdot F'''\right)\\ +2f_{,\alpha\beta}F^{\frac{n}{2}-2}\left(\left(\frac{n}{2}-1\right)F'F'+F\cdot F''\right)\end{array}\right]_y, \qquad (721)$$

$$\frac{\delta W}{\delta f}=0=\frac{\delta}{\delta f}\int_V d^n x\left(\sqrt{g\cdot F^n}\cdot R^{**}\right)=-(n-1)\cdot\sqrt{g}\cdot g^{\alpha\beta}\Big|_{\mathbf{y}}$$

$$\times\left[\begin{matrix}+F^{\frac{n}{2}-3}f_{,\alpha}f_{,\beta}\left(\begin{matrix}\left(\frac{n}{2}-1\right)\left(\left(\frac{n}{2}-2\right)(F')^3+2FF'F''\right)+F^2\cdot F''' \\ -\frac{1}{4}\left(\begin{matrix}F\left(4(F'F''+FF''')+2F'\cdot F''(n-6)\right) \\ +\left(\frac{n}{2}-2\right)F'\left(4FF''+F'\cdot F'(n-6)\right)\end{matrix}\right)\end{matrix}\right) \\ +2f_{,\alpha\beta}F^{\frac{n}{2}-2}\left(\left(\frac{n}{2}-1\right)F'F'+F\cdot F''-\frac{4FF''+F'\cdot F'(n-6)}{4}\right)\end{matrix}\right]_{\mathbf{y}},\qquad(722)$$

$$\frac{\delta W}{\delta f}=0=\frac{\delta}{\delta f}\int_V d^n x\left(\sqrt{g\cdot F^n}\cdot R^{**}\right)=-(n-1)\cdot\sqrt{g}\cdot g^{\alpha\beta}\Big|_{\mathbf{y}}$$

$$\times\left[\begin{matrix}+F^{\frac{n}{2}-3}f_{,\alpha}f_{,\beta}\left(\begin{matrix}\left(\frac{n}{2}-1\right)\left(\left(\frac{n}{2}-2\right)(F')^3+2FF'F''\right)+F^2\cdot F''' \\ -\frac{1}{4}\left(\begin{matrix}F\left(4(F'F''+FF''')+2F'\cdot F''(n-6)\right) \\ +\left(\frac{n}{2}-2\right)F'\left(4FF''+F'\cdot F'(n-6)\right)\end{matrix}\right)\end{matrix}\right) \\ +\frac{1}{2}(n+2)f_{,\alpha\beta}F^{\frac{n}{2}-2}F'F'\end{matrix}\right]_{\mathbf{y}},\qquad(723)$$

$$\frac{\delta W}{\delta f}=0=\frac{\delta}{\delta f}\int_V d^n x\left(\sqrt{g\cdot F^n}\cdot R^{**}\right)=-(n-1)\cdot\sqrt{g}\cdot g^{\alpha\beta}\Big|_y$$

$$\times\left[\begin{array}{c}+\frac{F^{\frac{n}{2}-3}f_{,\alpha}f_{,\beta}}{4}\left(\begin{array}{l}\left(4\left(\frac{n}{2}-2\right)\left(\frac{n}{2}-1\right)(F')^3+8\left(\frac{n}{2}-1\right)FF'F''\right)+4F^2\cdot F'''\\-4F^2\cdot F'''-4FF'F''-2FF'\cdot F''(n-6)\\-\left(\frac{n}{2}-2\right)\left(4FF'F''+F'\cdot F'F'(n-6)\right)\end{array}\right)\\+\frac{1}{2}(n+2)f_{,\alpha\beta}F^{\frac{n}{2}-2}F'F'\end{array}\right]_y$$

$$=-(n-1)\cdot\sqrt{g}\cdot g^{\alpha\beta}\Big|_y$$

$$\times\left[\begin{array}{c}+\frac{F^{\frac{n}{2}-3}f_{,\alpha}f_{,\beta}}{4}\left(\begin{array}{l}4\left(\frac{n}{2}-2\right)\left(\frac{n}{2}-1\right)(F')^3+8\left(\frac{n}{2}-1\right)FF'F''\\-4FF'F''-2FF'\cdot F''(n-6)\\-4\left(\frac{n}{2}-2\right)FF'F''-F'\cdot F'F'\left(\frac{n}{2}-2\right)(n-6)\end{array}\right)\\+\frac{1}{2}(n+2)f_{,\alpha\beta}F^{\frac{n}{2}-2}F'F'\end{array}\right]_y$$

$$=-(n-1)\cdot\sqrt{g}\cdot g^{\alpha\beta}\Big|_y$$

$$\times\left[\begin{array}{c}+\frac{F^{\frac{n}{2}-3}f_{,\alpha}f_{,\beta}}{4}\left(\begin{array}{l}\left(\frac{n}{2}-2\right)\left(4\left(\frac{n}{2}-1\right)-(n-6)\right)(F')^3\\+\left(8\left(\frac{n}{2}-1\right)-4-2(n-6)-4\left(\frac{n}{2}-2\right)\right)FF'F''\end{array}\right)\\+\frac{1}{2}(n+2)f_{,\alpha\beta}F^{\frac{n}{2}-2}F'F'\end{array}\right]_y, \qquad (724)$$

$$\frac{\delta W}{\delta f}=0=\frac{\delta}{\delta f}\int_V d^n x\left(\sqrt{g\cdot F^n}\cdot R^{**}\right)=-(n-1)\cdot\sqrt{g}\cdot g^{\alpha\beta}\Big|_y$$
$$\times\left[+\frac{F^{\frac{n}{2}-3}f_{,\alpha}f_{,\beta}}{4}\begin{pmatrix}\left(\frac{n}{2}-2\right)(n+2)(F')^3\\+(4n-8-4-2n+6-2n+8)FF'F''\end{pmatrix}\right.$$
$$\left.+\frac{1}{2}(n+2)f_{,\alpha\beta}F^{\frac{n}{2}-2}F'F'\right]_y \qquad . \qquad (725)$$
$$=-(n-1)\cdot\left[\frac{\sqrt{g}\cdot g^{\alpha\beta}}{2}F'F^{\frac{n}{2}-3}\begin{pmatrix}\frac{f_{,\alpha}f_{,\beta}}{2}\left(\frac{1}{2}(n-4)(n+2)(F')^2+2FF''\right)\\+(n+2)f_{,\alpha\beta}FF'\end{pmatrix}\right]_y$$

Now we incorporate all terms we have previously omitted due to our assumption of a metric of constants:

$$\frac{\delta W}{\delta f}=0=\frac{\delta}{\delta f}\int_V d^n x\left(\sqrt{g\cdot F^n}\cdot R^{**}\right)$$
$$=-\left[\begin{array}{l}(n-1)\cdot\left(F^{\frac{n}{2}-1}F'\left(\sqrt{g}\cdot g^{\alpha\beta}\right)_{,\alpha\beta}+2\left(F^{\frac{n}{2}-1}F'\right)_{,\alpha}\left(\sqrt{g}\cdot g^{\alpha\beta}\right)_{,\beta}\right)\\-\frac{1}{2}\left(\sqrt{g}\cdot F^{\frac{n}{2}-1}g^{\alpha\beta}\cdot F'g^{cd}\begin{pmatrix}g_{\alpha c,\beta}-g_{\beta\alpha,c}-g_{c\beta,\alpha}\\+\frac{n}{2}\left(2g_{\alpha\beta,c}-g_{\alpha c,\beta}-g_{\beta c,\alpha}\right)\end{pmatrix}\right)_{,d}\\+\frac{\sqrt{g}\cdot g^{\alpha\beta}}{2}f_{,d}g^{cd}\begin{pmatrix}g_{\alpha c,\beta}-g_{\beta\alpha,c}-g_{c\beta,\alpha}\\+\frac{n}{2}\left(2g_{\alpha\beta,c}-g_{\alpha c,\beta}-g_{\beta c,\alpha}\right)\end{pmatrix}\frac{d\left[F^{\frac{n}{2}-1}F'\right]}{df}\\-\frac{(n-1)}{2}f_{,\alpha}\cdot F^{\frac{n}{2}-2}\left(4FF''+F'\cdot F'(n-6)\right)\left(\sqrt{g}\cdot g^{\alpha\beta}\right)_{,\beta}\end{array}\right]_y$$
$$-(n-1)\cdot\left[\frac{\sqrt{g}\cdot g^{\alpha\beta}}{2}F'F^{\frac{n}{2}-3}\begin{pmatrix}\frac{f_{,\alpha}f_{,\beta}}{2}\left(\frac{1}{2}(n-4)(n+2)(F')^2+2FF''\right)\\+(n+2)f_{,\alpha\beta}FF'\end{pmatrix}\right]_y, \qquad (726)$$

$$\frac{\delta W}{\delta f}=0=\frac{\delta}{\delta f}\int_V d^n x\left(\sqrt{g}\cdot F^n\cdot R^{**}\right)$$

$$=-\left[\begin{array}{c}(n-1)\cdot F^{\frac{n}{2}-2}\left(FF'\left(\sqrt{g}\cdot g^{\alpha\beta}\right)_{,\alpha\beta}+2\left(\left(\frac{n}{2}-1\right)F'F'+FF''\right)f_{,\alpha}\left(\sqrt{g}\cdot g^{\alpha\beta}\right)_{,\beta}\right)\\ -F^{\frac{n}{2}-2}\frac{\left(\left(\frac{n}{2}-1\right)F'F'+FF''\right)f_{,d}}{2}g^{cd}\left(\sqrt{g}\cdot g^{\alpha\beta}\left(\begin{array}{c}g_{\alpha c,\beta}-g_{\beta\alpha,c}-g_{c\beta,\alpha}\\ +\frac{n}{2}\left(2g_{\alpha\beta,c}-g_{\alpha c,\beta}-g_{\beta c,\alpha}\right)\end{array}\right)\right)\\ -\frac{F^{\frac{n}{2}-1}F'}{2}\left(\sqrt{g}\cdot g^{\alpha\beta}\cdot g^{cd}\left(\begin{array}{c}g_{\alpha c,\beta}-g_{\beta\alpha,c}-g_{c\beta,\alpha}\\ +\frac{n}{2}\left(2g_{\alpha\beta,c}-g_{\alpha c,\beta}-g_{\beta c,\alpha}\right)\end{array}\right)\right)_{,d}\\ +\frac{\sqrt{g}\cdot g^{\alpha\beta}}{2}f_{,d}g^{cd}F^{\frac{n}{2}-2}\left(\begin{array}{c}g_{\alpha c,\beta}-g_{\beta\alpha,c}-g_{c\beta,\alpha}\\ +\frac{n}{2}\left(2g_{\alpha\beta,c}-g_{\alpha c,\beta}-g_{\beta c,\alpha}\right)\end{array}\right)\left(\left(\frac{n}{2}-1\right)F'F'+FF''\right)\\ -\frac{(n-1)}{2}f_{,\alpha}\cdot F^{\frac{n}{2}-2}\left(4FF''+F'\cdot F'(n-6)\right)\left(\sqrt{g}\cdot g^{\alpha\beta}\right)_{,\beta}\end{array}\right]_{y} \quad (727)$$

$$-(n-1)\cdot\left[\frac{\sqrt{g}\cdot g^{\alpha\beta}}{2}F'F^{\frac{n}{2}-3}\left(\begin{array}{c}\frac{f_{,\alpha}f_{,\beta}}{2}\left(\frac{1}{2}(n-4)(n+2)(F')^2+2FF''\right)\\ +(n+2)f_{,\alpha\beta}FF'\end{array}\right)\right]_{y}$$

Factoring out first derivatives of f yields:

$$\frac{\delta W}{\delta f}=0=\frac{\delta}{\delta f}\int_V d^n x\left(\sqrt{g}\cdot F^n\cdot R^{**}\right)$$

$$=-\left[\begin{array}{c}(n-1)\cdot F^{\frac{n}{2}-2}\left(FF'\left(\sqrt{g}\cdot g^{\alpha\beta}\right)_{,\alpha\beta}+2\left(\left(\frac{n}{2}-1\right)F'F'+FF''\right)f_{,\alpha}\left(\sqrt{g}\cdot g^{\alpha\beta}\right)_{,\beta}\right)\\ -F^{\frac{n}{2}-2}\frac{\left(\left(\frac{n}{2}-1\right)F'F'+FF''\right)f_{,\alpha}}{2}g^{ij}\sqrt{g}\cdot g^{\alpha\beta}\left(\begin{array}{c}g_{i\beta,j}-g_{ij,\beta}-g_{\beta j,i}\\ +\frac{n}{2}\left(2g_{ij,\beta}-g_{i\beta,j}-g_{j\beta,i}\right)\end{array}\right)\\ -\frac{F^{\frac{n}{2}-1}F'}{2}\left(\sqrt{g}\cdot g^{\alpha\beta}\cdot g^{ij}\left(\begin{array}{c}g_{i\beta,j}-g_{ij,\beta}-g_{\beta j,i}\\ +\frac{n}{2}\left(2g_{ij,\beta}-g_{i\beta,j}-g_{j\beta,i}\right)\end{array}\right)\right)_{,\alpha}\\ +\frac{\sqrt{g}\cdot g^{\alpha\beta}}{2}f_{,\alpha}g^{ij}F^{\frac{n}{2}-2}\left(\begin{array}{c}g_{i\beta,j}-g_{ij,\beta}-g_{\beta j,i}\\ +\frac{n}{2}\left(2g_{ij,\beta}-g_{i\beta,j}-g_{j\beta,i}\right)\end{array}\right)\left(\left(\frac{n}{2}-1\right)F'F'+FF''\right)\\ -\frac{(n-1)}{2}f_{,\alpha}\cdot F^{\frac{n}{2}-2}\left(4FF''+F'\cdot F'(n-6)\right)\left(\sqrt{g}\cdot g^{\alpha\beta}\right)_{,\beta}\end{array}\right]_{y}$$

$$-(n-1)\cdot\left[\frac{\sqrt{g}\cdot g^{\alpha\beta}}{2}F'F^{\frac{n}{2}-3}\left(\begin{array}{c}\frac{f_{,\alpha}f_{,\beta}}{2}\left(\frac{1}{2}(n-4)(n+2)(F')^2+2FF''\right)\\ +(n+2)f_{,\alpha\beta}FF'\end{array}\right)\right]_{y}, \quad (728)$$

$$\frac{\delta W}{\delta f}=0=\frac{\delta}{\delta f}\int_V d^n x\left(\sqrt{g\cdot F^n}\cdot R^{**}\right)$$

$$=-\left[\begin{array}{c}(n-1)\cdot F^{\frac{n}{2}-1}F'\left(\sqrt{g}\cdot g^{\alpha\beta}\right)_{,\alpha\beta}\\ -\frac{F^{\frac{n}{2}-1}F'}{2}\left(\sqrt{g}\cdot g^{\alpha\beta}\cdot g^{ij}\left(\begin{array}{c}g_{i\beta,j}-g_{ij,\beta}-g_{\beta j,i}\\ +\frac{n}{2}\left(2g_{ij,\beta}-g_{i\beta,j}-g_{j\beta,i}\right)\end{array}\right)\right)_{,\alpha}\\ f_{,\alpha}\left(\begin{array}{c}(n-1)\cdot F^{\frac{n}{2}-2}2\left(\left(\frac{n}{2}-1\right)F'F'+FF''\right)\left(\sqrt{g}\cdot g^{\alpha\beta}\right)_{,\beta}\\ -F^{\frac{n}{2}-2}\frac{\left(\left(\frac{n}{2}-1\right)F'F'+FF''\right)}{2}g^{ij}\sqrt{g}\cdot g^{\alpha\beta}\left(\begin{array}{c}g_{i\beta,j}-g_{ij,\beta}-g_{\beta j,i}\\ +\frac{n}{2}\left(2g_{ij,\beta}-g_{i\beta,j}-g_{j\beta,i}\right)\end{array}\right)\\ +F^{\frac{n}{2}-2}\frac{\left(\left(\frac{n}{2}-1\right)F'F'+FF''\right)}{2}g^{ij}\sqrt{g}\cdot g^{\alpha\beta}\left(\begin{array}{c}g_{i\beta,j}-g_{ij,\beta}-g_{\beta j,i}\\ +\frac{n}{2}\left(2g_{ij,\beta}-g_{i\beta,j}-g_{j\beta,i}\right)\end{array}\right)\\ -\frac{(n-1)}{2}\cdot F^{\frac{n}{2}-2}\left(4FF''+F'\cdot F'(n-6)\right)\left(\sqrt{g}\cdot g^{\alpha\beta}\right)_{,\beta}\end{array}\right)\end{array}\right]_{y}, \quad (729)$$

$$-(n-1)\cdot\left[\frac{\sqrt{g}\cdot g^{\alpha\beta}}{2}F'F^{\frac{n}{2}-3}\left(\begin{array}{c}\frac{f_{,\alpha}f_{,\beta}}{2}\left(\frac{1}{2}(n-4)(n+2)(F')^2+2FF''\right)\\ +(n+2)f_{,\alpha\beta}FF'\end{array}\right)\right]_{y}$$

$$\frac{\delta W}{\delta f}=0=\frac{\delta}{\delta f}\int_V d^n x\left(\sqrt{g\cdot F^n}\cdot R^{**}\right)$$

$$=-\left[\begin{array}{c}(n-1)\cdot F^{\frac{n}{2}-1}F'\left(\sqrt{g}\cdot g^{\alpha\beta}\right)_{,\alpha\beta}\\ -\frac{F^{\frac{n}{2}-1}F'}{2}\left(\sqrt{g}\cdot g^{\alpha\beta}\cdot g^{ij}\left(\begin{array}{c}g_{i\beta,j}-g_{ij,\beta}-g_{\beta j,i}\\ +\frac{n}{2}\left(2g_{ij,\beta}-g_{i\beta,j}-g_{j\beta,i}\right)\end{array}\right)\right)_{,\alpha}\\ +f_{,\alpha}\frac{(n-1)}{2}\cdot F^{\frac{n}{2}-2}\left(\sqrt{g}\cdot g^{\alpha\beta}\right)_{,\beta}\left((2n-4)F'F'+4FF''-4FF''-F'\cdot F'(n-6)\right)\end{array}\right]_{y}$$

$$-(n-1)\cdot\left[\frac{\sqrt{g}\cdot g^{\alpha\beta}}{2}F'F^{\frac{n}{2}-3}\left(\begin{array}{c}\frac{f_{,\alpha}f_{,\beta}}{2}\left(\frac{1}{2}(n-4)(n+2)(F')^2+2FF''\right)\\ +(n+2)f_{,\alpha\beta}FF'\end{array}\right)\right]_{y}$$

$$=-\left[\begin{array}{c}(n-1)\cdot F^{\frac{n}{2}-1}F'\left(\sqrt{g}\cdot g^{\alpha\beta}\right)_{,\alpha\beta}\\ -\frac{F^{\frac{n}{2}-1}F'}{2}\left(\sqrt{g}\cdot g^{\alpha\beta}\cdot g^{ij}\left(\begin{array}{c}g_{i\beta,j}-g_{ij,\beta}-g_{\beta j,i}\\ +\frac{n}{2}\left(2g_{ij,\beta}-g_{i\beta,j}-g_{j\beta,i}\right)\end{array}\right)\right)_{,\alpha}\\ +f_{,\alpha}\frac{(n-1)}{2}\cdot F^{\frac{n}{2}-2}\left(\sqrt{g}\cdot g^{\alpha\beta}\right)_{,\beta}F'F'(n+2)\end{array}\right]_{y}$$

$$-(n-1)\cdot\left[\frac{\sqrt{g}\cdot g^{\alpha\beta}}{2}F'F^{\frac{n}{2}-3}\left(\begin{array}{c}\frac{f_{,\alpha}f_{,\beta}}{2}\left(\frac{1}{2}(n-4)(n+2)(F')^2+2FF''\right)\\ +(n+2)f_{,\alpha\beta}FF'\end{array}\right)\right]_{y} \quad . \quad (730)$$

Now we bring back in the R-term:

$$
\begin{aligned}
&\frac{\delta W}{\delta f}=0\\
&=-\left[F^{\frac{n}{2}-2}F'\left(\begin{array}{c}(n-1)\cdot\left(F\left(\sqrt{g}\cdot g^{\alpha\beta}\right)_{,\alpha\beta}+f_{,\alpha}\frac{F'}{2}\cdot\left(\sqrt{g}\cdot g^{\alpha\beta}\right)_{,\beta}(n+2)\right)\\ -\frac{F}{2}\left(\sqrt{g}\cdot g^{\alpha\beta}\cdot g^{ij}\left(\begin{array}{c}g_{i\beta,j}-g_{ij,\beta}-g_{\beta j,i}\\ +\frac{n}{2}\left(2g_{ij,\beta}-g_{i\beta,j}-g_{j\beta,i}\right)\end{array}\right)\right)_{,\alpha}\end{array}\right)-\sqrt{g}\cdot R\frac{dF^{\frac{n}{2}}}{df}\right]_{\mathbf{y}}\\
&\quad-(n-1)\cdot\left[\frac{\sqrt{g}\cdot g^{\alpha\beta}}{2}F'F^{\frac{n}{2}-3}\left(\begin{array}{c}\frac{f_{,\alpha}f_{,\beta}}{2}\left(\frac{1}{2}(n-4)(n+2)(F')^{2}+2FF''\right)\\ +(n+2)f_{,\alpha\beta}FF'\end{array}\right)\right]_{\mathbf{y}}\\
&=-\left[\begin{array}{c}F^{\frac{n}{2}-2}F'\left(\begin{array}{c}(n-1)\cdot\left(F\left(\sqrt{g}\cdot g^{\alpha\beta}\right)_{,\alpha\beta}+f_{,\alpha}\frac{F'}{2}\cdot\left(\sqrt{g}\cdot g^{\alpha\beta}\right)_{,\beta}(n+2)\right)\\ -\frac{F}{2}\left(\sqrt{g}\cdot g^{\alpha\beta}\cdot g^{ij}\left(\begin{array}{c}g_{i\beta,j}-g_{ij,\beta}-g_{\beta j,i}\\ +\frac{n}{2}\left(2g_{ij,\beta}-g_{i\beta,j}-g_{j\beta,i}\right)\end{array}\right)\right)_{,\alpha}\\ +(n-1)\cdot\frac{\sqrt{g}\cdot g^{\alpha\beta}}{2F}\left(\begin{array}{c}\frac{f_{,\alpha}f_{,\beta}}{2}\left(\frac{1}{2}(n-4)(n+2)(F')^{2}+2FF''\right)\\ +(n+2)f_{,\alpha\beta}FF'\end{array}\right)\end{array}\right)\\ -\sqrt{g}\cdot\frac{n}{2}\cdot RF^{\frac{n}{2}-1}F'\end{array}\right]_{\mathbf{y}}
\end{aligned}
\quad , \quad (731)
$$

$$\frac{\delta W}{\delta f}=0$$

$$=-\left[F^{\frac{n}{2}-3}F'(n-1)\left(\begin{array}{c}\frac{\sqrt{g}\cdot g^{\alpha\beta}}{2}\frac{f_{,\alpha}f_{,\beta}}{2}\left(\frac{1}{2}(n-4)(n+2)(F')^2+2FF''\right)\\ +\frac{(n+2)}{2}FF'\overbrace{\left(\sqrt{g}\cdot g^{\alpha\beta}f_{,\alpha\beta}+f_{,\alpha}\left(\sqrt{g}\cdot g^{\alpha\beta}\right)_{,\beta}\right)}^{=\sqrt{g}\cdot\Delta f}\\ +F^2\left(\sqrt{g}\cdot g^{\alpha\beta}\right)_{,\alpha\beta}\\ -\frac{F^2}{2(n-1)}\left(\sqrt{g}\cdot g^{\alpha\beta}\cdot g^{ij}\left(\begin{array}{c}g_{i\beta,j}-g_{ij,\beta}-g_{\beta j,i}\\ +\frac{n}{2}\left(2g_{ij,\beta}-g_{i\beta,j}-g_{j\beta,i}\right)\end{array}\right)\right)_{,\alpha}\end{array}\right)\\ -\sqrt{g}\cdot\frac{n}{2}\cdot RF^{\frac{n}{2}-1}F'\right]_y$$

$$=-\left[F^{\frac{n}{2}-3}F'(n-1)\left(\begin{array}{c}\frac{\sqrt{g}\cdot g^{\alpha\beta}}{2}\frac{f_{,\alpha}f_{,\beta}}{2}\left(\frac{1}{2}(n-4)(n+2)(F')^2+2FF''\right)\\ +\frac{(n+2)}{2}FF'\sqrt{g}\cdot\Delta f\\ +F^2\left(\sqrt{g}\cdot g^{\alpha\beta}\right)_{,\alpha\beta}-\sqrt{g}\cdot\frac{n}{2}\cdot\frac{R\cdot F^2}{(n-1)}\\ -\frac{F^2}{2(n-1)}\left(\sqrt{g}\cdot g^{\alpha\beta}\cdot g^{ij}\left(\begin{array}{c}g_{i\beta,j}-g_{ij,\beta}-g_{\beta j,i}\\ +\frac{n}{2}\left(2g_{ij,\beta}-g_{i\beta,j}-g_{j\beta,i}\right)\end{array}\right)\right)_{,\alpha}\end{array}\right)\right]_y. \quad (732)$$

It totally suffices to demand that the inner factor gives zero in order to fulfill the variational condition and thus, we obtain:

$$\frac{\delta W}{\delta f}=0 \Rightarrow \Rightarrow$$

$$0=\left[\begin{array}{c}\frac{\sqrt{g}}{2}\left(\frac{f_{,\alpha}f_{,\beta}}{2}g^{\alpha\beta}\left(\frac{1}{2}(n-4)(n+2)(F')^2+2FF''\right)+(n+2)FF'\Delta f\right)\\ +F^2\left(\begin{array}{c}\left(\sqrt{g}\cdot g^{\alpha\beta}\right)_{,\alpha\beta}-\frac{1}{2(n-1)}\left(\sqrt{g}\cdot g^{\alpha\beta}\cdot g^{ij}\left(\begin{array}{c}g_{i\beta,j}-g_{ij,\beta}-g_{\beta j,i}\\ +\frac{n}{2}\left(2g_{ij,\beta}-g_{i\beta,j}-g_{j\beta,i}\right)\end{array}\right)\right)_{,\alpha}\\ -\frac{n}{2}\cdot\frac{\sqrt{g}\cdot R}{(n-1)}\end{array}\right)\end{array}\right]_y, \quad (733)$$

$$\xrightarrow{F=C+a\cdot f}$$

$$0=\left[\begin{array}{c}\frac{\sqrt{g}}{2}(n+2)\left(\frac{f_{,\alpha}f_{,\beta}}{4}g^{\alpha\beta}(n-4)a^{2}+(C+a\cdot f)\cdot a\cdot\Delta f\right)\\ +(C+a\cdot f)^{2}\left(\begin{array}{c}\left(\sqrt{g}\cdot g^{\alpha\beta}\right)_{,\alpha\beta}-\frac{n}{2}\cdot\frac{\sqrt{g}\cdot R}{(n-1)}\\ -\frac{1}{2(n-1)}\left(\sqrt{g}\cdot g^{\alpha\beta}\cdot g^{ij}\left(\begin{array}{c}g_{i\beta,j}-g_{ij,\beta}-g_{\beta j,i}\\ +\frac{n}{2}\left(2g_{ij,\beta}-g_{i\beta,j}-g_{j\beta,i}\right)\end{array}\right)\right)_{,\alpha}\end{array}\right)\end{array}\right]_{y}. \qquad (734)$$

22 Appendix D (by N. Schwarzer): Derivation of a Few Important Equations

22.1 Derivation of the Ricci Tensor and Scalar for a Scaled Metric Tensor

22.1.1 Derivation of the Ricci Tensor for a Scaled Metric Tensor

In the following we perform the derivation of the Ricci tensor for a scaled metric tensor of type:

$$G_{\alpha\beta} = g_{\alpha\beta} \cdot F[f]. \tag{735}$$

Due to its importance in this book, we are going to present the derivation in some detail. Our starting point shall be the well-known derivation of the Ricci tensor as follows:

$$R_{\alpha\beta} = \begin{pmatrix} -\frac{1}{2}\left(g_{\alpha\beta,ab} + g_{ab,\alpha\beta} - g_{\alpha b,a\beta} - g_{\beta b,a\alpha}\right)g^{ab} \\ +\frac{1}{2}\left(\frac{1}{2}g_{ac,\alpha} \cdot g_{bd,\beta} + g_{\alpha c,a} \cdot g_{\beta d,b} - g_{\alpha c,a} \cdot g_{\beta b,d}\right)g^{ab}g^{cd} \\ -\frac{1}{4}\left(g_{\alpha c,\beta} + g_{\beta c,\alpha} - g_{\alpha\beta,c}\right)\left(2g_{bd,a} - g_{ab,d}\right)g^{ab}g^{cd} \end{pmatrix}. \tag{736}$$

From there we extract the corresponding result for the scaled metric tensor:

$$R^{*}{}_{\alpha\beta} = \begin{pmatrix} -\frac{1}{2}\left(G_{\alpha\beta,ab} + G_{ab,\alpha\beta} - G_{\alpha b,a\beta} - G_{\beta b,a\alpha}\right)G^{ab} \\ +\frac{1}{2}\left(\frac{1}{2}G_{ac,\alpha}\cdot G_{bd,\beta} + G_{\alpha c,a}\cdot G_{\beta d,b} - G_{\alpha c,a}\cdot G_{\beta b,d}\right)G^{ab}G^{cd} \\ -\frac{1}{4}\left(G_{\beta c,\alpha} + G_{\alpha c,\beta} - G_{\alpha\beta,c}\right)\left(2G_{bd,a} - G_{ab,d}\right)G^{ab}G^{cd} \end{pmatrix}$$

$$= \begin{pmatrix} -\frac{1}{2F}\begin{pmatrix} F\cdot g_{\alpha\beta,ab} + F_{,b}g_{\alpha\beta,a} + F_{,a}g_{\alpha\beta,b} + F_{,ab}g_{\alpha\beta} \\ +F\cdot g_{ab,\alpha\beta} + F_{,\alpha\beta}g_{ab} + F_{,\alpha}g_{ab,\beta} + F_{,\beta}g_{ab,\alpha} \\ -F\cdot g_{\alpha b,a\beta} - F_{,a\beta}g_{\alpha b} - F_{,a}g_{\alpha b,\beta} - F_{,\beta}g_{\alpha b,a} \\ -F\cdot g_{\beta b,a\alpha} - F_{,a\alpha}g_{\beta b} - F_{,a}g_{\beta b,\alpha} - F_{,\alpha}g_{\beta b,a} \end{pmatrix} g^{ab} \\ +\frac{1}{2F^{2}}\begin{pmatrix} \frac{1}{2}\left(F\cdot g_{ac,\alpha} + F_{,\alpha}g_{ac}\right)\cdot\left(F\cdot g_{bd,\beta} + F_{,\beta}g_{bd}\right) \\ +\left(F\cdot g_{\alpha c,a} + F_{,a}g_{\alpha c}\right)\cdot\left(F\cdot g_{\beta d,b} + F_{,b}g_{\beta d}\right) \\ -\left(F\cdot g_{\alpha c,a} + F_{,a}g_{\alpha c}\right)\cdot\left(F\cdot g_{\beta b,d} + F_{,d}g_{\beta b}\right) \end{pmatrix} g^{ab}g^{cd} \\ -\frac{1}{4F^{2}}\begin{pmatrix} F\cdot g_{\alpha c,\beta} + F_{,\beta}g_{\alpha c} + F\cdot g_{\beta c,\alpha} \\ +F_{,\alpha}g_{\beta c} - F\cdot g_{\alpha\beta,c} - F_{,c}g_{\alpha\beta} \end{pmatrix}\left(2\left(F\cdot g_{bd,a} + F_{,a}g_{bd}\right) - F\cdot g_{ab,d} - F_{,d}g_{ab}\right)g^{ab}g^{cd} \end{pmatrix}. \quad (737)$$

Various transformations via:

$$
\begin{aligned}
R^{*}_{\alpha\beta} &= -\frac{1}{2F}\begin{pmatrix} F\cdot g_{\alpha\beta,ab}+F_{,b}g_{\alpha\beta,a}+F_{,a}g_{\alpha\beta,b}+F_{,ab}g_{\alpha\beta} \\ +F\cdot g_{ab,\alpha\beta}+F_{,\alpha\beta}g_{ab}+F_{,\alpha}g_{ab,\beta}+F_{,\beta}g_{ab,\alpha} \\ -F\cdot g_{\alpha b,a\beta}-F_{,a\beta}g_{\alpha b}-F_{,a}g_{\alpha b,\beta}-F_{,\beta}g_{\alpha b,a} \\ -F\cdot g_{\beta b,a\alpha}-F_{,a\alpha}g_{\beta b}-F_{,a}g_{\beta b,\alpha}-F_{,\alpha}g_{\beta b,a} \end{pmatrix} g^{ab} \\
&+\frac{1}{2F^2}\begin{pmatrix} \frac{1}{2}\left(F\cdot g_{ac,\alpha}+F_{,\alpha}g_{ac}\right)\cdot\left(F\cdot g_{bd,\beta}+F_{,\beta}g_{bd}\right) \\ +\left(F\cdot g_{\alpha c,a}+F_{,a}g_{\alpha c}\right)\cdot\left(F\cdot g_{\beta d,b}+F_{,b}g_{\beta d}\right) \\ -\left(F\cdot g_{\alpha c,a}+F_{,a}g_{\alpha c}\right)\cdot\left(F\cdot g_{\beta b,d}+F_{,d}g_{\beta b}\right) \end{pmatrix} g^{ab}g^{cd} \\
&-\frac{1}{4F^2}\begin{pmatrix} F\cdot g_{\alpha c,\beta}+F_{,\beta}g_{\alpha c}+F\cdot g_{\beta c,\alpha} \\ +F_{,\alpha}g_{\beta c}-F\cdot g_{\alpha\beta,c}-F_{,c}g_{\alpha\beta} \end{pmatrix}\left(2\left(F\cdot g_{bd,a}+F_{,a}g_{bd}\right)-F\cdot g_{ab,d}-F_{,d}g_{ab}\right)g^{ab}g^{cd} \\
&= -\frac{1}{2F}\begin{pmatrix} F\cdot g_{\alpha\beta,ab} \\ +F\cdot g_{ab,\alpha\beta} \\ -F\cdot g_{\alpha b,a\beta} \\ -F\cdot g_{\beta b,a\alpha} \end{pmatrix} g^{ab} - \frac{1}{2F}\begin{pmatrix} F_{,b}g_{\alpha\beta,a}+F_{,a}g_{\alpha\beta,b}+F_{,ab}g_{\alpha\beta} \\ +F_{,\alpha\beta}g_{ab}+F_{,\alpha}g_{ab,\beta}+F_{,\beta}g_{ab,\alpha} \\ -F_{,a\beta}g_{\alpha b}-F_{,a}g_{\alpha b,\beta}-F_{,\beta}g_{\alpha b,a} \\ -F_{,a\alpha}g_{\beta b}-F_{,a}g_{\beta b,\alpha}-F_{,\alpha}g_{\beta b,a} \end{pmatrix} g^{ab} \\
&+\frac{1}{2F^2}\begin{pmatrix} \frac{1}{2}\left(F\cdot g_{ac,\alpha}\cdot\left(F\cdot g_{bd,\beta}+F_{,\beta}g_{bd}\right)+F_{,\alpha}g_{ac}\cdot\left(F\cdot g_{bd,\beta}+F_{,\beta}g_{bd}\right)\right) \\ +\left(F\cdot g_{\alpha c,a}\cdot\left(F\cdot g_{\beta d,b}+F_{,b}g_{\beta d}\right)+F_{,a}g_{\alpha c}\cdot\left(F\cdot g_{\beta d,b}+F_{,b}g_{\beta d}\right)\right) \\ -\left(F\cdot g_{\alpha c,a}\cdot\left(F\cdot g_{\beta b,d}+F_{,d}g_{\beta b}\right)+F_{,a}g_{\alpha c}\cdot\left(F\cdot g_{\beta b,d}+F_{,d}g_{\beta b}\right)\right) \end{pmatrix} g^{ab}g^{cd} \\
&-\frac{1}{4F^2}\begin{pmatrix} F\cdot g_{\alpha c,\beta}\left(2\left(F\cdot g_{bd,a}+F_{,a}g_{bd}\right)-F\cdot g_{ab,d}-F_{,d}g_{ab}\right) \\ +F_{,\beta}g_{\alpha c}\left(2\left(F\cdot g_{bd,a}+F_{,a}g_{bd}\right)-F\cdot g_{ab,d}-F_{,d}g_{ab}\right) \\ +F\cdot g_{\beta c,\alpha}\left(2\left(F\cdot g_{bd,a}+F_{,a}g_{bd}\right)-F\cdot g_{ab,d}-F_{,d}g_{ab}\right) \\ +F_{,\alpha}g_{\beta c}\left(2\left(F\cdot g_{bd,a}+F_{,a}g_{bd}\right)-F\cdot g_{ab,d}-F_{,d}g_{ab}\right) \\ -F\cdot g_{\alpha\beta,c}\left(2\left(F\cdot g_{bd,a}+F_{,a}g_{bd}\right)-F\cdot g_{ab,d}-F_{,d}g_{ab}\right) \\ -F_{,c}g_{\alpha\beta}\left(2\left(F\cdot g_{bd,a}+F_{,a}g_{bd}\right)-F\cdot g_{ab,d}-F_{,d}g_{ab}\right) \end{pmatrix} g^{ab}g^{cd}
\end{aligned}
\quad , \qquad (738)
$$

$$
\begin{aligned}
R^{*}_{\alpha\beta} = & -\frac{1}{2F}\begin{pmatrix} F\cdot g_{\alpha\beta,ab} \\ +F\cdot g_{ab,\alpha\beta} \\ -F\cdot g_{\alpha b,a\beta} \\ -F\cdot g_{\beta b,a\alpha} \end{pmatrix} g^{ab} - \frac{1}{2F}\begin{pmatrix} F_{,b}g_{\alpha\beta,a} + F_{,a}g_{\alpha\beta,b} + F_{,ab}g_{\alpha\beta} \\ +F_{,\alpha\beta}g_{ab} + F_{,\alpha}g_{ab,\beta} + F_{,\beta}g_{ab,\alpha} \\ -F_{,a\beta}g_{\alpha b} - F_{,a}g_{\alpha b,\beta} - F_{,\beta}g_{\alpha b,a} \\ -F_{,a\alpha}g_{\beta b} - F_{,a}g_{\beta b,\alpha} - F_{,\alpha}g_{\beta b,a} \end{pmatrix} g^{ab} \\
& + \frac{1}{2F^{2}}\begin{pmatrix} \frac{1}{2}\begin{pmatrix} \left(F\cdot g_{ac,\alpha}\cdot F\cdot g_{bd,\beta} + F\cdot g_{ac,\alpha}\cdot F_{,\beta}g_{bd}\right) \\ +\left(F_{,\alpha}g_{ac}\cdot F\cdot g_{bd,\beta} + F_{,\alpha}g_{ac}\cdot F_{,\beta}g_{bd}\right) \end{pmatrix} \\ +\begin{pmatrix} \left(F\cdot g_{\alpha c,a}\cdot F\cdot g_{\beta d,b} + F\cdot g_{\alpha c,a}\cdot F_{,b}g_{\beta d}\right) \\ +\left(F_{,a}g_{\alpha c}\cdot F\cdot g_{\beta d,b} + F_{,a}g_{\alpha c}\cdot F_{,b}g_{\beta d}\right) \end{pmatrix} \\ -\begin{pmatrix} \left(F\cdot g_{\alpha c,a}\cdot F\cdot g_{\beta b,d} + F\cdot g_{\alpha c,a}\cdot F_{,d}g_{\beta b}\right) \\ +\left(F_{,a}g_{\alpha c}\cdot F\cdot g_{\beta b,d} + F_{,a}g_{\alpha c}\cdot F_{,d}g_{\beta b}\right) \end{pmatrix} \end{pmatrix} g^{ab}g^{cd} \\
& - \frac{1}{4F^{2}}\begin{pmatrix} \begin{pmatrix} 2\left(F\cdot g_{\alpha c,\beta}F\cdot g_{bd,a} + F\cdot g_{\alpha c,\beta}F_{,a}g_{bd}\right) \\ -F\cdot g_{\alpha c,\beta}F\cdot g_{ab,d} - F\cdot g_{\alpha c,\beta}F_{,d}g_{ab} \end{pmatrix} \\ +\begin{pmatrix} 2\left(F_{,\beta}g_{\alpha c}F\cdot g_{bd,a} + F_{,\beta}g_{\alpha c}F_{,a}g_{bd}\right) \\ -F_{,\beta}g_{\alpha c}F\cdot g_{ab,d} - F_{,\beta}g_{\alpha c}F_{,d}g_{ab} \end{pmatrix} \\ +\begin{pmatrix} 2\left(F\cdot g_{\beta c,\alpha}F\cdot g_{bd,a} + F\cdot g_{\beta c,\alpha}F_{,a}g_{bd}\right) \\ -F\cdot g_{\beta c,\alpha}F\cdot g_{ab,d} - F\cdot g_{\beta c,\alpha}F_{,d}g_{ab} \end{pmatrix} \\ +\begin{pmatrix} 2\left(F_{,\alpha}g_{\beta c}F\cdot g_{bd,a} + F_{,\alpha}g_{\beta c}F_{,a}g_{bd}\right) \\ -F_{,\alpha}g_{\beta c}F\cdot g_{ab,d} - F_{,\alpha}g_{\beta c}F_{,d}g_{ab} \end{pmatrix} \\ -\begin{pmatrix} 2\left(F\cdot g_{\alpha\beta,c}F\cdot g_{bd,a} + F\cdot g_{\alpha\beta,c}F_{,a}g_{bd}\right) \\ -F\cdot g_{\alpha\beta,c}F\cdot g_{ab,d} - F\cdot g_{\alpha\beta,c}F_{,d}g_{ab} \end{pmatrix} \\ -\begin{pmatrix} 2\left(F_{,c}g_{\alpha\beta}F\cdot g_{bd,a} + F_{,c}g_{\alpha\beta}F_{,a}g_{bd}\right) \\ -F_{,c}g_{\alpha\beta}F\cdot g_{ab,d} - F_{,c}g_{\alpha\beta}F_{,d}g_{ab} \end{pmatrix} \end{pmatrix} g^{ab}g^{cd} \quad ,
\end{aligned}
\tag{739}
$$

$$
\begin{aligned}
R^{*}_{\alpha\beta} = & -\frac{1}{2F}\begin{pmatrix} F\cdot g_{\alpha\beta,ab} \\ +F\cdot g_{ab,\alpha\beta} \\ -F\cdot g_{\alpha b,a\beta} \\ -F\cdot g_{\beta b,a\alpha} \end{pmatrix} g^{ab} - \frac{1}{2F}\begin{pmatrix} F_{,b}g_{\alpha\beta,a} + F_{,a}g_{\alpha\beta,b} + F_{,ab}g_{\alpha\beta} \\ +F_{,\alpha\beta}g_{ab} + F_{,\alpha}g_{ab,\beta} + F_{,\beta}g_{ab,\alpha} \\ -F_{,a\beta}g_{\alpha b} - F_{,a}g_{\alpha b,\beta} - F_{,\beta}g_{\alpha b,a} \\ -F_{,a\alpha}g_{\beta b} - F_{,a}g_{\beta b,\alpha} - F_{,\alpha}g_{\beta b,a} \end{pmatrix} g^{ab} \\
& + \frac{1}{2F^2}\left(\frac{1}{2}\left(\left(F\cdot g_{ac,\alpha}\cdot F\cdot g_{bd,\beta}\right)\right) + \left(\left(F\cdot g_{\alpha c,a}\cdot F\cdot g_{\beta d,b}\right)\right) - \left(\left(F\cdot g_{\alpha c,a}\cdot F\cdot g_{\beta b,d}\right)\right)\right) g^{ab}g^{cd} \\
& + \frac{1}{2F^2}\begin{pmatrix} \frac{1}{2}\left(\left(F\cdot g_{ac,\alpha}\cdot F_{,\beta}g_{bd}\right) + \left(F_{,\alpha}g_{ac}\cdot F\cdot g_{bd,\beta} + F_{,\alpha}g_{ac}\cdot F_{,\beta}g_{bd}\right)\right) \\ +\left(\left(F\cdot g_{\alpha c,a}\cdot F_{,b}g_{\beta d}\right) + \left(F_{,a}g_{\alpha c}\cdot F\cdot g_{\beta d,b} + F_{,a}g_{\alpha c}\cdot F_{,b}g_{\beta d}\right)\right) \\ -\left(\left(F\cdot g_{\alpha c,a}\cdot F_{,d}g_{\beta b}\right) + \left(F_{,a}g_{\alpha c}\cdot F\cdot g_{\beta b,d} + F_{,a}g_{\alpha c}\cdot F_{,d}g_{\beta b}\right)\right) \end{pmatrix} g^{ab}g^{cd} \\
& - \frac{1}{4F^2}\begin{pmatrix} \left(2\left(F\cdot g_{\alpha c,\beta}F\cdot g_{bd,a}\right) - F\cdot g_{\alpha c,\beta}F\cdot g_{ab,d}\right) \\ +\left(2\left(F\cdot g_{\beta c,\alpha}F\cdot g_{bd,a}\right) - F\cdot g_{\beta c,\alpha}F\cdot g_{ab,d}\right) \\ -\left(2\left(F\cdot g_{\alpha\beta,c}F\cdot g_{bd,a}\right) - F\cdot g_{\alpha\beta,c}F\cdot g_{ab,d}\right) \end{pmatrix} g^{ab}g^{cd} \\
& - \frac{1}{4F^2}\begin{pmatrix} \left(2\left(F\cdot g_{\alpha c,\beta}F_{,a}g_{bd}\right) - F\cdot g_{\alpha c,\beta}F_{,d}g_{ab}\right) \\ +\left(2\left(F_{,\beta}g_{\alpha c}F\cdot g_{bd,a} + F_{,\beta}g_{\alpha c}F_{,a}g_{bd}\right) - F_{,\beta}g_{\alpha c}F\cdot g_{ab,d} - F_{,\beta}g_{\alpha c}F_{,d}g_{ab}\right) \\ +\left(2\left(F\cdot g_{\beta c,\alpha}F_{,a}g_{bd}\right) - F\cdot g_{\beta c,\alpha}F_{,d}g_{ab}\right) \\ +\left(2\left(F_{,\alpha}g_{\beta c}F\cdot g_{bd,a} + F_{,\alpha}g_{\beta c}F_{,a}g_{bd}\right) - F_{,\alpha}g_{\beta c}F\cdot g_{ab,d} - F_{,\alpha}g_{\beta c}F_{,d}g_{ab}\right) \\ -\left(2\left(F\cdot g_{\alpha\beta,c}F_{,a}g_{bd}\right) - F\cdot g_{\alpha\beta,c}F_{,d}g_{ab}\right) \\ -\left(2\left(F_{,c}g_{\alpha\beta}F\cdot g_{bd,a} + F_{,c}g_{\alpha\beta}F_{,a}g_{bd}\right) - F_{,c}g_{\alpha\beta}F\cdot g_{ab,d} - F_{,c}g_{\alpha\beta}F_{,d}g_{ab}\right) \end{pmatrix} g^{ab}g^{cd}
\end{aligned}
\quad , \quad (740)
$$

$$
\begin{aligned}
R^{*}{}_{\alpha\beta} = & -\frac{1}{2F}\begin{pmatrix} F\cdot g_{\alpha\beta,ab} \\ +F\cdot g_{ab,\alpha\beta} \\ -F\cdot g_{\alpha b,a\beta} \\ -F\cdot g_{\beta b,a\alpha} \end{pmatrix} g^{ab} \\
& +\frac{1}{2F^2}\left(\frac{1}{2}\big(\big(F\cdot g_{ac,\alpha}\cdot F\cdot g_{bd,\beta}\big)\big)+\big(\big(F\cdot g_{\alpha c,a}\cdot F\cdot g_{\beta d,b}\big)\big)-\big(\big(F\cdot g_{\alpha c,a}\cdot F\cdot g_{\beta b,d}\big)\big)\right) g^{ab}g^{cd} \\
& -\frac{1}{4F^2}\begin{pmatrix} \big(2\big(F\cdot g_{\alpha c,\beta}F\cdot g_{bd,a}\big)-F\cdot g_{\alpha c,\beta}F\cdot g_{ab,d}\big) \\ +\big(2\big(F\cdot g_{\beta c,\alpha}F\cdot g_{bd,a}\big)-F\cdot g_{\beta c,\alpha}F\cdot g_{ab,d}\big) \\ -\big(2\big(F\cdot g_{\alpha\beta,c}F\cdot g_{bd,a}\big)-F\cdot g_{\alpha\beta,c}F\cdot g_{ab,d}\big) \end{pmatrix} g^{ab}g^{cd} \\
& -\frac{1}{2F}\begin{pmatrix} F_{,b}g_{\alpha\beta,a}+F_{,a}g_{\alpha\beta,b}+F_{,ab}g_{\alpha\beta} \\ +F_{,\alpha\beta}g_{ab}+F_{,\alpha}g_{ab,\beta}+F_{,\beta}g_{ab,\alpha} \\ -F_{,a\beta}g_{\alpha b}-F_{,a}g_{\alpha b,\beta}-F_{,\beta}g_{\alpha b,a} \\ -F_{,a\alpha}g_{\beta b}-F_{,a}g_{\beta b,\alpha}-F_{,\alpha}g_{\beta b,a} \end{pmatrix} g^{ab} \\
& +\frac{1}{2F^2}\begin{pmatrix} \frac{1}{2}\big(\big(F\cdot g_{ac,\alpha}\cdot F_{,\beta}g_{bd}\big)+\big(F_{,\alpha}g_{ac}\cdot F\cdot g_{bd,\beta}+F_{,\alpha}g_{ac}\cdot F_{,\beta}g_{bd}\big)\big) \\ +\big(\big(F\cdot g_{\alpha c,a}\cdot F_{,b}g_{\beta d}\big)+\big(F_{,a}g_{\alpha c}\cdot F\cdot g_{\beta d,b}+F_{,a}g_{\alpha c}\cdot F_{,b}g_{\beta d}\big)\big) \\ -\big(\big(F\cdot g_{\alpha c,a}\cdot F_{,d}g_{\beta b}\big)+\big(F_{,a}g_{\alpha c}\cdot F\cdot g_{\beta b,d}+F_{,a}g_{\alpha c}\cdot F_{,d}g_{\beta b}\big)\big) \end{pmatrix} g^{ab}g^{cd} \\
& -\frac{1}{4F^2}\begin{pmatrix} \big(2\big(F\cdot g_{\alpha c,\beta}F_{,a}g_{bd}\big)-F\cdot g_{\alpha c,\beta}F_{,d}g_{ab}\big) \\ +\big(2\big(F_{,\beta}g_{\alpha c}F\cdot g_{bd,a}+F_{,\beta}g_{\alpha c}F_{,a}g_{bd}\big)-F_{,\beta}g_{\alpha c}F\cdot g_{ab,d}-F_{,\beta}g_{\alpha c}F_{,d}g_{ab}\big) \\ +\big(2\big(F\cdot g_{\beta c,\alpha}F_{,a}g_{bd}\big)-F\cdot g_{\beta c,\alpha}F_{,d}g_{ab}\big) \\ +\big(2\big(F_{,\alpha}g_{\beta c}F\cdot g_{bd,a}+F_{,\alpha}g_{\beta c}F_{,a}g_{bd}\big)-F_{,\alpha}g_{\beta c}F\cdot g_{ab,d}-F_{,\alpha}g_{\beta c}F_{,d}g_{ab}\big) \\ -\big(2\big(F\cdot g_{\alpha\beta,c}F_{,a}g_{bd}\big)-F\cdot g_{\alpha\beta,c}F_{,d}g_{ab}\big) \\ -\big(2\big(F_{,c}g_{\alpha\beta}F\cdot g_{bd,a}+F_{,c}g_{\alpha\beta}F_{,a}g_{bd}\big)-F_{,c}g_{\alpha\beta}F\cdot g_{ab,d}-F_{,c}g_{\alpha\beta}F_{,d}g_{ab}\big) \end{pmatrix} g^{ab}g^{cd}
\end{aligned}
\quad , \tag{741}
$$

the extraction of the F^2-terms:

$$
\begin{aligned}
R^{*}{}_{\alpha\beta} = R_{\alpha\beta} - \frac{1}{2F}
&\begin{pmatrix}
F_{,b}g_{\alpha\beta,a} + F_{,a}g_{\alpha\beta,b} + F_{,ab}g_{\alpha\beta} \\
+F_{,\alpha\beta}g_{ab} + F_{,\alpha}g_{ab,\beta} + F_{,\beta}g_{ab,\alpha} \\
-F_{,a\beta}g_{\alpha b} - F_{,a}g_{\alpha b,\beta} - F_{,\beta}g_{\alpha b,a} \\
-F_{,a\alpha}g_{\beta b} - F_{,a}g_{\beta b,\alpha} - F_{,\alpha}g_{\beta b,a}
\end{pmatrix} g^{ab} \\
+\frac{1}{2F^2}
&\begin{pmatrix}
\frac{1}{2}\left(\left(F\cdot g_{ac,\alpha}\cdot F_{,\beta}g_{bd}\right)+\left(F_{,\alpha}g_{ac}\cdot F\cdot g_{bd,\beta}+F_{,\alpha}g_{ac}\cdot F_{,\beta}g_{bd}\right)\right) \\
+\left(\left(F\cdot g_{\alpha c,a}\cdot F_{,b}g_{\beta d}\right)+\left(F_{,a}g_{\alpha c}\cdot F\cdot g_{\beta d,b}+F_{,a}g_{\alpha c}\cdot F_{,b}g_{\beta d}\right)\right) \\
-\left(\left(F\cdot g_{\alpha c,a}\cdot F_{,d}g_{\beta b}\right)+\left(F_{,a}g_{\alpha c}\cdot F\cdot g_{\beta b,d}+F_{,a}g_{\alpha c}\cdot F_{,d}g_{\beta b}\right)\right)
\end{pmatrix} g^{ab}g^{cd} \\
-\frac{1}{4F^2}
&\begin{pmatrix}
\left(2\left(F\cdot g_{\alpha c,\beta}F_{,a}g_{bd}\right)-F\cdot g_{\alpha c,\beta}F_{,d}g_{ab}\right) \\
+\left(2\left(F_{,\beta}g_{\alpha c}F\cdot g_{bd,a}+F_{,\beta}g_{\alpha c}F_{,a}g_{bd}\right)-F_{,\beta}g_{\alpha c}F\cdot g_{ab,d}-F_{,\beta}g_{\alpha c}F_{,d}g_{ab}\right) \\
+\left(2\left(F\cdot g_{\beta c,\alpha}F_{,a}g_{bd}\right)-F\cdot g_{\beta c,\alpha}F_{,d}g_{ab}\right) \\
+\left(2\left(F_{,\alpha}g_{\beta c}F\cdot g_{bd,a}+F_{,\alpha}g_{\beta c}F_{,a}g_{bd}\right)-F_{,\alpha}g_{\beta c}F\cdot g_{ab,d}-F_{,\alpha}g_{\beta c}F_{,d}g_{ab}\right) \\
-\left(2\left(F\cdot g_{\alpha\beta,c}F_{,a}g_{bd}\right)-F\cdot g_{\alpha\beta,c}F_{,d}g_{ab}\right) \\
-\left(2\left(F_{,c}g_{\alpha\beta}F\cdot g_{bd,a}+F_{,c}g_{\alpha\beta}F_{,a}g_{bd}\right)-F_{,c}g_{\alpha\beta}F\cdot g_{ab,d}-F_{,c}g_{\alpha\beta}F_{,d}g_{ab}\right)
\end{pmatrix} g^{ab}g^{cd},
\end{aligned}
\quad (742)
$$

$$
\begin{aligned}
R^{*}_{\alpha\beta} = R_{\alpha\beta} &- \frac{1}{2F}\left(\begin{array}{l}
F_{,\alpha\beta}g_{ab} + F_{,b}g_{\alpha\beta,a} \\
+F_{,a}g_{\alpha\beta,b} - F_{,a}g_{\alpha b,\beta} - F_{,a}g_{\beta b,\alpha} \\
+F_{,ab}g_{\alpha\beta} \\
-F_{,a\beta}g_{\alpha b} \\
-F_{,a\alpha}g_{\beta b} \\
+F_{,\alpha}g_{ab,\beta} - F_{,\alpha}g_{\beta b,a} \\
+F_{,\beta}g_{ab,\alpha} - F_{,\beta}g_{\alpha b,a}
\end{array}\right)g^{ab} \\
&+\frac{1}{2F}\left(\begin{array}{l}
\frac{1}{2}\left(\left(g_{ac,\alpha}\cdot F_{,\beta}g_{bd}\right)+\left(F_{,\alpha}g_{ac}\cdot g_{bd,\beta}\right)\right)+\left(\left(g_{\alpha c,a}\cdot F_{,b}g_{\beta d}\right)+\left(F_{,a}g_{\alpha c}\cdot g_{\beta d,b}\right)\right) \\
-\left(\left(g_{\alpha c,a}\cdot F_{,d}g_{\beta b}\right)+\left(F_{,a}g_{\alpha c}\cdot g_{\beta b,d}\right)\right)
\end{array}\right)g^{ab}g^{cd} \\
&-\frac{1}{4F}\left(\begin{array}{l}
\left(2\left(g_{\alpha c,\beta}F_{,a}g_{bd}\right)-g_{\alpha c,\beta}F_{,d}g_{ab}\right)+\left(-F_{,\beta}g_{\alpha c}g_{ab,d}\right) \\
+\left(2\left(g_{\beta c,\alpha}F_{,a}g_{bd}\right)-g_{\beta c,\alpha}F_{,d}g_{ab}\right)+\left(-F_{,\alpha}g_{\beta c}g_{ab,d}\right) \\
-\left(2\left(g_{\alpha\beta,c}F_{,a}g_{bd}\right)-g_{\alpha\beta,c}F_{,d}g_{ab}\right)-\left(-F_{,c}g_{\alpha\beta}g_{ab,d}\right)
\end{array}\right)g^{ab}g^{cd} \\
&+\frac{1}{4F^{2}}\left(F_{,\alpha}\cdot F_{,\beta}\,(3n-6)+g_{\alpha\beta}F_{,c}F_{,d}g^{cd}\,(4-n)\right)
\end{aligned}
\quad , \quad (743)
$$

incorporation of the g^{cd}:

$$
\begin{aligned}
R^{*}_{\alpha\beta} = R_{\alpha\beta} &- \frac{1}{2F}\left(\begin{array}{l}
F_{,\alpha\beta}n + F_{,b}g_{\alpha\beta,a}g^{ab} + F_{,a}g_{\alpha\beta,b}g^{ab} \\
-F_{,a}g_{\alpha b,\beta}g^{ab} - F_{,a}g_{\beta b,\alpha}g^{ab} + g_{\alpha b,\beta}F_{,a}g^{ab} + g_{\beta b,\alpha}F_{,a}g^{ab} \\
+F_{,ab}g_{\alpha\beta}g^{ab} \\
-F_{,\alpha\beta} - F_{,\alpha\beta} + F_{,\alpha}g_{ab,\beta}g^{ab} - F_{,\alpha}g_{\beta b,a}g^{ab} - F_{,\alpha}g_{ab,\beta}g^{ab} \\
+F_{,\beta}g_{ab,\alpha}g^{ab} - F_{,\beta}g_{\alpha b,a}g^{ab} - F_{,\beta}g_{ab,\alpha}g^{ab} \\
-g_{\alpha\beta,a}\cdot F_{,b}g^{ab} - F_{,a}\cdot g_{\beta\alpha,b}g^{ab} + g_{\alpha c,\beta}\cdot F_{,d}g^{cd} + F_{,a}\cdot g_{\beta b,\alpha}g^{ab} \\
-\frac{1}{2}ng_{\alpha c,\beta}F_{,d}g^{cd} - \frac{1}{2}ng_{\beta c,\alpha}F_{,d}g^{cd} \\
-\left(g_{\alpha\beta,b}F_{,a}g^{ab}\right) + \frac{1}{2}ng_{\alpha\beta,c}F_{,d}g^{cd} + \frac{1}{2}F_{,c}g_{\alpha\beta}g_{ab,d}g^{cd}g^{ab}
\end{array}\right), \\
&+\frac{1}{4F^{2}}\left(F_{,\alpha}\cdot F_{,\beta}\,(3n-6)+g_{\alpha\beta}F_{,c}F_{,d}g^{cd}\,(4-n)\right)
\end{aligned}
\qquad (744)
$$

$$R^*_{\alpha\beta}=R_{\alpha\beta}-\frac{1}{2F}\begin{pmatrix}F_{,\alpha\beta}(n-2)+F_{,ab}g_{\alpha\beta}g^{ab}\\+F_{,b}g_{\alpha\beta,a}g^{ab}-g_{\alpha\beta,a}\cdot F_{,b}g^{ab}\\+F_{,a}g_{\alpha\beta,b}g^{ab}-F_{,a}g_{\alpha b,\beta}g^{ab}-F_{,a}g_{\beta b,\alpha}g^{ab}+F_{,a}\cdot g_{\beta b,\alpha}g^{ab}\\+g_{\alpha b,\beta}F_{,a}g^{ab}+g_{\beta b,\alpha}F_{,a}g^{ab}-g_{\alpha\beta,b}F_{,a}g^{ab}-F_{,a}\cdot g_{\beta\alpha,b}g^{ab}\\+F_{,\alpha}g_{ab,\beta}g^{ab}-F_{,\alpha}g_{\beta b,a}g^{ab}-F_{,\alpha}g_{ab,\beta}g^{ab}\\+F_{,\beta}g_{ab,\alpha}g^{ab}-F_{,\beta}g_{\alpha b,a}g^{ab}-F_{,\beta}g_{ab,\alpha}g^{ab}\\+g_{\alpha c,\beta}\cdot F_{,d}g^{cd}-\frac{1}{2}ng_{\alpha c,\beta}F_{,d}g^{cd}-\frac{1}{2}ng_{\beta c,\alpha}F_{,d}g^{cd}\\+\frac{1}{2}ng_{\alpha\beta,c}F_{,d}g^{cd}+\frac{1}{2}F_{,c}g_{\alpha\beta}g_{ab,d}g^{cd}g^{ab}\end{pmatrix}, \tag{745}$$

$$+\frac{1}{4F^2}\left(F_{,\alpha}\cdot F_{,\beta}(3n-6)+g_{\alpha\beta}F_{,c}F_{,d}g^{cd}(4-n)\right)$$

$$R^*_{\alpha\beta}=R_{\alpha\beta}-\frac{1}{2F}\begin{pmatrix}F_{,\alpha\beta}(n-2)+F_{,ab}g_{\alpha\beta}g^{ab}\\+F_{,a}g^{ab}\begin{pmatrix}g_{\alpha\beta,b}-g_{\alpha b,\beta}-g_{\beta b,\alpha}+g_{\beta b,\alpha}\\+g_{\alpha b,\beta}+g_{\beta b,\alpha}-g_{\alpha\beta,b}-g_{\beta\alpha,b}\end{pmatrix}\\+F_{,\alpha}g_{ab,\beta}g^{ab}-F_{,\alpha}g_{\beta b,a}g^{ab}-F_{,\alpha}g_{ab,\beta}g^{ab}\\+F_{,\beta}g_{ab,\alpha}g^{ab}-F_{,\beta}g_{\alpha b,a}g^{ab}-F_{,\beta}g_{ab,\alpha}g^{ab}\\+g_{\alpha c,\beta}\cdot F_{,d}g^{cd}-\frac{1}{2}ng_{\alpha c,\beta}F_{,d}g^{cd}-\frac{1}{2}ng_{\beta c,\alpha}F_{,d}g^{cd}\\+\frac{1}{2}ng_{\alpha\beta,c}F_{,d}g^{cd}+\frac{1}{2}F_{,c}g_{\alpha\beta}g_{ab,d}g^{cd}g^{ab}\end{pmatrix}, \tag{746}$$

$$+\frac{1}{4F^2}\left(F_{,\alpha}\cdot F_{,\beta}(3n-6)+g_{\alpha\beta}F_{,c}F_{,d}g^{cd}(4-n)\right)$$

finally gives us:

$$R^*_{\alpha\beta}=\begin{pmatrix}R_{\alpha\beta}-\frac{1}{2F}\begin{pmatrix}F_{,\alpha\beta}(n-2)+F_{,ab}g_{\alpha\beta}g^{ab}\\+F_{,a}g^{ab}\left(g_{\beta b,\alpha}-g_{\beta\alpha,b}\right)-F_{,\alpha}g^{ab}g_{\beta b,a}-F_{,\beta}g^{ab}g_{\alpha b,a}\\+F_{,d}g^{cd}\left(g_{\alpha c,\beta}-\frac{1}{2}ng_{\alpha c,\beta}-\frac{1}{2}ng_{\beta c,\alpha}+\frac{1}{2}ng_{\alpha\beta,c}+\frac{1}{2}g_{\alpha\beta}g_{ab,c}g^{ab}\right)\end{pmatrix}\\+\frac{1}{4F^2}\left(F_{,\alpha}\cdot F_{,\beta}(3n-6)+g_{\alpha\beta}F_{,c}F_{,d}g^{cd}(4-n)\right)\end{pmatrix}. \tag{747}$$

22.2 Derivation of the Ricci Scalar for a Scaled Metric Tensor

For the derivation of the corresponding Ricci scalar of a scaled metric tensor we can directly use the result for the Ricci tensor as given above in equation (747) and start the evaluation as follows:

$$
\begin{aligned}
& R^{*}\cdot F = R^{*}_{\alpha\beta}G^{\alpha\beta}\cdot F = R^{*}_{\alpha\beta}g^{\alpha\beta} \\
& = R_{\alpha\beta}g^{\alpha\beta} - \frac{g^{\alpha\beta}}{2F}\begin{pmatrix} F_{,\alpha\beta}(n-2)+F_{,ab}g_{\alpha\beta}g^{ab} \\ +F_{,a}g^{ab}\left(g_{\beta b,\alpha}-g_{\beta\alpha,b}\right)-F_{,\alpha}g^{ab}g_{\beta b,a}-F_{,\beta}g^{ab}g_{\alpha b,a} \\ +F_{,d}g^{cd}\left(g_{\alpha c,\beta}-\frac{1}{2}ng_{\alpha c,\beta}-\frac{1}{2}ng_{\beta c,\alpha}+\frac{1}{2}ng_{\alpha\beta,c}+\frac{1}{2}g_{\alpha\beta}g_{ab,c}g^{ab}\right) \end{pmatrix} \\
& +\frac{g^{\alpha\beta}}{4F^{2}}\left(F_{,\alpha}\cdot F_{,\beta}(3n-6)+g_{\alpha\beta}F_{,c}F_{,d}g^{cd}(4-n)\right) \\
& = R-\frac{1}{2F}\begin{pmatrix} g^{\alpha\beta}F_{,\alpha\beta}(n-2)+nF_{,ab}g^{ab} \\ +F_{,a}g^{ab}g^{\alpha\beta}\left(g_{\beta b,\alpha}-g_{\beta\alpha,b}\right)-F_{,\alpha}g^{\alpha\beta}g^{ab}g_{\beta b,a}-F_{,\beta}g^{\alpha\beta}g^{ab}g_{\alpha b,a} \\ +F_{,d}g^{cd}\left(g^{\alpha\beta}g_{\alpha c,\beta}-\frac{g^{\alpha\beta}}{2}ng_{\alpha c,\beta}-\frac{g^{\alpha\beta}}{2}ng_{\beta c,\alpha}+\frac{g^{\alpha\beta}}{2}ng_{\alpha\beta,c}+\frac{n}{2}g_{ab,c}g^{ab}\right) \end{pmatrix} \\
& +\frac{g^{\alpha\beta}F_{,\alpha}\cdot F_{,\beta}}{4F^{2}}\left((3n-6)+n(4-n)\right) \\
& = R-\frac{1}{2F}\begin{pmatrix} 2g^{\alpha\beta}F_{,\alpha\beta}(n-1) \\ +F_{,d}g^{cd}g^{\alpha\beta}\left(g_{\beta c,\alpha}-g_{\beta\alpha,c}\right)-F_{,\alpha}g^{\alpha\beta}g^{cd}g_{\beta d,c}-F_{,\beta}g^{\alpha\beta}g^{cd}g_{\alpha d,c}+F_{,d}g^{cd}g^{\alpha\beta}g_{\alpha c,\beta} \\ +F_{,d}g^{cd}g^{\alpha\beta}\left(-\frac{1}{2}ng_{\alpha c,\beta}-\frac{1}{2}ng_{\beta c,\alpha}+ng_{\alpha\beta,c}\right) \end{pmatrix} \\
& -(n-1)\frac{g^{\alpha\beta}F_{,\alpha}\cdot F_{,\beta}}{4F^{2}}(n-6) \\
& = R-\frac{1}{2F}\begin{pmatrix} 2g^{\alpha\beta}F_{,\alpha\beta}(n-1) \\ +F_{,d}g^{cd}g^{\alpha\beta}\left(g_{\beta d,\alpha}-g_{\beta\alpha,d}+g_{\alpha c,\beta}\right)-F_{,\alpha}g^{\alpha\beta}g^{cd}g_{\beta d,c}-F_{,\beta}g^{\alpha\beta}g^{cd}g_{\alpha d,c} \\ +F_{,d}g^{cd}g^{\alpha\beta}\left(-\frac{1}{2}ng_{\alpha c,\beta}-\frac{1}{2}ng_{\beta c,\alpha}+ng_{\alpha\beta,c}\right) \end{pmatrix} \\
& -(n-1)\frac{g^{\alpha\beta}F_{,\alpha}\cdot F_{,\beta}}{4F^{2}}(n-6)
\end{aligned} \quad . \tag{748}
$$

Thereby, we have started to evaluate not the Ricci scalar of the scaled metric directly, but the corresponding term (see rectangle in the equation below) from the quantum Einstein field equations, reading:

$$R^*_{\alpha\beta}-\frac{R^*}{2}G_{\alpha\beta}=R^*_{\alpha\beta}-\frac{R^*}{2}G_{\alpha\beta}=R^*_{\alpha\beta}-R^*_{\alpha\beta}G^{\alpha\beta}\cdot\frac{G_{\alpha\beta}}{2}=R^*_{\alpha\beta}-\boxed{R^*_{\alpha\beta}G^{\alpha\beta}\cdot F}\cdot\frac{g_{\alpha\beta}}{2}\,. \quad (749)$$
$$\text{with: } R^*_{\alpha\beta}G^{\alpha\beta}\cdot F=R^*_{\alpha\beta}g^{\alpha\beta}$$

The evaluation yields:

$$R^*=\left(\begin{array}{l} R-\frac{1}{2F}\left(\begin{array}{l} 2g^{\alpha\beta}F_{,\alpha\beta}(n-1) \\ +F_{,d}g^{cd}g^{\alpha\beta}\left(g_{\beta c,\alpha}-g_{\beta\alpha,c}+g_{\alpha c,\beta}\right)-F_{,\alpha}g^{\alpha\beta}g^{cd}\left(g_{\beta d,c}+g_{\beta d,c}\right) \\ +F_{,d}g^{cd}g^{\alpha\beta}\left(-\frac{1}{2}ng_{\alpha c,\beta}-\frac{1}{2}ng_{\beta c,\alpha}+ng_{\alpha\beta,c}\right) \end{array}\right) \\ -(n-1)\frac{g^{\alpha\beta}F_{,\alpha}\cdot F_{,\beta}}{4F^2}(n-6) \end{array}\right) \quad (750)$$

$$=\left(\begin{array}{c} R-\frac{1}{2F}\left(\begin{array}{l} 2g^{\alpha\beta}F_{,\alpha\beta}(n-1) \\ +F_{,d}g^{cd}g^{\alpha\beta}\left(g_{\beta c,\alpha}-g_{\beta\alpha,c}+g_{\alpha c,\beta}\right)-2F_{,\alpha}g^{\alpha\beta}g^{cd}g_{\beta d,c} \\ +F_{,d}g^{cd}g^{\alpha\beta}\left(-\frac{1}{2}ng_{\alpha c,\beta}-\frac{1}{2}ng_{\beta c,\alpha}+ng_{\alpha\beta,c}\right) \end{array}\right) \\ -(n-1)\frac{g^{\alpha\beta}F_{,\alpha}\cdot F_{,\beta}}{4F^2}(n-6) \end{array}\right). \quad (751)$$

Further simplification is possible when using the symmetry properties of the metric tensor:

$$=\left(\begin{array}{l} R-\frac{1}{2F}\left(\begin{array}{l} 2g^{\alpha\beta}F_{,\alpha\beta}(n-1) \\ +F_{,d}g^{cd}g^{\alpha\beta}\left(g_{\beta c,\alpha}-g_{\beta\alpha,c}+g_{\alpha c,\beta}\right)-2F_{,d}g^{\alpha\beta}g^{cd}g_{\alpha c,\beta} \\ +F_{,d}g^{cd}g^{\alpha\beta}\left(-\frac{1}{2}ng_{\alpha c,\beta}-\frac{1}{2}ng_{\beta c,\alpha}+ng_{\alpha\beta,c}\right) \end{array}\right) \\ -(n-1)\frac{g^{\alpha\beta}F_{,\alpha}\cdot F_{,\beta}}{4F^2}(n-6) \end{array}\right). \quad (752)$$
$$=\left(\begin{array}{c} R-\frac{1}{2F}\left(2g^{\alpha\beta}F_{,\alpha\beta}(n-1)+F_{,d}g^{cd}g^{\alpha\beta}\left((n-1)g_{\alpha\beta,c}-ng_{\alpha c,\beta}\right)\right) \\ -(n-1)\frac{g^{\alpha\beta}F_{,\alpha}\cdot F_{,\beta}}{4F^2}(n-6) \end{array}\right)$$

In some cases it is more practical to introduce the Laplace operator, e.g., when intending to work out the connection with classical Quantum Theory:

$$R^{*}_{\alpha\beta}G^{\alpha\beta}\cdot F = R^{*}_{\alpha\beta}g^{\alpha\beta}$$

$$= \begin{pmatrix} R - \frac{1}{2F}\left((n-1)\left(\overbrace{2g^{\alpha\beta}F_{,\alpha\beta} + F_{,d}g^{cd}g^{\alpha\beta}g_{\alpha\beta,c}}^{=2\Delta F - 2F_{,d}g^{cd}{}_{,c}}\right) - nF_{,d}g^{cd}g^{\alpha\beta}g_{\alpha c,\beta}\right) \\ -(n-1)\frac{g^{\alpha\beta}F_{,\alpha}\cdot F_{,\beta}}{4F^2}(n-6) \end{pmatrix}$$

$$= \begin{pmatrix} R - \frac{1}{2F}\left(2(n-1)\left(\Delta F - F_{,d}g^{cd}{}_{,c}\right) - nF_{,d}g^{cd}g^{\alpha\beta}g_{\alpha c,\beta}\right) \\ -(n-1)\frac{g^{\alpha\beta}F_{,\alpha}\cdot F_{,\beta}}{4F^2}(n-6) \end{pmatrix} \quad . \tag{753}$$

$$= \begin{pmatrix} R - \frac{1}{2F}\left(2(n-1)\Delta F - F_{,d}\left(2(n-1)g^{cd}{}_{,c} + ng^{cd}g^{\alpha\beta}g_{\alpha c,\beta}\right)\right) \\ -(n-1)\frac{g^{\alpha\beta}F_{,\alpha}\cdot F_{,\beta}}{4F^2}(n-6) \end{pmatrix}$$

22.3 Quantum Einstein Field Equations for Scaled Metric Tensor in Various Forms

In combining the results of the previous two subsections, we can now present the scaled metric field equations in various forms.

$$R^{*}_{\alpha\beta} - \frac{R^{*}}{2}G_{\alpha\beta}$$

$$= \begin{pmatrix} R_{\alpha\beta} - \frac{R}{2}g_{\alpha\beta} + \frac{g_{\alpha\beta}}{2}\begin{pmatrix} \frac{1}{2F}\begin{pmatrix} 2g^{ab}F_{,ab}(n-1) \\ +F_{,d}g^{cd}g^{ab}\left(g_{bc,a} - g_{ba,c} + g_{ac,b}\right) \\ -F_{,a}g^{ab}g^{cd}\left(g_{bd,c} + g_{ad,c}\right) \\ +F_{,d}g^{cd}g^{ab}\left(-\frac{1}{2}ng_{ac,b} - \frac{1}{2}ng_{bc,a} + ng_{ab,c}\right) \end{pmatrix} \\ +(n-1)\frac{g^{ab}F_{,a}\cdot F_{,b}}{4F^2}(n-6) \end{pmatrix} \\ -\frac{1}{2F}\begin{pmatrix} F_{,\alpha\beta}(n-2) + F_{,ab}g_{\alpha\beta}g^{ab} \\ +F_{,a}g^{ab}\left(g_{\beta b,\alpha} - g_{\beta\alpha,b}\right) - F_{,\alpha}g^{ab}g_{\beta b,a} - F_{,\beta}g^{ab}g_{\alpha b,a} \\ +F_{,d}g^{cd}\left(g_{\alpha c,\beta} - \frac{1}{2}ng_{\alpha c,\beta} - \frac{1}{2}ng_{\beta c,\alpha} + \frac{1}{2}ng_{\alpha\beta,c} + \frac{1}{2}g_{\alpha\beta}g_{ab,c}g^{ab}\right) \\ -\frac{1}{2F}\left(F_{,\alpha}\cdot F_{,\beta}(3n-6) + g_{\alpha\beta}F_{,c}F_{,d}g^{cd}(4-n)\right) \end{pmatrix} \end{pmatrix}, \tag{754}$$

$$R^{*}_{\alpha\beta}-\frac{R^{*}}{2}G_{\alpha\beta}$$

$$=\left(\begin{array}{l}R_{\alpha\beta}-\frac{R}{2}g_{\alpha\beta}+\frac{g_{\alpha\beta}}{2}\left(\begin{array}{l}\frac{1}{2F}\left(\begin{array}{l}2g^{ab}F_{,ab}(n-1)\\+F_{,d}g^{cd}g^{ab}\left(g_{bc,a}-g_{ba,c}+g_{ac,b}\right)-F_{,a}g^{ab}g^{cd}\left(g_{bd,c}+g_{ad,c}\right)\\+F_{,d}g^{cd}g^{ab}\left(-\frac{1}{2}ng_{ac,b}-\frac{1}{2}ng_{bc,a}+ng_{ab,c}\right)\end{array}\right)\\+(n-1)\frac{g^{ab}F_{,a}\cdot F_{,b}}{4F^{2}}(n-6)\end{array}\right)\\-\frac{1}{2F}\left(F_{,ab}g_{\alpha\beta}g^{ab}+F_{,d}g^{cd}\frac{1}{2}g_{\alpha\beta}g_{ab,c}g^{ab}-\frac{1}{2F}g_{\alpha\beta}F_{,c}F_{,d}g^{cd}(4-n)\right)\\-\frac{1}{2F}\left(\begin{array}{l}F_{,\alpha\beta}(n-2)\\+F_{,a}g^{ab}\left(g_{\beta b,\alpha}-g_{\beta\alpha,b}\right)-F_{,\alpha}g^{ab}g_{\beta b,a}-F_{,\beta}g^{ab}g_{\alpha b,a}\\+F_{,d}g^{cd}\left(g_{\alpha c,\beta}-\frac{1}{2}ng_{\alpha c,\beta}-\frac{1}{2}ng_{\beta c,\alpha}+\frac{1}{2}ng_{\alpha\beta,c}\right)\\-\frac{1}{2F}\left(F_{,\alpha}\cdot F_{,\beta}(3n-6)\right)\end{array}\right)\end{array}\right),\quad(755)$$

$$R^{*}_{\alpha\beta}-\frac{R^{*}}{2}G_{\alpha\beta}$$

$$=\left(\begin{array}{l}R_{\alpha\beta}-\frac{R}{2}g_{\alpha\beta}+\frac{g_{\alpha\beta}}{2}\left(\begin{array}{l}\frac{1}{2F}\left(\begin{array}{l}2g^{ab}F_{,ab}(n-1)\\+F_{,d}g^{cd}g^{ab}\left(g_{bc,a}-g_{ba,c}+g_{ac,b}\right)-2F_{,a}g^{ab}g^{cd}g_{bd,c}\\+F_{,d}g^{cd}g^{ab}\left(-\frac{1}{2}ng_{ac,b}-\frac{1}{2}ng_{bc,a}+ng_{ab,c}\right)\end{array}\right)\\+\frac{g^{ab}F_{,a}\cdot F_{,b}}{4F^{2}}((n-1)(n-6))\end{array}\right)\\-\frac{g_{\alpha\beta}}{4F}\left(2F_{,ab}g^{ab}+F_{,d}g^{cd}g_{ab,c}g^{ab}\right)+\frac{g_{\alpha\beta}}{4F^{2}}F_{,c}F_{,d}g^{cd}(4-n)\\-\frac{1}{2F}\left(\begin{array}{l}F_{,\alpha\beta}(n-2)\\+F_{,a}g^{ab}\left(g_{\beta b,\alpha}-g_{\beta\alpha,b}\right)-F_{,\alpha}g^{ab}g_{\beta b,a}-F_{,\beta}g^{ab}g_{\alpha b,a}\\+F_{,d}g^{cd}\left(g_{\alpha c,\beta}-\frac{1}{2}ng_{\alpha c,\beta}-\frac{1}{2}ng_{\beta c,\alpha}+\frac{1}{2}ng_{\alpha\beta,c}\right)\\-\frac{1}{2F}\left(F_{,\alpha}\cdot F_{,\beta}(3n-6)\right)\end{array}\right)\end{array}\right),\quad(756)$$

$$R^*_{\alpha\beta}-\frac{R^*}{2}G_{\alpha\beta}$$

$$=\left(\begin{array}{l}R_{\alpha\beta}-\frac{R}{2}g_{\alpha\beta}+\frac{g_{\alpha\beta}}{2}\left(\begin{array}{l}\frac{1}{2F}\left(\begin{array}{l}2g^{ab}F_{,ab}(n-2)\\+F_{,d}g^{cd}g^{ab}\left(g_{bc,a}-g_{ba,c}+g_{ac,b}\right)\\-2F_{,d}g^{ab}g^{cd}g_{bc,a}-F_{,d}g^{cd}g_{ab,c}g^{ab}\\+F_{,d}g^{cd}g^{ab}\left(-\frac{1}{2}ng_{ac,b}-\frac{1}{2}ng_{bc,a}+ng_{ab,c}\right)\end{array}\right)\\+\frac{g^{ab}F_{,a}\cdot F_{,b}}{4F^2}((n-1)(n-6)+2(4-n))\end{array}\right)\\-\frac{1}{2F}\left(\begin{array}{l}F_{,\alpha\beta}(n-2)\\+F_{,a}g^{ab}\left(g_{\beta b,\alpha}-g_{\beta\alpha,b}\right)-F_{,\alpha}g^{ab}g_{\beta b,a}-F_{,\beta}g^{ab}g_{\alpha b,a}\\+F_{,d}g^{cd}\left(g_{\alpha c,\beta}-\frac{1}{2}ng_{\alpha c,\beta}-\frac{1}{2}ng_{\beta c,\alpha}+\frac{1}{2}ng_{\alpha\beta,c}\right)\\-\frac{1}{2F}\left(F_{,\alpha}\cdot F_{,\beta}(3n-6)\right)\end{array}\right)\end{array}\right), \tag{757}$$

$$R^*_{\alpha\beta}-\frac{R^*}{2}G_{\alpha\beta}$$

$$=\left(\begin{array}{l}R_{\alpha\beta}-\frac{R}{2}g_{\alpha\beta}+\frac{g_{\alpha\beta}}{2}\left(\begin{array}{l}\frac{1}{2F}\left(\begin{array}{l}2g^{ab}F_{,ab}(n-2)-2F_{,d}g^{cd}g^{ab}g_{ab,c}\\+F_{,d}g^{cd}g^{ab}\left(-ng_{ac,b}+ng_{ab,c}\right)\end{array}\right)\\+\frac{g^{ab}F_{,a}\cdot F_{,b}}{4F^2}((n-1)(n-6)+2(4-n))\end{array}\right)\\-\frac{1}{2F}\left(\begin{array}{l}F_{,\alpha\beta}(n-2)\\+F_{,a}g^{ab}\left(g_{\beta b,\alpha}-g_{\beta\alpha,b}\right)-F_{,\alpha}g^{ab}g_{\beta b,a}-F_{,\beta}g^{ab}g_{\alpha b,a}\\+F_{,d}g^{cd}\left(g_{\alpha c,\beta}-\frac{1}{2}ng_{\alpha c,\beta}-\frac{1}{2}ng_{\beta c,\alpha}+\frac{1}{2}ng_{\alpha\beta,c}\right)\\-\frac{1}{2F}\left(F_{,\alpha}\cdot F_{,\beta}(3n-6)\right)\end{array}\right)\end{array}\right), \tag{758}$$

$$
\begin{aligned}
&R^{*}_{\alpha\beta}-\frac{R^{*}}{2}G_{\alpha\beta} \\
&=\left(\begin{array}{l}
R_{\alpha\beta}-\frac{R}{2}g_{\alpha\beta}+\frac{g_{\alpha\beta}}{2}\left(\begin{array}{l}
\frac{1}{2F}\left(\begin{array}{c}
(n-2)\left(\overbrace{2g^{ab}F_{,ab}+F_{,d}g^{cd}g^{ab}g_{ab,c}}^{=2\Delta F-2F_{,d}g^{cd}{}_{,c}}\right) \\
-nF_{,d}g^{cd}g^{ab}g_{ac,b}
\end{array}\right) \\
+\frac{g^{ab}F_{,a}\cdot F_{,b}}{4F^{2}}((n-1)(n-6)+2(4-n))
\end{array}\right) \\
-\frac{1}{2F}\left(\begin{array}{l}
F_{,\alpha\beta}(n-2) \\
+F_{,a}g^{ab}\left(g_{\beta b,\alpha}-g_{\beta\alpha,b}\right)-F_{,\alpha}g^{ab}g_{\beta b,a}-F_{,\beta}g^{ab}g_{\alpha b,a} \\
+F_{,d}g^{cd}\left(g_{\alpha c,\beta}-\frac{1}{2}ng_{\alpha c,\beta}-\frac{1}{2}ng_{\beta c,\alpha}+\frac{1}{2}ng_{\alpha\beta,c}\right) \\
-\frac{1}{2F}\left(F_{,\alpha}\cdot F_{,\beta}(3n-6)\right)
\end{array}\right)
\end{array}\right),
\end{aligned}
\tag{759}
$$

$$
\begin{aligned}
&R^{*}_{\alpha\beta}-\frac{R^{*}}{2}G_{\alpha\beta} \\
&=\left(\begin{array}{l}
R_{\alpha\beta}-\frac{R}{2}g_{\alpha\beta}+\frac{g_{\alpha\beta}}{2}\left(\begin{array}{l}
\frac{1}{2F}\left(2(n-2)\Delta F-F_{,d}\left(2(n-2)g^{cd}{}_{,c}+ng^{cd}g^{ab}g_{ac,b}\right)\right) \\
+\frac{g^{ab}F_{,a}\cdot F_{,b}}{4F^{2}}((n-1)(n-6)+2(4-n))
\end{array}\right) \\
-\frac{1}{2F}\left(\begin{array}{l}
F_{,\alpha\beta}(n-2) \\
+F_{,a}g^{ab}\left(g_{\beta b,\alpha}-g_{\beta\alpha,b}\right)-F_{,\alpha}g^{ab}g_{\beta b,a}-F_{,\beta}g^{ab}g_{\alpha b,a} \\
+F_{,d}g^{cd}\left(g_{\alpha c,\beta}-\frac{1}{2}ng_{\alpha c,\beta}-\frac{1}{2}ng_{\beta c,\alpha}+\frac{1}{2}ng_{\alpha\beta,c}\right) \\
-\frac{1}{2F}\left(F_{,\alpha}\cdot F_{,\beta}(3n-6)\right)
\end{array}\right)
\end{array}\right).
\end{aligned}
\tag{760}
$$

Incorporation of the scaling function F[f] gives:

$$R^*_{\alpha\beta}-\frac{R^*}{2}G_{\alpha\beta}$$

$$=\begin{pmatrix} R_{\alpha\beta}-\frac{R}{2}g_{\alpha\beta}+\frac{g_{\alpha\beta}}{2}\begin{pmatrix}\frac{1}{2F}\begin{pmatrix}(n-2)\left(2g^{ab}\left(F''f_{,a}f_{,b}+F'f_{,ab}\right)+F_{,d}g^{cd}g^{ab}g_{ab,c}\right)\\ -nF_{,d}g^{cd}g^{ab}g_{ac,b}\end{pmatrix}\\ +\frac{g^{ab}F_{,a}\cdot F_{,b}}{4F^2}((n-1)(n-6)+2(4-n))\end{pmatrix}\\ -\frac{1}{2F}\begin{pmatrix}\left(F''f_{,\alpha}f_{,\beta}+F'f_{,\alpha\beta}\right)(n-2)\\ +F_{,a}g^{ab}\left(g_{\beta b,\alpha}-g_{\beta\alpha,b}\right)-F_{,\alpha}g^{ab}g_{\beta b,a}-F_{,\beta}g^{ab}g_{\alpha b,a}\\ +F_{,d}g^{cd}\left(g_{\alpha c,\beta}-\frac{1}{2}ng_{\alpha c,\beta}-\frac{1}{2}ng_{\beta c,\alpha}+\frac{1}{2}ng_{\alpha\beta,c}\right)\\ -\frac{1}{2F}\left(F_{,\alpha}\cdot F_{,\beta}(3n-6)\right)\end{pmatrix}\end{pmatrix}$$

$$=\begin{pmatrix} R_{\alpha\beta}-\frac{R}{2}g_{\alpha\beta}+\frac{g_{\alpha\beta}}{2}\begin{pmatrix}\frac{F'}{2F}\begin{pmatrix}(n-2)\left(\overbrace{2g^{ab}f_{,ab}+f_{,d}g^{cd}g^{ab}g_{ab,c}}^{=2\Delta f-2f_{,d}g^{cd}{}_{,c}}\right)\\ -nf_{,d}g^{cd}g^{ab}g_{ac,b}\end{pmatrix}\\ +\frac{g^{ab}f_{,a}\cdot f_{,b}}{4F^2}\left((F')^2\begin{pmatrix}(n-1)(n-6)\\+2(4-n)\end{pmatrix}+4FF''(n-2)\right)\end{pmatrix}\\ -\frac{1}{2F}\begin{pmatrix}F'f_{,\alpha\beta}(n-2)\\ +F_{,a}g^{ab}\left(g_{\beta b,\alpha}-g_{\beta\alpha,b}\right)-F_{,\alpha}g^{ab}g_{\beta b,a}-F_{,\beta}g^{ab}g_{\alpha b,a}\\ +F_{,d}g^{cd}\left(g_{\alpha c,\beta}-\frac{1}{2}ng_{\alpha c,\beta}-\frac{1}{2}ng_{\beta c,\alpha}+\frac{1}{2}ng_{\alpha\beta,c}\right)\\ -\frac{1}{2F}\left(F_{,\alpha}\cdot F_{,\beta}(3n-6)-2FF''(n-2)\right)\end{pmatrix}\end{pmatrix}, \quad (761)$$

$$R^*_{\alpha\beta}-\frac{R^*}{2}G_{\alpha\beta}$$

$$=\begin{pmatrix} R_{\alpha\beta}-\frac{R}{2}g_{\alpha\beta}+\frac{g_{\alpha\beta}}{2}\frac{F'}{2F}\left((n-2)\left(2g^{ab}f_{,ab}+f_{,d}g^{cd}g^{ab}g_{ab,c}\right)-nf_{,d}g^{cd}g^{ab}g_{ac,b}\right)\\ +\frac{(n-2)}{4F^2}\left(f_{,\alpha}\cdot f_{,\beta}\left(3(F')^2-2FF''\right)+\frac{g_{\alpha\beta}}{2}g^{ab}f_{,a}\cdot f_{,b}\left((F')^2(n-7)+4FF''\right)\right)\\ -\frac{F'}{2F}\begin{pmatrix}f_{,\alpha\beta}(n-2)\\ +f_{,a}g^{ab}\left(g_{\beta b,\alpha}-g_{\beta\alpha,b}\right)-f_{,\alpha}g^{ab}g_{\beta b,a}-f_{,\beta}g^{ab}g_{\alpha b,a}\\ +f_{,d}g^{cd}\left(g_{\alpha c,\beta}-\frac{1}{2}ng_{\alpha c,\beta}-\frac{1}{2}ng_{\beta c,\alpha}+\frac{1}{2}ng_{\alpha\beta,c}\right)\end{pmatrix}\end{pmatrix}. \quad (762)$$

23 References

[1] D. Hilbert, "Die Grundlagen der Physik", Teil 1, Göttinger Nachrichten, 1915, pp. 395-407

[2] N. Schwarzer, "The World Formula: A Late Recognition of David Hilbert's Stroke of Genius", Jenny Stanford Publishing, 2020, ISBN: 9789814877206

[3] A. Einstein, "Grundlage der allgemeinen Relativitätstheorie", Annalen der Physik (ser. 4), 49, pp. 769-822

[4] N. Schwarzer, "The Math of Body, Soul, and the Universe", Jenny Stanford Publishing, 2022, ISBN: 9789814968249

[5] N. Schwarzer, "The Theory of Everything – Quantum and Relativity is Everywhere – A Fermat Universe", Pan Stanford Publishing, 2020, ISBN-10: 9814774472

[6] C. A. Sporea, "Notes on f(R) Theories of Gravity", 2014, arxiv.org/pdf/1403.3852.pdf

[7] N. Schwarzer, "How Systems Evolve – A Red Pill Course on Evolution", self-published, Amazon Digital Services, 2022, Kindle, ASIN: B0B6RGBKQ4; see also [70]

[8] N. Schwarzer, "Mathematical Psychology – Tools for the understanding, simulation, mitigation and potential prevention of mass formation psychoses", self-published, Amazon Digital Services, 2022, Kindle, ASIN: B0BC5T1HB4

[9] N. Schwarzer, "Towards the Metric Quark – Derivation of More Spin-1/2-Objects from the Einstein-Hilbert Action", self-published, Amazon Digital Services, 2023, Kindle, ASIN: B0BWDQDBFW

[10] N. Schwarzer, "The Metric Electron – Derivation of Spin-1/2-Objects from the Einstein-Hilbert Action", self-published, Amazon Digital Services, 2023, Kindle, ASIN: B0BTPQM26X

[11] N. Schwarzer, "The Foreseeable Quantum Enigma Disaster", Part 10b of "Medical Socio-Economic Quantum Gravity", self-published, Amazon Digital Services, 2021, Kindle, ASIN: B096PJRNTK

[12] N. Schwarzer, "The Quantum Gravity War – How will the Nearby Unification of Physics Change the Future Warfare?", Jenny Stanford Publishing, 2024, ISBN: 9789814968584

[13] N. Schwarzer, "Quantum Gravity Systems Science", self-published, Amazon Digital Services, 2023, Kindle, ASIN: B0BRL8SKSB

[14] N. Schwarzer, "Why Tucker Carlson is Perfectly Right about the WEF-People – A Mathematical Proof", self-published, Amazon Digital Services, 2023, Kindle, ASIN: B0BSTGKC6F

[15] N. Schwarzer, "Derivation of the Harmonic Oscillator from the Einstein-Hilbert Action – The Metric Harmonic Oscillator", self-published, Amazon Digital Services, 2023, Kindle, ASIN: B0BSXJCJVB

[16] N. Schwarzer, "Derivation of the Schrödinger Hydrogen from the Einstein-Hilbert Action – The Metric Schrödinger Hydrogen", self-published, Amazon Digital Services, 2023, Kindle, ASIN: B0BT9Y54XP

[17] N. Schwarzer, "Derivation Newton's Gravity Law from Quantum Einstein Field Equations", Amazon Digital Services, 2023, Kindle, ASIN: B0BXPS2ZJH

[18] N. Schwarzer, "Derivation Hawking Radiation of Black Holes from Quantum Einstein Field Equations", Amazon Digital Services, 2023, Kindle, ASIN: B0BXQ5ZG4Z

[19] N. Schwarzer, "Quantum Gravity – Brief Derivation of Quantum Einstein Field Equations", self-published, Amazon Digital Services, 2023, Kindle, ASIN: B0BWTWZHP5

[20] N. Schwarzer, "Quantum Gravity and Einstein's Famous Moon Problem – The Theory of Perspectivity", self-published, Amazon Digital Services, 2023, Kindle, ASIN: B0BXPQQWZN

[21] N. Schwarzer, "Shouldn't Quantum Gravity Solve the 3-Generation Problem?", self-published, Amazon Digital Services, 2023, Kindle, ASIN: B0BXPXP33L

[22] N. Schwarzer, "About Apparent Asymmetries in the 3-Generations", self-published, Amazon Digital Services, 2023, Kindle, ASIN: B0BXPYGMNF

[23] N. Schwarzer, "The Universe's Half-Spin Deception Game", self-published, Amazon Digital Services, 2023, Kindle, ASIN: B0BXPKWPQZ

[24] H. Haken, H. Chr. Wolf, "Atom- und Quantenphysik" (in German), 4th edition, Springer Heidelberg, 1990, ISBN: 0-387-52198-4

[25] E. Schrödinger, "Quantisierung als Eigenwertproblem (erste Mitteilung)", Ann. Phys., Vol. 384, No. 4., 1926, pp. 361-376

[26] E. Schrödinger, "Quantisierung als Eigenwertproblem (zweite Mitteilung)", Ann. Phys., Vol. 384, No. 6., 1926, pp. 489-527

[27] P. A. M. Dirac, "The Quantum Theory of the Electron", Published 1 February 1928, DOI: 10.1098/rspa.1928.0023

[28] C. Cohen-Tannoudji, B. Diu, F. Laloë, Franckv "Quantenmechanik 1&2", 2. Auflage, Walter de Gruyter, Berlin - New York, 1999, ISBN-10: 3110626004

[29] N. Schwarzer, "Metric Interpretation of Quantum Theory – How can Quantum Gravity Help Us to Understand Quantum Strangeness", self-published, Amazon Digital Services, 2023, Kindle, ASIN: B0BXQRKPV7

[30] N. Schwarzer, "The Metric Potential – How can Quantum Gravity Help Us to Understand the Origin of Forces", self-published, Amazon Digital Services, 2023, Kindle, ASIN: B0BXSHQZX1

[31] N. Schwarzer, "Strings, Branes, Friedmanns and the Matryoshka Universe", Part 7d of "Medical Socio-Economic Quantum Gravity", self-published, Amazon Digital Services, 2022, Kindle, ASIN: B09V37VXMB

[32] N. Schwarzer, "Towards Quantum Einstein Field Equations", Part 7 of "Medical Socio-Economic Quantum Gravity", self-published, Amazon Digital Services, December 2020, Kindle, ASIN: B08NG6H22X

[33] N. Schwarzer, "Einstein had it, but he did not see it – Part XXIV: A Variety of Solutions and the Dirac-Schwarzschild-Particle", self-published, Amazon Digital Services, 2018, Kindle, ASIN: B0796GZX8R

[34] N. Schwarzer, "Einstein had it, but he did not see it – Part XLIII: A Selection of Einstein-Dirac-Particles", self-published, Amazon Digital Services, 2018, Kindle, ASIN: B07DPVRZ8B

[35] N. Schwarzer, "Einstein had it... Part LXIII: Einstein-Field-Equations = Dirac² (+) Klein-Gordon", self-published, Amazon Digital Services, 2018, Kindle, ASIN: B07JVFV4HP

[36] N. Schwarzer, "Einstein had it, but he did not see it – Part LXXX: Short Note on the Killing of Dirac", self-published, Amazon Digital Services, 2018, Kindle, ASIN: B07NKZVF61

[37] N. Schwarzer, "Science Riddles – Riddle No. 16: How to Understand the Dirac Equation?", self-published, Amazon Digital Services, 2019, Kindle, ASIN: B07VFW2Z3F

[38] N. Schwarzer, "Science Riddles – Riddle No. 17: How Einstein becomes first order and goes Dirac – Can we factorize the Einstein-Field-Equations?", self-published, Amazon Digital Services, 2019, Kindle, ASIN: B07VV9FG7K

[39] N. Schwarzer, "Einstein had it, but he did not see it – Part LXXXI: More Dirac Killing", self-published, Amazon Digital Services, 2019, Kindle, ASIN: B07WW1G6N7

[40] N. Schwarzer, "Science Riddles – Riddle No. 23: Why does the Dirac Equation work so well?", self-published, Amazon Digital Services, 2019, Kindle, ASIN: B0818V32JY

[41] N. Schwarzer, "My Horcruxes – A Curvy Math to Salvation", Part 9 of "Medical Socio-Economic Quantum Gravity", self-published, Amazon Digital Services, 2021, Kindle, ASIN: B096SPB5MW

[42] N. Schwarzer, "The "New" Dirac Equation: Originally Postulated, Now Geometrically Derived and – along the way – Generalized to Arbitrary Dimensions and Arbitrary Metrics Space-Times", self-published, Amazon Digital Services, 2020, Kindle, ASIN: B085636Q88

[43] N. Schwarzer, "The Metric Dirac Equation Revisited and the Geometry of Spinors", Part 8 of "Medical Socio-Economic Quantum Gravity", self-published, Amazon Digital Services, April 2021, Kindle, ASIN: B08Y96TL3D

[44] N. Schwarzer, "The Dirac Miracle", Part 8a of "Medical Socio-Economic Quantum Gravity", self-published, Amazon Digital Services, May 2021, Kindle, ASIN: B0963Z1Z74

[45] N. Schwarzer, "How Einstein gives Dirac, Klein-Gordon and Schrödinger", self-published, Amazon Digital Services, 2017, Kindle, ASIN: B071K2Y4V2

[46] N. Schwarzer, "Dirac's Little Spinor Problem – Can Quantum Gravity Explain the Occurrence of this Postulated Object?", self-published, Amazon Digital Services, 2023, Kindle, ASIN: B0C3QWJSQS

[47] N. Schwarzer, "How Fermions Get Massy – Can Quantum Gravity Answer this Question?", self-published, Amazon Digital Services, 2023, Kindle, ASIN: B0C5TM9ZCG

[48] https://en.wikipedia.org/wiki/Dirac_equation_in_curved_spacetime

[49] N. Schwarzer, "What Was Before Time? – How Quantum Gravity Might Give an Answer", Amazon Digital Services, 2023, Kindle, ASIN: B0CB3ZKKNL

[50] N. Schwarzer, "Quantum Gravity Thermodynamics – And it May Get Hotter", self-published, Amazon Digital Services, 2019, Kindle, ASIN: B07XC2JW7F

[51] see [1], pp. 249 and 280 and
N. Schwarzer, "Science Riddles – Riddle No. 20: Second Law of Thermodynamics – Where is its Fundamental Origin?", self-published, Amazon Digital Services, 2019, Kindle, ASIN: B07Y79BTT9

[52] N. Schwarzer, "Science Riddles – Riddle No. 21: Evolution – Where is its Fundamental Origin?", self-published, Amazon Digital Services, 2019, Kindle, ASIN: B07YKL37DL

[53] N. Schwarzer, "Quantum Gravity Thermodynamics II – Derivation of the Second Law of Thermodynamics and the Metric Driving Force of Evolution", self-published, Amazon Digital Services, 2019, Kindle, ASIN: B07XWPXF3G

[54] N. Schwarzer, "Brief Proof of Hilbert's World Formula – Dirac, Klein-Gordon, Schrödinger, Einstein, Evolution and the 2nd Law of Thermodynamics all from one origin", self-published, Amazon Digital Services, 2020, Kindle, ASIN: B08585TRB8

[55] N. Schwarzer, "Virus – How Did Evolution Invent this Thing and How Could It Be Made a Weapon Against Humanity?", to be published, Amazon Digital Services, Kindle

[56] N. Schwarzer, "Mathematical Psychology – The World of Thoughts as a Quantum Space-Time with a Gravitational Core", Jenny Stanford Publishing, ISBN: 9789815129274

[57] N. Schwarzer, "Why the 4 Dimensions? – Can Quantum Gravity Explain this Peculiar Fact?", self-published, Amazon Digital Services, 2023, Kindle, ASIN: B0C5TKF9MB

[58] Peter W. Higgs, "Broken Symmetries and the Masses of Gauge Bosons", Phys. Rev. Lett. 13, 508, Published 19 October 1964

[59] https://en.wikipedia.org/wiki/Higgs_boson

[60] https://en.wikipedia.org/wiki/Flatness_problem

[61] N. Schwarzer, "Einstein had it, but he did not see it – Part LXXVI: Quantum Universes – We don't need no… an Inflation", self-published, Amazon Digital Services, 2019, Kindle, ASIN: B07NLH3JJV

[62] N. Schwarzer, "Einstein had it, but he did not see it – Part LXXVII: Matter is Nothing and so Nothing Matters", self-published, Amazon Digital Services, 2019, Kindle, ASIN: B07NQKKC31

[63] T. Bodan, N. Schwarzer, "The Photon – Connector of Space and Time: The "Simplest" Particle in a Theory of Everything", self-published, Amazon Digital Services, 2016, Kindle, ASIN: B06XGC4NDM

[64] A. E. Green, W. Zerna, "Theoretical Elasticity", London: Oxford University Press, 1968

[65] H. Neuber, "Kerbspannungslehre", in German, 3rd edition, Springer-Verlag, Berlin, Heidelberg, New York, Tokyo, 1985, ISBN: 3-540-13558-8

[66] N. Schwarzer, "Science Riddles – Riddle No. 20: Second Law of Thermodynamics – Where is its Fundamental Origin?", self-published, Amazon Digital Services, 2019, Kindle, ASIN: B07Y79BTT9

[67] N. Schwarzer, "3rd Epistle to Elementary Particle Physicists – Beyond the Standard Model – Metric Solutions for Neutrino, Electron, Quark", self-published, Amazon Digital Services, 2019, Kindle, ASIN: B07XJJ535T

[68] N. Schwarzer, "Einstein had it, but he did not see it – Part LXXXV: In Conclusion", self-published, Amazon Digital Services, 2019, Kindle, ASIN: B07Y37LNRW

[69] P. W. Atkins, "Physical Chemistry", Oxford University Press, 4th edition 1990, ISBN: 0-19-855283-1

[70] N. Schwarzer: "How Systems Evolve – A Red Pill Course on Evolution", self-published, Amazon Digital Services, 2022, Kindle, ASIN: B0B6RGBKQ4
See also:
N. Schwarzer, "The Mathematical Darwin – How Quantum Gravity Explains The Miracle of Evolution", upcoming book-project (fig. 4)
Synopsis: Since its first publication in 1859 "On the Origin of Species by Means of Natural Selection, or The Preservation of Favoured Races in the Struggle for Life" was never out of debate (and sale). Now we are ready for a "remake" because we have the math that proves Darwin – almost completely – right. We can show that evolution is a fundamental driving force of and within the universe and that it is just the other side of the coin which also holds the second law of thermodynamics. We will show that evolution reveals itself as a part of a comprehensive Quantum Gravity Theory in which Einstein's General Theory and Quantum Theory are unified. Many examples will be considered and special attention will be put on the co-evolution of man and machine.

[71] C. L. Bennett et al, "Nine-Year Wilkinson Microwave Anisotropy Probe (WMAP) Observations: Final Maps and Results", 2013, arxiv.org/pdf/1212.5225.pdf

[72] https://en.wikipedia.org/wiki/Accelerating_expansion_of_the_universe

[73] W. Laskowski, W. Pohlit, "Biophysik", Vol. 1 (in German), dtv Wissenschaftliche Reihe, 1974, ISBN: 342304229X

[74] S. Noiré (Co-authoring N. Schwarzer, T. Bodan), "The Womanizer and his Image of Diseases – About love fields and isolation traps – a victim reveals", illustrated version, self-published, BoD Classic, ISBN: 9783756841370

[75] https://demonstrations.wolfram.com/DiracMatricesInHigherDimensions/

[76] A. Bejan, "The Physics of Life: The Evolution of Everything", St. Martin's Press, 2016, ISBN-10: 1250078822

[77] W. Pauli, "Über den Zusammenhang des Abschlusses der Elektronengruppen im Atom mit der Komplexstruktur der Spektren", 1925, Zeitschrift für Physik 31: 765-783, Bibcode:1925ZPhy...31..765P. doi:10.1007/BF02980631

[78] K. Schwarzschild, "Über das Gravitationsfeld einer Kugel aus inkompressibler Flüssigkeit nach der Einsteinschen Theorie" ["On the gravitational field of a ball of incompressible fluid

following Einstein's theory"], 1916, Sitzungsberichte der Königlich-Preussischen Akademie der Wissenschaften (in German), Berlin: 424-434

[79] K. Schwarzschild, "On the gravitational field of a mass point according to Einstein's theory", (translation and foreword by S. Antoci and A. Loinger), arXiv:physics/9905030v1

[80] R. P. Kerr, "Gravitational field of a spinning mass as an example of algebraically special metrics", Phys. Rev. Lett., 1963, 11, 26.

[81] R. P. Kerr, "Gravitational collapse and rotation, Quasi-stellar Sources and Gravitational Collapse", including the Proceedings of the First Texas Symposium on Relativistic Astrophysics, Edited by I. Robinson, A. Schild and E.L. Schucking. The University of Chicago Press, Chicago and London, 1965, 99-109

[82] N. Schwarzer, "Quantum Gravity Psychology – Is there a Fundamental Theory of the Mind?", self-published, Amazon Digital Services, 2023, Kindle, ASIN: B0C5TLCG3S

[83] N. Schwarzer, "Einstein had it, but he did not see it – Part XXXI: A Cosmologic Pairwise Entanglement of Dimensions and the Holographic Principle", self-published, Amazon Digital Services, 2018, Kindle, ASIN: B07B5GBQ4N

[84] N. Schwarzer, "Einstein had it… Part XXXV: The 2-Body Problem and the GTR-Origin of Mass", self-published, Amazon Digital Services, 2018, Kindle, ASIN: B07CWP3V1S

[85] N. Schwarzer, T. Bodan, "Sherlock, Watson, Einstein – Part 1: The Mystery of Entanglement and the Spooky Action at a Distance", self-published, Amazon Digital Services, 2018, Kindle, ASIN: B079Z92GGM

[86] N. Schwarzer, T. Bodan, "Sherlock, Watson, Stalin Part 2: The Hell of Gender, Merkel, Communism and the Dictatorship of Parasites", self-published, Amazon Digital Services, Kindle, ASIN: B07BJ9PZWM

[87] N. Schwarzer, "Einstein had it… Part XXXIII: Elementary Particle Universes (without spin and shear components)", self-published, Amazon Digital Services, 2018, Kindle, ASIN: B07CGFHBC2

[88] N. Schwarzer, "The Einstein Quantum Computer – Mathematical Principle and Transition to the Classical Discrete and Quantum Computer Design", self-published, Amazon Digital Services, 2018, Kindle, ASIN: B07D9J5VLV

[89] N. Schwarzer, "Einstein had it… Part XXXVI: The Classical and Principle Misinterpretation of the Einstein-Field-Equations AND How it Might be Done Correctly", self-published, Amazon Digital Services, 2018, Kindle, ASIN: B07D68G9M9

[90] N. Schwarzer, "Einstein had it, but he did not see it – Part XXXIX: EQ or The Einstein Quantum Computer", self-published, Amazon Digital Services, 2018, Kindle, ASIN: B07D9MBRS3

[91] M. M. Vopsen, "The information catastrophe", AIP Advances 10, 085014, 2020, https://doi.org/10.1063/5.0019941

[92] N. Schwarzer, "Is there an ultimate, truly fundamental and universal Computer Machine?", Self-published, Amazon Digital Services, 2019, Kindle, ASIN: B07V52RB2F

[93] N. Schwarzer, "Einstein had it, but he did not see it – Part LIII: They Are Everywhere! Why there are so Many Sigmoid-Dependencies in this World", self-published, Amazon Digital Services, 2018, Kindle, ASIN: B07G8F9X6T

[94] J. Schwarzer, "On the co-evolution of man and machine Or Elon Musk's neuralink, Aristotle's ethic, Hamilton, Hilbert, Einstein and the fate of Man – A 16 year-old's school project", self-published, Amazon Digital Services, 2023, Kindle, ASIN: B0BSDMZ97S

[95] N. Schwarzer, "The Quantum Gravity Well – From Particle Physics to Psychology – a True Theory of Everything?", self-published, Amazon Digital Services, 2023, Kindle, ASIN: B0C3QV9S1Y

[96] W. Greiner, "Quantenmechanik – Teil 1 Einführung", vol. 4 of "Theoretische Physik", 1992, 5th edition, in German, Verlag Harri Deutsch, Frankfurt am Main, ISBN: 3-8171-1206-8

[97] F. Schwabl, "Quantenmechanik",1990, Springer-Verlag Berlin Heidelberg New York, 2nd edition, in German, ISBN: 3-540-52359-6

[98] N. Schwarzer, "Understanding Quantum Tunneling – How can Quantum Gravity Help us to Get the Gist about this Strange Phenomenon", Amazon Digital Services, 2023, Kindle, ASIN: B0C241FQH3

[99] N. Schwarzer, "The Death of Schrödinger's Cat – How Quantum Gravity Solves the Problem of the Wave Function Collapse", Amazon Digital Services, 2023, Kindle, ASIN: B0C2444JZX

[100] M. D. Pollock, "ON THE DIRAC EQUATION IN CURVED SPACE-TIME", ACTA PHYSICA POLONICA B, Vo. 41, 2010, No. 8, 1827-1846 and: https://en.wikipedia.org/wiki/Dirac_equation_in_curved_spacetime

[101] H. Goenner, "Einführung in die spezielle und allgemeine Relativitätstheorie" (in German), Spektrum Akad. Verlag, 1996, Heidelberg, Berlin, Oxford, ISBN: 3-86025-333-6

[102] J. Schwarzer, "3 Generations of Consciousness", YouTube, 2022, https://youtu.be/0AlFzIqwxN8

[103] https://en.wikipedia.org/wiki/Generation_(particle_physics)

[104] J. D. Bekenstein, "Black holes and entropy", 1973, Phys. Rev. D 7:2333-2346

[105] J. D. Bekenstein, "Information in the Holographic Universe", Scientific American, Volume 289, Number 2, August 2003, p. 61

[106] N. Schwarzer, "The Wave Particle Dualism – How Does Quantum Gravity Explain this Apparent Peculiarity", self-published, Amazon Digital Services, 2023, Kindle, ASIN: B0C5TMS8W5

[107] N. Schwarzer, T. Chudoba, F. Richter, "Investigation of ultra-thin coatings using Nanoindentation", Surface and Coatings Technology, 2006, Vol 200/18-19 pp 5566 - 5580, online at doi: http://dx.doi.org/10.1016/j.surfcoat.2005.07.075

[108] N. Schwarzer, "FilmDoctor", software package, www.siomec.com; e.g., see:
Fusion Reactor Optimization: https://youtu.be/eBWfikGxhuQ
How to get the depth profile of a complex coating system – Part I: https://youtu.be/EmXfSFiwdVE

How to get the depth profile of a complex coating system – Part II: https://youtu.be/sn82LOGMOao

[109] F. Richter, T. Chudoba, N. Schwarzer, G. Hecht, “Neue Möglichkeiten zur Charakterisierung dünner Schichten mit Indentermethoden – Novel Possibilities for Thin Film Characterisation Using Indentation Methods”, Materialwissenschaften und Werkstofftechnik 32, 2001, pp. 621-627

[110] O. Wändstrand, N. Schwarzer, T. Chudoba, Å. Kassman-Rudiphi, “Load-carrying capacitity of Ni-plated media in spherical indentation: experimental and theoretical results”, Surface Engineering Vol. 18, 2002, No. 2, pp. 98-104

[111] N. Schwarzer, I. Hermann, T. Chudoba, F. Richter, “Contact Modelling in the Vicinity of an Edge”, Surface and Coatings Technology 146-147, 2001, pp. 371-377

[112] T. Chudoba, N. Schwarzer, F. Richter, “Steps towards a mechanical modeling of layered systems”, Surface and Coatings Technology 154, 2002, pp. 140-151

[113] N. Schwarzer, “Experimental and analytical modelling of mechanical contact problems of Coating Systems”, proceedings of the Philips-NIMR Technology roadmap workshop, Nijenrode University, 6-7 June 2002, published in “Towards a Detailed NIR-Roadmap on Wear Resistant/Low Friction Coatings” from A.J.W.A. Vermeulen, Philips Centre for Industrial Technology, CTB598-02-3099 2002-07-30, Juli 2002

[114] N. Schwarzer, “About the theory of thin coated plates”, published at: http://nbn-resolving.de/urn:nbn:de:bsz:ch1-200200050

[115] V. Linss, I. Hermann, N. Schwarzer, U. Kreissig, F. Richter, “Mechanical properties of thin films in the ternary triangle B-C-N”, Surf. Coat. Technol. 163-164, 2003, pp. 220-226 (ISSN 0257-8972)

[116] N. Schwarzer, “Modelling of the mechanics of thin films using analytical linear elastic approaches”, Habilitationsschrift der TU-Chemnitz 2004, FB Physik Fester Körper, http://archiv.tu-chemnitz.de/pub/2004/0077

[117] V. Linss, N. Schwarzer, T. Chudoba, M. Karniychuk, F. Richter, “Mechanical Properties of a Graded BCN Sputtered Coating with VaryingYoung's Modulus: Deposition, Theoretical Modelling and Nanoindentation”, Surf. Coat. Technol., 195, 2005, pp. 287-297, https://doi.org/10.1016/j.surfcoat.2004.06.010

[118] T. Chudoba, N. Schwarzer, V. Linss, F. Richter, “Determination of Mechanical Properties of Graded Coatings using Nanoindentation”, proceedings of the ICMCTF 2004 in San Diego, California, USA, also in Thin Solid Films, 2004, 469-470C, pp. 239-247

[119] N. Schwarzer, “Elastic Surface Deformation due to Indenters with Arbitrary symmetry of revolution”, J. Phys. D: Appl. Phys., 37, 2004, pp. 2761-2772

[120] N. Schwarzer, G. M. Pharr, “On the evaluation of stresses during nanoindentation with sharp indenters”, proceedings of the ICMCTF 2004 in San Diego, California, USA, also in Thin Solid Films, Vol. 469-470C, 2004, pp. 194-200

[121] N. Schwarzer, T. Chudoba, G. M. Pharr, “On the evaluation of stresses for coated materials during nanoindentation with sharp indenters”, Surf. Coat. Technol, Vol 200/14-15, 2006, pp. 4220-4226, doi:10.1016/j.surfcoat.2005.01.011

[122] R. Puschmann, N. Schwarzer, F. Richter, S. Frühauf, S. E. Schulz, "An applicable concept for the indentation of thin porous films", proceedings of the NanoMech 5, 7-9 September 2004 in Hückelhoven, Germany, also in Z. Metallkd. 96, 2005, 11, 1-6

[123] N. Schwarzer, F. Richter, "On the determination of film stress from substrate bending: Stoney's formula and its limits", online archive of the Technical University of Chemnitz: http://nbn-resolving.de/urn:nbn:de:swb:ch1-200600111

[124] N. Schwarzer, "Determining intrinsic stresses in layered materials via nanoindentation – the question of in principle feasibility", online archive of the Technical University of Chemnitz: http://archiv.tu-chemnitz.de/pub/2006/0018 (the publication contains also a prototype of the software FilmDoctor for the modelling of contact loads for layered materials for half spaces consisting of up to 11 layers)

[125] N. Schwarzer, P. Heuer-Schwarzer, "Qualitative failure analysis on laminate structures of windsurfing boards using analytical linear elastic modelling", online archive of the Technical University of Chemnitz: http://archiv.tu-chemnitz.de/pub/2006/0010

[126] N. Schwarzer, "Analysing Nanoindenation Unloading Curves using Pharr's Concept of the Effective Indenter Shape", proceedings of the ICMCTF 2005 in San Diego, California, USA, also in Thin Solid Films 494, 2006, pp. 168-172

[127] N. Schwarzer, "The extended Hertzian theory and its uses in analysing indentation experiments", Phil. Mag. 86(33-35) 21 Nov-11 Dec 2006, pp. 5153-5767, Special Issue: "Instrumented Indentation Testing in Materials Research and Development"

[128] N. Schwarzer, "Intrinsic stresses – Their influence on the yield strength and their measurement via nanoindentation", online archives of the Saxonian Institute of Surface Mechanics www.siomec.de/pub/2007/001

[129] N. Schwarzer, P. Heuer, "Didactically optimized training tools for mechanical thin film design", online archives of the Saxonian Institute of Surface Mechanics www.siomec.de/pub/2007/002

[130] N. Schwarzer, P. Heuer, "Failure analysis on laminate structures of windsurfing boards using thin film modelling techniques", Poster, online archives of the Saxonian Institute of Surface Mechanics www.siomec.de/pub/2007/003

[131] N. Schwarzer, P. Heuer, "Failure analysis on laminate structures of windsurfing boards using thin film modelling techniques", online archives of the Saxonian Institute of Surface Mechanics www.siomec.de/pub/2007/004

[132] N. Schwarzer, P. Heuer, "Qualitative failure analysis on laminate structures of windsurfing boards using analytical linear elastic modelling", online archives of the Saxonian Institute of Surface Mechanics www.siomec.de/pub/2007/005

[133] D. Heuer, T. Chudoba, N. Schwarzer, P. Heuer-Schwarzer, "Nanoindentation in teeth: the influence of experimental conditions on local mechanical properties", Poster, online archives of the Saxonian Institute of Surface Mechanics www.siomec.de/pub/2007/006

[134] N. Schwarzer, "Modelling of Contact Problems of Rough Surfaces", online archives of the Saxonian Institute of Surface Mechanics www.siomec.de/pub/2007/007

[135] D. Heuer, T. Chudoba, N. Schwarzer, "Contact Mechanics in Dentistry: A systematic investigation of modern composite materials used for fillings", online archives of the Saxonian Institute of Surface Mechanics www.siomec.de/pub/2007/008

[136] N. Schwarzer, "An Extension of the Oliver and Pharr Method to Ultra-Thin Structures, Coatings, Functionally Graded Coatings and Multilayer Systems", online archives of the Saxonian Institute of Surface Mechanics www.siomec.de/pub/2007/010

[137] N. Schwarzer, "Modelling of Mechanical Loads on Thin Film Solar Cell Structures", online archives of the Saxonian Institute of Surface Mechanics www.siomec.de/pub/2007/011

[138] N. Schwarzer, "Modelling of Mechanical Loads on Multi-Layer Varnishes", online archives of the Saxonian Institute of Surface Mechanics www.siomec.de/pub/2007/012

[139] N. Schwarzer, "Analysing system for the investigation of ultra-thin coatings – determination of their Young's modulus, hardness and yield strength using the method of the 'Effective Indenter'", online-documents of the Saxonian Institute of Surface Mechanics, www.siomec.de/doc/2007/001

[140] N. Schwarzer, "How to measure intrinsic stresses via nanoindentation – an example", online-documents of the Saxonian Institute of Surface Mechanics, www.siomec.de/doc/2007/002

[141] N. Schwarzer, "About the understanding of loading and unloading curves in nanoindentation", online-documents of the Saxonian Institute of Surface Mechanics, www.siomec.de/doc/2007/003

[142] N. Schwarzer, "Effect of lateral displacement on the surface stress distribution for cone and sphere contact", Phil. Mag. 86(33-35) 21 Nov-11 Dec 2006, pp. 5231-5237, Special Issue: "Instrumented Indentation Testing in Materials Research and Development"

[143] James G. Kohl, Irwin L. Singer, Norbert Schwarzer, and Victor Y. Yu, "Effect of Bond Coat Modulus on the Durability of Silicone Duplex Coatings", Progress in Organic Coatings, 56, 2006, pp. 220-226

[144] N. Schwarzer, "Modelling of Contact Problems of Rough Surfaces", online archive of the Technical University of Chemnitz: http://archiv.tu-chemnitz.de/pub/2006/0016

[145] N. Schwarzer, "The extended Hertzian Approach for lateral loading", online archive of the Technical University of Chemnitz: http://archiv.tu-chemnitz.de/pub/2006/0015

[146] N. Schwarzer, L. Geidel, "Automated Analysing of Thin Film Nanoindentation Data Using the Concept of the Effectively Shaped Indenter", proceedings of the ICMCTF 1-5 May 2006 in San Diego, California, USA, also in Surface and Coatings Technology, Volume 201, Issue 7, December 2006, pp. 4377-4383, doi:10.1016/j.surfcoat.2006.08.043.

[147] F. Richter, M. Herrmann, F. Molnar, T. Chudoba, N. Schwarzer, M. Keunecke, K. Bewilogua, X.W. Xiang, H.-G. Boyen, P. Ziemann, "On the evaluation of stresses in coated materials during nanoindentation with sharp indenters Substrate influence in Young's modulus determination of thin films by indentation methods: Cubic boron nitride as an example", Surf. Coat. Technol. 201, 2006, pp. 3577-3587

[149] M. Herrmann, N. Schwarzer, F. Richter, "Determination of Young's modulus and yield stress of porous low-k materials by nanoindentation", proceedings of the ICMCTF 1-5 May 2006

in San Diego, California, USA and in Surface & Coatings Technology 201, 2006, pp. 4305-4310

[150] N. Schwarzer, M. Fuchs, "Comprehensive analysis of thin film nanoindentation data via internetportal – A principal feasibility study", Thin Solid Films, 2006, Vol. 515, Issue 3, November 2006, pp. 1080-1086, doi: 10.1016/j.tsf.2006.07.165. (proceedings of the ICMCTF 1-5 May 2006 in San Diego, California, USA)

[151] N. Schwarzer, "Short note on the potential use of a rotating indenter with respect to the next generation of nanoindenters", Int. J. Surface Science and Engineering, Vol.1 2007 2/3, pp. 239-258

[152] N. Schwarzer, "An Extension of the Oliver and Pharr Method to Ultra-Thin Structures, Coatings, Functionally Graded Coatings and Multilayer Systems", online archives of the Saxonian Institute of Surface Mechanics www.siomec.de/pub/2007/010

[153] M. Herrmann, F. Richter, N. Schwarzer, "About the problem of properly designed FE-Models for mechanical contact problems on layered materials", proceedings of the ICMCTF 23-27 April 2007 in San Diego, California, USA

[154] Riaz Akhtar, Michael J Sherratt, Nick Bierwisch, Brian Derby, Paul M Mummery, Rachel, E.B Watson, and Norbert Schwarzer, "Nanoindentation of Histological Specimens using an Extension of the Oliver and Pharr Method", Volume 1097, Pages GG01-09 of the Materials Research Society Proceedings 2008

[155] R. Akhtar, N. Schwarzer, M.J. Sherratt, R.E.B. Watson, H. K. Graham, A. W. Trafford, P.M. Mummery and B. Derby, "Nanoindentation of Histological Specimens: Mapping the Elastic Properties of Soft Tissues", JMR Special Focus Issue on "Indentation Methods in Advanced Materials Research", J. Mater. Res., Vol. 24, No. 3, March 2009, pp. 638-646

[156] N. Schwarzer, "Some Basic Equations for the Next Generation of Surface Testers Solving the Problem of Pile-up, Sink-in and Making Area-Function-Calibration obsolete", JMR Special Focus Issue on "Indentation Methods in Advanced Materials Research", J. Mater. Res., Vol. 24, No. 3, March 2009, pp. 1032-1036

[157] M. Sebastiani, E. Bemporad, F. Carassiti and N. Schwarzer, "Residual stress measurement at the micrometer scale: Focused ion beam (FIB) milling and nanoindentation testing", Philosophical Magazine, 2010, pp. 1-16, iFirst, http://dx.doi.org/10.1080/14786431003800883

[158] A. Gies, N. Schwarzer, J. Becker, H. Rudigier, "Untersuchung der mechanischen Eigenschaften von DLC-Schichtsystemen mittels Nanoindentation und deren Modellierung", Tagungsband der ThGOT 2010, ISBN: 978-3-00-025424-6, pp. 104-109

[159] N. Schwarzer, "Multiaxial mechanical surface tests of the next generation and better modeling and analysis of the same", Tagungsband der ThGOT 2010, ISBN: 978-3-00-025424-6, pp. 77-90

[160] N. Schwarzer, Q.-H. Duong, N. Bierwisch, G. Favaro, M. Fuchs, P. Kempe, B. Widrig, J. Ramm, "Optimization of the Scratch Test for Specific Coating Designs", Surface and Coatings Technology, volume 206, issue 6, year 2011, pp. 1327-1335

[161] A. C. Fischer-Cripps, St. J. Bull, N. Schwarzer, "Critical review of Claims for Ultra-Hardness in Nanocomposite Coatings", Philosophical Magazine, Volume 92, Issue 13, 2012, pp. 1601-1630

[162] N. Schwarzer, "Short note on the effect of pressure induced increase of Young's modulus", Philosophical Magazine, accepted January 2012, Vol. 92, issue 13, pp. 1631-1648

[163] J.G. Kohl, N.X. Randall, N. Schwarzer, T.T. Ngo, J.M. Shockley, and R.P. Nair, "An Investigation of Scratch Testing of Silicone Elastomer Coatings with a Thickness Gradient", Journal of Applied Polymer Science, Vol. 124, 2012, pp. 2978-2986

[164] N. Schwarzer, "Analyse und Simulation der mechanischen Eigenschaften beschichteter Polymere unter Berücksichtigung der meist zeitabhängigen Materialparameter", 19. NDVaK, 19. u. 20.10. 2011 Dresden, Germany, ISBN: 978-3-9812550-3-4, pp. 96-101

[165] N. Schwarzer, "Completely Analytical Tools for the Next Generation of Surface and Coating Optimization", Coatings 2014, 4, pp. 263-291, doi:10.3390/coatings4020263

[166] James G. Kohl, Nick Bierwisch, Truc T. Ngo, Gregory Favaro, Eric Renget, Norbert Schwarzer, "Determining the viscoelastic behavior of polyester fiberglass composite by continuous micro-indentation and friction properties", Wear 350-351, 2016, pp. 63-67

[167] M. Zawischa, S. Makowski, N. Schwarzer, V. Weihnacht, "Scratch resistance of superhard carbon coatings – A new approach to failure and adhesion evaluation", Surface and Coatings Technology, Volume 308, 25 December 2016, pp. 341-348, https://doi.org/10.1016/j.surfcoat.2016.07.109

[168] James G. Kohl, Norbert Schwarzer, Truc T. Ngo, Gregory Favaro, Eric Rengnet and Nick Bierwisch, "Determining the viscoelastic properties obtained by depth sensing microindentation of epoxy and polyester thermosets using a new phenomenological method", Mater. Res. Express 2, 2015, 015301

[169] N. Schwarzer, "About Holistic Optimization – Examples From In- and Outside the World of Coatings", 2015, Proceedings of the 58th Annual Technical SVC Conference

[170] N. Schwarzer, "From Hertz via Higgs to a Paradox Failure Mechanism", The 2015 Fall issue of the SVC Bulletin, https://www.flipsnack.com/svcdigitalpublications/2015-fall-winter-bulletin.html , 46-49

[171] Simon Vogt, Thomas Greß, Franz Ferdinand Neumayer, Norbert Schwarzer, Adrian Harris, Wolfram Volk, "Method for highly spatially resolved determination of residual stress by using nanoindentation", Production Engineering, 2019, 13:133-138, https://doi.org/10.1007/s11740-018-0857-5

[172] Adam van Casteren, Peter W. Lucas, David S. Strait, Shaji Michael, Nick Bierwisch, Norbert Schwarzer, Khaled J. Al-Fadhalah, Abdulwahab S. Almusallam, Lidia A. Thai, Sreeja Saji, Ali Shekeban and Michael V. Swain, "Metallic proxies remain unsuitable for assessing the mechanics of microwear formation: reply to comment on van Casteren et al. (2018)", Royal Society Open Science (2019), https://doi.org/10.1098/rsos.190572

[173] N. Schwarzer, "From interatomic interaction potentials via Einstein field equation techniques to time dependent contact mechanics", Materials Research Express, 1, 1, IOP Publishing, 19-Mar-14, http://dx.doi.org/10.1088/2053-1591/1/1/015042

[174] Adam van Casteren, Peter W. Lucas, David S. Strait, Shaji Michael, Nick Bierwisch, Norbert Schwarzer, Khaled J. Al-Fadhalah, Abdulwahab S. Almusallam, Lidia A. Thai, Sreeja Saji, Ali Shekeban and Michael V. Swain, "Evidence that metallic proxies are unsuitable for assessing the mechanics of microwear formation and a new theory of the meaning of microwear", Royal Society Open Science, 2018, https://doi.org/10.1098/rsos.171699

[175] J. G. Kohl, N. Schwarzer, T. T. Ngo, G. Favaro, N. Bierwisch, "Relating viscoelastic properties obtained by depth sensing microindentation to scratch behavior of epoxy and polyester thermosets", Wear, submitted August 2012

[176] Florian Seibert, Max Döbeli, Doris M. Fopp-Spori, Kerstin Glaentz, Helmut Rudigier, Norbert Schwarzer, Beno Widrig, Jürgen Ramm, "Comparison of arc evaporated Mo-based coatings versus Cr1N1 and ta-C coatings by reciprocating wear test", Wear, February 2013, 298-299, Complete, pp. 14-22

[177] N. Schwarzer, "Endlessly Touchable – Structural Solutions for the Next Generation of Tribo-Protective Coatings", Micromaterials and Nanomaterials, July 2013

[178] Liskiewicz TW, Beake BD, Schwarzer N, Davies MI, "Short note on improved integration of mechanical testing in predictive wear models", Surface & Coatings Technology, 2013, Vol. 237, 212-218.

[179] N. Schwarzer, "About Stress Field, Time and Temperature Dependent Contact Mechanics for Thin Film Structures – Especially the Role of Intrinsic Stresses", Proceedings of TechCon 2014, accepted August 2014 & N. Schwarzer, "About Stress Field, Time and Temperature Dependent Contact Mechanics for Thin Film Structures – Especially the Role of Intrinsic Stresses", Proceedings of TechCon 2015, D. M. Mattox invited talk for SVC 2015, accepted August 2015

[180] N. Schwarzer, "SCALE INVARIANT MECHANICAL SURFACE OPTIMIZATION APPLYING ANALYTICAL TIME DEPENDENT CONTACT MECHANICS FOR LAYERED STRUCTURES", Proceedings of SMT-28, Tampere 2014, 287, Tampere University of Technology Tampere, Finland, June 16-18, 2014, http://www.gbv.de/dms/tib-ub-hannover/78861911x.pdf

[181] N. Schwarzer, "Scale invariant mechanical surface optimization applying analytical time dependent contact mechanics for layered structures", Chapter 22 in "Applied Nanoindentation in Advanced Materials", Atul Tiwari (Editor), Sridhar Natarajan (Co-Editor), ISBN: 978-1-119-08449-5, 2017, www.wiley.com/WileyCDA/WileyTitle/productCd-1119084490.html

[182] N. Schwarzer, "Epistle to Elementary Particle Physicists – A Chance You Might not Want to Miss", self-published, Amazon Digital Services, 2019, Kindle, ASIN: B07XDMLDQQ

[183] N. Schwarzer, "2nd Epistle to Elementary Particle Physicists – Brief Study about Gravity Field Solutions for Confined Particles", self-published, Amazon Digital Services, 2019, Kindle, ASIN: B07XN8GLXD

[184] S. Lifson, A. T. Haggler, and P. Dauber, 1979, J. Amer. Chem. Soc. 101, 5111

[185] R. Brooks, R. E. Bruccoleri, B. D. Olafson, D. J. States, S. Swaminathan, and M. Karplus, 1983, J. Comput. Chem. 4, 187

[186] W. F. van Gunsteren and H. J. C. Berendsen, Groningen Molecular Simulation (GROMOS) library manual, 1987

[187] C. Oostenbrink, A. Villa, A.E. Mark and W.F. van Gunsteren, "A biomolecular force field based on the free enthalpy of hydration and solvation: the GROMOS force-field parameter sets 53A5 and 53A6", Journal of Computational Chemistry 25, 2004, pp. 1656-1676

[188] M. Clark, R. D. Cramer III, and N. van Opdenbosch, 1989, J. Comput. Chem. 10, 982

[189] S. L. Mayo, B. D. Olafson, and W. A. Goddard III, 1990, J. Phys. Chem. 94, 8897

[190] V. S. Allured, C. M. Kelly, and C. R. Landis, 1991, J. Amer. Chem. Soc. 113, 1

[191] A. K. Rappe, C. J. Casewit, K. S. Colwell, W. A. Goddard III, and W. M. Skiff, 1992, J. Amer. Chem. Soc. 114, 10024

[192] G. Nemethy, K. D. Gibsen, K. A. Palmer, C. N. Yoon, G. Paterlini, A. Zagari, S. Rumsey, and H. A. Sheraga, 1992, J. Phys. Chem. 96, 6472

[193] W. D. Cornell, P. Cieplak, C. I. Bayly, I. R. Gould, K. M. Merz Jr., D. M. Ferguson, D. C. Spellmeyer, T. Fox, J. W. Caldwell, and P. A. Kollman, 1995, J. Amer. Chem. Soc. 117, 5179

[194] W. Damm, A. Frontera, J. Tirado-Rives, and W. L. Jorgensen, 1997, J. Comput. Chem. 18, 1955

[195] S. L. Mayo, B. D. Olafson, and W. A. Goddard, 1990, J. Phys. Chem. 94, 8897

[196] S. D. Morley, R. J. Abraham, I. S. Haworth, D. E. Jackson, M. R. Saunders, and J. G. Vinter, 1991, J. Comput. Aided Mol. Des. 5, 475

[197] Maple, J. R., Dinur, U., Hagler, A. T., "Derivation of Force Fields for Molecular Mechanics and Dynamics from ab initio Energy Surfaces", Proc. Natl. Acad. Sci. 1988, 85, 5350.

[198] Maple J. R., Hwang, M. J., Stockfisch, T. P., Dinur U., Waldman, M., Ewig, C. S., Hagler, A. T., "Derivation Of Class II Force Fields .1. Methodology and Quantum Force Field for the Alkyl Functional Group and Alkane Molecules", J. Comput. Chem. 1994, 15, 162.

[199] Peng, Z., Ewig, C. S., Hwang, M-J., Waldman, M., Hagler, A. T., "Derivation of Class II Force Fields, 4. Van der Waals Parameters of Alkali Metal Cations and Halide Anions", J. Phys. Chem. A 1997, 101, 7243.

[200] Nguyen Van Hung, Nguyen Cong Toan, Nguyen Bao Trung, Ngo Hoang Giang, "Calculation of Morse Potential for Diamond Crystals. Applification to Anharmonic Effective Potential", VNU Journal of Science, Mathematics - Physics 24, 2008, pp. 125-131

[201] M. Lie, N. X. Chen, "Möbius inversion transform for diamond-type materials and phonon dispersions", Physical Review B, 52, No. 2, 1995, pp. 997-1003

[202] C. Domb, M. S. Green, "Phase Transitions and Critical Phenomena", vol. 6, 1976, Academic Press, London, New York, San Francisco, ISBN: 0-12-220306-2

[203] N. Schwarzer, "The Covariance Principle is Dead, Long Live the Covariance Principle – How Quantum Gravity Kills a Cornerstone of Physics", self-published, Amazon Digital Services, 2023, Kindle, ASIN: B0C96L41LS

[204] N. Schwarzer, "Super-Brains via Neuron Entanglement", self-published, Amazon Digital Services, 2020, Kindle, ASIN: B085PWJ249

[205] N. Schwarzer, "Societons and Ecotons – The Photons of the Human Society – Control them and Rule the World", Part 1 of "Medical Socio-Economic Quantum Gravity", self-published, Amazon Digital Services, 2020, Kindle, ASIN: B0876CLT7C

[206] N. Schwarzer, "How can we measure the Size of a Thought", to be published, Amazon Digital Services, Kindle; short version 2019 on: www.worldformulaapps.com/wp-content/uploads/2019/12/What-is-the-Size-of-a-Thought.pdf

[207] J. Kruger, D. Dunning, "Unskilled and unaware of it. How difficulties in recognizing one's own incompetence lead to inflated self-assessments", Journal of Personality and Social Psychology, Vol. 77, No. 6, 1999, pp. 1121-1134

[208] D. Dunning, "Chapter five – The Dunning–Kruger Effect: On Being Ignorant of One's Own Ignorance", Advances in Experimental Social Psychology, Vol. 44, 2011, pp. 247-296, doi:10.1016/B978-0-12-385522-0.00005-6

[209] Planck, Max K., "Scientific Autobiography and Other Papers", New York: Philosophical library, 1950

[210] N. Schwarzer, "Science Riddles – Riddle No. 11: What is Mass?", self-published, Amazon Digital Services, 2021, Kindle, ASIN: B07SSF1DFP

[211] N. Schwarzer, "THE PSA-PAPER – The Fundamental Theory", Feb. 2024, classified RASA strategy publication

[212] N. Schwarzer, "The Quantum Black Hole", Part 7b of "Medical Socio-Economic Quantum Gravity", self-published, Amazon Digital Services, 2021, Kindle, ASIN: B092MNP8DN

[213] N. Schwarzer, "Humanitons – The Intrinsic Photons of the Human Body – Understand them and Cure Yourself", Part 2 of "Medical Socio-Economic Quantum Gravity", self-published, Amazon Digital Services, 2020, Kindle, ASIN: B088QGK6ST

[214] Ch. W. Misner, K. S. Thorne, J. A. Wheeler, "Gravitation", W. H. Freeman and Company, New York 1997 (20th edition), ISBN: 0-7167-0344

[215] N. Schwarzer, "Hugh Everett's Multiverse – How Quantum Gravity Revives an old Idea and – Perhaps – Gives it a Second Life", self-published, Amazon Digital Services, 2023, Kindle, ASIN: B0CCQWC6YX

[216] M. Desmet, "The psychology of totalitarianism", Chelsea Green Publishing, 2022, ISBN: 978-1645021728

[217] M. Desmet, "Lacan's Logic of Subjectivity – A Walk on the Graph of Desire", QWL Press 2019, Ghent, ISBN: 9789089318695

[218] D. Martin, "The Fauci/COVID-19 Dossier", 205 pages, This document is prepared for humanity by Dr. David E. Martin, as pdf: www.davidmartin.world/wp-content/uploads/2021/01/The_Fauci_COVID-19_Dossier.pdf

[219] N. Schwarzer, "Einstein had it, but he did not see it – Part LXXIX: Dark Matter Options", self-published, Amazon Digital Services, 2019, Kindle, ASIN: B07PDMH2JB

[220] F. Schwarzer, J. Schwarzer, "How to get the depth profile of a complex coating system – an interview", YouTube, 2023, https://youtu.be/EmXfSFiwdVE

[221] F. Schwarzer, J. Schwarzer, "Scratch test analysis with the Calotte Module of FilmDoctor – an interview", YouTube, 2023, https://youtu.be/sn82LOGMOao

[222] Dr. David Martin and Julius Schwarzer, "Rejuvenating The World: A Discussion with Dr. David Martin & Julius Schwarzer", YouTube, 2024, https://youtu.be/aEh3_HTCzqA

23.1 Picture References

[A] Artwork by Livia Schwarzer

[B] Artwork by Filia Schwarzer

[C] Artwork by Laura Kube, 2023

[D] Artwork by Julius Schwarzer, 2012

24 The Motivator – With Math

Now we are going to repeat the story, but this time, all the necessary equations and a few references will be given.

In order to allow stand-alone readability for this section, we restart the numeration of the equations with (1) and avoid any cross-referencing with the rest of the book.

24.1 Introduction – Part Two

"Come on, Watson," Holmes urged me, "You must have heard about these gentlemen."

"Of course, I have heard about the first one, this Einstein," I said. "A pipe-smoking genius who likes to stick out his tongue the moment he sees a camera pointing at him. However, much to my regret I have never heard, saw, or read anything about the other two… what were their names again, Holmes?" I asked.

"Podolsky and Rosen," Holmes answered. "To be precise, it is Professor Podolsky and Professor Rosen. The three are famous because of two publications they wrote together many years ago [M1, M2]."

"When exactly?" I asked.

"It was in 1935 when the main paper, which is this one here [M1] was published," answered Holmes. "The other paper [M2] was about something you might already have heard of. Physicists call it 'wormholes' and mean connections between distant parts of space…"

"Like short cuts. I mean the ones the science fiction people are always using for their interstellar travels?" I asked.

"Precisely, Watson!" said Holmes.

"But when the story is so old, why does it matter today… why… ah…," I hesitated.

"Why I'd like to make it to one of my cases?" asked Holmes and gave the answer straight away, "Because the apparently ancient problem is still unsolved and many scientists of today keep busying themselves with it."

"Like who?"

"Like these, for instance," answered Holmes and pointed out a few more papers on his desk [M3, M4, M5].

I merely nodded and in order to fight my desire to ask the question nagging so forcefully inside me, I bent forward and took the cup of tea in front of me. But my friend only knew me too well. He took his pipe out of his mouth and gave me one of his rare smiles.

"My dear Watson," he said, "of course I could enlighten you myself about the mystery these three most respected gentlemen are confronted with, but as there is a much better way, I suggest we just sit and enjoy the rest of our tea, while we are waiting."

"Waiting for what?" asked I.

"For the three gentlemen to arrive, of course!" Holmes replied.

I looked startled. My mouth must have fallen open, but Holmes – as polite as he was – generously ignored this quite obvious lack of self-control and pulled out his pocket watch. This brought me back to my senses.

"You still have this old thing?" I asked.

Holmes did not answer immediately. He turned the watch in his hand before he finally said, "Old habits, Doctor Watson, just old habits." Then he held up his pipe. "It is the same with this thing here." He twisted his hand slowly, as if to intensely observe the object in his hand. After all, an object he should know so well as he has had it for so many years. "As you have suggested about 70 years or so ago, I had stopped smoking."

I nodded vigorously. "And right I was in making such a suggestion."

"Yes, of course, you were right, Watson!" said Holmes, "but still, I think as if something very important would be missing if I wasn't having it in my mouth or in my hand."

I thought about this for a while. Then I nodded.

"More tea, Holmes?" I asked.

"Yes please!" Holmes answered, "… and now you may also place the question bothering you so profoundly."

"Well," I started, "how can we expect a visitor who is known to be dead?"

"Yes, Watson, Einstein is dead and so are the other two…" He paused and looked at me quite calmly. I could not make out a trace of irony neither in his words, nor in their intonation, nor in his face. Then, after it had felt like an endless stretch of time, he finally added, "… and so are we, Watson."

I stared at him.

"Yes Watson, we are dead, too. Well, to be precise, we never actually were alive, but this does not mean that we did not exist, you see?"

I shrugged.

"All is relative, Watson! All is relative!" Holmes said. Then he added, "Besides Watson, Einstein is allowed to bring his pipe. He has it with him all the time, probably died with it… I don't know, but don't worry, he will not smoke in here."

"So, he is going to use it for show then, just like you?" I asked.

“Not for show, my dear Doctor. As I just said, old habits, nothing but habits. What would we be without them.”

Then the bell rang.

24.2 The Strangest Client There Could Be – Part Two

“So, you are not the clients yourself, but you are here on someone else’s behalf?” Still I could not comprehend it. It was the strangest thing I had ever heard and it already had taken us a while to get this far in the conversation.

Oh yes, the three men were quite eager to share information, but Holmes needed a while to make them all three understand that the information they were willing to give was not necessarily the one he and I needed to solve the problem or to help them solving it. At the beginning, it even seemed almost impossible to define the problem, even finding out who wanted that problem to be solved so very badly that he had organized this unusual gathering.

But let us move backwards in time a little bit.

It was obvious that the three men had no time to lose. They had barely entered the room when Einstein started to speak in a rather un-British briskly manner, “Only a very few moments of eternity were given to us, Mister Holmes,” Einstein said after we had formally greeted each other and the usual amenities were – almost – hastily exchanged. Einstein looked imploringly at my friend when he added, “Only a few moments to solve the very problem we have already failed to solve a long time ago.” He spoke all this rather quickly, with the pipe in his mouth and his hat still on.

Podolsky added, “We have not been able to do so while we were alive, but this time with you, Mister Holmes…”

“And what or who gave you the idea that I could be of assistance?” Holmes asked.

“You and Doctor Watson!” insisted Podolsky, who was the only one of the three visitors who had bothered to take off his hat.

“My apologies, gentlemen,” Holmes responded. “I meant to say ‘us’ of course.” He corrected himself and gave a pronounced nod in my direction, his eyes winking. I understood the message and now it was I who repeated his question:

“Still Professor Einstein, we would both like to know what gave you the idea that WE could be of assistance?”

“It wasn’t us,” said Einstein almost a bit impatiently, “but, please, can we leave it for later? I don’t think it matters here and I’d rather…”

“On the contrary, Professor Einstein,” Holmes interrupted him and Einstein looked startled, “it may matter very much… very much indeed.”

There was a pause and I asked again, “So, who or what was it then? Who made you think…”

“The universe itself did!” he said with the same kind of impatience he had shown before.

"The universe? The universe has talked to you? How?" asked Holmes, who was the first who recovered from this peculiar answer. But when Einstein merely shrugged, he gave the answer himself. "It is what you call a gut's feeling, Professor, is it not?"

Einstein nodded and Rosen muttered an almost silent "Yes!" Then he added, "We concluded this, because all three of us have the same 'gut's feeling', as you just characterized it. We also think…" But Holmes took his hand in the air and silenced him easily. His gesture was very soft, but such a power emanated from the man that even a simple thing like a raised hand could change the climate in a room completely and instantly.

"So, the universe doesn't know!" he said distinctly. Everybody in the room stared at the man in the chair, who was deep in thought. Nobody said a word. Then quite suddenly, Holmes stood up and walked to the window. He looked out into the street for a while and when he finely turned and faced us, he was almost beaming with excitement.

"Now we know who the client is, gentlemen," he said.

"Who?" I asked, but Holmes did not answer. Therefore, I faced the three professors and repeated my question in their direction, "So, you are not the clients yourself, but you are here on someone else's behalf?"

The three men nodded. Then we heard Holmes' voice. He had faced the window again and it was as if he spoke to himself. Nevertheless, his voice was loud and clear. It sounded through the room as if being amplified from every corner.

"The universe only knows what has been found out by the things it hosts. It is like a computer which only has a chance of knowing what one of his programs has evaluated. The universe is nothing but a huge computer and we are its programs [M6].

This most specific problem you have brought with you, my dear Professors, must be of great importance to the universe, because it went to quite some length to construct a simulation like this."

"Simulation?" Einstein asked.

"Well, you could also give it another name, my dear Professor. What was it again, the one technique you always applied to understand such strange things like a bending space-time, like riding on a photon, or other rather strange adventures like this?"

"Are you by any chance referring to a 'thought experiment', Mister Holmes?" suggested Einstein.

"Yes, Professor, we might as well assume that we are all just parts – essential parts I hope – of a little thought experiment. So, now we, Mister Watson and I, are ready to listen to your story, gentlemen."

24.3 The Spooky Action at a Distance – Part Two

"Originally," Rosen started, "we did this experiment…"

"This thought experiment!" rendered Einstein more precisely, thereby smiling.

"Yes, of course," Rosen added quickly, "at the time there was no chance… no chance to do such experiments in the real world. Thus, naturally, it only was a thought experiment. We constructed a certain quantum theoretical system with two objects in it…"

"Any particular objects?"

"No," Rosen said. "Just quantum objects, which is to say, small things which would follow the rules of Quantum Theory, that's the only condition."

"Almost!" hinted Einstein.

"Ah yes," Rosen agreed quickly again, "the two objects were assumed to be connected."

"Entangled!" Einstein said.

"You see, it is a very special form of connection. You simply assume that the two particles would have properties and that one of these properties kind of ties them together."

"Entangles them!" Einstein corrected. But Rosen seemed unperturbed by his interruptions, this obvious lack of education of his famous colleague. He took the hints as if they were coming from his own mind.

"All other properties are free to do as they please, only these specially connected…"

"Entangled!" said Einstein in the same almost indifferent tone as before. As much as he insisted on the correction as much one would have expected him to say it a bit sharper when presenting it the second time, but there was no change detected. And as before, Rosen took it as if it were just coming from another part of his brain.

"… properties are kind of unfree," he finished his sentence.

"Could you please specify this for us?" asked Holmes.

"Of course!" Rosen replied. "Simply imagine the two objects having nothing more than the two properties, we might like to name them black or white and positions in space for each of the objects."

"Space-time!" Einstein said.

"Pardon?" I asked.

"Space-time!" repeated Einstein, "it is positions in space and time and not just in space alone."

Rosen went on, "Quantum Theory allows us to formulate the system in such a way that we could make the black-and-white property coupled…"

"Entangled!" said Einstein.

"… while we leave the positions of the objects arbitrary."

"Which does mean what?" Holmes asked.

"It means that the quantum objects, let us imagine we have two particles, could move apart quite easily. Nothing prevents them from traveling away from each other… farther and farther until they may be lightyears apart from each other. Then something or someone measures the black-and-white property of one of the two particles. Let us assume the coupling…"

"The entanglement!" said Einstein.

"… would bind the black for one particle to a white of the other. Then in the very instant that you have seen the color-state of the one which was measured, you will immediately also know the color-state of the other… even though it is several lightyears away and the two particles have no way to communicate with each other."

"Well, what is so strange about that?" asked Holmes. "If I were taking coins out of a box, cut them into halves and were tossing them to my left and right, each piece in one direction and the other into the other one, then of course, if someone catches a coin half on my left, he would immediately know how its partner on the right does look like."

"Yes, Mister Holmes!" Podolsky said like somebody who had a lot of practice in making allowances to people who are less familiar with the funny laws of Quantum Theory. "But that is the macroscopic reality you are referring to. We, however, are talking about the quantum world. Here the particle does not have a defined black-or-white property until this very property is measured. You might put it like 'the particle has been black AND white until the very moment of its measurement'. It is like a fast-spinning coin and you have no way of knowing which side it will show until you will have caught it."

Holmes nodded. "I see!" he said softly.

"So, there is the same likelihood for the color white just as there also is for the color black to be measured at the first particle. There is also the same likelihood for the color white just as there is for the color black to be measured at the other particle… only, that…"

But suddenly Podolsky was interrupted by Holmes. His deep voice almost shook the room, "… only that the measurement should not have taken place with the first particle, before the second one is due with its own, otherwise the one measurement will already have fixed the result for the other, isn't it so, Professor?"

The three scientists looked stunned.

"How do you know?" they asked in unison.

Holmes almost laughed. "Well, anything else would be no surprise, nothing out of the ordinary. So, I simply picked the most mysterious constellation as it was the most logic under the circumstances. After all, this gathering would not be, if it wasn't for some very great mystery."

Einstein smiled. "Yes!" he said simply, "but we aren't there yet. The real mystery is still to come."

"What is it?" I asked.

"What do you think makes these particles behave so strangely, I mean, what makes them appear so rigidly bound together?"

"There must be a connection!" I said simply. "This entanglement, as you named it?"

"Ok," Einstein said, "remember, the two particles are lightyears apart from each other. Nevertheless, the very instant the black-white property has been determined for one of the two particles, the thing, I mean the color, is also fixed for the other one. So, I ask you, what kind of connection could this be? What would allow such information to travel several billions of miles in no time at all?"

I was flabbergasted. "Oh!" was all I could say. But then I recovered quickly and said, "However, this was all just a thought experiment, right? I mean, it was your thought experiment?"

"Yes, that is quite true," admitted Einstein, "we came up with this funny idea, because we wanted to show the world, which is to say the scientific community, that Quantum Theory cannot be complete. There had to be something wrong or missing. If strange stuff like this was the outcome, surely there was a snag."

"And did the scientific community understand and accept your objection?" Holmes asked. His voice was almost a bit amused, as if he already knew the answer… and in fact, I was sure that he did.

"That's hardly the point. The community well understood what our problem was. So, all fine there. However, a few years back, from today, I mean, some clever scientists found a way to realize what we considered the perfect thought experiment, simply because we thought it would never be possible to perform it in practice. Yes, they did the one thing we were sure would never be possible at all. Not possible in the real world, but to our great dismay they exactly found what we had predicted. They measured the strangest thing there is. We name it

The Spooky Action at a Distance."

24.4 But Where Is the Problem – Part Two

Everybody stared at Holmes. He had started to laugh. It was as if somebody had made a wonderful joke, but he was the only one who had gotten the gist.

"My apologies, gentlemen," Holmes said after a while, "but I could not help noticing that the very man who helped to create the paradox of this spooky action at a distance in Quantum Theory also is the one who has to be held responsible for making this a problem at all."

Everybody looked puzzled. Everybody, except Einstein, who understood, smiled, and said, "Yes, I got the point, Mister Holmes, but the constancy of the speed of light is not to be negotiated here. It is a solid and well-proven fact. Nothing can travel faster than light. This speed is about three hundred thousand kilometers per second and that is THE UNIVERSAL limit, end of story."

"And?" I threw in, in order to show that I simply could not see why there was a problem at all.

Einstein made an almost exasperated gesture, but Podolsky politely answered, "The thing is quite simple, Doctor Watson, if there is nothing faster than light, how can it be that the signal from one of the two particles…"

"… entangled particles…," added Einstein, his huge eyes rolling impatiently, but Podolsky took it just like Rosen had done and went on as if the add-on from Einstein, no matter how impolitely brought forward by the famous man, was just his own thought, "… reaches the other in an instant, with absolutely no time being elapsed? That is the riddle, Doctor Watson."

24.5 Finding the Starting Point – Part Two

"So," I said, "then we are here because there is an effect, which allows some spooky action to travel huge distances in no time…"

"… no time at all…" Einstein said, but I ignored him.

"… but there is also a law which does not allow velocities faster than the speed of light."

"Precisely!" cried Rosen excitedly.

"So that's just an antagonism," I said. "One theory cannot be correct or must at least be somehow incomplete, right?"

"That is what we thought, too," Einstein said. "We guessed that it is Quantum Theory, which must be wrong, because it is too fantastic anyway. But then there were the experiments and the spooky action at a distance was observed in the real world."

I smiled at the famous scientist. I simply had to smile, I could not help it, because I knew that I would play the role of the devil's advocate now and I savored the moment before finally saying what had to be said anyway, "Well then, the solution is quite simple, is it not?" I started. "It must be the other theory, which is wrong, am I correct?"

"Yes," said Einstein a little bit sharper than necessary. "However, even ignoring the fact that this very theory is from me – we call it General Theory of Relativity, by the way – this theory has also been proven in reality. So, the riddle is, how can there exist two theories of such rather obvious antagonism inside the same world?"

"And you have no idea where to start in your efforts of finding an answer to that question?" I asked.

"No!" Einstein said with determination and Rosen and Podolsky nodded.

"But isn't that obvious?" Holmes said suddenly.

Everybody looked at him, the three professors puzzled, me merely interested. I knew my friend too well to be surprised by his power of deduction. He could draw conclusions when others were unable to even see the pieces, not to speak about the connection between. In the light of the topic of the day, I would even suggest that his brain must have already made good use of the spooky action at a distance… taking the speed of his thoughts.

"You are here, gentlemen," Holmes said slowly. "I cannot see anybody from the quantum section. So, the starting point must have to do with you."

"Then we should start with our old thought experiment, the one thing they nowadays call the Einstein-Podolsky-Rosen or EPR paradox, right?" asked Rosen.

"No!" Holmes said sharply, "this is not what I had in mind. The fact that you are here does not automatically lead to the conclusion that your starting point from almost 90 years ago was correct. No, this only leads to the suggestion that you are able – in principle – to provide something which you must have forgotten to throw into the story 90 years ago. Now we need to find out what this missing piece could have been."

"But you just said that this is obvious?" Podolsky said and there was a tiny note of accusation in his voice.

"And so I still think it is," Holmes answered unperturbed.

"So why not just enlighten us then?" Einstein asked.

"Because you already know it yourself, don't you, Professor?" Holmes said. "You are just afraid of what might be the outcome, is it not so?"

Einstein inhaled deeply before he answered, "When doing the publication about this paradox, I was convinced that our doubts would settle the matter and that Quantum Theory in fact was incomplete. Then, a short while after my death, a certain Mister Bell [M7] found a way to design such an experiment and he also showed a mathematical way to prove that our own approach to address the problem was wrong… foolish even. The man succeeded, the experiments, which were performed later on [M8 – M13], were most successful and as said, we three looked like fools."

"Pardon me?" I asked. "But if the matter was decided, I mean with tests and everything, then what problem is there to be solved?"

Einstein shrugged.

"Because the universe was not satisfied with the answer Mister Bell and all his colleagues had given [M7 – M13]. The whole thing was still incomplete. So please, Professor Einstein, tell us everything!" Holmes urged.

Now, finally Einstein took off his hat. He drew a chair and sat down. Podolsky and Rosen followed his example.

"Everybody is of the opinion that I regretted the introduction of the cosmological constant as 'THE biggest blunder of my life'. In fact, I had made two such 'biggest blunders' and today I consider the second one even bigger than the first."

"You shouldn't be too hard on yourself," said Rosen, "after all, it was us who came up with the idea of the hidden parameters in the first place." Podolsky nodded vigorously in agreement.

Einstein made an impatient hand gesture.

"Not really… I was so over consumed with the idea that there must be a God who does not throw dice that I would have taken anything. I would have grasped at straws to save my deep beliefs about the world. When you came with the idea that there are hidden parameters, assuring the connection of the two particles within our EPR paradox, I was so relieved that I couldn't see the obvious flaw… simply because I did not want to see… did not want to even think about it."

"What flaw?" asked Holmes.

"A theory like Quantum Theory, which was able to perfectly describe the world around us, including the tiny electrons and their peculiar behavior, should have deserved some more credit. It was not right of us to throw it overboard only because we did not like what this theory told us about our God."

"Instead of making up paradoxes in order to prove these theories wrong," added Rosen, "we should have gone deeper into the matter, question everything, but most of all, we should have questioned our beliefs, because… after all… they are only just…"

"… beliefs," ended Podolsky the sentence for him. Then, in an almost excusing tone, he went on, "I would like to add that these beliefs are planted so deeply into ourselves, into our brains, even into our genetic code [M14, M15] that our desire for certain results, certain vectors in recognition may indeed be quite understandable, but nevertheless it is wrong and unforgivable to make them dogmatic. The greatest idiots in the world are always those with dogmata and ideologies instead of knowledge and a healthy desire for a deep understanding of the true nature of all things."

They were all quiet for a while.

"So, what is the current state of the matter?" I finally asked.

Einstein took a deep breath, "Quantum theoreticians simply say that the two particles are not separated. No matter how far they are apart from each other, it is one system. If, however, it is one system, then the black-white-parameter is well-defined inside the system and as long as the system stays intact, the parameter may have oscillated as much as it likes, but it is always oscillating according to the boundary conditions of the whole system. Thus, if the boundary conditions demand the second particle to be black if the first was white, then this will be so all the time."

"All the time the system stays intact!" Rosen added softly.

"And what could destroy this state of intactness?" I asked.

"A measurement, for instance," said Podolsky.

There was silence for a while. Then a question occurred to me and I asked, "You do not like the explanation of the quantum guys, do you?"

"No!" said Rosen flatly and Holmes elaborated on his behalf, "You do not like it, because it is against the common sense to even consider something like a non-local[30] black-and-white parameter, right?"

The three professors nodded and Holmes went on, "You have fought so hard to establish the fact that nothing can travel faster than the speed of light and that everything is relative and well-determined, which also includes that it has a certain position, that you almost detest the idea about a system, being spread over lightyears distances and still being just one system, one entity. Almost, … like being one cosmos."

They nodded again and Holmes went on, "You consider the fact unbearable that something like position loses its meaning, that locality does not matter with a thing like the black-and-white parameter."

"Yes!" said Einstein. Suddenly he got up and started pacing the room. "And this has nothing to do with my God," he said. "It has to do with the fact that I cannot imagine such a thing anymore. I cannot construct a proper association, I cannot do a…," he made a pause and I suggested, "… a thought experiment, Professor?"

Einstein nodded.

---

[30] Here the reader is reminded of our section about the destruction of the particle idea (see "There are no Particles"). There, it is been shown that the moment something, which could be anything, should be localized, it automatically becomes a wave. The particle effect is just the superpositions of many waves to a compactified form. In case this something is two "somethings" which "happens" at two different positions, we still have just one wave-structure, but with two compactified areas.

"Professor Einstein," Holmes started, "what kind of solution would console you?"

Einstein thought for a while and I assumed that he was seeking an answer, but he was already far beyond that point.

"It is very interesting that you think that an answer consoling us would also be the one satisfying the universe, Mister Holmes."

"Well, this is obvious, is it not?" Holmes replied.

"Yes indeed!" said Einstein, "because otherwise there wouldn't be much reason for us...," he pointed towards his colleagues, "... to be here at all."

"Precisely!" Holmes cried out, delightedly.

"But there is only one thing which would make me feel more comfortable with the EPR paradox, and that is impossible."

"Nevertheless, we are here and even more strangely, we are here together." Holmes made a waving movement with his right hand. "That fact alone should tell us that something or someone must be of the opinion that there is a chance."

But Einstein was skeptical. "The only thing that could convince me would be to show that the funny non-localized-parameter stuff would also reside in my own theory."

Now Rosen cut in and said, "This, however, requires the unification of Quantum Theory and the General Theory of Relativity. So, we are talking about a task that whole generations of scientists have failed to achieve."

"But why not try and then see how far we will get?" I suggested and I felt Holmes' gaze riveted on me. The warm smile of my dear friend was filling the room and his eyes were full of confidence. But there also was something else in his features. At first, I could not quite place it, but then I recognized it. Holmes was amused.

24.6 Approaching the Problem – Part Two

"Tell us about the other 'biggest blunder', Professor Einstein," Holmes said, and the addressed cleared his throat at once. Obviously, this man had no problem admitting that he had made mistakes.

"Well, Mister Holmes, even though I have no idea how this could help us, that particular story is quite simple. After I had established my field equations [M16], it was discovered that they, I mean the equations, give solutions to the whole universe. To my great surprise, however, the solutions demanded the universe to move... or evolve. I had always considered the universe as something static, something just being... existing. Now my own theory, my General Theory of Relativity suggested the universe to be most dynamic. This could not be. I could not accept this and so, I was looking for a way out and this is a way I found. I discovered that my equations were still correct if I added a constant term. I named this constant the 'cosmological constant' and gave it the symbol Λ, which is the big L from the Greek alphabet." Einstein made a short pause. Apparently, he was waiting for a remark, but nobody made any comment.

"Then Hubble, a famous astronomer in the United States, discovered the so-called red shift in the light of distant stars [M17], which means he discovered that galaxies are moving away from us. The greater the distance to the moving galaxies, the greater there is the velocity with which these galaxies flee. It became immediately clear to me that this could only mean that the universe in fact was a dynamic entity and so I removed the cosmological constant. Which is to say I erased the one term I had just added from my equations and announced its introduction as the 'biggest blunder' of my life."

"And do you still think that it was a blunder?" Holmes asked.

"That depends…" Einstein said. Then he chuckled and said, "It is kind of relative, you know?"

"I see!" Holmes said. "Please, Professor Einstein. Would you be so kind and show us the full equation, I mean the one with this 'biggest blunder' in it?"

Einstein obliged. He stood up, strolled to the blackboard right next to the window and started to draw his famous equation[31].

$$R^{\alpha\beta} - \frac{1}{2} R g^{\alpha\beta} + \Lambda g^{\alpha\beta} = -\kappa T^{\alpha\beta} \tag{1}$$

"And now," Holmes demanded, "please, tell us what the various terms mean, Professor!"

"Certainly!" Einstein replied and pointed to the left-hand side of his equation. "This here is all just about curvature. It tells us how a certain space-time warps in order to find a stable state… a minimum in action, we physicists say."

"So, this whole thing comes out of a condition?" Holmes asked.

"Yes, of course," answered Einstein as if this was just obvious as anything, "the whole is nothing but the result of demanding that a space of a certain number of dimensions can only exist in states of minima."

"You mean, like finding the deepest point in a valley or even in a chain of mountains?" I asked.

"Precisely!" Einstein said.

"Interesting!" said Holmes, "and was it ever shown in a rigorous manner that your equations result from such a minimum principle?"

"Yes, of course!" Einstein said promptly. "A great mathematician of my time, David Hilbert, had shown this quite clearly [M18]. The only condition he needed was to demand a minimum for the curvature of the space. The curvature is this term here, this R. We call it the Ricci scalar. Then, when doing this minimum-evaluation-thing with the Ricci scalar, my equations come out automatically. Here is the essential evaluation as Hilbert has done it."[32] Einstein turned to the blackboard and started to evaluate. "This here is the so-called Hilbert or Einstein-Hilbert action:

---

[31] Here we have: $R^{\alpha\beta}$, $T^{\alpha\beta}$ the Ricci- and the energy momentum tensor, respectively, while the parameters Λ and κ are constants (usually called cosmological and coupling constant, respectively). These are the well-known Einstein field equations in n dimensions with the indices α and β running from 1 to n. The theory behind it is called "General Theory of Relativity".

[32] Please note that Einstein's physics is not the only possible outcome from Hilbert's variational task. This was shown in this book in the subsection "The Unconventional Physics Realm".

$$\delta_g W = 0 = \delta_g \int_V d^n x \left(\sqrt{-g} R\right). \tag{2}$$

In this equation g denotes the determinant of the metric tensor, W and V are giving the action and the volume of the n-dimensional space, respectively, R is denoting the scalar curvature or Ricci scalar. The evaluation under the integral goes like this:

$$\begin{aligned} \delta_g W &= 0 = \delta_g \int_V d^n x \sqrt{-g} R = \delta_g \int_V d^n x \sqrt{-g} g^{\alpha\beta} R_{\alpha\beta} \\ &= \int_V d^n x \delta_g \left(\sqrt{-g} g^{\alpha\beta} R_{\alpha\beta}\right) = \int_V d^n x \left[\delta_g \left(\sqrt{-g} g^{\alpha\beta}\right) R_{\alpha\beta} + \sqrt{-g} g^{\alpha\beta} \delta_g \left(R_{\alpha\beta}\right)\right] \\ &= \int_V d^n x \sqrt{-g} \left[-G^{\kappa\lambda} \delta_g g_{\kappa\lambda} + g^{\alpha\beta} \delta_g R_{\alpha\beta}\right] \end{aligned} \tag{3}$$

with $G^{\alpha\beta} = R^{\alpha\beta} - \frac{1}{2} R g^{\alpha\beta}$ denoting the Einstein tensor. It has to be noted that the result does not change when we add a constant Λ, giving us $G^{\alpha\beta} + \Lambda g^{\alpha\beta} = R^{\alpha\beta} - \frac{1}{2} R g^{\alpha\beta} + \Lambda g^{\alpha\beta}$ (c.f. (1)). In order to achieve this result from the last line in (3), however, we have to see that the term $\delta_g R_{\alpha\beta}$ gives zero under the integral. Hilbert has shown that in fact this is the case and thus,…"

"Well, Professor Einstein," Holmes interrupted, "I have to admit that I'm familiar with this classical part, of course, but you should know that nowadays people are a bit skeptical about the setting of $\delta_g R_{\alpha\beta}$ to zero."

"What do you mean?" asked Einstein bewildered.

"As you have just mentioned," my friend started, "it was assumed by Hilbert [M18] that the second term in the last line in the integrand of (3), which is to say the term $g^{\alpha\beta} \delta_g R_{\alpha\beta}$, could be ignored. Or, in order to say it a bit more correctly, Hilbert assumed to be able to make the term a surface integral and thus, to be allowed to disregard it. However, as demonstrated in later publications [M19, M20], such a surface term could still influence the outcome of a more holistic approach if, for example, our space is considered as a manifold, which is embedded in a space of higher dimension."

"Who made such a strange suggestion?" Rosen asked.

"Ahh…" answered Holmes, "this does not matter here. I would only like to show you how the originally neglected term could still play a role." He stood up, went to the blackboard and took a piece of chalk. "After performing the variation in (3) we have:

$$\begin{aligned} 0 &= \int_V d^n x \sqrt{-g} \left[R^{\kappa\lambda} - \frac{1}{2} R g^{\kappa\lambda} + \Lambda g^{\kappa\lambda}\right] \delta_g g_{\kappa\lambda} + \int_V d^n x \sqrt{-g} \left[\left(\left(g^{\alpha\beta} \delta_g \Gamma^{\rho}_{\alpha\beta} - g^{\alpha\rho} \delta_g \Gamma^{\gamma}_{\alpha\gamma}\right)_{;\rho}\right)\right] \\ &= \int_V d^n x \sqrt{-g} \left[R^{\kappa\lambda} - \frac{1}{2} R g^{\kappa\lambda} + \Lambda g^{\kappa\lambda}\right] \delta_g g_{\kappa\lambda} + \int_{\partial V} d^{n-1} x \sqrt{-h} \left[\left(g^{\alpha\beta} \delta_g \Gamma^{\rho}_{\alpha\beta} - g^{\alpha\rho} \delta_g \Gamma^{\gamma}_{\alpha\gamma}\right) V_\rho\right] \end{aligned}. \tag{4}$$

By applying the following identities:

$$\left(g^{\kappa}_{\alpha}g^{\lambda}_{\sigma}g^{\rho}_{\beta} + g^{\kappa}_{\sigma}g^{\lambda}_{\beta}g^{\rho}_{\alpha} - g^{\kappa}_{\alpha}g^{\lambda}_{\beta}g^{\rho}_{\sigma}\right) \equiv GX^{\kappa\lambda\rho}_{\alpha\beta\sigma} \quad (5)$$

$$\delta_g\left(\Gamma^{\gamma}_{\alpha\beta}\right) = \frac{1}{2}g^{\gamma\sigma}\left(\delta g_{\sigma\alpha;\beta} + \delta g_{\sigma\beta;\alpha} - \delta g_{\alpha\beta;\sigma}\right) = \frac{g^{\gamma\sigma}}{2}GX^{\kappa\lambda\rho}_{\alpha\beta\sigma}\delta g_{\kappa\lambda;\rho} \quad (6)$$

we can give (4) in the following form:

$$\begin{aligned}
0 &= \int_V d^n x\sqrt{-g}\left[R^{\kappa\lambda} - \frac{1}{2}Rg^{\kappa\lambda} + \Lambda g^{\kappa\lambda}\right]\delta g_{\kappa\lambda} + \int_{\partial V} d^{n-1}x\sqrt{-h}\left[\left(g^{\alpha\beta}\delta_g\Gamma^{\rho}_{\alpha\beta} - g^{\alpha\rho}\delta_g\Gamma^{\gamma}_{\alpha\gamma}\right)V_\rho\right] \\
&= \int_V d^n x\sqrt{-g}\left[R^{\kappa\lambda} - \frac{1}{2}Rg^{\kappa\lambda} + \Lambda g^{\kappa\lambda}\right]\delta g_{\kappa\lambda} \\
&+ \int_{\partial V} d^{n-1}x\frac{\sqrt{-h}}{2}\left(\overbrace{\left(g^{\alpha\beta}g^{\rho\sigma}GX^{\kappa\lambda\omega}_{\alpha\beta\sigma} - g^{\alpha\rho}g^{\gamma\sigma}GX^{\kappa\lambda\omega}_{\alpha\gamma\sigma}\right)}^{\equiv MB^{\kappa\lambda\omega\rho}}\delta g_{\kappa\lambda;\omega}\right)V_\rho
\end{aligned} \quad . \quad (7)$$

Integration by parts allows us to get rid of the $\delta_g g_{\kappa\lambda;\omega}$-term, assuming higher dimensions coupling into our manifold and after some evaluation as shown in [M19, M20] we obtain a set of new field equations like this:

$$\begin{aligned}
&\text{a) } \delta g_{\kappa\lambda}, V: \quad 0 = R^{\kappa\lambda} - \frac{1}{2}Rg^{\kappa\lambda} + \Lambda g^{\kappa\lambda} + \frac{1}{2}B\cdot V_\rho MB\gamma^{\kappa\lambda\omega\rho}v_\omega \\
&\text{b) } \delta g_{\kappa\lambda}, \partial^2 V: \quad 0 = \Re^{\kappa\lambda} - \frac{1}{2}\Re h^{\kappa\lambda} + \Lambda h^{\kappa\lambda} + \frac{1}{2}A\cdot U_\rho MB^{\kappa\lambda\omega\rho}u_\omega
\end{aligned} \quad . \quad (8)$$

Here A and B are constants, V_σ, v_σ and U_σ, u_σ denote the normal vectors on the surface ∂V and the sub-surface $\partial^2 V$, respectively. With respect to the complete evaluation, I have to refer to the papers [M19] or [M20]."

"But Holmes," I blurted out, "how did you come across such papers? After all, this is not necessarily your typical field of interest."

"On first sight you might be right," Holmes said, "but there was this case about two Jews being murdered by the Nazis during World War II, which I was most interested in [M21, M22]. This case led to some interesting papers and finally to the new solution I have just presented."

"Do they have a special name?" asked Rosen.

"Who?" asked Holmes back.

"Not who, Mister Holmes! I meant the equations," elaborated Rosen. "Do they have a special name?"

"Yes, I think I read it somewhere in those papers," Holmes said and went back to his chair in front of his desk. There he quickly found what he was looking for. "Here it is," he cried, "Extended Einstein field equations. That's the name the author gave them. And you might find it most interesting that he named the additional terms 'surface' or 'entanglement' terms."

"That is quite fascinating," Einstein said. Then he asked, "Will this help us to solve our problem?"

"Perhaps it does," Holmes said. "But before we come to that, please let me summarize what we have learned about your theory, which is to say the General Theory of Relativity. In essence, there is nothing you need to postulate, except that you say:

A) Here is some space and the space has n dimensions. Perhaps if we set n=4 then we might speak about our ordinary space-time. Then…
B) … you say that this space could be warped somehow and you want to know the nature or geometry of this funny deformation.
C) So, you simply take the thing you called curvature and denote it with an R in your equation…"

"The Ricci scalar!" said Einstein.

"… and then you do the Hilbert-trick and out comes your equation, right?"

"In a nutshell, yes!" said Einstein.

"What's the thing on the other side, this T-something?"

"That is matter," Einstein answered, thereby pointing out the $-\kappa T^{\alpha\beta}$ -term on the right-hand side of (1). "We called it the energy-momentum tensor and gave it the symbol T."

"Does it also come out of a minimum principle?" asked Holmes.

"No! Not really…" Einstein seemed uncertain, almost a bit uncomfortable, "To tell the truth, it had to be postulated."

"Why?" asked Holmes.

"Well, because we all thought there must be some matter in the universe and as the Ricci scalar alone does not show any matter, we had to bring it in somehow, didn't we?"

"In this case, discard it!" Holmes demanded.

The three professors looked flabbergasted.

"What?" Podolsky even cried out, "But where else should matter come from then?"

"I don't care at the moment," answered Holmes, "but I most certainly do not want to bias my starting approach with something I have no idea where it comes from. Therefore, until there is no proper explanation about the origin of this T-term, would you please treat it as if it were not there… Thank you, Professor!"

"And what shall I put there instead?" Einstein asked almost a bit mockingly.

"A zero, of course!" Holmes answered simply.

Einstein obliged and drew a pronounced '0' after the '='-sign on the right-hand side of his equation. "But you should know that we then only have a vacuum equation, Mister Holmes."

$$R^{\alpha\beta} - \frac{1}{2} R g^{\alpha\beta} + \Lambda g^{\alpha\beta} = 0 \tag{9}$$

"A vacuum is not necessarily nothing!"[33] Holmes said simply and then added in a suddenly very busy tone, "And now we find the right solution for the 'spooky action mystery'!"

24.7 The Quantum Side of the General Theory of Relativity – Part Two

"But there are already hundreds of solutions to Einstein's equations, if not thousands," Rosen said. "How shall we know in which direction to go? Where do we start?"

"Simple!" said Holmes, "there must be a set of solutions clearly sporting typical properties of quantum solutions."

"Like what?" Einstein asked, almost a bit snappishly.

"Like the solutions of Professor Dirac [M23]. I mean the ones where the time coordinate oscillates and can have two signs, namely one for matter and..."

"... the other one for antimatter," completed Podolsky. "My dear Mister Holmes, but the Dirac-solutions are solely solutions to the Dirac equation [M23]. Here, you might have missed that point, but trust me, it is an important one; we are talking about the Einstein field equations. This is something completely different. Or let me put it differently: We have two completely different equations, namely the one of Dirac and the one of Einstein and you expect us to find a solution satisfying both. This is impossible."

"Really?" said Holmes amused. Then he stood up and walked to the blackboard. "How about this one here?"

$$g_{\alpha\beta} = \begin{pmatrix} H & 0 & 0 & 0 \\ 0 & A\cdot f[\tau]\cdot g_{\varrho}'[c_{\varrho}\varrho]^2 & 0 & 0 \\ 0 & 0 & g_{22} = B\cdot f[\tau]\cdot g_{\theta}'[c_{\theta}\theta]^2 g_{\varrho}[c_{\varrho}\varrho]^2 & 0 \\ 0 & 0 & 0 & g_{33} \end{pmatrix} \tag{10}$$

$$g_{33} = \frac{D\cdot g_{\varphi}'[c_{\varphi}\varphi]^2}{B\cdot g_{\theta}'[c_{\theta}\theta]^2}\cdot g_{22}\cdot \sin\left[g_{\theta}[c_{\theta}\theta]\right]^2$$

"The letters A, B, D, H stand for arbitrary constants," explained Holmes, "and as you will probably find quite easy to prove, the solution to your vacuum equation (9), Professor Einstein, can be given by the following setting:"

$$f[\tau] = C_2\cdot e^{\pm 2\cdot i\cdot\tau\cdot\sqrt{\frac{H\cdot\Lambda}{3}}};\quad \frac{c_{\varrho}^2}{A} = \frac{c_{\theta}^2}{B} \tag{11}$$

---

[33] C.f. subsection „What Is Matter?"

With this, he wrote a rather simple expression on the blackboard and turned to his famous guests. At first, they stared at him. Then, almost in perfect unison, they started to calculate. Einstein on the blackboard. Rosen and Podolsky on the table in front of them using sheets of paper and pencils apparently ready to be used for just this purpose. There was a bit back and forth and a few exchanges between the blackboard and the table before the three faced Holmes and Einstein said, "Yes Holmes, that is a solution to my equation… and surprisingly…"

"It has the quantum properties of the Dirac-solutions," added Rosen.

"And still a lot of degrees of freedom to be adjusted to the experimental observations," finished Podolsky.

"But how did you come up with it?" I asked.

"It was given to me." Holmes said and with this he opened a drawer and pulled out one sheet of paper. He handed it to Einstein, who only needed about three seconds to take in its content and then he asked, "Schwarzer? Never heard of this guy [M24, M25, M26]. Did you?" he addressed his colleagues.

Both shook their heads.

Holmes smiled. "That is quite understandable, gentlemen. After all, the person is a bit too recent to be known by someone who is as… well, as antique as you are. And what is more, he is not even a real scientist and this bit of paper isn't even known among the scientific community today."

"But the solution is correct, we just checked," Podolsky said incredulously, "how come nobody knows this man or this solution?"

"Explaining this would lead us too far away. For now, it needs to suffice to tell you that nowadays the whole science and education system is often more a circus than a well-organized and fair community. Politicians and ideologists have taken over at too many places and made it a farce, a mere show for an ever-dimmer audience to pretend importance and meaning where extremely often there is not a trace of neither of the two[34]."

"I'm very sorry to hear that!" said Einstein and his colleagues mumbled their agreement.

"Well, as bad as this may be, this can hardly be our concern right now. We have a different problem we want to solve and with the Dirac-like solution I think we do have a good start." Holmes sounded extremely positive, but Einstein was not convinced.

"I'm sorry, but I simply do not see how this thing could be of help with the EPR paradox. It is great, of course, that we now have seen a solution to my equation, sporting quantum properties, but the EPR-thing is a different story, is it not?"

He looked straight at Holmes and when he found him smiling, he added, "You already have an idea, Mister Holmes?"

"Let's call it a suggestion, Professor Einstein… just a suggestion. But before I will convey this to you, I need to ask you something. When reading papers about the General Theory of Relativity, one always finds hints about so-called test masses. What exactly does that mean?"

---

[34] See our section about „Planck's Principle…"!

"Ha!" laughed Einstein loudly, "It means nothing else but that we have no true two-body solution. That's just it."

Holmes smiled. "Good!" he said, "This is what I thought it is and it suits me very well!"

"I'm sorry, gentlemen!" I cut in, "but I have to confess myself completely lost. What on earth is a two-body solution?"

"Oh, that is not a very complicated thing, Doctor," said Rosen. "We have very nice solutions to the Einstein field equations as long as we are just talking about one body. Mister Holmes' solution here, for instance…"

"It is not mine," hinted Holmes, "I found it in a paper of a guy named Schwarzer and even he doesn't claim it as his and therefore names it the 'Dirac-Schwarzschild solution'."

"… well, anyway, it also is a one-body or one-center solution and it seems to be so complicated to construct true two-center or two-body solutions in the General Theory of Relativity that we scientists resorted to the model of the test-mass. There we simply assume that a small mass, which is to say we have a mass small in comparison with another object, determining the warping of the space, comes near the space-warping object and, because the test-mass is so small, we can ignore the warping, which itself is producing. That's just it."

"So, you are cheating!" I said.

"Let's better call it approximating," Rosen suggested, smiling.

"Yes, and here, so I think, is the point," Holmes cut in. "This approximation hinders you from getting the full picture. Without that little scientific skullduggery, you would probably be able to get the holistic view and find the missing something. Thus, you exactly lack THE one thing you obviously need to understand the physical meaning about the peculiar effect which right now still appears as mystery."

"The quantum guys will most certainly disagree at this point, Mister Holmes!" said Podolsky. "You know that their argument is that the entangled system is just one entity and that the black-white-property is non-local, which means, it is of such character that the position of the two particles and the huge distance they have from each other does not matter. For them, I mean quantum people, there is no communication and the EPR-treatment is not correct. In other words, they consider the problem solved, which is to say they consider it non-existent."

Holmes only laughed. "In other words: they ignore the observable distance, cannot give any explanation why this distance does not matter with the black-white-property, and forbid you to think about that, because they say it is the wrong question, right?"

"Yes, Mister Holmes," said Einstein, "it is a little hard to put it like you just did, but in essence, you… kind of… got the point."

"In other words," Holmes added, "the quantum believers say to us that the quantum world is too special to be understood anyway and that we simply have to accept its strangeness. Seeking an explanation for its strangeness is prohibited or at least counter argued with the hint, that Quantum Theory is nothing one can truly understand and therefore it is foolish to the extent to even try to do so."

"But you are not willing to accept this, Mister Holmes, are you?" asked Rosen.

"No!" Holmes said flatly, "99% of my cases I only solved because I never did accept such restrictions, such human made limitations and I will not start to do it now."

"But Holmes," I blurted out. I simply could not restrain myself, "You say Quantum Theory should not forbid us to seek for reasonable explanations, but at the same time you accept the strange things that General Theory of Relativity gives out. Where is the difference?"

"Two reasons, my dear Watson, just two reasons:

a) The General Theory of Relativity comes out of a nice and clear mathematical principle. I'm referring to the minimum principle of Hamilton in the form of Professor Hilbert's action integral. It goes like that: you start here and you end there… there is no fuzzy stuff in between, except for this postulated matter term. However, as you remember, I had just thrown it out of Einstein's theory, because I do not consider it genuine. Sorry for that, Professor Einstein!
b) I do not know about you, my dear Watson, but a curved space-time, my brain can handle, but non-local parameters and arbitrary, almost religiously dogmatic rules like this about 'it is just one system' or 'this is nothing the human brain could understand anyway' or even 'don't ask questions', I cannot bear."

"This is all very nice, Mister Holmes," interrupted Einstein, "but I don't see how to proceed from where we are now in order to come forward with this EPR-paradox or EPR-non-paradox. What is the 'sanity and reason' way to grasp the effect of entanglement?"

"I'm very glad you brought it down to this straightforward question, Professor," answered Holmes and smiled. "In my very humble opinion the key for the correct handling of entanglement or the EPR paradox…" Rosen snorted and Holmes added, "Well, EPR-paradox or EPR-non-paradox then… Anyway, I think it all comes down to the two-body or two-center problem. Without a consistent model for this thing, we will helplessly run in circles and this, I assure you, is not my preferred type of exercise."

"But so far nobody has such a solution, Mister Holmes!" cut in Podolsky, "and please rest assured, we all have tried our best to find one."

"Hm!" said Holmes thoughtfully, his hands under his chin, "then every one of you and the whole rest of the community, too, must have looked in the wrong direction."

"With all due respect, Mister Holmes, but how on earth would you know where we have already looked?" Rosen almost sounded a little bit offended.

"Oh, that is obvious," answered Holmes lightly, "pairs of bodies exist in reality, and so there must be a way to mathematically describe such systems. Because, obviously the universe can handle them. However, as nobody has found the corresponding solution, they must all have sought at the wrong places… I mean as long as we exclude those blind dunderheads who would not see the solution even when having it right in front of them."

Einstein laughed, while Rosen and Podolsky looked rather sour.

"Ok, Mister Holmes," asked Rosen after Einstein had finished laughing, "and where should we start to search for the perfect solution then?"

"In the infinite space of a few more dimensions, preferably orthogonal ones, of course!" answered Holmes simply. Everybody in the room stared at my friend. We expected him to elaborate, but Holmes stood up, strolled to the blackboard again and simply extended his previous drawing.

"I'm only giving you the simple form. There are many other options and degrees of freedom," he explained, "but this one here should suffice to make it all clear:"

$$g_{\alpha\beta} = \begin{pmatrix} g_{00} & 0 & 0 & 0 & 0 & 0 & 0 & 0 \\ 0 & g_{11} & 0 & 0 & 0 & 0 & 0 & 0 \\ 0 & 0 & g_{22} & 0 & 0 & 0 & 0 & 0 \\ 0 & 0 & 0 & g_{33} & 0 & 0 & 0 & 0 \\ 0 & 0 & 0 & 0 & g_{44} & 0 & 0 & 0 \\ 0 & 0 & 0 & 0 & 0 & g_{55} & 0 & 0 \\ 0 & 0 & 0 & 0 & 0 & 0 & g_{66} & 0 \\ 0 & 0 & 0 & 0 & 0 & 0 & 0 & g_{77} \end{pmatrix} \tag{12}$$

$$g_{00} = 1; g_{11} = A \cdot f[\tau] \cdot g_{\varrho}'\left[c_{\varrho}\varrho\right]^2; g_{44} = 1; g_{55} = a \cdot g[t] \cdot g_x'[c_x x]^2$$

$$g_{22} = B \cdot f[\tau] \cdot g_{\theta}'[c_{\theta}\theta]^2 g_{\varrho}\left[c_{\varrho}\varrho\right]^2; g_{66} = b \cdot g[t] \cdot g_y'\left[c_y y\right]^2 g_x[c_x x]^2$$

$$g_{33} = \frac{D \cdot g_{\varphi}'\left[c_{\varphi}\varphi\right]^2}{B \cdot g_{\theta}'[c_{\theta}\theta]^2} \cdot g_{22} \cdot \sin\left[g_{\theta}[c_{\theta}\theta]\right]^2; g_{77} = \frac{d \cdot g_z'[c_z z]^2}{b \cdot g_y'\left[c_y y\right]^2} \cdot g_{66} \cdot \sin\left[g_y\left[c_y y\right]\right]^2$$

While adding term after term, the whole still looking relatively simple, he elaborated his idea. "Until today, you have always tried to mathematically treat the second body in the same space where you already had placed the first one. You thought that the coordinates of your mathematical space have to do with the spatial directions of your own experience, but this is not the case. You must separate the mathematical space from the real one, because the mathematical one does not stand for a spatial reality at all, it only gives the degrees of freedom of a certain object."

"Good Lord," said Einstein with dawning comprehension.

"As long as you have only one object," Holmes went on, "you have the usual coordinates for space and time. Then, when talking about our normal space, for instance, you have the three dimensions for space and an additional one for time and these are also the degrees of freedom for your single object. However, when adding a second object, you should be aware of the fact that this object brings in its own degrees of freedom and this..."

"... this automatically increases the number of dimensions I have to treat the two-body system with!" Einstein finished the sentence for my friend.

"Precisely, Professor!" Holmes said. "And here are the two functions f and g solving your very vacuum field equations." He started to write again:

$$f[\tau] = C_2 \cdot e^{\pm 2 \cdot i \cdot \tau \cdot \sqrt{\frac{\Lambda_\tau}{3}}}; \quad \frac{c_{\varrho}^2}{A} = \frac{c_{\theta}^2}{B}; \quad g[x] = C_3 \cdot e^{\pm 2 \cdot i \cdot x \cdot \sqrt{\frac{\Lambda_x}{3}}}; \quad \frac{c_x^2}{a} = \frac{c_y^2}{b} \tag{13}$$

"You see that we have split up the cosmological constant into two parts as follows:

$$\Lambda = \Lambda_\tau + \Lambda_x \tag{14}$$

where we need to set the two parts of the cosmological constant to be equal, which is to say, we have

$$\Lambda_\tau = \Lambda_x \text{."} \qquad (15)$$

Meanwhile Holmes had finished his drawing and stepped back from the blackboard. Einstein took one look at the new equation and cried, “So, it is eight dimensions now! You have simply doubled the number of degrees of freedom when going from one object to two.”

“That is correct, Professor,” said Holmes, “but may I attract your attention to the important fact that there is a peculiar coupling among the objects when solving the whole system. Which is to say, when introducing my little eight-dimensional approach here…”

“Metric!” said Einstein.

“Pardon me?” asked I.

“Metric!” repeated Einstein and then elaborated, “We call these systems, I mean the ones Mister Holmes has just drawn on the blackboard, metrics, Doctor Watson. No need to worry about. It is only an expression which shows us that we deal with a certain mathematical tool which describes the state of a space.”

“Ah!” said I and thought that his explanation was rather compact for a layman like me, but nevertheless, I thought to have gotten the gist about that thing on the blackboard and that was fine with me at the moment. After all, I would be able to ask Holmes about it later, when the “wise men” were gone.

While I was still mulling over the “tool” on the blackboard, the professors had set to work on Holmes’ new “metric solution” and as they now had to consider eight dimensions instead of only four, they were considerably more “busy”. Whereas I am using this funny description, which means the word “busy”, here to point out that there was much more excitement among them than before.

Finally, Rosen cried, “They are coupled…”

“Entangled!” said Einstein.

“… via their cosmological constant.” And Podolsky immediately added, “And as this cosmological constant is – well, just as the name says – a constant everywhere…”

“In the whole cosmos the system spans!” said Einstein.

“… we have found the coupling parameter holding the two objects firmly bound together.”

“But it depends on the property the constant couples in. That is quite important!” elaborated Einstein, who had jumped to his feet and started to pace the room while he was speaking to us. “While here it is the two time-coordinates for each of the two objects and I’m sure, Mister Holmes, you have only chosen this, because this allowed us to reuse the results from the first one-body solution you had named a Dirac-Schwarzschild solution…? ...” Einstein stopped for a moment in his pacing and looked at my friend imploringly.

Holmes simply nodded.

“… we can couple…”

“Entangle!” I heard myself say.

“What?” Einstein was totally taken aback.

"Entangle, Professor Einstein!" I repeated. "I guess this is a bit more precise than 'couple', because I have realized that, so far, you had insisted quite intensely in using this term rather than anything else, have you not?"

There was a funny pause. Podolsky and Rosen chuckled and Holmes was almost beaming at me. Einstein took the pipe out of his mouth, smiled briefly, and then started pacing again.

"Where was I?" he said. "Ah yes. In this extremely simple example, it just happened to be that the entanglement was on time, because Mister Holmes was so kind to choose a close enough solution to the one-center case we had before – thank you for your thoughtfulness and consideration, Mister Holmes. But we clearly see that we could do the entanglement on every other pair or probably even triple or any tuple of degrees of freedom within this extended metric. The entanglement via the cosmological constant always assures the coupling. Be it with time, to perhaps couple matter and antimatter, be it spin, charge… it does not matter. I see no limits here. Every property… at least as far as I can see right now… could be made an entangled one by the cosmological constant… or is it constants?"

Einstein suddenly stopped. He faced us, looked everybody in the room straight in the eyes, and cried, "But this means…"

"… that we were right!" shouted Podolsky and Rosen in unison.

"Yes, my friends!" Einstein took over again. "There are 'hidden' parameters providing the entanglement, only that these are not truly hidden. As cosmological constants they are omnipresent and everywhere in the cosmos. They always assure the entanglement. There is no specific quantum-mystery after all. It was already inside my equations. All the time. The cosmological constant is the 'not so very much hidden parameter' and as it is just present… present like a celestial God, I'd almost say, the connection of the entanglement is guaranteed. Everywhere and everywhen, I'd say. But then… wait a moment." Einstein almost ran to Holmes' desk, stopped sharply before the edge, stared at my friend, and cried, "You knew!"

"Rest assured, Professor Einstein," my friend answered very calmly, "I did nothing more than guess."

"Sorry gentlemen, but what is the sudden problem?" I asked.

But Einstein ignored me. He continued to stare at Holmes and said, "There are infinitely many of them. Entanglement is everywhere. Entangled systems, which are all little cosmoi, I mean, with their own private cosmological constant… Our whole universe is just brim-full with them."

"Yes!" said Holmes, "that is what comes out of your own equations, Professor Einstein."

"And the measurement?" asked Rosen. "What about the measurement, I mean when we determine such an entangled parameter with one particle? What happens to the system… to the…"

"… the little cosmos?" Einstein asked. "You mean, what happens to the little entangled system, which is forming its own small cosmos… a cosmos of two, of three, and so on? You would like to know what happens to this cosmos?"

Rosen nodded. That was his question.

"It will be destroyed during the measurement!" Holmes said softly.

There was silence for quite a while. Then Einstein mumbled, "The collapse of the wave function is the death of a minuscule universe… well, its coupling!"

"Sorry, gentlemen, but what is what?" I asked.

Podolsky looked at me in surprise, but then his eyes showed pity and he started to elaborate. "The quantum people saw that there is a problem in interpreting their funny combined black-and-white property, which should be either the one or the other but in fact is always both… I mean it is both until the measurement. Then the quantum state becomes clear and its mathematical description takes on one specific form. As they named this mathematical description a 'wave function', they are speaking about the 'collapse of the wave function' in the moment of the measurement. Now your friend was just suggesting that this mysterious collapse is nothing but the destruction of a mini-cosmos. Fascinating I'd say!"

Rosen, who had been very contemplative during Podolsky's short talk, said, "This is all very nice, but taking what Einstein had said about truly understanding the nature of a thing by literally grasping it with one's thoughts, I still have no inkling as to where this whole quantum behavior is coming from?"

He looked imploringly at Holmes, who, to everyone's astonishment, was smiling again. "Well," he started, "now we are coming back to the two murdered Jews and their funny idea of a sub-structured universe [M21, M22]…"

"So far," Einstein interrupted him, "you have not told us that they had the idea of the universe being sub-structured."

"That is correct, Professor Einstein," Holmes admitted. "It did not matter so far. Now, however, in the light of Professor Podolsky's question, it is the perfect moment to present this idea. Do you remember, Professor Einstein, that in your so-called 'annus mirabilis' in 1905, you published three important papers?"

"How could I have forgotten?" Einstein said. "One paper was about the photoelectric effect, one was about my Special Theory of Relativity, and the third was about…" He became very pale and fell temporarily silent. Then he said in a hoarse whisper, "… you have it again, Mister Holmes!"

"Maybe, Professor, maybe…" Holmes said and then he explained. "In 1905 you also published a paper about the so-called 'Brownian motion'. Now simply use what you know about this Brownian motion and imagine something, which is being placed in a grainy universe, a particle perhaps. The grains shall be of extremely small scale. Moreover, they should permanently jitter. This jitter comes out of your own equation, because there are solutions exactly giving the right objects, exactly showing such properties [M27]. By the way, the two Jews I had mentioned before had called them 'Friedmanns' [M21], these objects or grains. This is the basis for all the quantum stuff that we can observe. It is nothing but a scale effect, because we cannot see beyond the scale of the minuscule grains and so our measurements only give what the physicists call the 'classical results'. In other words: your theory contains the Quantum Theory and with the additional assumption of the substructure you obtain the Heisenberg uncertainty and the speed of light limit and more [M27]."

"So, I had it," said Einstein very quietly. "I had it all… years ago, I could have solved it. I was so close."

For a very long time, nobody spoke a word.

24.8 Good Bye – Part Two

"What do you think, Holmes? Where did they go?" I asked.

"On!" Holmes answered simply and then he added, "However, in these modern times, some might rather formulate it as 'they went to the next level'."

Then he smiled and said, "I am of the opinion, Watson, that right now, it would be the perfect time for a nice cup of tea, don't you think?"

24.9 Epilog – Part Two

I was almost bursting with the question, but pulled myself together and it was only after the tea was on the table, cups were filled, the honey was properly stirred into the hot, steaming liquid, that I dared to bring it forward.

"Holmes?" I started as casually as I could, but the moment the word was out of my mouth I knew that it already had betrayed me. Nevertheless, my friend sat rather quietly. He laid the spoon on the tray, lifted the saucer and cup, but did not drink. Instead, he only inhaled the steam wafting from the tea.

"Fine Indian tea," he said as if he had not heard me. But then, he added, "Yes, Watson I know. You have asked yourself two questions…"

"Ahhh…" I mumbled, because until this moment I had not noticed myself, that it actually was two questions, but when thinking about it…

"For one, which is the more obvious, you have asked yourself, why it was only us, I mean the three famous scientists plus you and me. Clearly, it would only have been most reasonable from the universe to also invite the very man who had recently come up with the idea, which turned out to provide the necessary prerequisite to solve the riddle, would it not, Watson?"

I nodded. Holmes sipped from his tea and while he was savoring it, I asked myself which part of his famous brain right now was busy with the analysis of the liquid in his mouth.

"Lime!" he said.

I concluded at once that he must refer to the honey, because the tea truly came from India and could not have anything to do with a lime tree.

"How can you taste this, Holmes?" I asked, "I mean this is a particular strong tea and the honey in it doesn't give much away, apart from its sweetness, but still you…"

"It says on the label, Watson!" my friend answered, pointing to the glass of honey on the tray. I felt embarrassed, but inwardly I had to smile nevertheless. My friend was not the type to harp and so he just drunk a bit more from his tea and then he said, "There can be only one reason why he was not here today and I'm sure you already know, don't you, my dear Doctor Watson?"

Holmes did not wait for an answer.

“It is the subsequent, which is to say the second, question immediately following out of the answer of the first, which hinders you to make this very first answer manifest, is it not?”

I felt my mouth going dry. My friend was right, of course. It was not too difficult to give the only logic answer to the question why the man who had come up with the essential ingredient has not been with us today. I mean, in the same way as we were there. But what did this mean? Immediately, just as Holmes had predicted, that second question popped up in my head and I said, “But what are we then, Holmes?”

Holmes smiled and one could see that he was rather pleased with himself.

“I’d say that we are pretty good programs, Watson… or, as Einstein would probably have said, we are quite fascinating thought experiments, doing our job inside a nice ‘computer’. A computer, which in fact happens to be a rather beautiful mind. It probably does belong to the one person you were missing and that itself is nothing but a thought experiment on an even greater level.”

“And this awareness does make you feel… comfortable?” I asked.

But Holmes did not answer. He simply studied me. I felt like yet another interesting case to him, … like a small one, however, almost insignificant. Then I heard my friend say, “Just drink a bit more of that wonderful tea and then you only need to ask yourself one thing. It will immediately make you feel much, much better:

24.10 Personal Statement

The reader should note that the story in this book is also a result of the intention of the authors to find ways for better uncertainty budget estimates in socioeconomic models. Even though the three authors had different starting points, it became clear to them independently that the current physics does not give satisfactory answers to many socioeconomic questions. The same can be said to many very practical fields and applications. However, as Quantum Theory is THE one field of science, which only is in existence, because of very principal (if not to say universal) uncertainties, it was soon clear that a general solution of the problem requires the quantization of arbitrary spaces and space-times. Thus, when starting this work, the authors also knew that the connection of Einstein’s General Theory of Relativity and Quantum Theory would be a necessary by-product. This statement should not be interpreted as any kind of disrespect regarding the quest of bringing Einstein’s and Quantum Theory together. It simply is a fact that the motivation of the authors for starting this whole work did not lay in theoretical physics, but in practical problem-solving and socioeconomics. It should be pointed out that the application of the “Theory of Everything” to socioeconomics seems to have even more drastic consequences than its application in physics [M28, M29].

For one, it can be mathematically shown that most politicians and their mainstream-media-cronies are totally and utterly – and probably purposefully – wrong. As this holds true for almost everything they are doing, one could state that they are in fact nothing but dim-doers instead of do-gooders [M30].

24.11 References for the Motivator Story

[M1] A. Einstein, B. Podolsky and N. Rosen, "Can quantum mechanical description of physical reality be considered complete?", 1935, Phys. Rev. 47, 777

[M2] A. Einstein and N. Rosen, "The Particle Problem in the General Theory of Relativity", 1935, Phys. Rev. 48, 73

[M3] J. Maldacena, L. Susskind, "Cool horizons for entangled black holes", 2013, arXiv:1306.0533v2

[M4] L. Susskind, "Copenhagen vs Everett, Teleportation, and ER=EPR", 2016, arXiv:1604.02589v2

[M5] G. N. Remmen, N. Bao, J. Pollack, "Entanglement Conservation, ER=EPR, and a New Classical Area Theorem for Wormholes", 2016, arXiv:1604.08217v1

[M6] N. Schwarzer, "Einstein had it, but he did not see it – Part VII: Konrad Zuse's Computing Universe", self-published, Amazon Digital Services, 2017, Kindle, ASIN: B0752Z99DL

[M7] J. S. Bell, "On the Einstein-Podolsky-Rosen paradox", Physics vol. 1, No. 3, 1964, pp. 195-200

[M8] C. A. Kocher, E. D. Commins, "Polarization Correlation of Photons Emitted in an Atomic Cascade", Physical Review Letters vol 18, No. 15, 1967, pp. 575-577, doi:10.1103/PhysRevLett.18.575

[M9] S. J. Freedman, J. F. Clauser, "Experimental Test of Local Hidden-Variable Theories", Physical Review Letters vol. 28, No. 14, 1972, pp. 938-941, doi:10.1103/PhysRevLett.28.938

[M10] A. Aspect, J. Dalibard, G. Roger, "Experimental Test of Bell's Inequalities Using Time-Varying Analyzers", Physical Review Letters vol. 49, No. 25, 1982, pp. 1804-1807, doi:10.1103/PhysRevLett.49.1804

[M11] G. Weihs, T. Jennewein, C. Simon, H. Weinfurter, A. Zeilinger, "Violation of Bell's Inequality under Strict Einstein Locality Conditions", Physical Review Letters vol. 81, No. 23, 1998, pp. 5039-5043, doi:10.1103/PhysRevLett.81.5039, arxiv:quant-ph/9810080v1

[M12] M. A. Rowe, D. Kielpinski, V. Meyer, C. A. Sackett, W. M. Itano, C. Monroe, D. J. Wineland, "Experimental violation of a Bell's inequality with efficient detection", Nature vol. 409, No. 6822, 2001, pp. 791-4, doi:10.1038/35057215.

[M13] J. F. Clauser, M. A. Horne, A. Shimony, R. A. Holt, "Proposed Experiment to Test Local Hidden-Variable Theories", Physical Review Letters vol. 23, No. 15, 1969, pp 880-884, doi:10.1103/PhysRevLett.23.880

[M14] N. Schwarzer, "Quantized Relativized Theology – Where is God?", self-published, Amazon Digital Services, 2016, Kindle, ASIN: B01M0XPXTT

[M15] N. Schwarzer, "Humanized Artificial Intelligence – Spiritual Computers – Make Them Believe and They'll Start to Think", self-published, Amazon Digital Services, 2017, Kindle, ASIN: B072MNRLJP

[M16] A. Einstein, "Grundlage der allgemeinen Relativitätstheorie", Annalen der Physik (ser. 4), 49, pp. 769-822

[M17] E. Hubble, "A relation between distance and radial velocity among extra-galactic nebulae", 1929, Proceedings of the National Academy of Sciences 15 (3): pp. 168–173, Bibcode: 1929PNAS...15..168H. doi:10.1073/pnas.15.3.168

[M18] D. Hilbert, "Die Grundlagen der Physik", 1915, Teil 1, Göttinger Nachrichten, pp. 395-407

[M19] N. Schwarzer, "Einstein had it, but he did not see it – Part XX: Higher Order Covariant Variation of the Einstein-Hilbert-Action", self-published, Amazon Digital Services, 2017, Kindle, ASIN: B0788VWKD4

[M20] N. Schwarzer, "Einstein had it, but he did not see it – Part XXI: A Very Simple Theory of Everything", self-published, Amazon Digital Services, 2017, Kindle, ASIN: B078QTVKJS

[M21] T. Bodan, "7 Days – How to explain the world to my dying child", Books on Demand, 2021, ISBN: 978-3-7526-3972-8; in German: "7 Tage – Wie erkläre ich meinem sterbenden Kind die Welt", Books on Demand, 2021, ISBN: 978-3-7534-4187-0

[M22] T. Bodan, "The Eighth Day – Two Jews against the Third Reich", Books on Demand, 2021, ISBN: 978-3-7534-1725-7

[M23] P. A. M. Dirac, "The Quantum Theory of the Electron", Published 1 February 1928, doi: 10.1098/rspa.1928.0023

[M24] N. Schwarzer, "Einstein had it, but he did not see it – Part XXIV: A Variety of Solutions and the Dirac-Schwarzschild-Particle", self-published, Amazon Digital Services, 2018, Kindle, ASIN: B0796GZX8R

[M25] N. Schwarzer, "Einstein had it, but he did not see it – Part XXVI: The Nature of SPIN 1/2", self-published, Amazon Digital Services, 2018, Kindle, ASIN: B0798R3P8D

[M26] N. Schwarzer, "Einstein had it, but he did not see it – Part XXVIII: ¿ Anti Gravity ?", self-published, Amazon Digital Services, 2018, Kindle, ASIN: B079M1VG62

[M27] N. Schwarzer, "Einstein had it, but he did not see it – Part XXV: The Vacuum Friedmann Cosmos and the Origin of a Quantal World", self-published, Amazon Digital Services, 2018, Kindle, ASIN: B078VRLJSS

[M28] T. Bodan, N. Schwarzer, "Quantum Economy", self-published, Amazon Digital Services, 2017, Kindle, ASIN: B01N80I0NG

[M29] T. Bodan, "EU vs. Britain – The other Monkey Trial", self-published, Amazon Digital Services, Kindle, 2017, ISBN-13: 978-1521304792, ASIN: B071Z6S2XV

[M30] D. Martin, "The Fauci/COVID-19 Dossier", 205 pages, This document is prepared for humanity by Dr. David E. Martin, as pdf: www.davidmartin.world/wp-content/uploads/2021/01/The_Fauci_COVID-19_Dossier.pdf

[A] Artwork by Livia Schwarzer, 2024

[B] Artwork by Filia Schwarzer, 2024